HIGH-PRESSURE RESEARCH

Applications in Geophysics

ACADEMIC PRESS RAPID MANUSCRIPT REPRODUCTION

HIGH-PRESSURE RESEARCH

Applications in Geophysics

Edited by

MURLI H. MANGHNANI
University of Hawaii at Manoa
Honolulu, Hawaii

SYUN-ITI AKIMOTO
University of Tokyo
Tokyo, Japan

ACADEMIC PRESS, INC. NEW YORK SAN FRANCISCO LONDON 1977
A Subsidiary of Harcourt Brace Jovanovich, Publishers

ACADEMIC PRESS, INC.
111 Fifth Avenue, New York, New York 10003

United Kingdom Edition published by
ACADEMIC PRESS, INC. (LONDON) LTD.
24/28 Oval Road, London NW1

Library of Congress Cataloging in Publication Data

Main entry under title:

High-pressure research.

Papers presented at a U.S.-Japan joint seminar held in Honolulu, 6-9 July 1976, sponsored by the U.S. National Science Foundation and the Japan Society for the Promotion of Science.
Includes bibliographical references and index.
1. Geophysics--Congresses. 2. Geochemistry--Congresses. 3. High pressure research--Congresses. 4. Geodynamics--Congresses. I. Manghnani, Murli H., Date II. Akimoto, Syun-iti, III. United States. National Science Foundation. IV. Nippon Gakujutsu Shinkōkai.
QE501.3.H53 551 77-1689
ISBN 0-12-468750-4

PRINTED IN THE UNITED STATES OF AMERICA

CONTENTS

SESSION III
Equations of State and Shock Wave Experiments

SESSION IV
Instrumentation, Pressure Calibration, and Standardization

LIST OF CONTRIBUTORS

AHRENS, T. J. Seismological Laboratory, Division of Geological and Planetary Sciences, California Institute of Technology, Pasadena, CA 91125

AKENI, K. Institute of Geophysics, Faculty of Science, Kyoto University, Kyoto, Japan

AKIMOTO, S. Institute for Solid State Physics, University of Tokyo, Roppongi, Minato-ku, Tokyo 106, Japan

ANDERSON, O. L. Institute of Geophysics and Planetary Physics, University of California, Los Angeles, CA 90024

AOKI, K. Institute of Mineralogy, Petrology and Economic Geology, Tohoku University, Aoba, Sendai 980, Japan

BASSETT, W. A. Department of Geological Sciences, The University of Rochester, Rochester, N.Y. 14627

BELL, P. M. Geophysical Laboratory, Carnegie Institution of Washington, Washington, D.C. 20008

BLOCK, S. Institute for Materials Research, National Bureau of Standards, Washington, D.C. 20234

BOEHLER, R. Institute of Geophysics and Planetary Physics, University of California, Los Angeles, Los Angeles, CA 90024

BOETTCHER, A. L. Institute of Geophysics and Planetary Physics and The Department of Earth and Space Sciences, University of California at Los Angeles, Los Angeles, CA 90024

BRODY, E. M. The Institute of Optics, The University of Rochester, Rochester, N.Y. 14627

BUKOWINSKI, M. S. T. Institute of Geophysics and Planetary Physics, University of California, Los Angeles, Los Angeles, CA 90024

DAVIES, G. F. Department of Geological Sciences, University of Rochester, Rochester, N.Y. 14627

DEMAREST, H. H., JR. Department of the Geophysical Sciences and Materials Research Laboratory, University of Chicago, Chicago, Ill. 60637

DUBA, A. G. University of California, Lawrence Livermore Laboratory, Livermore, CA 94550

FISHER, E. S. Argonne National Laboratory, Argonne, Ill. 60439

FORMAN, R. A. Institute for Materials Research, National Bureau of Standards, Washington, D.C. 20234

FUJII, N. Geophysical Institute, University of Tokyo, Tokyo, Japan

FUKAO, Y. Department of Earth Sciences, Nagoya University, Chikusa, Nagoya 464, Japan

GETTING, I. C. Institute of Geophysics and Planetary Physics, University of California, Los Angeles, Los Angeles, CA 90024

GOETZE, C. Department of Earth and Planetary Sciences, Massachusetts Institute of Technology, Cambridge, Mass. 02139

GOTO, T. The Research Institute for Iron, Steel and Other Metals, Tohoku University, Katahira, Sendai 980, Japan

GRADY, D. E. Sandia Laboratories, Albuquerque, N.M. 87115

HAMANO, Y. Geophysical Institute, University of Tokyo, Tokyo, Japan

HEARD, H. C. University of California, Lawrence Livermore Laboratory, Livermore, CA 94550

ICHINOSE, K. Toshiba Research and Development Center, Komukai, Kawasaki 210, Japan

INOUE, K. Asada Fundamental Research Laboratory, Kobe Steel Ltd., Nada-ku, Kobe 657, Japan

ITO, E. Institute for Thermal Spring Research, Okayama University, Misasa, Tottori-ken 682-02, Japan

ITO, H. Geological Survey of Japan, Hisamoto, Kawasaki 213, Japan

ITO, K. Department of Earth Sciences, Faculty of Science, Kobe University, Nada, Kobe 657, Japan

JAMIESON, J. C. Department of the Geophysical Sciences, University of Chicago, Chicago, Ill. 60637

JEANLOZ, R. Seismological Laboratory, Division of Geological and Planetary Sciences, California Institute of Technology, Pasadena, CA 91125

KATAHARA, K. W. Hawaii Institute of Geophysics, University of Hawaii, Honolulu, Hawaii 96822

KAWAI, N. Department of Physics, Faculty of Engineering Science, Osaka University, Toyonaka, Osaka 560, Japan

KENNEDY, G. C. Institute of Geophysics and Planetary Physics, University of California, Los Angeles, Los Angeles, CA 90024

KINOMURA, N. Institute of Scientific and Industrial Research, Osaka University, Yamadakami, Suita, Osaka 565, Japan

KITAMURA, M. Institute of Mineralogy, Petrology and Economic Geology, Tohoku University, Aoba, Sendai 980, Japan

KNOPOFF, L. Institute of Geophysics and Planetary Physics, University of California, Los Angeles, Los Angeles, CA 90024

KOIZUMI, M. Institute of Scientific and Industrial Research, Osaka University, Yamadakami, Suita, Osaka 565, Japan

KUMAZAWA, M. Department of Earth Science, Nagoya University, Nagoya, Japan

KUME, S. College of General Education, Osaka University, Machikaneyama-cho, Toyonaka, Osaka 560, Japan

KUSHIRO, I. Geological Institute, University of Tokyo, Tokyo, Japan

LIEBERMANN, R. C. Research School of Earth Sciences, Australian National University, Canberra, A.C.T. 2600, Australia

LIU, L. Research School of Earth Sciences, Australian National University, Canberra, A.C.T. 2600, Australia

MANGHNANI, M. H. Hawaii Institute of Geophysics, University of Hawaii, Honolulu, Hawaii 96822

MAO, H. K. Geophysical Laboratory, Carnegie Institution of Washington, Washington, D.C. 20008

MARTIN, R. J., III Cooperative Institute for Research in Environmental Sciences, University of Colorado/NOAA, Boulder, Colorado 80309

MATSUI, Y. Institute for Thermal Spring Research, Okayama University, Misasa, Tottori-ken 682-02, Japan

MATSUSHIMA, S. Institute of Earth Science, School of Liberal Arts, Kyoto University, Kyoto, Japan

MING, L. C. Hawaii Institute of Geophysics, University of Hawaii, Honolulu, Hawaii 96822

MISHIMA, O. Department of Physics, Faculty of Engineering Science, Osaka University, Toyonaka, Osaka 560, Japan

MIZUTANI, H. Geophysical Institute, University of Tokyo, Hongo, Bunkyo-ku, Tokyo 113, Japan

NAKAGAWA, Y. The Research Institute for Iron, Steel and Other Metals, Tohoku University, Sendai 980, Japan

O'CONNELL, R. J. Department of Geological Sciences, Harvard University, Cambridge, Mass. 02138

OLINGER, B. Los Alamos Scientific Laboratory, Los Alamos, N.M. 87545

OTA, R. Institute of Geophysics and Planetary Physics, University of California, Los Angeles, CA 90024

PIERMARINI, G. J. Institute for Materials Research, National Bureau of Standards, Washington, D.C. 20234

RINGWOOD, A. E. Research School of Earth Sciences, Australian National University, Canberra, A.C.T. 2600, Australia

SATO, Y. Meteorological College, Asahi-cho, Kashiwa, Chiba 277, Japan

SAWAMOTO, H. Department of Earth Science, Nagoya University, Nagoya, Japan

SCHOCK, R. N. University of California, Lawrence Livermore Laboratory, Livermore, CA 94550

SOGA, N. Cooperative Institute for Research in Environmental Sciences, University of Colorado/NOAA Boulder, Colorado 80309

SPETZLER, H. Cooperative Institute for Research in Environmental Sciences, University of Colorado/NOAA, Boulder, Colorado 80309

STROMBERG, H. D. University of California, Lawrence Livermore Laboratory, Livermore, CA 94550

SUITO, K. Department of Material Physics, Faculty of Engineering Science, Osaka University, Toyonaka, Osaka-fu 560, Japan

SYONO, Y. The Research Institute for Iron, Steel and other Metals, Tohoku University, Sendai 980, Japan

TOGAYA, M. Department of Physics, Faculty of Engineering Science, Osaka University, Toyonaka, Osaka 560, Japan

WYLLIE, P. J. Department of the Geophysical Sciences, University of Chicago, Chicago, Ill. 60637

YAGI, T. Institute for Solid State Physics, University of Tokyo, Roppongi, Minato-ku, Tokyo 106, Japan

YAMAMOTO, K. Water Research Institute, Nagoya University, Chikusa-ku, Nagoya 464, Japan

PREFACE

In the past decade significant advances have been made in the fields of high pressure–high temperature research related to the geophysics and geochemistry of the earth's interior. New innovations in the design of pressure–temperature apparatus (e.g., the diamond anvil pressure cell and laser heating system in the U.S.A., the multiple anvil sliding system (MASS) and split sphere apparatus in Japan) and breakthroughs in pressure in calibration techniques (e.g., ruby fluorescence scale) have enabled studies at pressures and temperatures not attained before in the laboratory. Recent data especially have provided new and relatively more accurate knowledge of the structure–property relations for earth materials and their analogs. Such knowledge, together with shock wave data, has enabled substantial elucidation of the composition and structure of the earth's interior.

In the light of these breakthroughs, the time was most opportune for scientists in high pressure–high temperature research to get together and exchange ideas on the latest state-of-the-art developments, their experimental results, and their latest interpretations with regard to the significance of these results to the geophysical sciences in general.

Under the U.S.–Japan Cooperative Science Program sponsored by the U.S. National Science Foundation (NSF) and the Japan Society for the Promotion of Science (JSPS), a four-day U.S.–Japan joint seminar was held in Honolulu, Hawaii, 6–9 July 1976. It was fortunate that, besides 26 participants from the U.S.A. and 19 from Japan, three participants and observers from Australia and one from France were also able to attend and contribute an international atmosphere. It was unfortunate that the high pressure research colleagues invited from the U.S.S.R. could not participate.

Four formal sessions, consisting of morning and afternoon subsessions, were held. After the formal presentation of papers, informal discussions were held covering topics of the day's sessions. No attempt was made to formally record the spontaneous discussions so as not to hinder free exchange of ideas. The four topics covered in the program and the chairmen of the morning and afternoon subsessions are listed in the program.

Of the forty-two papers presented at the seminar, thirty-nine appear as contributed papers and three as abstracts in this volume. The editors appreciate the task of a number of reviewers in maintaining the high quality of the papers. Deep appreciation is expressed to the U.S. National Science Foundation and the Japan Society for the Promotion of Science for their fianancial support and guidance for the seminar, to the Logistics Office of the East-West Center, in particular, Mr. J. McMahon, for arranging the many phases of this seminar and for the use of the Asia Room at the Center, to Dr. H. P. McKaughan, Dean of the Graduate School and Director of the Office of Research Administration, University of Hawaii, for financial support for the publication of the proceedings, to Dr. J. P. Craven, Dean of Marine Programs, University of Hawaii, for financial assistance for the seminar, and to Ms. Cynthia Garver for assistance in the editorial work. Sincere thanks are due Mrs. J. Taga, who also assisted in the editorial work and untiringly typed a number of the manuscripts, and Miss K. Lee for drafting.

The seminar committee wishes to thank all the participants, contributors, and session chairmen for making the first U.S.–Japan Seminar on High Pressure Research Application in Geophysics a tremendous success.

MURLI H. MANGHNANI

SYUN-ITI AKIMOTO

PROGRAM

SESSION I. Geophysics and Geochemistry of the Crust and Upper Mantle

CHAIRMEN: *M. Kumazawa and T. J. Ahrens*
K. Ito and P. J. Wyllie

SESSION II. Phase Transitions Related to Earth's Deep Interior

CHAIRMEN: *J. C. Jamieson and N. Kawai*
W. A. Bassett and I. Kushiro

SESSION III. Equations of State and Shock Wave Experiments

CHAIRMEN: *R. G. McQueen and H. Mizutani*
O. L. Anderson and Y. Syono

SESSION IV. Instrumentation, Pressure Calibration, and Standardization

CHAIRMEN: *J. Osugi and G. J. Piermarini*
S. Kume and T. J. Shankland

SEMINAR COMMITTEE

UNITED STATES

Murli H. Manghnani
(coordinator)

John C. Jamieson
(co-sponsor)

Don L. Anderson
(co-sponsor)

JAPAN

Syun-iti Akimoto
(coordinator)

Hitoshi Mizutani
(co-sponsor)

SESSION I

GEOPHYSICS AND GEOCHEMISTRY OF THE CRUST AND UPPER MANTLE

A BRIEF SUMMARY OF OUR PRESENT DAY UNDERSTANDING OF THE EFFECT OF VOLATILES AND PARTIAL MELT ON THE MECHANICAL PROPERTIES OF THE UPPER MANTLE

C. GOETZE
Department of Earth and Planetary Sciences
Massachusetts Institute of Technology
Cambridge, Massachusetts 02139

Abstract

The attenuation and velocity in the LVZ are linked through the approximate dispersion relation

$$\frac{1}{Q} \simeq \frac{\pi}{2} \cdot \frac{d\ell n\ (modulus)}{d\ell n\ (frequency)}$$

which holds for all linear attenuation effects regardless of mechanism. The attenuation level alone cannot be used to predict the depression in velocity, but it is reasonable to seek a common explanation for both. Models based on liquid melt pockets in an elastic solid predict attenuation/modulus effects which range widely in magnitude and characteristic frequency, depending critically on the melt geometry. This geometry is at present not known. There is therefore at present no hope of inverting the seismic data to yield melt fraction. Ubiquitous and substantial subsolidus attenuation/modulus effects have been documented for many metals, salts and ceramics; however, it is not possible at present to predict the attenuation level due to this source under upper mantle conditions. The seismic data alone do not permit a choice between a melt or subsolidus explanation for the LVZ.

The creep properties of mantle material are certain to be affected by partial melt only if the melt fraction exceeds about 25-30%. Extrapolated laboratory creep data on unmelted olivines are compatible with geophysical evidence of the rheology of the upper mantle. Evidence has accumulated that hydrogen within the olivine lattice ("wet olivine") leads to mechanical softening. If enough hydrogen can be accommodated in the lattice, it will further complicate the line of reasoning proposed by Wyllie and other petrologists about the melt fraction in the LVZ based on the crystallographic site of mantle water. It appears to this author that no reasonable case can be made at present for any correlation other than empirical between regions of high attenuation, regions of partial melt, and the zone of decoupling of the continents.

I. INTRODUCTION

The idea that the abnormally low shear velocity and high attenuation in the Low Velocity Zone result from partial melting of the upper mantle and that this causes the reduction in creep strength that allows the plates to decouple from the asthenosphere is an intuitively attractive idea [*Press*, 1959; *Anderson and Sammis*, 1970; *Lambert and Wyllie*, 1970; *Solomon*, 1972]. It has proved to be a very effective catalyst, bringing together a number of disciplines on a single geophysical problem of outstanding importance [*Wyllie and Schreyer*, 1976]. However, the fact that seismologists are now using the top of the LVZ to map the asthenosphere/lithosphere boundary in considerable detail [*Leeds et al.*, 1974; *Leeds*, 1975] while others are computing the melt microstructure and lithospheric thickness within different portions of the LVZ by combining seismic and petrologic data [*Green and Liebermann*, 1976] and yet others are determining mantle temperatures from the degree of melting [*Solomon*, 1972, 1975] indicates to this author that the idea LVZ = partial melt = zone of decoupling is calcifying into fact through repetition rather than the introduction of new evidence. It is therefore timely to review the progress made, if any, in providing a firmer basis for what was initially admitted to be only an intuitive idea [*Press*, 1959].

1. Are the attenuation and velocity effects traceable to the same cause?

Experimentally, attenuation effects, of whatever cause, appear to be linear (independent of amplitude) if strain amplitudes are below 10^{-5}-10^{-6}. Seismic amplitudes are comfortably below this value. It has long been known that a single time-dependent function, such as $1/Q(\nu)$ where ν is the frequency, completely characterizes a linear rheology. This follows directly from linearity, not from any understanding of the mechanism [*Zener*, 1948; *Bergen*, 1960]. During sinusoidal oscillation, for example, the real (M') and imaginary (M'') parts of any modulus[1] which relate stress and strain can be written as

$$
\begin{aligned}
M'(\nu) &= \int_0^\infty R(r)\,\frac{\nu^2}{\nu^2+r^2}\,dr + M(0) \\
M''(\nu) &= \int_0^\infty R(r)\,\frac{\nu r}{\nu^2+r^2}\,dr
\end{aligned}
\tag{1}
$$

[1]Attenuation in the LVZ appears to occur largely or exclusively in the shear modulus and not in the bulk modulus [*Anderson et al.*, 1965].

where $R(r)$ is a non-negative function which must be found by integral inversion from an experimentally determined $M'(\nu)$ or $M''(\nu)$ [*Zener*, 1948; *Bergen*, 1960; *Goetze*, 1969]. The modulus appropriate to the sonic velocities is given by

$$M_{sonic} = \sqrt{(M')^2 + (M'')^2}$$

while (2)

$$\frac{1}{Q}(\nu) = \frac{M''(\nu)}{M'(\nu)}$$

Approximate relations between M' and M'' have come into use in polymer rheology [*Schwarzl and Staverman*, 1952; *Leaderman*, 1958] of which the most useful is

$$\frac{1}{Q}(\nu) = \frac{\pi}{2} \cdot \frac{d\ell n\ M'(\nu)}{d\ell n\ \nu} \qquad (3)$$

which relates the attenuation and dispersion at each frequency.

The sharpest frequency dependence occurs if $R(r)$ is a delta function, in which case the behavior is that of the "standard linear solid" shown in Figure 1a. Two parameters, $\Delta M/M(0)$ and ν_0, completely specify this behavior. A more distributed $R(r)$ which is compatible with typical LVZ characteristics ($\Delta M/M = 20\%$, $1/Q = .025$[2]) is shown in Figure 1b. At ultrasonic frequencies the dispersion and hence attenuation (1) are believed to be small, although this has seldom been tested above a few hundred degrees. In the range of seismic frequencies, attenuation is taken to be near $1/Q = .025$, requiring a frequency dependence of the modulus of 3-4%/decade in frequency. This region is also known to creep at an effective viscosity which we will assume, for discussion purposes, to be $\eta = 10^{21}$ poise. This requires a final drop of stiffness near $\nu \cong \mu/\eta \cong 10^{-9}$hz [*Zener*, 1948]. In between, modulus and dispersion can have arbitrary positive values provided they can be derived from a positive $R(r)$ according to (1), which is to say they must be fairly gently varying.

---

[2]These figures are taken from a comparison of shear velocities [*Biswas and Knopoff*, 1974] and shear attenuation [*Lee and Solomon*, 1975] with depth under various regions in the continental United States:

	Velocity drop entering LVZ	Peak attenuation ($1/Q$)
Western U.S.	10-16%	.04-.06
South Central U.S.	7-10%	.01-.03
North Central U.S.	3-7%	
"Typical Values"	10%	.025

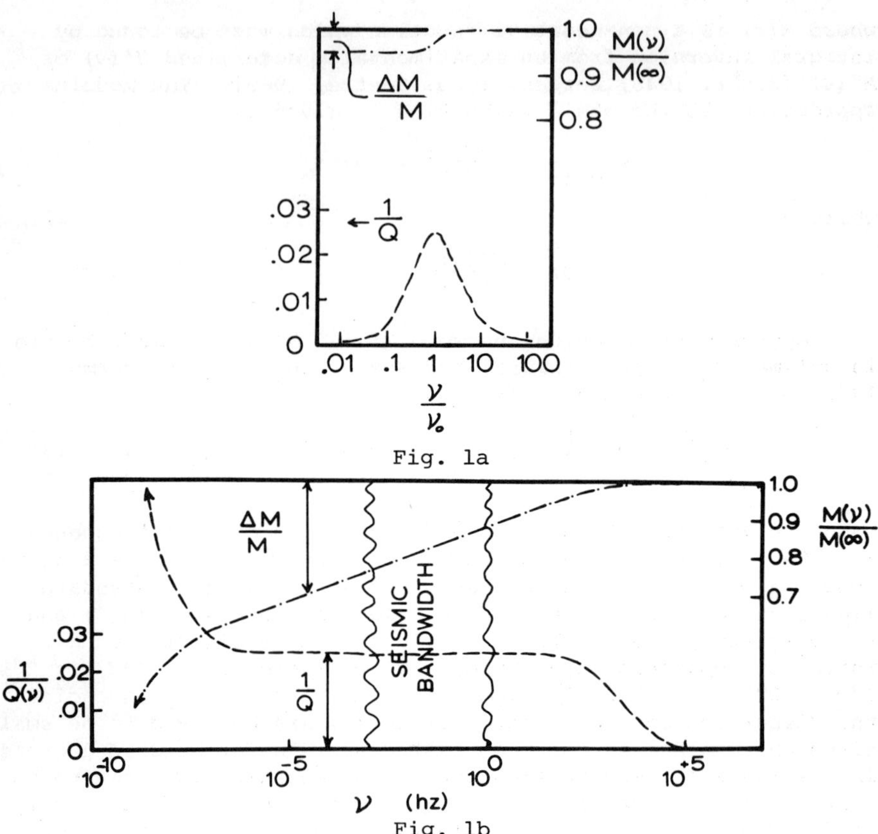

Fig. 1a

Fig. 1b

Fig. 1. Frequency dependence of the attenuation and shear modulus for (a) a standard linear solid, and (b) a more realistic model for the LVZ (see text).

At a single frequency, modulus defect ($\Delta M/M$) and attenuation ($1/Q$) are not uniquely related. However, even if R were a delta function, the observed attenuation would not be compatible with a relaxation effect of $\Delta M/M(0) < 5\%$ (Figure 1a). If this were the case, then over half of the relaxation in modulus might result from mineralogical layering [*Clark and Ringwood*, 1964] and be completely unrelated to the attenuation. On the other hand, by extending the attenuation and dispersion to very high frequencies, frequencies which can never be samples by seismic methods, $\Delta M/M$ can be made arbitrarily large. The fact that a modulus defect is expected in an attenuating region and the fact that the observed modulus defect and attenuation appear to be confined to the same region in the earth [*Anderson et al.*, 1965; *Solomon*, 1972; *Biswas and Knopoff*, 1974, vs. *Lee and Solomon*, 1975] makes a fairly strong case that they result from the same relaxation mechanism. Direct seismic evidence that the modulus and attenuation are linked through the dispersion relation (1) or (3) would prove this point. However, in view of the difficulty in obtaining depth and frequency resolution simultaneously by seismic methods, this evidence is not likely to be forthcoming.

2. Can partial melting account for the observed attenuation and modulus defect?

This question has been approached theoretically [*Oldroyd*, 1956; *Biot*, 1956; *Walsh*, 1969; *Mavko and Nur*, 1975; *O'Connell and Budiansky*, submitted to JGR] and experimentally [*Köster and Bangert*, 1951; *Shimozuru*, 1956; *Knopoff*, 1959; *Mizutani and Kanamori*, 1964; *Umekawa et al.*, 1965; *Pokorny*, 1965; *Spetzler and Anderson*, 1968; *Nur*, 1971; *Arzi*, 1972, 1974; *Benzing*, 1974; *Stocker and Gordon*, 1975]. The starting point for all theoretical models has been a non-attenuating elastic homogeneous matrix containing inclusions of viscous fluid. It should be borne in mind that in real materials near their melting point the attenuation of the solid components will be added to the attenuation predicted by these models. Models differ in the geometry of the fluid inclusions. A few simple models illustrate the possible range of effects.

The exact solution for $1/Q(\nu)$ and $\Delta M/M(\nu)$ for a dilute dispersion of ellipsoidal-shaped inclusions was first given by *Walsh* [1969]. The behavior for inclusions of a single aspect ratio (α) is that of the standard linear solid (Figure 1a) with parameters

$$\frac{\Delta M}{M}(0) = \frac{C}{2\alpha}$$

$$\nu_0 = \frac{\alpha\mu}{20\eta}$$

where C is the melt concentration by volume, μ the shear modulus of the matrix, and η the fluid viscosity[3].

For spherical or equi-axed inclusions $\Delta M/M(0) \sim C/2$ and $\nu_0 = 10^9$-10^{11}hz. The relaxation is very small and no attenuation is expected in a reasonable frequency range. Only an unreasonably small aspect ratio ($\alpha \sim 10^{-9}$) can bring ν_0 into the seismic range. However, $\Delta M/M$ can be made sensitive to very minute melt concentrations for aspect ratios which could reasonably correspond to 1% melt distributed along smooth grain boundaries ($\alpha \sim 10^{-2}$). However, under these conditions the Walsh model becomes unrealistic for several reasons.

---

[3] In practice, a distribution of aspect ratios is, of course, expected which would spread out the relaxation effect over a larger range of frequencies, perhaps much larger, than corresponds to the standard linear solid.

Firstly the relaxation in modulus which results from a small amount of grain boundary melting is caused by the relaxation of the shear stress supported by the grain boundaries [*Ké*, 1947]. *Zener* [1941], who first recognized that this relaxation could not depend on the melt fraction at small melt fraction, computed the relaxation strength to be 35% in the shear modulus, 20% in Young's modulus, and 0% in the bulk modulus with a minor dependence on Poisson's ratio. These numbers have been experimentally verified a number of times [*Zener,* 1948]. An improved grain boundary melting model, therefore, has a fixed relaxation strength $\Delta\mu/\mu$ = 35% and the same characteristic frequency as the Walsh model [*Stocker and Gordon,* 1975]. The Walsh model simply breaks down because the melt cavities are not widely dispersed as assumed.

Secondly the melt now has the opportunity to transfer from one cavity to another ("melt squirt"). *Mavko and Nur* [1975] and *O'Connell and Budiansky* [submitted to JGR] have extended the theory to take this into account. Since the flow of melt now depends on the third and fourth power of the dimensions and geometry of the interconnections, the uncertainties in any analysis increase tremendously and the characteristic frequencies for which relaxation occurs will undoubtedly show an immense spread. *Mavko and Nur* [1975] conclude that characteristic times extend to a matter of years while the shortest times near 10^{-10}sec occur for equi-axed isolated melt pockets.

The singular importance of the geometry of the melt was not appreciated in much of the early experimental work, and the interpretation of these results in terms of the models is therefore very ambiguous. *Spetzler and Anderson* [1968], *Goetze* [1969] and *Stocker and Gordon* [1975] have reported acoustic measurements in partially melted aggregates of various non-silicates which are believed to show wetting of the grain boundaries. However, even in these the microstructures were insufficiently characterized to allow more than qualitative comparison with theory. They did substantiate, however, that a modulus defect as high as 30% could be obtained from a few percent melt. The only acoustic experiments on partially melted silicates (granite) have been reported by *Benzing* [1974], but the few results obtained were somewhat puzzling and the distribution of glass not well characterized. *Arzi's* [1974] work on creep of partially melted granite makes it evident that the initial distribution of melt is controlled by previous fracturing, moisture content, distribution of hydrous phases and other kinetic considerations and is not the distribution expected after annealing in the partially melted condition (see Discussion).

The principal purpose of these studies has been to obtain a velocity-melt fraction curve which could be used to estimate the melt fraction in the LVZ and other regions of presumed partial melt. Figure 2 shows the experimental data assembled by *Stocker and Gordon* in 1975 for materials reported to wet (o) and not wet (●) the grain boundaries. Any of these experimental points is compatible with the Walsh theory if the aspect ratio is treated as a disposable parameter. It appears to this author that the effect of very small melt fractions on the sound velocity is practically unconstrained.

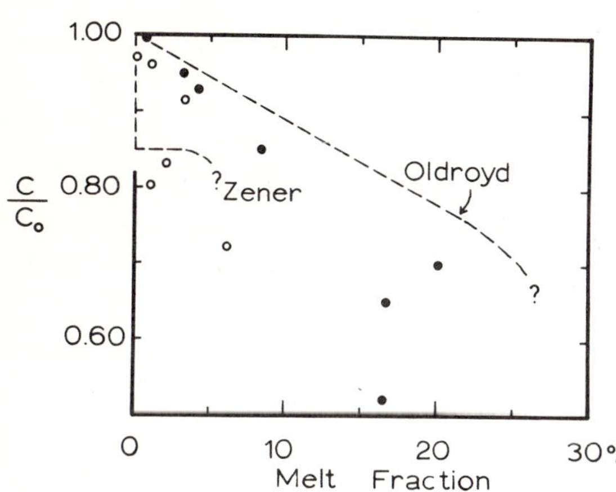

Fig. 2. Relative extensional velocity as a function of melt content after Stocker and Gordon [1975]. Curve labelled Oldroyd is the theoretical prediction for spherical melt cavities; curve labelled Zener is the theoretical prediction for grain boundary melting. Circles are experimental points from a number of different studies in which wetting behavior was reported for (o) and non-wetting for (•).

3. Is partial melting the only explanation for the observed attenuation and modulus defect?

Rocks have long been known to exhibit a high attenuation at room pressure [*Bradley and Fort*, 1966]. *Birch* [1942] and *Gordon and Davis* [1968] showed that this attenuation was suppressed with the application of 1-3 kb pressure. There is now general agreement that this attenuation effect is traceable in some way to the microscopic fractures which are common to all natural rocks at room pressure and has nothing to do with mantle attenuation. Nevertheless this troublesome effect eliminates the possibility of getting attenuation data on rock aggregates, meaningful for the mantle, without the application of pressure. To date no one has attempted to obtain attenuation data at simultaneous high temperature and pressure. The subsolidus attenuation data believed to be relevant to the mantle has therefore been largely restricted to metals, salts, and those hot-pressed aggregates of oxides which do not disaggregate upon heating.

Experimental data covering much of the subsolidus temperature range have been reported for aggregates of Al, Cu, Ag, Ni, Fe, Au, various metal alloys, NaCl, KCl, Al_2O_3, Y_2O_3, Zr_2O_3, Mullite, ThO_2, BeO, $MgAl_2O_4$, Mg_2SiO_4, SiC, TiC, as well as two-phase mixtures of Al_2O_3-Cr_2O_3, Al_2O_3-La_2O_3, Al_2O_3-SiO_2, BeO-MgO, KCl-AgCl, NaCl-KCl, and NaCl-NaF. Much of this literature was summarized by *Gordon and Nelson* [1966] and *Jackson and Anderson* [1970], although emphasis in these reviews was placed on the attenuation effects and not on the modulus defect. They are related, of course, through equation (1). Additional references may be found in *Goetze* [1969].

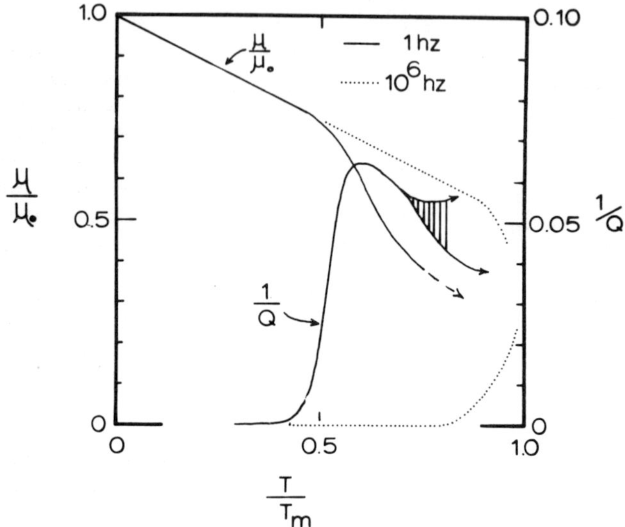

Fig. 3. Generalized curve showing the subsolidus relaxation of the shear modulus and corresponding attenuation (see text).

Qualitatively a very consistent picture emerges from this literature which is summarized in Figure 3. At low frequencies all aggregates show a very high attenuation as they approach the melting point (Tm)[4]. The broad peak in attenuation which is thought to result from the relaxation of shear stresses along the grain boundaries occurs, as rule of thumb, at $T/Tm = 0.4 - 0.7$ at a frequency of 1 hz and near $T/Tm = 1$ at 10^6 hz. Attenuation peak heights vary from $1/Q = .02 - .1$. In some cases the attenuation falls markedly after the first peak is reached; in others the peak is not clearly discernible. The effect of adding more constituents to the aggregate is to broaden the region of attenuation both in temperature and frequency. No exceptions to this general picture are known to me.

The characteristic frequencies of these relaxation effects are thermally activated and therefore strongly temperature sensitive. Rising temperature at constant frequency (Figure 3) yields

---

[4]This rule of thumb was developed principally for elements and simple compounds. The two-phase data listed above suggests that the corresponding temperature for multi-component systems such as rocks that best predicts the relaxation data is closer to the melting point of the component minerals than to the solidus of the rock.

a plot which is qualitatively similar to falling frequency at constant temperature (Figure 1). The data can be presented in either form.

Some of the older oxide data have been criticized because glass was found in the grain boundaries, and this was thought to have caused the relaxation of stiffness. However, the improvement in aggregates since that time and the generality of the effect even among very pure metals and salts leaves little doubt that such relaxations are characteristic of a wide range of materials and can probably be expected to occur in lherzolite as well. Whether they do so at seismic frequencies and mantle temperatures can only be established by experimentation.

A number of consequences follow from Figure 3:

(1) Subsolidus attenuation levels are large, typically $1/Q$ = .02 - .1, as compared to .01 - .05 in the LVZ, so peak subsolidus attenuation levels need not be reached to account for the attenuation and modulus defect observed in the LVZ.

(2) In common with partial melting and practically all mechanisms which can be expected to give a pronounced relaxation at high pressure, the acoustic effects shown in Figure 3 are negligible for the bulk modulus and most pronounced for the shear modulus, as is observed to be the case in the earth [*Goetze*, 1969; *Anderson et al.*, 1965].

(3) The extrapolation of $d\mu/dT$ measured at low temperatures or high frequencies or both, to seismic frequencies and LVZ temperatures, has frequently been used to discredit the idea that the modulus defect in the LVZ could result from thermal gradients under subsolidus conditions [*Birch*, 1969; *Anderson and Sammis*, 1970]. It can be seen from Figure 3 that this is likely to be grossly in error when subsolidus relaxation effects occur.

(4) The complete relaxation of shear stresses supported along the grain boundaries results in a relaxation of shear stiffness of approximately 35% [*Zener*, 1941]. This grain boundary relaxation has been identified as a principal contributor to the subsolidus acoustic effects shown in Figure 2. It is the same relaxation which was just postulated to occur as a result of partial melting along the grain boundaries. It may prove difficult to discern in the case of the LVZ whether this relaxation occurs in the subsolidus or with melting, or perhaps differently in different regions.

(5) Laboratory experiments will have to be done at very low frequencies in order to be useful in the interpretation of seismic data.

In conclusion, I do not believe I am saying anything controversial in concluding that the seismic data alone can be accounted for equally well by subsolidus or partial melting effects. *Jackson and Anderson* [1970], for example, conclude that the subsolidus relaxation effects shown in Figure 3 "are the most probable mechanism of seismic attenuation in the mantle, although partial melting may predominate in the LVZ of the upper mantle."

4. Is partial melting required to explain the decoupling of the plates?

Early objections to continental drift centered on the supposed finite strength of the mantle [*Jeffreys*, 1929, 1972], but experimental creep data on olivine has not confirmed this earlier intuition. A number of numerical experiments have shown that the "hardest" experimental flow laws of the mineral olivine are compatible with our meager knowledge of the stresses, strain rates, and temperatures at which plates decouple from the lower mantle [*Froidevaux and Schubert*, 1975; *Schubert et al.*, 1976]. Effects such as transformational superplasticity [*Dein and Sammis*, 1976], water weakening [*Carter and Ave'Lallement*, 1970; *Blacic*, 1972], or partial melting, if effective in weakening the flow law appropriate to the mantle, appear to predict embarrassingly weak behavior. However, such arguments, even if they could be made firm, cannot at present be used to exclude the possibility of a small degree of partial melt. To do this an understanding is needed of how partial melt influences the creep rate. Intuition has suggested to many that a small degree of partial melt should weaken the LVZ dramatically [*Press*, 1959; *Ringwood*, 1969; *Ritsema*, 1972].

5. Does partial melting even influence the creep rheology?

The modest increases in viscosity which result from a suspension of solids in a viscous fluid have been investigated in a wide variety of fields [*Roscoe*, 1952]. A little reflection will show that a critical point is reached in the range of melt concentrations from 20-40% in which particles will not be able to flow past each other but must be deformed or fractured[5]. According to the formulation of Roscoe, this point is reached at a volume melt concentration of 26%. In practice it is found to depend on the grain size distribution and the grain shapes. I will assume that we are concerned, in the LVZ, with melt concentrations well below this critical value for which the creep strength is comparable with, though perhaps lower than, that of the rock matrix.

At room pressure and low melt concentrations, partially melted ceramic aggregates are found experimentally to deform by a dilatant mechanism involving the introduction of air or gas to form enough total porosity for the solids to move past each other.

---

[5]This discussion does not consider the possibility of unusual grain geometries; for example, a perfect alignment of platy silicates in which slip on the foliation plane could presumably occur at very small melt fraction.

strength is provided by capillary attraction [*Allison et al.*, 1959]. Adequate simulation of mantle conditions therefore requires the application of external pressure in addition to the control of temperature, stress, and strain.

Arzi [1972, 1974] has measured the creep properties of partially melted granite under hydrostatic water pressure of 2 kb. Figure 4, which is modified from Arzi's thesis, shows the results of three successful runs on Westerly granite. Small (1-3%) transient strains resulted from the melt, but the terminal creep strengths which are plotted in Figure 4 did not differ significantly from the unmelted rock until 10% melt was exceeded. The solid line gives the viscosity computed from Roscoe's formula. The shaded band appears to be the region separating "rock-like" from "fluid-like" behavior of the aggregate. Microscopic inspection indicated that melt separated some of the grain boundaries but not others [*Arzi*, 1974].

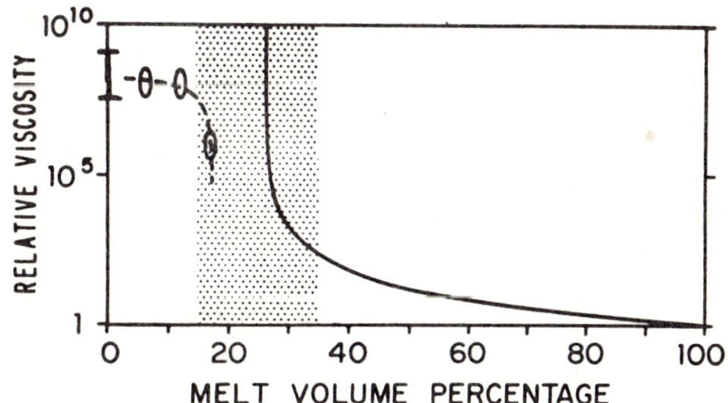

Fig. 4. The effect of partial melting on the creep properties of Westerly granite (adapted from Arzi [1974]).

Auten et al. [1974] and *Auten and Gordon* [1975] chose AlGe as a model material to explore creep under conditions of partial melt because the melt is known to be strongly wetting in this system. Again, a small transient was observed upon melting 1% of the specimen followed by creep at the subsolidus creep rate.

A few poorly controlled experiments in this laboratory [unpublished] were conducted under confining pressure at 1300-1400°C and 1 kbar differential stress on a partially melted dunite (Twin Sisters dunite) and lherzolite nodule (Dish Hill, Texas). Creep strengths were again similar to those of solid olivine.

The few experiments that have been done, therefore, appear to argue that small degrees of partial melt do not affect the creep properties. The only kind of mechanism which could provide a substantial creep enhancement at low melt concentration must involve the stress-induced melting and redeposition of matrix material under stress[6]. The current theories of nonhydrostatic thermodynamics indicate that such melting can happen [*Paterson*, 1973; *Robin*, 1974], but at present, to my knowledge, there is no data on which to base an estimate of the strain rate to be expected in the LVZ. Figure 5 shows a sketch of the relation between differential stress and strain rate in the LVZ. It is for illustrative purposes only. The hatched band shows the range of strain rates to be expected from unmelted olivine based on an extrapolation from experimental data to 1100°C [*Kohlstedt and Goetze*, 1974; *Kohlstedt et al.*, 1976]. A melt and redeposition mechanism is likely to be linear in stress, as shown by the lines A, B, C, and will therefore probably diverge markedly at lower stress and strain rate. Data already in hand (the dunite and lherzolite just mentioned) indicate that the melt and redeposition mechanism is unlikely to proceed at a rate faster than A. If this mechanism contributes a strain rate slower than C, then it will not be pertinent to the decoupling of continents since the solid components will creep faster in the strain rate range of interest. However, the rheology shown by B is quite possible; it would have important consequences for mantle rheology, and is very difficult to explore experimentally. Experiments are therefore unlikely to be conclusive as to whether a small melt concentration does or does not influence the rheology in the LVZ. The evidence on hand is against it.

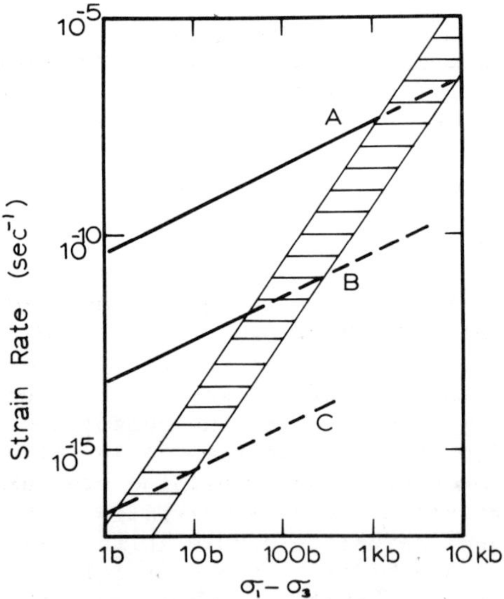

Fig. 5. Three curves illustrating a possible enhancement of the creep rate through partial melting in the LVZ (see text).

---

[6]Enhancement from stress concentrations resulting from the melt/solid microstructure are not regarded here as substantial.

6. Is partial melting known to be present in the LVZ?

The high temperature of the dry solidus of peridotite was originally seen as an obstacle to accepting the concept of a partially melted LVZ. Subsequently a number of authors, most recently and extensively *Mysen and Boettcher* [1975], have shown that the solidus of peridotite in the presence of excess water falls below the dry solidus by many hundreds of degrees. *Lambert and Wyllie* [1970] and *Wyllie* [1971] have discussed the solidus and melt fraction to be expected in the mantle under the assumptions (a) that there exists a small but poorly known amount of H_2O widely distributed in the mantle (for example 10^{-3} by weight) and (b) that this water must either reside in hydrous minerals such as amphibole, phlogopite, etc., occur as a vapor, or be dissolved in a melt. Using data on the water-saturated solidus of peridotite and data on the stability limits of appropriate hydrous minerals, Lambert and Wyllie predicted that perhaps 0.1-1.0% melt should exist along most currently accepted geotherms. *Green* [1973] later prepared samples containing 0.2% water added to pyrolite composition and appears to have verified Lambert and Wyllie's arguments. This simple picture has now become somewhat clouded by experiments on the effect of CO_2 on the peridotite solidus [*Mysen and Boettcher,* 1975; *Wyllie,* 1977]. Both these volatiles, in excess quantities, lower the solidus into the range at least of the oceanic geotherms, but a consensus on the solidus appropriate for the LVZ is not at hand [*Wyllie and Schreyer,* 1976].

It is important to appreciate that a common assumption in these petrologic studies, and particularly in the arguments of Lambert and Wyllie just stated, is that no hydrogen will be found as a cation impurity within the "anhydrous" solid phases such as olivine and pyroxenes. Adsorbed water on surfaces, submicroscopic inclusions of fluids or hydrous minerals, coupled with the analytic difficulties in detecting hydrogen, make the search for trace hydrogen as a distributed cation impurity in anhydrous phases notoriously difficult. It is this water which appears to have an important effect on the rheology of quartz and olivine and perhaps other phases as well [*Griggs,* 1967; *Carter and Ave'Lallement,* 1970; *Blacic,* 1972].

The subject of water as an impurity in quartz has been studied largely because of its effect on the acoustic properties of transducers [*Brunner et al.,* 1961; *Katz,* 1962; *Dodd and Fraser,* 1965; *Wilkins and Sabine,* 1973] and because of its effect on high temperature rheology [*Griggs and Blacic,* 1965; *Griggs,* 1967; *Baeta and Ashbee,* 1970; *Hobbs et al.,* 1972; *Griggs,* 1974]. The problem of water as an impurity in MgO has been the subject of intense investigation in recent years [*Glass and Searle,* 1967; *Henderson and Wertz,* 1968; *Henderson and Silbey,* 1971; *Chen et al.,* 1975; *Briggs,* 1975a, 1975b] principally as a result of studies to determine the effect of radiation on color centers. Much of this work has been involved with analytic techniques, identification of the crystallographic site of the hydrogen and its mobility.

At this stage almost no data is on hand to estimate the equilibrium water content under different conditions of temperature and fugacity of water, and perhaps also hydrogen and oxygen. Very large water contents have been observed in hydrothermally grown quartz crystals (H/Si up to 1%), but these may not be equilibrium water contents. Water has also been introduced in the melt, by high temperature electrolysis in the solid state [*Katz*, 1962] and by equilibrating at high temperatures under modest water pressures (25 bars, 1000°C in quartz, *Katz* [1962]; 2 kb, 1400°C in MgO, *Cheng* [unpublished]). Water contents in the range H/Si = 1 - 10 x 10^{-4} have been reached in these exploratory experiments. All that can be concluded from these studies is that MgO and quartz appear to develop an equilibrium impurity content of water which is enhanced by increasing water pressure. It also appears to this author that the bubbles in olivine crystals of lherzolite xenoliths from the mantle, identified by *Green and Radcliffe* [1975] as having precipitated on dislocations and dislocation subboundaries, do convincingly show that some volatile has been precipitated from the olivine lattice, either before or during its ascent from the mantle. Whether water concentrations which are believed to be typical of the mantle can be accommodated in this way remains to be explored experimentally. However, these studies do illustrate that water, as well as other trace elements, can be accommodated as impurities, or through slight non-stoichiometry, in ways that are not being tested by current petrologic techniques. These facts weaken the argument that water, if not accommodated within an identifiable hydrous phase, must form vapor or melt.

II. DISCUSSION

Today the situation is therefore as follows:

(1) The attenuation and modulus defect in the LVZ probably result from a common mechanism, but neither seismic, experimental, nor theoretical evidence can be used to infer whether the effect is due to partial melting or occurs below the solidus.

(2) The rheology may be influenced by a small degree of partial melt, but experiments so far have shown the contrary. The rheology shown by the mantle is compatible with experimental data on unmelted olivine.

(3) The temperature/pressure trajectory of the solidus in the upper mantle is still under considerable discussion, but it appears that either some or most geotherms will intersect regions of partial melt. Nevertheless, uncertainties about the melt fraction, about the microstructure of the melt pockets, and about the acoustic effects to be expected do not allow one to conclude

that partial melting is the cause of the attenuation and modulus defect observed seismically and therefore to correlate the observed boundaries of the LVZ with the solidus.

Directions in which one could look for progress on these questions might include the following:

(a) If the LVZ results from partial melting alone, it should have a very sharp inception with depth, perhaps a few kilometers wide. Subsolidus relaxation effects are thermally activated and the drop in modulus would occur over a few tens of kilometers. This point is illustrated in Figure 6, in which a velocity structure compatible with Figure 1b has been sketched for a geothermal gradient of 9°/km, using 85 kcal/mole[7] for the activation energy. The velocity gradient with depth is approximately proportional to the dispersion (hence attenuation), the temperature gradient, and the activation energy of the relaxation effect. Figure 6 serves simply to illustrate what may be a typical "sharpness" predicted for the subsolidus velocity curve. The sharpness in the earth could perhaps be investigated by seismic reflection techniques.

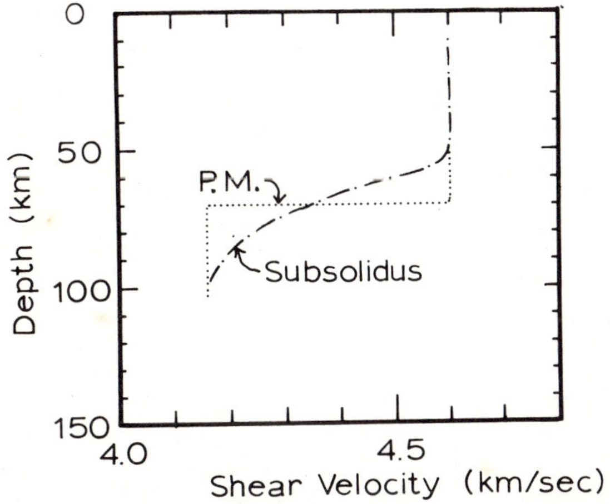

Fig. 6. Curve illustrating the "sharpness" of the LVZ lid as predicted by a subsolidus mechanism and by a partial melting mechanism. The subsolidus curve is computed from Figure 1b and a geothermal gradient of 9°/km. It occurs at an arbitrarily selected depth.

---

[7]Grain boundary diffusivity is believed to control the relaxation effects. The rule of thumb used by *Stocker and Ashby* [1973] for predicting the corresponding activation energy gives 83 kcal/mole for olivine. Grain-size sensitive hot-pressing data for olivine powders which are believed to be controlled by grain boundary diffusivity [*Goetze and Schwenn*, in preparation] give 85±29 kcal/mole.

(b) Further acoustic experimental work on partially melted materials is of little value until a consensus is reached about the solid/melt microstructure to be expected in the LVZ. This subject is complicated. Direct melting of natural rocks in the laboratory [*Krank and Oja*, 1960; *Unger*, 1967; *Mehnert et al.*, 1973; *Arzi*, 1974; *Benzing*, 1974] results in a melt distribution controlled by pre-existing fractures and the location of low melting point phases. Annealing for laboratory times appears to be unable to redistribute the melt into a geometry controlled by surface energies [*Arzi*, 1974; experience in this laboratory, unpublished]. The same may be said of the partial melting commonly observed in xenoliths that have intersected the solidus on their ascent to the surface [*Green et al.*, 1968]. One test of equilibrium of the melt distribution is to allow the microstructure to coarsen by perhaps a factor of three or more while in a partially melted condition. If the melt geometry remains unaltered during this coarsening, then it is in a dynamic equilibrium. To my knowledge, no studies have been reported which meet this test. It is hoped that the experience of experimental petrologists in dealing with partially melted specimens may yet contribute to this problem. A number of allusions have been made about the melt microstructure observed in equilibration experiments [*Green*, 1973; *Green and Liebermann*, 1976]. Experience in the ceramics industry indicates that the relevant surface energies are quite sensitive to small changes in temperature and composition of the melt. What is needed, therefore, is a complete report giving compositions, temperatures, equilibration times, and micrographs of the resultant microstructure.

Another question that arises in discussing the melt distribution is how long the melt will remain distributed on the scale of the grain structure. In sea ice small spherical and tubular cavities trap brine within the ice crystals during freezing. These cavities migrate at perhaps 1-2 meters a year, leaving comparatively fresh ice behind [*Anderson and Weeks*, 1958]. Melt will coalesce into large bodies very rapidly if partial melt cavities within the LVZ move at comparable velocities.

Yet another complication may arise if brittle fracturing occurs while in the partially melted condition. Where the external pressure is carried by a low-viscosity fluid, the effective pressure is zero and the behavior approximates that under conditions of no pressure [*Brace*, 1972]. Stresses in the order of 100 bars could therefore lead to fracturing and considerable redistribution of melt deep within the mantle.

An investigation of these questions would not only illuminate the study of acoustic properties of partially melted rocks but also much more fundamental questions of magma generation and mobility.

Acknowledgments. Support from National Science Foundation Grant No. GA-36280A is gratefully acknowledged. The two partial melting experiments on lherzolite and dunite were done by D.L. Kohlstedt.

REFERENCES

Anderson, D.L., and W.F. Weeks, A theoretical analysis of sea ice strength, *Trans. Amer. Geophys. Union, 39,* 632-640, 1958.

Anderson, D.L., A. Ben-Menahem, and C.B. Archambeau, Attenuation of seismic energy in the upper mantle, *J. Geophys. Res., 70,* 1441-1448, 1965.

Anderson, D.L., and C. Sammis, Partial melting in the upper mantle, *Phys. Earth Planet. Inter., 3,* 41-50, 1970.

Allison, E.B., P. Brock, and J. White, The rheology of aggregates containing a liquid phase, *Trans. Br. Ceram. Soc., 58,* 495-531, 1959.

Arzi, A., Experimental study of partial melting in natural rocks and subsequent creep under low stresses (Abstract), *Trans. Amer. Geophys. Un., 53,* 513, 1972.

Arzi, A., Partial melting in rocks: rheology, kinetics, water diffusion, PhD thesis, Harvard University, 1974.

Auten, T.A., R.B. Gordon, and R.L. Stocker, Q and creep in the mantle, *Nature, 250,* 317-318, 1974.

Auten, T.A., and R.B. Gordon, Compressive creep rate of partially melted AlGa alloys, *Metall. Trans., 6A,* 584-586, 1975.

Baeta, R.D., and K.H.G. Ashbee, Mechanical deformation of quartz, *Phil. Mag., 22,* 601-635, 1970.

Benzing, W.M., III, Experimental stress wave propagation in media containing liquid inclusions, PhD thesis, University of Chicago, 1974.

Bergen, J.T., *Viscoelasticity, Phenomenological Aspects,* Academic Press, New York, 212 pp., 1960.

Biot, M.A., Theory of propagation of elastic waves in a fluid saturated porous solid, *J. Acoust. Soc. Amer., 28,* 168-191, 1956.

Birch, F. (editor), *Handbook of Physical Constants,* Geological Soc. of Amer., Special Paper #36, 1942.

Birch, F., Density and composition of the upper mantle: first order approximation as an olivine layer, in *The Earth's Crust and Upper Mantle,* edited by P.D. Hart, Geophys. Monogr., 13, 18-36, Amer. Geophys. Un., Washington, D.C., 1969.

Biswas, N.N., and L. Knopoff, The structure of the upper mantle under the United States from the dispersion of Raleigh waves, *Geophys. J. Roy. Astron. Soc., 36,* 515-539, 1974.

Blacic, J.D., Effect of water on the experimental deformation of olivine, in *Flow and Fracture of Rocks,* edited by Heard, Borg, Carter, and Raleigh, Geophys. Monogr., 16, 109-115, Amer. Geophys. Un., Washington, D.C., 1972.

Brace, W.F., Pore pressure in geophysics, in *Flow and Fracture of Rocks,* edited by Heard, Borg, Carter, and Raleigh, Geophys. Monogr., 16, 265-274, 1972.

Bradley, J.T., and A.N. Fort, Jr., Internal friction in rocks, in *Handbook of Physical Constants*, edited by S.P. Clark, Jr., Geol. Soc. Amer. Memoir #79, 175-195, 1966.

Briggs, A., The formation of hydrogen-filled cavities in MgO crystals annealing in reducing atmospheres, *J. Mat. Sci.*, *10*, 737-746, 1975a.

Briggs, A., Hydroxyl impurity and the formation and distribution of cavities in melt-grown MgO crystals, *J. Mat. Sci.*, *10*, 729-736, 1975b.

Brunner, Von G.O., H. Wondratschek and F. Laves, Ultrarotuntersuchung über den Einbau von H in natürlichem Quartz, *Zeit. für Elektrochemie*, *65*, 735-750, 1961.

Carter, N.L., and H.G. Ave'Lallement, High temperature flow of dunite and peridotite, *Geol. Soc. Amer. Bull.*, *81*, 2181-2202, 1970.

Chen, Y., M.M. Abraham, L.C. Templeton, and W.P. Unruh, Role of hydrogen and deuterium on the V^- center formation in MgO, *Phys. Rev. B.*, *11*, 881-890, 1975.

Clark, S.P., Jr., and A.E. Ringwood, Density distribution and constitution of the earth's mantle, *Rev. Geophys.*, *2*, 35-88, 1964.

Dein, J.L., and C.G. Sammis, A new transformational superplasticity theory with application to the earth's mantle (Abstract), *Trans. Amer. Geophys. Union*, *57*, 323, 1976.

Dodd, D.M., and D.B. Fraser, The 3000-3900 cm^{-1} absorption bands and anelasticity of crystalline α-quartz, *J. Phys. Chem. Solids*, *26*, 673-686, 1965.

Froidevaux, C., and G. Schubert, Plate motion and structure of the continental asthenosphere: a realistic model of the upper mantle, *J. Geophys. Res.*, *80*, 2553-2565, 1975.

Glass, A.M., and T.M. Searle, Reaction between vacancies and impurities in magnesium oxide II: Mn^{+4} and OH^- ion impurities, *J. Chem. Phys.*, *46*, 2092-2102, 1967.

Goetze, C., High temperature elasticity and anelasticity of polycrystalline salts, PhD thesis, Harvard University, 1969.

Gordon, R.B., and L.A. Davis, Velocity and attenuation of seismic waves in imperfectly elastic rock, *J. Geophys. Res.*, *73*, 3917-3935, 1968.

Gordon, R.B., and C.W. Nelson, Anelastic properties of the earth, *Rev. Geophys.*, *4*, 457-474, 1966.

Green, H.W., and C.V. Radcliffe, Fluid precipitates in rocks from the earth's mantle, *Geol. Soc. Amer. Bull.*, *86*, 846-852, 1975.

Green, D.H., J.W. Morgan, and K.S. Heier, Thorium, uranium, and potassium abundances in peridotite inclusions and their host basalts, *Earth Planet. Sci. Lett.*, *4*, 155-166, 1968.

Green, D.H., Experimental melting studies on a model upper mantle composition at high pressure under water-saturated and water under-saturated conditions, *Earth Planet. Sci. Lett.*, *19*, 37-53, 1973.

Green, D.H., and R.C. Liebermann, Phase equilibria and elastic properties of a pyrolite model for the oceanic upper mantle, *Tectonophysics, 32,* 61-92, 1976.

Griggs, D.T., and J.D. Blacic, Quartz: anomalous weakness of synthetic crystals, *Science, 147,* 292-295, 1965.

Griggs, D.T., Hydrolytic weakening of quartz and other silicates, *Roy. Astron. Soc. Geophys. J., 14,* 19-31, 1967.

Griggs, D.T., A model of hydrolytic weakening in quartz, *J. Geophys. Res., 79,* 1653-1661, 1974.

Henderson, B., and J.E. Wertz, Defects in alkaline earth oxides, *Advances in Phys., 17,* 749-778, 1968.

Henderson, B., and W.A. Silbey, Studies of OH^- and OD^- ions in magnesium oxide. I. Distribution and annealing of hydroxyl and deuteroxyl ions, *J. Chem. Phys., 55,* 1276-1285, 1971.

Hobbs, B.E., A.C. McLaren, and M.S. Paterson, Plasticity of single crystals of synthetic quartz, in *Flow and Fracture of Rocks,* edited by Heard, Borg, Carter, and Raleigh, Geophys. Monogr., 16, 29-53, Amer. Geophys. Un., Washington, D.C., 1972.

Jackson, D.D., and D.L. Anderson, Physical mechanisms of seismic-wave attenuation, *Rev. Geophys., 8,* 1-64, 1970.

Jeffreys, H., *The Earth,* 2nd edition, Cambridge University Press, Cambridge, 346 pp., 1929.

Jeffreys, H., Creep in the earth and planets, *Tectonophysics, 13,* 569-581, 1972.

Katz, A., Hydrogen in α-quartz, *Phillips Research Reports, 17,* 133-279, 1962.

Kê, T.S., Experimental evidence of the viscous behavior of grain boundaries in metals, *Phys. Rev., 71,* 533-546, 1947.

Kohlstedt, D.L., and C. Goetze, Low-stress high-temperature creep in olivine single crystals, *J. Geophys. Res., 79,* 2045-2051, 1974.

Kohlstedt, D.L., C. Goetze, and W.B. Durham, Experimental deformation of single crystal olivine with application to flow in the mantle, in *The Physics and Chemistry of Minerals and Rocks,* edited by S.K. Runcorn, 35-49, John Wiley & Sons, Ltd., London, 1976.

Knopoff, L., Velocity of sound in a two-component system, *J. Geophys. Res., 64,* 359-363, 1959.

Köster, W., and L. Bangert, Über den Einfluss von Blei auf Elastizitäts Moduli and Dämpfung von Kupfer, Silber, Messing, und Rotguss, *Zeit. Metall. R., 42,* 391-394, 1951.

Krank, E.H., and R.V. Oja, Experimental studies of anatexis, *International Geological Congress, 14,* 16-29, 1960.

Lambert, I.B., and P.J. Wyllie, Melting in the deep crust and upper mantle and the nature of the low velocity layer, *Phys. Earth Planet. Interiors, 3,* 316-322, 1970.

Leaderman, H., Viscoelastic phenomena, in *Rheology,* edited by F.R. Eirich, 1-61, Academic Press, New York, 1958.

Lee, W.B., and S.C. Solomon, Inversion schemes for surface wave attenuation and Q in the crust and the mantle, *Geophys. J. Roy. Astron. Soc., 43,* 47-71, 1975.

Leeds, A.R., L. Knopoff, and E.G. Kausel, Variations of upper mantle structure under the Pacific Ocean, *Science, 186,* 141-143, 1974.

Leeds, A.R., Lithospheric thickness in the western Pacific, *Phys. Earth Planet. Interiors, 11,* 61-64, 1975.

Mavko, G., and A. Nur, Melt squirt in the asthenosphere, *J. Geophys. Res., 80,* 1444-1448, 1975.

Mehnert, K.R., W. Busch, and G. Schneider, Initial melting at grain boundaries of quartz and feldspar in gneisses and granulites, *Neues Jahrb. Min. Abh., 4,* 165-183, 1973.

Mizutani, H., and H. Kanamori, Variation of elastic wave velocity and attenuative property near the melting temperature, *J. Phys. Earth, 12,* 43-49, 1964.

Mysen, B.O., and A.L. Boettcher, Melting of a hydrous mantle: I, Phase relations of natural peridotite at high temperatures with controlled activities of water, carbon dioxide, and hydrogen, *J. Petrol., 16,* 520-548, 1975.

Nur, A., Viscous phases in rocks and the low velocity zone, *J. Geophys. Res., 76,* 1270-1278, 1971.

Oldroyd, J.G., The effect of small viscous inclusions on the mechanical properties of an elastic solid, in *Deformation and Flow of Solids,* edited by R. Gammel, 304-313, Springer, Berlin, 1956.

Paterson, M.S., Nonhydrostatic thermodynamics and its geologic applications, *Rev. Geophys., 11,* 355-390, 1973.

Pokorny, M., Variation of velocity and attenuation of longitudinal waves during solid-liquid and liquid-solid transition in woods alloy, *Ceskoslovenska Akad. Ved. Stud. Geophys. et Geod., 9,* 250-258, 1965.

Press, F., Some implications on mantle and crustal structure from G waves and Love waves, *J. Geophys. Res., 64,* 565-568, 1959.

Ringwood, A.E., Composition and evolution of the upper mantle, in *The Earth's Crust and Upper Mantle,* edited by P.J. Hart, Geophys. Monogr., 13, 1-17, Am. Geophys. Un., Washington, D.C., 1969.

Ritsema, A.R., Synopsis of some comments on Sir Harold Jeffrey's paper on creep in the earth and planets, *Tectonophysics, 13,* 579-580, 1972.

Robin, P.F., Thermodynamic equilibrium across a coherent interface in a stressed crystal, *Amer. Mineral., 59,* 1286-1298, 1974.

Roscoe, R., The viscosity of suspension of rigid spheres, *Br. J. Appl. Phys., 3,* 267-269, 1952.

Schubert, G., C. Froidevaux, and D.A. Yuen, Oceanic lithosphere and asthenosphere: thermal and mechanical structure, *J. Geophys. Res., 81,* 3525-3540, 1976.

Schwarzl, F., and A.D. Staverman, Higher approximations of relaxation spectra, *Physica, 18*, 791-798, 1952.

Shimozuru, P., Study on the elasticity near the melting point. Part II: Velocity of the dilatational waves in sodium, *Bull. Earthquake Res. Inst., Tokyo Univ., 34*, 87-96, 1956.

Solomon, S.C., Seismic-wave attenuation and partial melting in the upper mantle of North America, *J. Geophys. Res., 77*, 1483-1502, 1972.

Solomon, S.C., Geophysical constraints on radial and lateral temperature variations in the upper mantle, Extended abstracts, *International Conference on Geothermometry and Geobarometry*, Penn. State University, Oct. 5-10, 1975.

Spetzler, H., and D.L. Anderson, The effect of temperature and partial melting on velocity and attenuation in a simple binary system, *J. Geophys. Res., 73*, 4333-4337, 1968.

Stocker, R.L., and M.F. Ashby, On the rheology of the upper mantle, *Rev. Geophys., 11*, 391-426, 1973.

Stocker, R.L., and R.B. Gordon, Velocity and internal friction in partial melts, *J. Geophys. Res., 80*, 4828-4836, 1975.

Umekawa, S., R. Kotfila, and O.D. Sherby, Elastic properties of a tungsten-silver composite above and below the melting point of silver, *J. Mech. Phys. Solids, 13*, 229-235, 1965.

Unger, J.D., Melting of granite under effective confining pressure, MS thesis, Massachusetts Institute of Technology, 1967.

Walsh, J.B., New analysis of attenuation in partially melted rock, *J. Geophys. Res., 74*, 4333-4337, 1969.

Wilkins, R.W.T., and W. Sabine, Water content of some nominally anhydrous silicates, *Am. Mineral., 58*, 508-516, 1973.

Wyllie, P.J., Role of water in magma generation and initiation of diapiric uprise in the mantle, *J. Geophys. Res., 76*, 1328-1338, 1971.

Wyllie, P.J., and W. Schreyer, Geophysical measurements and experimental petrology, *Tectonophysics, 32*, 1-6, 1976.

Wyllie, P.J., Experimental studies on the influence of CO_2 and H_2O in the upper mantle, in *High Pressure Research: Applications to Geophysics*, edited by M.H. Manghnani and S. Akimoto, Academic Press, New York, 77-106, 1977.

Zener, C., Theory of the elasticity of polycrystals with viscous grain boundaries, *Phys. Rev., 60*, 906-908, 1941.

Zener, C., *Elasticity and Anelasticity of Metals*, Chap. V, University of Chicago Press, Chicago, Ill., 1948.

PHASE TRANSFORMATION IN SILICATE MELTS UNDER UPPER-MANTLE CONDITIONS

I. KUSHIRO[1]

Geological Institute
University of Tokyo
Tokyo, Japan

Abstract

Viscosities of silicate melts of basaltic and andesitic compositions decrease with increasing pressure at constant temperature, the rate of decrease depending on the composition of the melt. The viscosities also decrease with increasing temperature by factors of 2 to 3 per 100°C at 1 atm. Volatile components also have significant effects on the viscosity of silicate melts: H_2O reduces the viscosity, the effect being greater for more silica-rich melts, whereas CO_2 and P_2O_5 may increase the viscosity.

Viscosity and density changes of silicate melts of several different compositions at high pressures suggest that the phase transformations take place in silicate melts. The pressure-induced phase transformations in silicate melts are continuous and gradual compared to those observed in crystalline silicates. It is suggested that magmas formed in the upper mantle would be denser and less viscous at greater depths than those formed at shallow depths.

I. INTRODUCTION

Most studies on the origin of magmas derived from the upper mantle have been concerned with the chemistry of magmas, or the relations between the chemical compositions of magmas and the physical conditions under which the magmas are formed, yet very few studies have been made on the physical properties of magmas

---

[1]Now at the *Geophysical Laboratory* of the *Carnegie Institution of Washington, Washington, D. C., U. S. A.*

under the upper mantle conditions. For a better understanding of the problems of magma genesis, including mechanisms of separation and ascent of magma, physical properties of magmas at high pressures must be determined.

As an extension of the studies on the nature of silicate melts, *Kushiro* [1976] and *Kushiro et al.* [1976] have determined the viscosities of synthetic and natural rock melts at high pressures. These two studies will be briefly reviewed, and possible phase transformations in silicate melts under upper mantle conditions will be outlined.

II. EFFECT OF PRESSURE ON VISCOSITY

Viscosities of silicate melts of many different compositions have been measured at 1 atm, whereas very few measurements have been made at high pressures, and the effect of pressure on the viscosity of silicate melts has not been determined in detail. Recently, *Kushiro* [*1976*] and *Kushiro et al.* [1976] determined the viscosities of both synthetic and natural rock melts at pressures between 5 and 30 kbar, and measured the change in viscosity of these melts with pressure and temperature. The effect of pressure on the viscosity was particularly well demonstrated in the $NaAlSi_2O_6$ melt: the viscosity decreases isothermally from 6.8×10^4 poise at 1 atm to 5.3×10^3 poise at 24 kbar at 1350°C [*Kushiro*, 1976]. It was also shown that the viscosity of melt of $Na_2Si_3O_7$ composition decreases by a factor of 3 from 1 atm to 20 kbar at 1175°C. From the results on melts of $NaAlSi_2O_6$ and $Na_2Si_3O_7$ compositions (Figure 1), it is concluded that the large viscosity drop in $NaAlSi_2O_6$ melt would most probably be due to the presence of Al. The results of *Kushiro* [1976] support the suggestion of *Waff* [1975] that Al transforms from 4- to 6-fold coordination in the melt at high pressures, and such transformation results in the depolymerization of the Si-O-Al-O-Si bonds and the decrease in melt viscosity.

Viscosity of another Al-free synthetic melt ($K_2O \cdot MgO \cdot 5SiO_2$ composition) has been determined at high pressures for comparison. In the system K_2O-MgO-SiO_2, the liquidus field of enstatite ($MgSiO_3$) expands relative to that of forsterite (Mg_2SiO_4) with increasing pressure [*Kushiro*, 1975]. This evidence may be related to the change of polymerization of the Si-O bonds with pressure. The composition $K_2MgSi_5O_{12}$ examined in those experiments lies near the extension of the forsterite-protoenstatite liquidus boundary at 1 atm. The method followed in the experiments was the same as that described by *Kushiro* [1976]. The starting materials were powdered glass made by the late

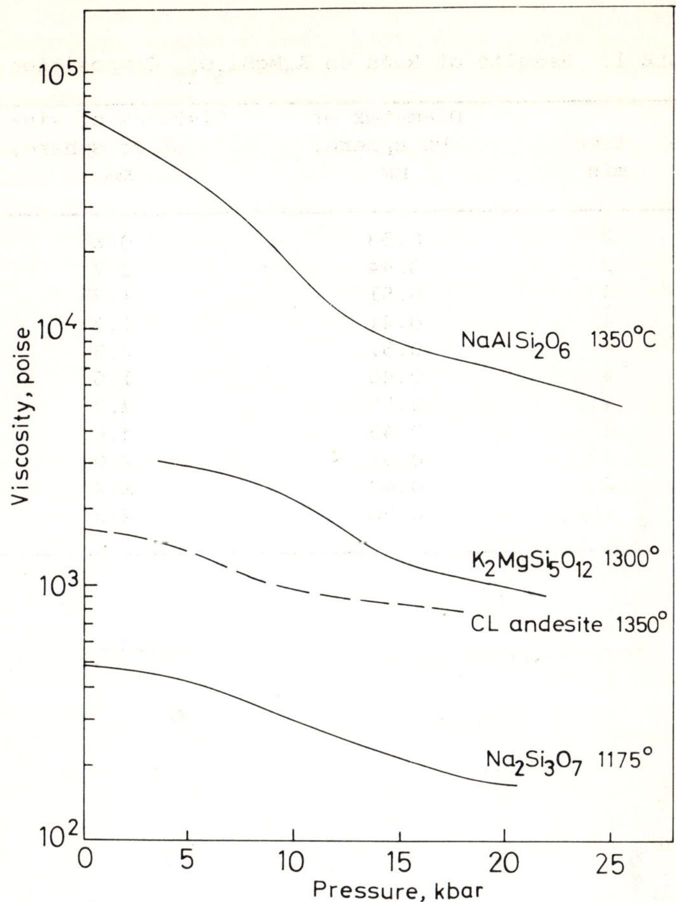

Fig. 1. Changes in viscosity with pressure of three synthetic silicate melts and one natural rock melt at constant temperature. Data from Kushiro [1976] for $NaAlSi_2O_6$ and $Na_2Si_3O_7$ melts, Kushiro et al. [1976] for Crater Lake (CL) andesite melt, and present study for $K_2MgSi_5O_{12}$ melt.

Dr. J. F. Schairer. The results of the experiments at 1300°C are given in Table 1, and the relations between time and distance of sinking of platinum spheres are shown in Figure 2. The lines connecting points obtained at 5 and 20 kbar extrapolate

TABLE 1. Results of Runs on $K_2MgSi_5O_{12}$ Composition

Pressure kbar	Time min	Diameter of Pt sphere, mm	Distance of sinking of Pt sphere, mm
5	2	0.53	0.85
5	2	0.44	0.7
5	4	0.51	1.75
5	4	0.41	1.3
10	4	0.51	2.2
10	4	0.40	1.6
15	4	0.55	4.7
15	4	0.45	3.6
20	2	0.51	2.0
20	2	0.40	1.2
20	4	0.50	4.5

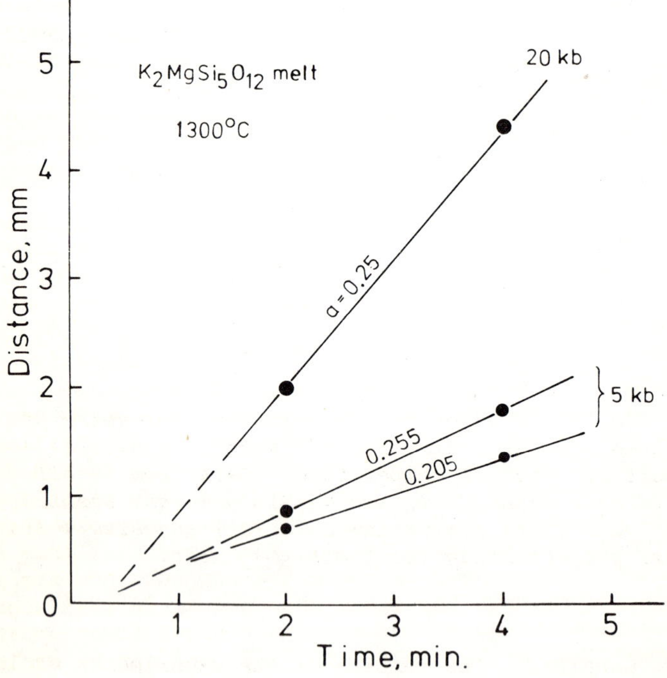

Fig. 2. The relations between time and distance of sinking of platinum spheres in a melt of $K_2MgSi_5O_{12}$ composition at 1300°C at 5 and 20 kbar. a indicates radii (mm) of platinum spheres.

to near 0 on the time axis, indicating that the platinum spheres begin to sink as soon as the temperature reaches 1300°C. The charges quenched after 2 and 4 min runs were clear glass, and no crystals were observed under the microscope.

The viscosity of the melt of $K_2MgSi_5O_{12}$ composition was calculated by the Stokes' law with the corrections for wall effects (Faxen correction) applied by *Shaw* [1963]. The data at 1300°C (Figure 1) are as follows: 2900 poise (5 kbar), 2200 (10 kbar), 1200 (15 kbar), and 1050 (20 kbar). (Figure 1). The viscosity decreases by a factor of 2.8 from 5 to 20 kbar. It is interesting to note that the rate of decrease in viscosity is nearly identical to that for the $Na_2Si_3O_7$ melt from 1 atm to 20 kbar.

The cause of the decrease in viscosity of these Al-free melts is not fully understood at present. One possible cause is a structural change of the melts associated with the coordination change of alkali ions, Mg^{2+} and Si^{4+}. Another possible cause is the formation of discrete ions at high pressures, such as $(Si_3O_9)^{6-}$, $(Si_4O_{12})^{8-}$, $(Si_6O_{15})^{6-}$ and $(Si_9O_{21})^{6-}$ suggested by *Bockris et al.* [1955]. Transformation of the Si-O chains to the above discrete ions should reduce the viscosity of silicate melts. Expansion of the liquidus field of enstatite relative to that of forsterite with increasing pressure in the system K_2O-MgO-SiO_2 could be explained by the formation of the $(Si_3O_9)^{6-}$ or $(Si_4O_{12})^{8-}$ ions in the melt.

The effect of pressure on viscosity is smaller for natural basalt and andesite melts than for synthetic silicate melts. The viscosity of Crater Lake calc-alkaline andesite melt decreases from 1660 poise at 5 kbar to 850 poise at 15 kbar at 1350°C, and that of Kilauea 1921 olivine tholeiite melt decreases from 22.4 poise at 1 atm to 15 poise at 20 kbar [*Kushiro et al.*, 1976]. The rate of decrease in viscosity is smaller for the Kilauea basalt melt than for the Crater Lake andesite melt. A possible reason for the difference is the lower content of Al_2O_3 and alkalies in the Kilauea basalt (13 and 2.8 wt. %, respectively) than in the Crater Lake andesite (18 and 5.7 wt. %).

At much higher pressures, Si^{4+} in silicate melts may shift from 4- to 6-fold coordination, as observed in crystalline silica, resulting in a drastic decrease in viscosity due to the depolymerization of the Si-O bonds. The infrared absorption spectra of the quenched glass of $Na_2Si_3O_7$ composition support this possibility (*Velde and Kushiro*, 1976). The glasses of this composition quenched at pressures up to 59 kbar do not show significant changes in spectra, but the glass quenched at 76 kbar clearly shows a new band at about 930 cm^{-1} and a significant decrease of the "polymer" band at 760 cm^{-1}. The new band can be explained by the presence of octahedrally coordinated Si^{4+}, and the decrease of the band at 760 cm^{-1} could be explained by the decrease of Si-O network in the melt.

III. EFFECT OF TEMPERATURE ON VISCOSITY

A number of experimental results demonstrate that viscosities of silicate melts decrease with increasing temperature at constant pressure. *Bottinga and Weill* [1973] summarized the available viscosity data on silicate melts at 1 atm and derived coefficients for calculating viscosities of silicate melts of given composition at given temperature. Some of the results obtained at 1 atm on both natural rock melts and synthetic silicate melts of geological interest are shown in Figure 3. The viscosities of these melts decrease by factors of 2 to 3 per 100°C. The data shown in Figure 3 are those at superliquidus conditions. At subliquidus temperatures, the viscosities increase with greater rates, and the temperature-viscosity curves usually show breaks at or near the liquidus temperatures [e.g., *Kozu and Kani*, 1935; *Shaw*, 1969; *Murase and McBirney*, 1973]. The logarithm viscosities of most of the synthetic and natural rock melts change linearly when they are plotted against $1/T$ (°K).

As shown in the previous section, the rates of decrease in viscosity with pressure of basalt and andesite melts are relatively small at constant temperature; however, the viscosities of the basalt and andesite melts decrease significantly with increasing pressure along their liquidi. For example, the viscosity of the Kilauea 1921 olivine tholeiite melt decreases from about 110 poise at 1 atm to about 8 poise at 30 kbar at temperatures slightly above the liquidus [*Kushiro et al.*, 1976]. Such decrease is mostly due to the effect of temperature, as shown in Figure 3.

IV. EFFECTS OF VOLATILE COMPONENTS ON VISCOSITY

It is widely accepted that volatile components play an important role in the generation of magmas. Among many volatile components, H_2O and CO_2 are most important in controlling the composition of magmas, and their effects on magma genesis have been extensively studied.

Volatile components also have a significant influence on the viscosity of magmas. The effect of H_2O on the viscosity of granitic or rhyolitic magmas has been extensively studied [e.g., *Saucier*, 1952; *Burnham*, 1963, 1967; *Friedman et al.*, 1963; *Shaw*, 1963].

Shaw [1972] derived an empirical equation to calculate the viscosity of granitic melts containing H_2O. The effect of H_2O on the viscosity of basic silicate melts, however, has not been studied in detail. *Kushiro et al.* [1976] found that the viscosity of Crater Lake calc-alkaline andesite decreases by a factor of about 20 by the addition of 4 wt. % H_2O at 15 kbar and 1350°C. *Scarfe* [1973] reported that the viscosities of basic andesite and tholeiitic basalt melts are lowered by factors of 4 and 2.7,

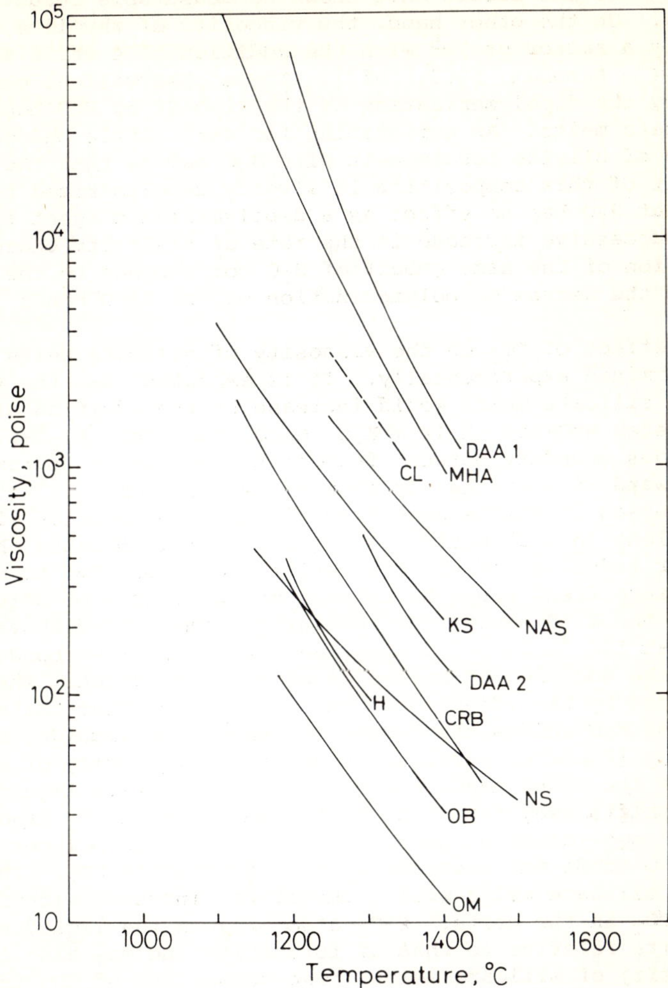

Fig. 3. Changes in viscosity with temperature of some synthetic and natural rock melts at 1 atm. Data from Kani and Kozu [1934] for DAA 1 ($Di_{20}An_{60}Ab_{20}$ wt %) and DAA 2 ($Di_{40}An_{40}Ab_{20}$); Murase and McBirney [1973] for MHA (Mount Hood andesite) and CRB (Columbia River basalt); Riebling [1966] for NAS (Na_2O 22.9, Al_2O_3 10.3, SiO_2 66.8); Bockris et al. [1955] for KS (K_2O 22.3, SiO_2 77.7; and NS (Na_2O 27.0, SiO_2 73.0); Shaw [1969] for H (Hawaiian tholeiite); and Scarfe [1973] for OB (olivine basalt) and OM (olivine melanephelinite). Dashed curve CL is a viscosity change of Crater Lake andesite at 5 kbar [Kushiro et al. 1976].

respectively, by the addition of 4 wt. % H_2O at 1 kbar and 1150°C, whereas an olivine basalt melt shows no measurable reduction in viscosity. On the other hand, the viscosity of rhyolite melt decreases by a factor of 10^5 with the addition of 4 wt. % at 1000°C [*Shaw*, 1963; *Burnham*, 1963, 1967]. These observations can be explained by the depolymerization of the Si-O-Si by H_2O bonds in the silicate melts. An explanation for very little change in viscosity of olivine basalt melt with H_2O may be that the anhydrous melt of this composition is already depolymerized to the extent that H_2O has no effect as a depolymerizing agent [*Scarfe*, 1973]. Successive increase in the rate of viscosity change by the addition of the same amount of H_2O corresponds to the increase in the degree of polymerization of the Si-O bonds in the melts.

The effect of CO_2 on the viscosity of silicate melts has not been determined experimentally. It is expected that the viscosity of silicate melts would increase by the addition of CO_2. In the system $MgO-SiO_2-X$ ($X = K_2O$, Na_2O, CaO, FeO, Al_2O_3, TiO_2), the liquidus boundary between forsterite and Ca-poor pyroxene shifts toward SiO_2 by the addition of oxides of monovalent cations, whereas it shifts away from SiO_2 by the addition of oxides of polyvalent ($\underline{n} \geqslant 3$) cations [*Kushiro*, 1975]. Because the expansion or reduction of the forsterite field relative to the Ca-poor pyroxene field would be related to the degree of polymerization of the Si-O bonds, the regularities mentioned above indicate that the degree of polymerization of the Si-O bonds is reduced by the addition of oxides of monovalent cations, whereas it is increased by the addition of oxides of polyvalent cations. The addition of H_2O shifts the forsterite-enstatite liquidus boundary toward SiO_2 [*Kushiro*, 1972] and reduces the viscosity of silicate melts. On the other hand, the forsterite-enstatite liquidus boundary shifts away from SiO_2 by the addition of CO_2 [*Eggler*, 1974]. This evidence suggests that the degree of polymerization of the Si-O bonds may increase by the addition of CO_2. The viscosity of silicate melts would, therefore, increase with increasing CO_2 in the melts. P_2O_5 also expands the liquidus field of enstatite relative to that of forsterite and may also increase the viscosity of silicate melts. The solubility of CO_2 in both synthetic and natural rock melts has been determined [e.g., *Eggler*, 1973; *Mysen*, 1975]. *Mysen* [1975] mentioned that the solubility of CO_2 increases with increasing basicity of the melts. The effect of CO_2 may, therefore, be to increase the viscosity in more basic melts.

V. DENSITY CHANGE OF SILICATE MELTS WITH PRESSURE

The density of silicate melts gives a clue to understand the structural change of the melts with pressure. Only a few direct measurements have been made on the density of silicate melts at

1 atm, and none have been previously reported at high pressures. *Murase and McBirney* [1973] measured the densities of basalt, andesite, and rhyolite melts at 1 atm and showed that the densities of these melts decrease slightly with increasing temperature above the liquidus.

The densities of glasses quenched at high pressures from melts of $NaAlSi_2O_6$, Kilauea 1921 basalt, and Crater Lake andesite compositions have been determined [*Kushiro*, 1976; *Kushiro et al.* 1976]: the densities of glasses of the Kilauea 1921 basalt and Crater Lake andesite increase with increasing pressure and decrease with increasing temperature of quenching. *Murase and McBirney* [1973] showed that the density of melt is lower than that of glass by 0.1-0.15 g/cc. Diopside glass at 20°C is also denser than the diopside melt by 0.16 g/cc [*Dane*, 1941]. Al-

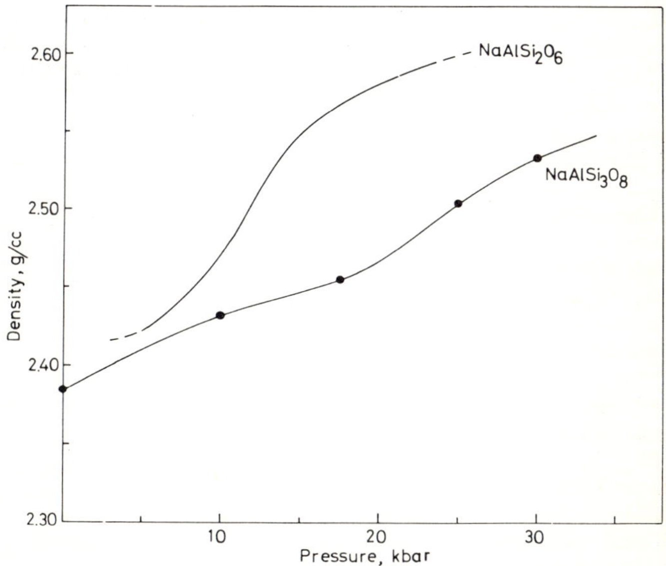

Fig. 4. Densities of $NaAlSi_2O_6$ (jadeite) and $NaAlSi_3O_8$ (albite) glasses as a function of pressure of quenching. The density of the $NaAlSi_2O_6$ glass is from Kushiro [1976], and that of the $NaAlSi_3O_8$ glass has been calculated from the refractive index of the glass given by Boyd and England [1963] using the Gladstone-Dale equation [Bowen and Schairer, 1935].

though the densities of glasses are higher than those of melts of the same compositions, basalt and andesite melts should show the density variations similar to glasses.

The density of glass of $NaAlSi_2O_6$ composition increases from 2.40 g/cm^3 at 5 kbar to 2.58 g/cc at 21 kbar, as shown in Figure 4. The density increase is about 7%. The rate of increase in density between 10 and 15 kbar is larger than in other pressure ranges investigated. As shown in Figure 1, the rate of decrease in viscosity is also larger, near 10 kbar. From these viscosity and density changes, it is suggested that the structural change caused by the shift of Al^{3+} from 4- to 6-fold coordination takes place efficiently at pressures near 10 kbar. For comparison, the density change of albite glass, based on the refractive index of albite glass given by *Boyd and England* [1963], is also shown in Figure 4. The rate of increase in density between 20 and 24 kbar appears to be larger than in other pressure ranges. The albite melt should also show a similar density change with pressure. *Mysen* [1975, 1976] observed a significant increase in the mole fraction of CO_3^{2-} relative to that of CO_2 in albite melt at pressures between 20 and 30 kbar and suggested that Al^{3+} shifts from 4- to 6-fold coordination in this pressure range.

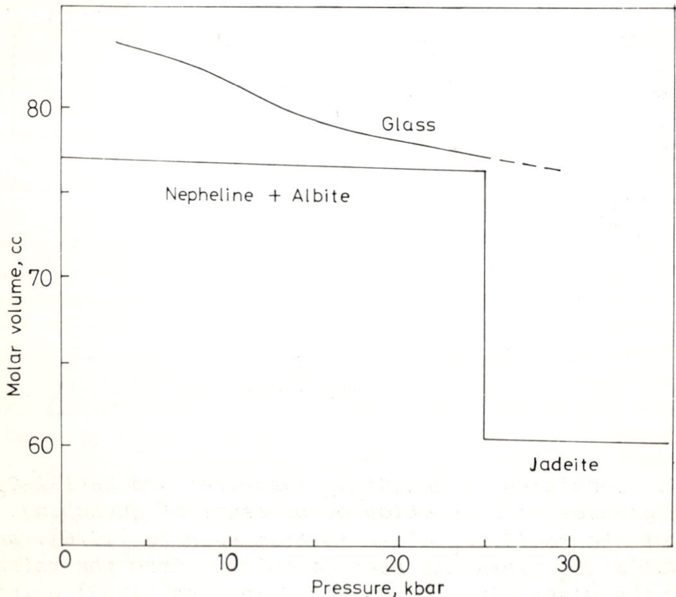

Fig. 5. Molar volumes of glass and crystalline phases of $NaAlSi_2O_6$ composition.

The density changes of jadeite and albite glasses are continuous and gradual compared to those of the crystalline phase. The change of molar volume of $NaAlSi_2O_6$ glass is compared to those of the crystalline phase (Figure 5): the molar volume of the crystalline phase changes discontinuously at the transition from nepheline + albite to jadeite, whereas the molar volume of the glass changes continuously with pressure. The difference in molar volume between the glass and the crystalline phases becomes smaller with increasing pressure in the stability field of nepheline + albite, and when jadeite becomes stable, the difference increases again.

VI. CONCLUSIONS

Changes in viscosity of some synthetic and natural rock melts and densities of their quenched glasses strongly indicate that phase transformations due to the structural change occur in silicate melts at high pressures. The structural changes in melts are probably similar to those in crystalline silicates; however, the transformation would be of higher order, as indicated by the continuous changes with pressure in viscosity and molar volume of the melts. The viscosities of magmas of basaltic and andesitic compositions decrease with increasing temperature at constant pressure and also decrease with increasing pressure at constant temperature, the rate of decrease depending on the compositions of magmas. In consequence, the viscosities of the magmas decrease significantly with increasing pressure along the liquidi. It is concluded that magmas formed in the upper mantle would be denser and less viscous at greater depths. Volatile components also have significant effects on the viscosity of magmas: H_2O reduces the viscosity, the effect being greater for more silica-rich compositions, whereas CO_2 and P_2O_5 may increase the viscosity. These conclusions are, however, based on a relatively small number of measurements on viscosity of silicate melts and density of glasses of limited compositions. More measurements on viscosity and density of silicate melts of wider compositional ranges and under wider ranges of pressure-temperature conditions are no doubt needed for a better understanding of the phase transformation and behavior of magmas in the upper mantle. Determination of the structure of silicate melts at high pressures is also highly desirable.

REFERENCES

Bockris, J. O'M. J. D. Mackenzie, and J. A. Kitchener, Viscous flow in silica and binary liquid silicates, *Trans. Faraday Soc.*, *51*, 1734-1748, 1955.

Bottinga, Y., and D. F. Weill, The viscosity of magmatic silicate liquids: A model for calculation, *Amer. J. Sci.*, *272*, 438-475, 1972.

Bowen, N. L., and J. F. Schairer, The system $MgO-FeO-SiO_2$, *J. Sci.*, *29*, 151-217, 1935.

Boyd, F. R., and J. L. England, Effect of pressure on the melting of diopside, $CaMgSi_2O_6$, and albite, $NaAlSi_3O_8$, in the range up to 50 kilobars, *J. Geophys. Res.*, *68*, 311-323,1963.

Burnham, C. W., Viscosity of a water-rich pegmatite melt at high pressure (abstract), *Geol. Soc. Amer. Spec. Paper 76*, 26, 1963.

Burnham, C. W., Hydrothermal fluids at the magmatic stage. In *Geochemistry of Hydrothermal Ore Deposits*, edited by H. L. Barnes, pp. 34-76, Holt, Rinehart & Winston, New York, 1967.

Dane, E. B., Densities of molten rocks and minerals, *Amer. J. Sci.*, *239*, 809-818, 1941.

Eggler, D. H., Role of CO_2 in melting processes in the mantle, *Carnegie Inst. Washington, Yearb.*, *72*, 457-467, 1973.

Eggler, D. H., Effect of CO_2 on the melting of peridotite, *Carnegie Inst. Washington, Yearb.*, *73*, 215-224, 1974.

Friedman, I., W. Long, and R. L. Smith, Viscosity and water content of rhyolite glass, *J. Geophys. Res.*, *68*, 6523-6535, 1963.

Kozu, S., and K. Kani, Viscosity measurements of the ternary system diopside-albite-anorthite at high temperatures, *Proc. Imp. Acad. Japan*, *11*, 383-385, 1935.

Kushiro, I., Effect of water on the compositions of magmas formed at high pressures, *J. Petrology*, *13*, 311-334, 1972.

Kushiro, I., On the nature of silicate melt and its significance in magma genesis: Regularities in the shift of the liquidus boundaries involving olivine, pyroxene, and silica minerals, *Amer. J. Sci.*, *275*, 411-431, 1975.

Kushiro, I., Changes in viscosity and structure of melt of $NaAlSi_2O_6$ composition at high pressures, *J. Geophys. Res.* *(in press)*, 1976.

Kushiro, I., H. S. Yoder, Jr., and B. O. Mysen, Viscosity of basalt and andesite melts at high pressures, *J. Geophys. Res.* *(in press)*, 1976.

Murase, T., and A. R. McBirney, Properties of some common igneous rocks and their melts at high temperatures, *Geol. Soc. Amer. Bull.*, *84*, 3563-3592, 1973.

Mysen, B. O., Solubility of volatiles in silicate melts at high pressures and temperature: The role of carbon dioxide and water in feldspar, pyroxene, and feldspathoidal melts, *Carnegie Inst. Washington, Yearb.*, *74*, 454-468, 1975.

Mysen, B. O., The role of volatiles in silicate melts: Solubility of carbon dioxide and water in feldspar, pyroxene and feldspathoid melts to 30 kb and 1625°C, *Amer. J. Sci.* *(in press)* 1976.

Riebling, E. F., Structure of sodium aluminosilicate melts containing at least 50 mole % SiO_2 at 1500°C, *J. Chem. Phys.*, *44*, 2857-2865, 1966.

Saucier, P. H., Quelques experiences sur la viscosité a haute température de verres ayant la composition d'un granite, Influence de la vapeur d'eau sous pression, *Soc. Franc. Mineral., 75,* 1-45, 1952.

Scarfe, C. M., Viscosity of basic magma at varying pressure, *Nature (London), 241,* 101-102, 1973.

Shaw, H. R., Obsidian-H_2O viscosities at 1000 and 2000 bars in the temperature range 700° to 900°C, *J. Geophys. Res., 68,* 6337-6343, 1963.

Shaw, H. R., Rheology of basalt in the melting range, *J. Petrology, 10,* 510-535, 1969.

Shaw, H. R., Viscosities of magmatic silicate liquids: An empirical method of prediction, *Amer. J. Sci., 272,* 870-893, 1972.

Velde, B. and I. Kushiro, Infrared spectra of high-pressure quenched silicate liquids. *Carnegie Inst. Washington, Yearb., 75 (in press),* 1976.

Waff, H. S., Pressure-induced coordination changes in magmatic liquids, *Geophys. Res. Lett., 2,* 193-196, 1975.

THE ELECTRICAL CONDUCTIVITY OF POLYCRYSTALLINE OLIVINE AND PYROXENE UNDER PRESSURE

R. N. SCHOCK, A. G. DUBA, H. C. HEARD, H. D. STROMBERG
University of California
Lawrence Livermore Laboratory
Livermore, California 94550

Abstract

Electrical conductivity (σ) measurements on polycrystalline olivine and pyroxene were made at pressures up to 5.0 GPa and at temperatures up to 1200°C to examine the effect of grain boundaries on conduction. Values of σ were observed within one-half order of magnitude of those for single crystals at lower pressures and under controlled oxygen fugacity (f_{O_2}). If this difference prevails over the entire temperature range, these values result in temperature profiles for the earth and moon that are lower than profiles from single-crystal data by no more than 200°C. However the small σ differences between single and polycrystals most probably are due to a higher f_{O_2} in the experimental apparatus rather than the presence of grain boundaries. The data indicate pressure has a small effect on conductivity and activation enthalpies associated with conduction.

I. INTRODUCTION

Recently we have measured the electrical conductivity (σ) of minerals thought to be representative of the interiors of terrestrial planets [*Duba, et al.*, 1972, 1974, 1975, 1976]. All of these measurements were made on single crystals and attention was given to the oxidation state of the crystal [*Duba, et al., Duba and Nicholls*, 1972]. In some cases, oxygen fugacity (f_{O_2}) was controlled directly, while in others σ was measured as a function of confining pressure. These measurements yielded very low conductivities for natural crystals which, when used with conductivity-depth profiles, resulted in higher temperatures at a given depth than hitherto obtained. Nevertheless, we have been cautious in generalizing these results because the electrical properties of single crystals may not be representative of those for polycrystalline planetary interiors. Grain boundaries influence σ in some instances, for example, by providing high diffusivity paths where the conduction mechanism is ionic or by creating resistive barriers where the mechanism is electronic.

In this report we present data on σ determined in polycrystalline olivine and pyroxene at pressures between 2.0 and 5.0 GPa[1] and at temperatures up to 1200°C. These conditions represent earth upper-mantle and deep lunar-interior conditions, and σ data obtained on polycrystalline material should be relevant to these regions.

II. EXPERIMENTAL TECHNIQUES

Olivine (Fo_{92}) from Mt. Leura, Australia [*Duba*, 1972] and pyroxene (En_{88}) from Bamle, Norway [*Ahrens and Gaffney*, 1971] were ground so as to pass through a 44-μm sieve. The powder then was sealed in a Ta capsule and hot-pressed in a girdle-anvil device [*Stromberg and Stephens*, 1964] at 1200°C and 5.0 GPa for 2 h. The capsule was then removed, and discs 3.2 mm in diameter and approximately 0.6-mm thick were sectioned with a diamond-wafer blade. These samples were translucent and green in a thin chip.

Measurements of σ were made in a girdle-anvil device modified to allow thermocouple wires to enter the sample chamber through the tungsten-carbide (WC) pistons. A section of the sample assembly in the girdle is shown in Figure 1. The sample (A) is placed between two Pt-foil electrodes (B) in contact with two Pt-Pt10Rh thermocouples. The thermocouple wires are insulated by alumina tubing (C), which has been inserted through the WC anvils. The tubing is held in the anvil by coating it with a small amount of powdered Fe_2O_3-epoxy mixture before insertion through the anvil. This assembly is then surrounded with unfired Al_2O_3 (D) and a Pt-foil resistance heater (E). Pressure is contained by unfired pyrophyllite (F). Sample temperatures up to 1200°C are achieved by this method, although a recent heater modification using Pt30Rh has extended the temperature range to 1450°C. Since the sample was never directly at the highest temperature zone of the furnace, temperature gradients of up to 50°C have been encountered across the sample. Temperatures reported are therefore average sample temperatures.

A simplified girdle-anvil device of this geometry will attain 8.0 GPa. We have limited ourselves to 5.0 GPa, which is achieved with 110 t force. Pressure calibration at room temperature was accomplished using the Bi I-II (2.55 GPa), Tl II-III (3.67 GPa), and Bi III-V (7.7 GPa) transitions [*Lloyd*, 1971]. At high temperature we used the α-γ Fe transition [*Strong, et al.*, 1973]. The well-known affinity of Fe for Pt in high temperatures [*Sosman and Hostetler*, 1915] makes it difficult to reproduce this measurement. The resulting accuracy in stated pressures at temperature is no better than ±0.25 GPa. All temperatures were corrected for the effect of pressure on thermocouple EMF by

---

[1] 1 Pa = $1 N/m^2 = 10^{-5}$ bar

extrapolating the data of *Getting and Kennedy* [1970] and *Lazarus, et al.* [1971] and are estimated to be accurate to ±2°C.

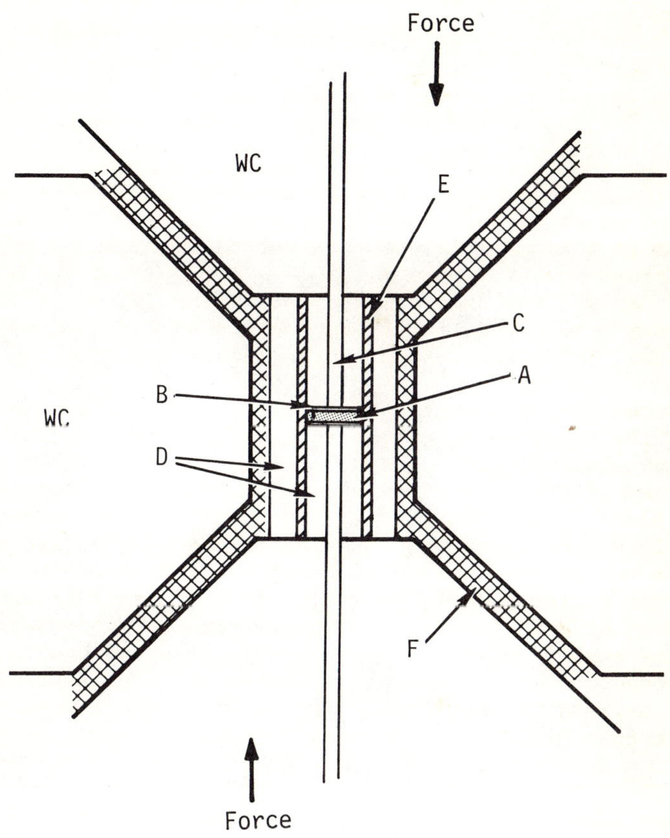

Fig. 1. Schematic drawing of sample assembly.

Conductance was measured using a 1.592-kHz 200-mV peak-to-peak excitation voltage in a bridge circuit. These values are accurate to 0.1%. Due to uncertainties about precise sample geometry after pressurization, conductance readings are accurate to no better than 10%, which results, however, in less than 1% error in log σ (for σ values reported here) when calculating activation enthalpies. The leakage conductance of the experimental configuration was checked by measuring the conductance of single-crystal Al_2O_3 (Linde) as a function of pressure and temperature. Conductivities below 10^{-7} S/m were measured. At equivalent temperatures, leakage conductance values were at least an order of magnitude lower than the sample conductance used to calculate σ values. Conductance data were taken only after σ had stabilized with time (1 to 2 h) at 900 to 1000°C and 2.0 GPa. We believe this represents equilibration of the sample to the unknown f_{O_2} in the apparatus.

An assembly sectioned after running at 5.0 GPa and up to 1200°C is shown in Figure 2. Post-run X-ray powder diffraction (detection level, 5%) revealed no changes in the starting material. The fog of diffuse darker material between the sample and the Pt-heaters is Al_2O_3, sintered by the hottest portion of the furnace.

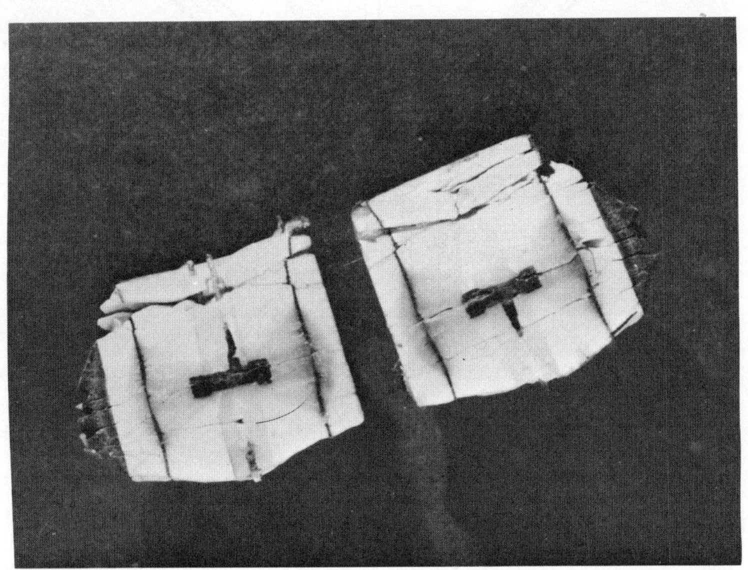

Fig. 2. Photograph of sectioned sample assembly.

III. RESULTS

The results for olivine are shown in Figure 3. The lines labeled A_1 and A_2 represent σ data gathered as a function of temperature at 2.0 and 5.0 GPa, respectively. The activation enthalpy (ΔH^*) in the Boltzmann expression

$$\ln\sigma = \ln\sigma_o + \Delta H^*/kT$$

varies between 1.17 ± 0.01 eV and 1.00 ± 0.01 eV over the pressure range 2.0 to 5.0 GPa. The line labeled B is data [*Duba*, 1972] on a single crystal (orientation unknown) from the same locality. These data were obtained at an unknown f_{O_2} in Ar.

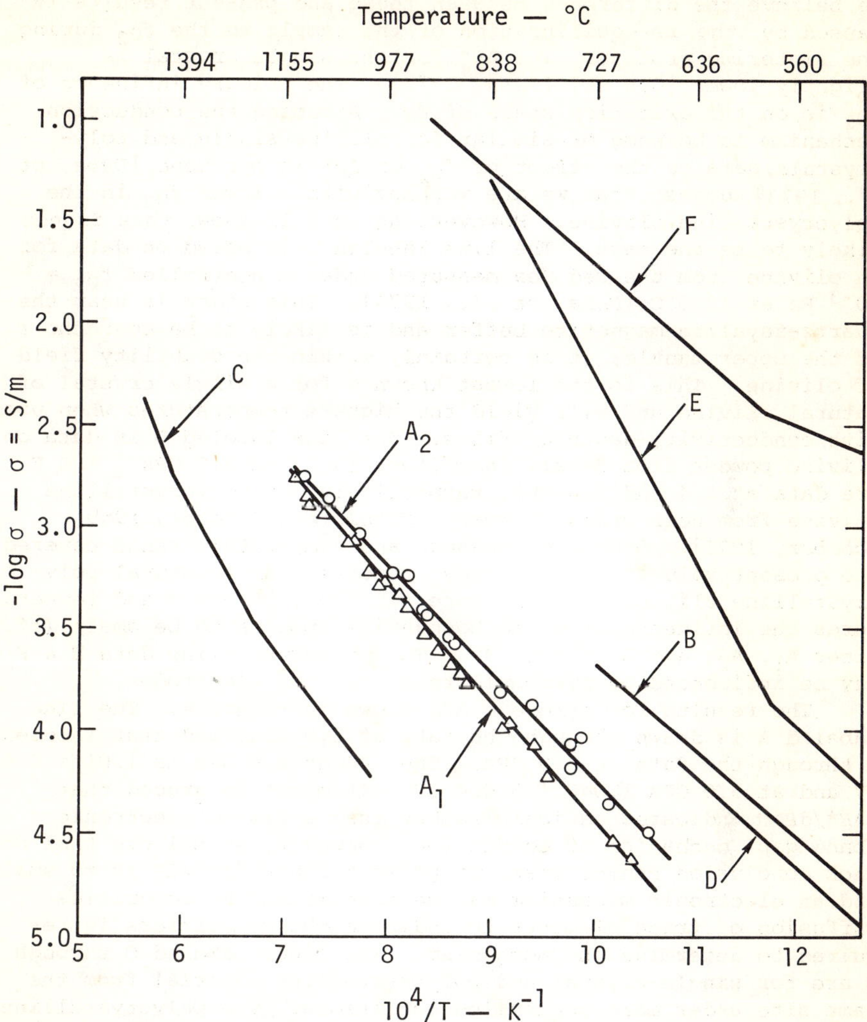

Fig. 3. Log σ versus temperature for (A_1) polycrystalline olivine (Fo92) f_{O_2} unknown, P = 2.0 GPa; (A_2) polycrystalline olivine (Fo92) f_{O_2} unknown, P = 5.0 GPa; (B) single-crystal olivine (Fo92) f_{O_2} unknown, P = 0.8 GPa [Duba, 1972]; (C) Single-crystal olivine (Fo91), $f_{O_2} = 10^{-3}$ Pa, P = 0.1 MPa [Duba et al., 1974]; (D) polycrystalline olivine (Fo90), f_{O_2} unknown, P = 2.2 GPa [Hamilton, 1965]; (E) polycrystalline olivine (Fo91), $f_{O_2} = 10^{-7}$ Pa, P = 2.4 GPa [Schult and Schober, 1969]; (F) polycrystalline olivine (Fo91), $f_{O_2} = 10^{-7}$ Pa, P = 5.0 GPa [Schober, 1971]. All f_{O_2} at 1200°C.

We believe the difference between these and present results is caused by the re-equilibration of the sample to the f_{O_2} during the sintering process. The ΔH^* for the single crystal is slightly lower [0.8 eV] [*Duba*, 1972] The primary influence of f_{O_2} is on the oxidation state of Fe. Assuming the conduction mechanism to be same or similar for olivine single and poly-crystals, data on the effect of f_{O_2} on ΔH^* in pyroxene [*Duba, et al.*, 1973] suggest that we are working with a lower f_{O_2} in the polycrystalline olivine. However, as we will show, this is not likely to be the case. The line labeled C is based on data for an olivine from the Red Sea measured under a controlled f_{O_2} = 10^{-3} Pa at 1200°C [*Duba, et al.*, 1974]. This state is near the quartz-fayalite-magnetite buffer and is likely to be near the f_{O_2} of the upper mantle; it is certainly within the stability field of olivine. This is the lowest known σ for a single crystal of natural olivine and will yield the highest temperatures when used with conductivity-depth profiles. The line labeled D is data on olivine powder from Hawaii [*Hamilton*, 1965] at 2.0 GPa. E & F are data at 2.4 and 5.0 GPa, respectively, on polycrystalline olivine from near Dreis, Germany [*Schult and Schober*, 1969; *Schober*, 1971]. Over the pressure and temperature range covered, the present values are the lowest reported for a natural poly-crystalline olivine. Furthermore, at these pressures and temperatures the influence of grain boundaries appears to be small (cf. lines A_1, A_2, and B, Figure 3). The polycrystalline data E & F may be influenced by pyrophyllite across the electrodes.

The results for pyroxene are shown in Figure 4. The line labeled A is drawn through the data at 2.0 GPa, and that labeled B through the data at 5.0 GPa. The ΔH^* at 2.0 GPa is 1.01 ± 0.01 eV and at 5.0 GPa 1.040 ± 0.004 eV. It might be argued that $d\Delta H^*/dP>0$ indicates an ionic rather than a simple electronic (band gap) mechanism of conduction. However, we believe that no such conclusion can be drawn at present since $d\Delta H^*/dP$ is so small, and an electronic mechanism may be complicated by impurities. Diffusion of trace elements to indicate charge carriers is required to determine the mechanism. The lines labeled C through E are for single-crystal and polycrystalline material from the same site under more controlled conditions. The polycrystalline data in Figure 4 at 2.0 and 5.0 GPa are in fair agreement with these measurements. It is probable that the difference is a result of the effect of f_{O_2}.

The σ data on polycrystalline pyroxene in Figure 4 are about one-half of an order of magnitude greater in the girdle-anvil cell than in a controlled atmosphere. Nevertheless, the slopes (ΔH^*) are approximately the same. *Duba, et al.*, [1973] in a study of the effect of f_{O_2} on other orthopyroxenes, showed that f_{O_2} had a small effect on slope. Their data on σ as a function of f_{O_2} indicate that σ may increase by 0.5 orders of magnitude with four orders of magnitude change in f_{O_2} (e.g., from 10^{-1} Pa to 10^{+3} Pa at 1000°C). Thus the polycrystalline pyroxene results at high pressure would seem to be slightly different than those under more controlled conditions due a higher f_{O_2}. These conclusions about f_{O_2} are applicable also to the olivine results.

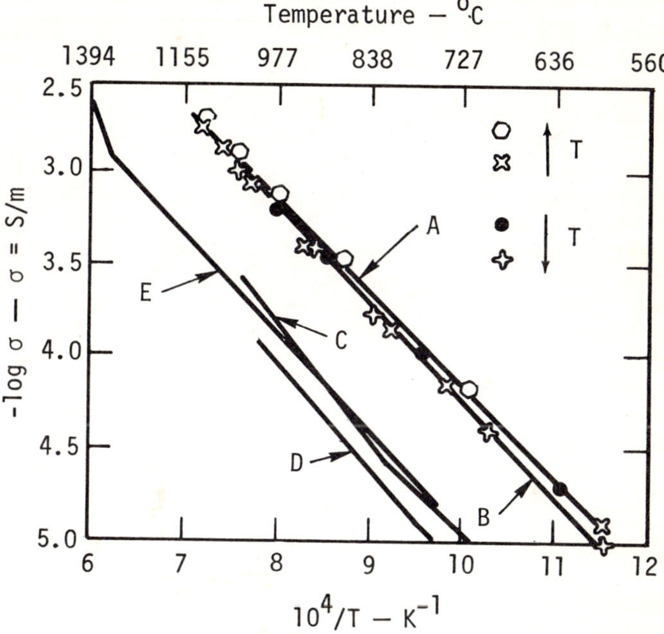

Fig. 4. Log σ versus temperature for (A) polycrystalline enstatite, f_{O_2} unknown, P = 2.0 GPa; (B) polycrystalline enstatite, f_{O_2} unknown, P = 5.0 GPa; (C) polycrystalline enstatite, $f_{O_2} = 10^{-3}$ Pa, P = 0.1 MPa [Heard, et al., 1975]; (D) single-crystal [010] enstatite, $f_{O_2} = 10^{-3}$ Pa, P = 0.1 MPa [Heard, et al., 1975]; (E) single-crystal [010] enstatite, f_{O_2} = 10^{-3}-10^{+1} Pa, P = 0.5 GPa [Heard, et al., 1975]. All f_{O_2} at 1200°C.

We may place an upper limit on f_{O_2} since we know from post-run conductivity levels that olivine remained in its stability field [*Nitsan*, 1974]. The stability field of olivine in Figure 5 indicates that if $f_{O_2} > 10^{-3}$ Pa, the olivine in these experiments (lines A, Figure 3) would have been outside the olivine stability field at temperatures high enough (>1000°C) to allow oxidation to proceed in the experimental times. It is also likely that f_{O_2} increases with temperature in our system, although the precise slope is dependent on the unknown buffers active in the girdle-anvil assembly.

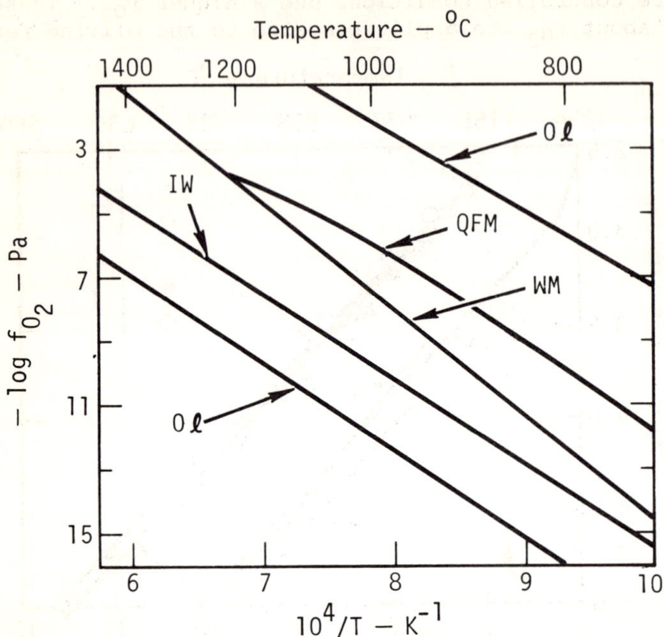

Fig. 5. Stability fields of olivine [*Nitsan, 1974*] *and iron-oxygen-silicon systems* [*Eugster and Wones, 1962*] *as a function of* f_{O_2} *and temperature. Boundaries are olivine (Ol), wüstite-magnetite (WM), and iron-wüstite (IW).*

In the experimental assembly the sample is, with the exception of Pt-foil, completely surrounded by Al_2O_3. Calculations based on thermodynamic data [*Coughlin*, 1954] show that if the decomposition of Al_2O_3 were the active buffer for the sample, f_{O_2} would be extraordinarily low (10^{-26} Pa at 1370°C). Clearly this is not the case since the olivine was not decomposed. Water in the pyrophyllite may be more active as a buffer since the reaction $2H_2O \rightleftarrows 2H_2 + O_2$ yields an $f_{O_2} = 10^{-4.7}$ at 1300°C. However, the most likely control is O_2 in the atmosphere, which diffuses into the sample area through the powdered pyrophyllite and Al_2O_3. Oxygen fugacity would thus be controlled by the permeability of these powders under pressure. This permeability is unknown.

IV. DISCUSSION

It has long been recognized [*Shakespeare*, 1622] that olivine is the major constituent of the earth's upper mantle and as such can be considered to control bulk electrical properties. The data in Figure 3 may therefore be used to construct a geotherm. The temperatures shown in Figure 6 are calculated using the geomagnetic data of *Banks* [1972] for depths below 300 km and the magnetotelluric data of *Swift* [1967] and *Kasameyer* [1974] for shallower depths, and the data from line C, Figure 3. Several other geotherms are shown for comparison. If the differences in σ between the polycrystalline and single-crystal data are assumed to hold at higher temperatures, the calculated geotherm would be about 200° lower. However, as already stated, we believe that the polycrystalline data are influenced by the uncontrolled f_{O_2} even though the samples are within the olivine stability field. We conclude that since the high-pressure polycrystalline and the single-crystal σ data (lines A and B, Figure 3) taken without f_{O_2} control are in reasonable agreement, there is no significant

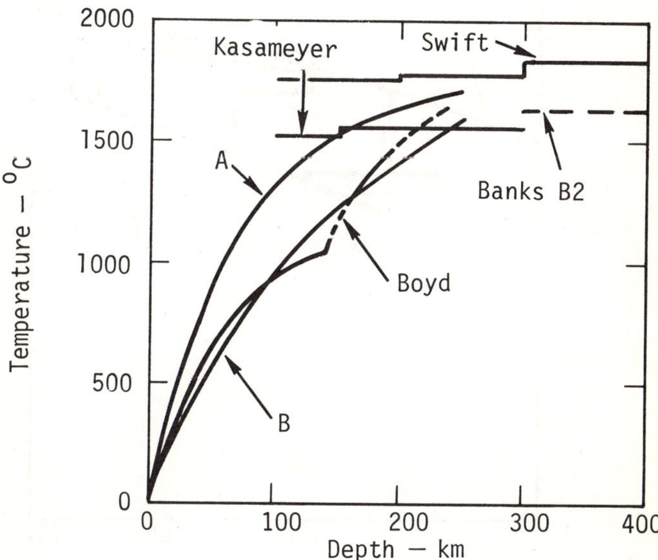

Fig. 6. Temperature limits with depth for the earth constructed using the σ data from lines A, Figure 3 and the electromagnetic induction data (limits of best fits) of Banks [1972], Swift [1967], and Kasameyer [1974]. The oceanic geotherm (A) and continental geotherm (B) of Ringwood [1966] and the pyroxene geotherm of Boyd [1973] are shown for comparison.

effect of grain boundaries. If polycrystalline data could be taken under controlled f_{O_2}, they would agree with the single-crystal σ data under controlled f_{O_2}. These conclusions assume no f_{O_2} effect on grain boundary conduction. Therefore, the geotherm in Figure 6 is applicable to a polycrystalline mantle even though the σ-temperature data are obtained on single crystals.

Ringwood and Essene [1971] have suggested that the interior of the moon is pyroxenite, thus the magnetometer data of *Sonett, et al.*, [1971] and *Dyal and Parkin* [1971] may be used with the data in Figure 4 to construct temperature limits for the moon as shown in Figure 7. Use of the polycrystalline σ data at high

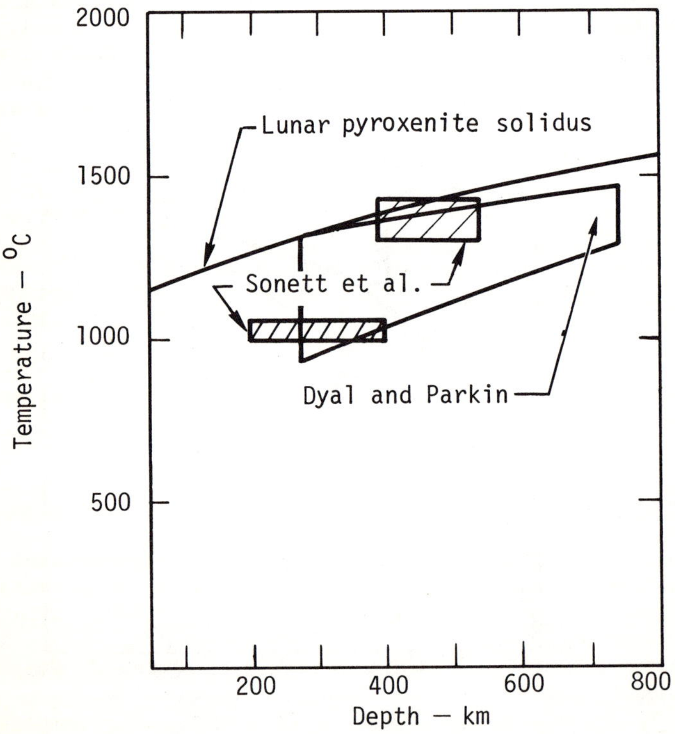

Fig. 7. Temperature limits with depth for the moon constructed using the σ data from line E in Figure 4 and the magnetic induction data of Sonett, et al., [*1971*] *and Dyal and Parkin* [*1971*]. *The lunar pyroxenite solidus* [*Ringwood and Essene, 1971*] *is shown for comparison.*

pressures would lower the temperature by some 150 to 200°C over the range 800 to 1400°C. This is the maximum possible effect. If used, they will yield a hot moon. Again, we believe that the polycrystalline data are influenced by f_{O_2} and that the single-crystal data (line D, Figure 4) best represent the lunar interior.

We conclude that the influence of grain boundaries on electrical conductivity of purported upper-mantle and lunar-interior materials is not significant in terms of constructed temperature-depth profiles and that single-crystal data may be used to accurately infer temperatures in the interiors of terrestrial bodies. In addition, the data up to 5.0 GPa indicate little pressure effect on the conductivity and activation enthalpies associated with conduction, and thereby reinforce our previous conclusions based on σ data for single crystals up to 8.0 GPa [*Duba et al.*, 1974].

Acknowledgments. C. P. Sonett was most helpful in discussing the results of these experiments and in providing generous hospitality. This work was performed under the auspices of the U. S. Energy Research and Development Administration under Contract No. W-7405-Eng-48 and was supported by the National Aeronautics and Space Administration.

REFERENCES

Ahrens, T. J. and E. S. Gaffney, Dynamic compression of enstatite, *J. Geophys. Res., 76,* 5504, 1971.

Banks, R. J., The overall conductivity distribution of the earth, *J. Geomagn. Geoelec., 24,* 337, 1972.

Boyd, F. R., A pyroxene geotherm, *Geochim. Cosmochim. Acta, 37,* 2533, 1973.

Coughlin, J. P., Contributions to the data on theoretical metallurgy. XII. Heats and free energies of formation of inorganic oxides, *U. S. Bur. Mines Bull., 542,* Washington, D. C., 1954.

Duba, A., The electrical conductivity of olivine, *J. Geophys. Res., 77,* 2483, 1972.

Duba, A., H. C. Heard, and R. N. Schock, The Lunar temperature profile, *Earth Planet. Sci. Lett., 15,* 301. 1972.

Duba, A., J. Ito, and J. C. Jamieson, The effect of ferric iron on the electrical conductivity of olivine, *Earth Planet. Sci. Lett., 18,* 279, 1972.

Duba, A., and I. A. Nicholls, The influence of oxidation state on the electrical conductivity of olivine, *Earth Planet. Sci. Lett., 18,* 2483, 1972.

Duba, A., J. N. Boland, and A. E. Ringwood, The electrical conductivity of pyroxene, *J. Geol., 81,* 727, 1973.

Duba, A., H. C. Heard, and R. N. Schock, Electrical conductivity of olivine at high pressure and under controlled oxygen fugacity, *J. Geophys. Res., 79,* 1667, 1974.

Duba, A., H.C. Heard, A. Piwinskii, and R. N. Schock, Electrical conductivity and the geotherm, in *Extended Abstracts of Intl. Conf. on Geothermometry and Geobarometry, Oct. 5-10, 1975,* Pennsylvania State Univ., College Park, Pa., 1975.

Duba, A., A. Piwinskii, H. C. Heard, and R. N. Schock, The electrical conductivity of forsterite, enstatite, and albite, in *The Physics and Chemistry of Rocks and Minerals,* edited by R. G. J. Strens, John Wiley & Sons, N. Y., 249-260, 1976.

Dyal, P., and C. W. Parkin, Electrical conductivity and temperature of the lunar interior from magnetic transient-response measurements, *J. Geophys. Res., 76,* 5947, 1971.

Eugster, H. P., and D. R. Wones, Stability of the ferruginous biotite, annite, *J. Petrology, 3,* 82, 1962.

Getting, I. C., and G. C. Kennedy, The effect of pressure on the E.M.F. of chromel-alumel and platinum-platinum 10% rhodium thermocouples, *J. Appl. Phys., 41,* 4552, 1970.

Hamilton, R. M., Temperature variation at constant pressures of the electrical conductivity of periclase and olivine, *J. Geophys. Res., 70,* 5679, 1965.

Heard, H. C., A. Duba, A. J. Piwinskii, and R. N. Schock, Electrical conductivity studies: refinement of the selenotherm (abstract), *Lunar Sci. Conf., 6,* 355, 1975.

Kasameyer, P. W., Low frequency magnetotelluric survey of New England, Ph.D. Thesis, Massachusetts Institute of Technology, Cambridge, Mass., 1974.

Lazarus, D., R. M. Jeffery, and J. D. Weiss, Relative pressure dependence of chromel/alumel and platinum/platinum 10% rhodium thermocouples, *J. Appl. Phys. Lett., 19,* 371, 1971.

Lloyd, E. C. (Ed.), The accurate characterization of the high pressure environment, *Nat. Bur. of Stand. Spec. Publ. 326,* 314, Washington, D. C., 1971.

Nitsan, U., Stability field of olivine with respect to oxidation and reduction, *J. Geophys. Res., 79,* 706, 1974.

Ringwood, A. E., and E. Essene, Petrogenesis of Apollo 11 basalts, internal constitution and origin of the moon, *Geochim. Cosmochim. Acta., Suppl. 1,* 769, 1971.

Ringwood, A. E., Mineralogy of the mantle, in *Advances in Earth Sciences,* edited by P. M. Hurley, MIT Press, Cambridge, Mass., 357, 1966.

Schober, M., The electrical conductivity of some samples of natural olivine at high temperatures and pressures, *Z. Geophysik, 37,* 283, 1971.

Schult, A., and M. Schober, Measurement of electrical conductivity of natural olivine at temperatures up to 950°C and pressures up to 42 kbar, *Z. Geophysik, 35,* 105, 1969.

Shakespeare, W., *The Tragedy of Othello, The Moore of Venice,* Act. V, Scene II, lines 171-174, 1622.

Sonett, C. P., D. S. Colburn, P. Dyal, C. W. Parkin, B. F. Smith, G. Schubert, and K. Schwartz, Luna electrical conductivity profile, *Nature, 230,* 359, 1971.

Sosman, R. B., and J. C. Hostetler, The reduction of iron oxides by platinum, with a note on the magnetic susceptibility of iron-bearing platinum, *J. Wash. Acad. Sci., 5,* 293, 1915.

Stromberg, H. D., and D. R. Stephens, Effects of pressure on the electrical resistance of certain metals, *J. Phys. Chem. Solids, 25,* 1015, 1964.

Strong, H. M., R. E. Tuft, and R. E. Hanneman, The iron fusion curve and the γ-δ-α triple point, *Mett. Trans., 4,* 2657, 1973.

Swift, C. M., A magnetotelluric investigation of an electrical conductivity anomaly in the southwestern United States, Ph.D. Thesis, Massachusetts Institute of Technology, Cambridge, Mass., 1967.

ANISOTROPIC CHANGES IN RESISTIVITY AND VELOCITY DURING ROCK DEFORMATION

N. FUJII, Y. HAMANO
Geophysical Institute
University of Tokyo
Tokyo, Japan

Abstract

The changes in electrical resistivity and elastic-wave velocity of partially saturated granite under uniaxial compression are measured in the directions parallel to and perpendicular to the loading axis up to about 75% of the fracture stress. The sample, of which about 50% of pore space is saturated by water, is first loaded to about 0.7 kbar and then loaded to the maximum stress of about 1.1 kbar.

The volumetric strain indicates almost no dilatancy in the first stress cycle, whereas the dilatant strain is reached to be about 0.35% in the second cycle. The changes in compressional wave velocities for both directions are nearly reversible with stress except in the second unloading cycle, where the velocity perpendicular to the loading axis decreases about 5% relative to that in the previous cycles. In contrast, the changes in resistivities for both directions are irreversible with stress cycles. The resistivity ratios to the initial value decrease with increasing stress and nearly constant with decreasing loading stress for both directions.

These observations and the differences of changes for two directions are the indications of the anisotropic behaviors of the connectivity of pores and newly formed cracks during rock deformation.

I. INTRODUCTION

The effects of dilatancy on bulk physical properties of low-porosity brittle rocks are important keys to understanding the state of stress in the crust. Although dilatancy is assumed to be responsible for *in situ* premonitory changes of seismic velocities, electrical resistivities and other rock properties [*Nur*, 1972; *Scholz et al.*, 1973; *Whitcomb et al.*, 1973] laboratory data are insufficient to account for these *in situ* observations. Dilatancy is essentially nonlinear phenomenon induced by shear stress so that the rock properties related to it undergoes some nonlinear and anisotropic changes.

Stress-induced velocity anisotropy in dry rock was comprehensively studied by *Nur and Simmons* [1969] under relatively low stress. The dilatancy-related changes in elastic wave velocities have been studied on dry samples [*Tocher*, 1957; *Matsushima*, 1962; *Thill*, 1973]. *Gupta* [1973] measured compressional and shear waves along three mutually perpendicular directions simultaneously for dry limestone at ambient pressure. However, strains were not recorded simultaneously in these studies. *Spetzler and Martin* [1974] and *Spetzler et al.* [1977] demonstrated the changes in compressional wave velocity with holographically observed strain distribusion. Shear wave birefringence with dilatancy was investigated under dry conditions [*Bonner*, 1974] and controlled pore pressure [*Hadley*, 1975].

Changes in electrical resistivity with stress were observed in rocks under uniaxial compression [*Yamazaki*, 1965, 1966; *Parkhomenko*, 1967] and various controlled confining pressures and pore pressures [*Brace et al.*, 1965; *Brace and Orange*, 1966, 1968*a*, 1968*b*]. These changes are generally understood as follows: since electrical conduction in rock is almost wholly through the pore fluids, the variation of resistivity reflects the changes in shape and connectivity of pore spaces during rock deformation. Electrical resistivity changes related to dilatancy were discussed by *Brace* [1975] based on the experimental observations of the effects of dilatant volume, degree of water saturation in pore spaces and resistivity of aqueous solutions. As the crack distribution becomes anisotropic under stress, the changes in resistivity are expected to be anisotropic.

In previous studies, little attention has been paid to the anisotropic changes in resistivity during rock deformation. In this paper, experimental data are reported on anisotropic changes in electrical resistivity and compressional wave velocity of partially water saturated granite under uniaxial compression.

II. EXPERIMENTAL METHODS

The rock sample studied was Inada granite from Japan. Mineral compositions by volume were 40% K-feldspar (≃4 mm), 20% plagioclase (≃1.5 mm), 30% quartz, (≃3 mm and 10% biotite (≃0.5 mm), in which the largest grain size was indicated in parentheses. A more detailed description of this sample was given in *Mogi* [1964]. Bulk density, porosity and fracture stress at 1 bar confining pressure were 2.63 g/cm^3, 1.1% and about 1.5 kbar, respectively. Before loading, the effect of the moisture content on resistivity at about 2 Hz was measured. Figure 1 shows the effect of water content by volume on resistivity at ambient temperature. The extreme left point in this figure corresponds to 100% water saturation, which was obtained by immersing the sample into distilled water under vacuum and for at least 2 hours. The extreme right point was obtained after heating the sample at 120°C for 2 hours and cooling it in a desiccator for a halfday. The break in the relationship between the resistivity and water content was found to be about 45% water saturation. Above this critical water content, the water is sufficient to coat the surface of grains (or pore spaces) uniformly so that the resistivity increases in proportion to the inverse square of the water content [*Keller*, 1953]. Below this content, the network of water-film channels through pore spaces either breaks at many places, or the water-film coating pore surfaces becomes too thin for the usual (intrinsic) electrical conduction in water.

The amount of partial saturation at 1 bar was selected to be about 50% so that the change in resistivity might be proportional to the inverse square of the amount of partial saturation, if pore spaces decrease uniformly. The sample was entirely coated with epoxy to keep the water content constant throughout the experiment.

The sample was in the form of rectangular parallelepiped (3 x 3 x 5 cm) and uniaxial load was applied along the longest

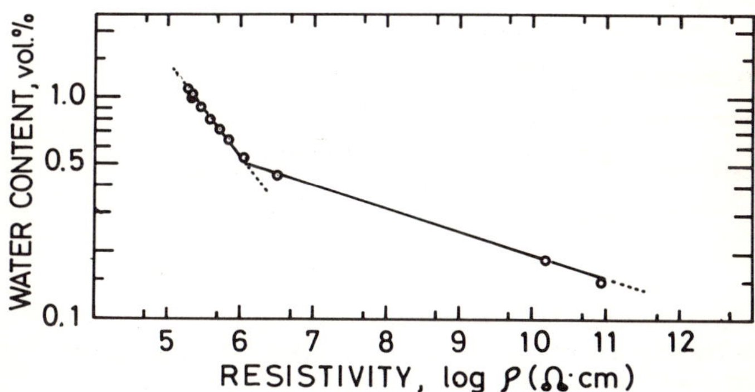

Fig. 1. The variation of electrical resistivity with water content (by volume %) for Inada granite at ambient temperature.

dimension. The original sample was almost isotropic within the experimental errors both for the resistivity and compressional wave velocity. The electrodes consisted of conductive epoxy painted directly to the surface of the sample. The areas of electrodes whose normal was parallel to and perpendicular to the uniaxial load were 3 x 3 cm and 1.5 x 2 cm, respectively. The effect of surface conduction on the resistivity measurement can be neglected for this electrode geometry as it is within the experimental error. 2-MHz barium-titanate compressional transducers were bonded to the electrodes for the velocity measurement by the pulse transmission method [*Birch*, 1960]. The other pair of surfaces was used for the strain measurement by bonding foil strain gauges. Thus strains, resistivities, and compressional wave velocities for two directions were measured simultaneously. Figure 2 shows a schematic diagram of the experimental set-up.

To clarify the effect of dilatancy, the sample is first loaded to 50% and then to about 75% of the fracture stress.

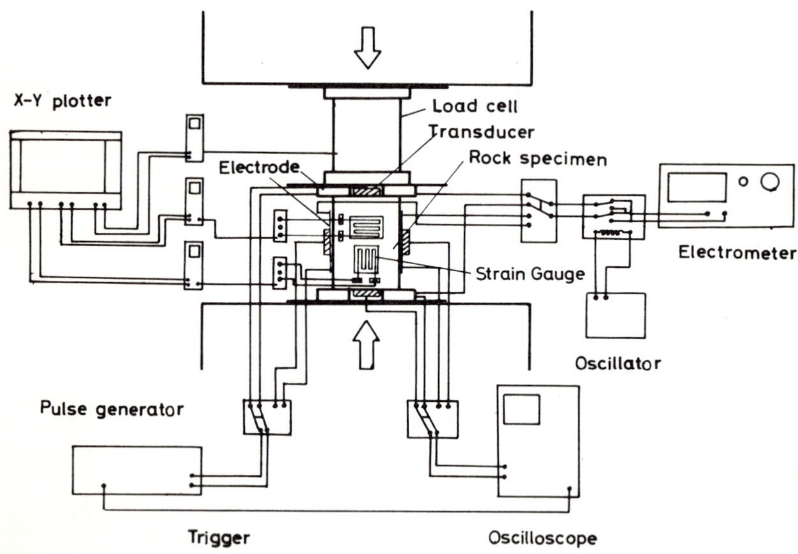

Fig. 2. Schematic diagram of the apparati used for simultaneous measurement on strains, resistivities, and compressional wave velocities for two directions.

III. RESULTS

Figure 3 shows the X-Y record of the axial and transverse strains. It took about 10 minutes for resistivity and velocity measurement at the points indicated by dots and triangles in this figure. Overall strain rate was about 1.5×10^{-5} sec^{-1} throughout this study. Almost reversible changes were obtained for both the first and second stress cycles in the axial strain ($\varepsilon_{\parallel}$). On the other hand, large deviations from quasi-linear change were observed for the second stress cycle in the transverse strain ($\varepsilon_{\perp}$). From this figure the volumetric strain $\varepsilon_{\parallel} + 2\varepsilon_{\perp}$ was calculated and is shown as a function of uniaxial compression in Figure 4. The dilatancy observed in the first cycle is less than that in the second cycle. The dilatant strain observed in the second unloading cycle was about 0.35% at 1 kbar

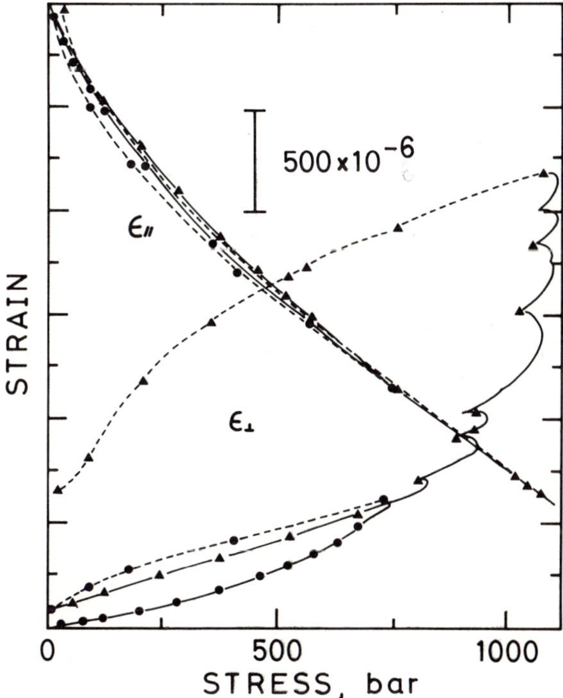

Fig. 3. Actual copy of X-Y recorder for axial ($\varepsilon_{\parallel}$) and transverse ($\varepsilon_{\perp}$) strains. Dots and triangles indicate the sampling points for resistivity and velocity measurement in the first and second stress cycles, respectively. Solid and dashed lines are loading and unloading cycles, respectively.

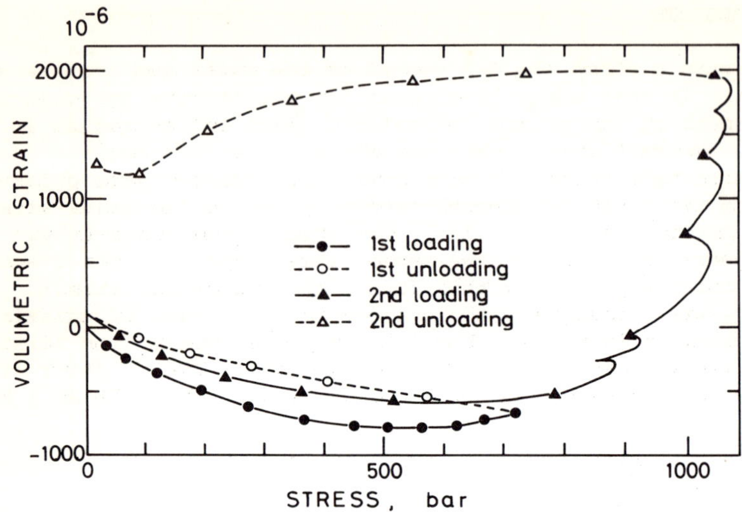

Fig. 4. Volumetric strain versus uniaxial compression. Closed circles and triangles connected by solid line indicate the sampling points for resistivity and velocity measurement in the first and second loading cycles, respectively. Open ones connected by dashed line correspond to unloading cycle.

and 0.25% at 500 bar. The parmanent volumetric strain after the second cycle was about 0.1%, and so the pore volume of this sample increased by about 10% of the initial porosity.

Figure 5 shows variations of resistivity ratios to the initial value for two directions as a function of uniaxial compression. The initial resistivity values for both directions (ρ_0) were 1.5 (±0.2) x 10^6 ohm•cm. The changes in resistivity for both directions clearly showed the hysteresis effect, *i.e.*, in the first unloading cycle resistivity was nearly the same value as that at the maximum stress, and it remained constant in the second loading cycle until the applied load reached the maximum stress in the previous cycle. The rate of decrease in axial resistivity ratio $(\rho/\rho_0)_{\parallel}$ was sharp (about -0.5/kbar) up to about a few hundred bars and became gradual (about -0.2/kbar) at higher stress level, while $(\rho/\rho_0)_{\perp}$ was almost constant or decreased slightly with stress in the initial low-stress region and then decreased rapidly (-0.4/kbar) in the higher stress region (>300 bar). This observation is qualitatively consistent with a previous investigation by *Brace* [1975], though he discussed only for axial resistivity variation under a few kbar confining pressure. After the first stress cycle, $(\rho/\rho_0)_{\parallel}$ was about 0.8 and nearly the same as

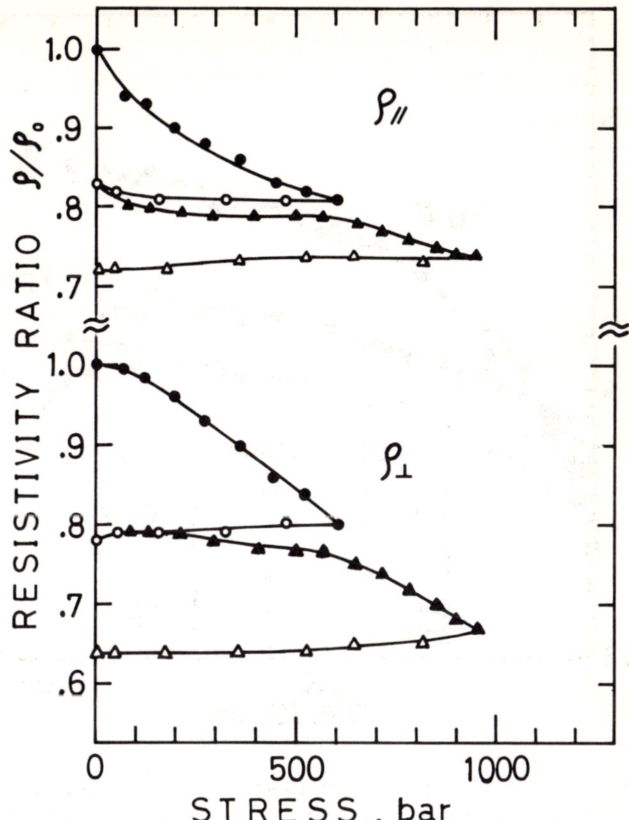

Fig. 5. Variation of resistivity ratio with uniaxial compression. ρ_0 *is the initial resistivity value.* $\rho_{\parallel}$ *and* $\rho_{\perp}$ *indicate the resistivity parallel to and perpendicular to the loading axis, respectively. Simbols are the same as in Fig. 4.*

$(\rho/\rho_0)_{\perp}$. $(\rho/\rho_0)_{\parallel}$ became 0.72 after the second stress cycle, whereas $(\rho/\rho_0)_{\perp}$ became 0.65.

In Figure 6, variations of compressional wave velocity (V_p) were shown as a function of uniaxial compression for both directions. V_p for both directions increased with increasing stress. This observation is consistent with a previous investigation [*Nur and Simmons*, 1969] at least below 300 bar. The velocity values below 50 bar were uncertain because of ambiguities in recognizing the first arrivals. The changes in V_p for both directions were almost reversible except for the transverse direction ($V_{p\perp}$) in the second stress cycle and at

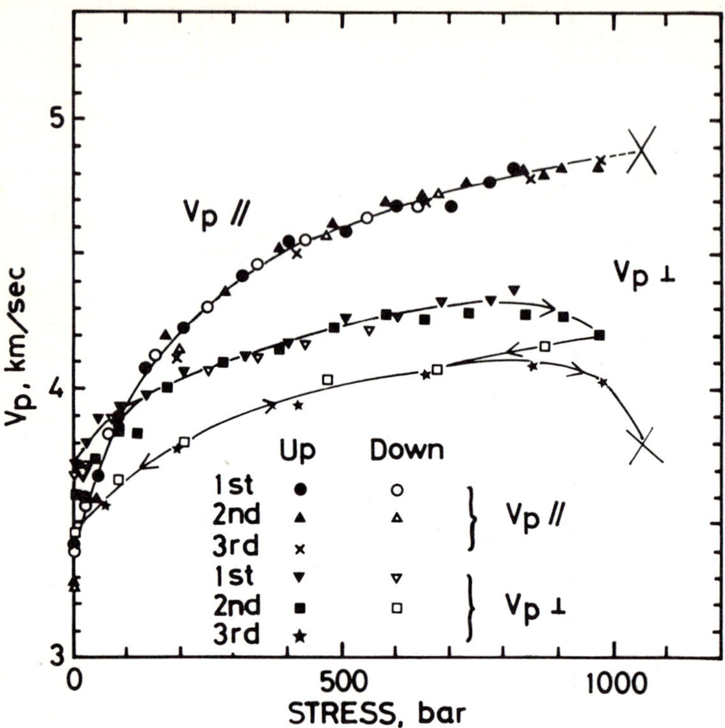

Fig. 6. Variation of compressional wave velocity with uniaxial compression. Stars indicate the values in third loading cycle, and a cross shows where the signal became weak. Other symbols are the same as in Fig. 4.

stress higher than 0.8 kbar. At this stress level, $V_{p\perp}$ started to decrease with increasing stress and the dilatancy became appreciable as shown in Figure 4. The amount of decrease (about 5%) seemed constant through the second unloading cycle. V_p measurements were made in the third stress cycle where the stress reached to about 1.1 kbar. Although $V_{p\perp}$ decreased rapidly near the point indicated by a cross, no decrease of $V_{p\parallel}$ was found up to the maximum stress studied.

IV. Discussion

Our results, summarized in Figure 7, compare the anisotropic changes in resistivity and velocity for two directions as a function of uniaxial compression.

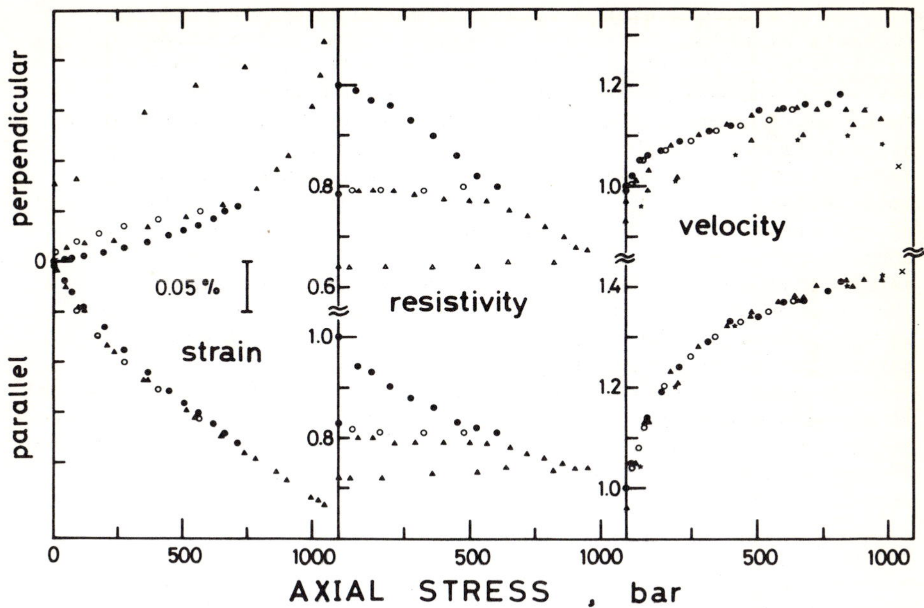

Fig. 7. Summary of variations of strain, resistivity, and compressional wave velocity with uniaxial compression for dirctions parallel to and perpendicular to the loading axis. Symbols are the same as in Fig. 6.

In the first stress cycle, the sample is compressed less than half of the fracture stress so that little dilatancy is observed. Velocity changes in the two directions can be interpreted in terms of crack closure with stress as discussed by *Nur and Simmons* [1969]. The decreases of resistivity for both directions are also explained by the effect of crack closure, which causes increase in water-film connectivity and partial saturation in pore spaces. Increase of partial saturation due to crack closure in the first cycle, estimated to be at most 0.08% for porosity decrease (Figure 4), corresponded to a decrease in resistivity of a few percent (Figure 1), provided the water-film channels were unchanged at the initial state. Observed decrease in resistivity ($\simeq$ 20%) was much larger than that estimate above. Although there is no reason to consider that the water channels remain isotropic, the observed changes in resistivity for the two directions are nearly the same in the first stress cycle. Observed hysteresis effect may be due to the characteristics between the water film and coated grain surfaces due to the affinity of water, *i.e.*, once the pore surface is coated by water, the water film is not disconnected until the pore is completely closed.

In the second stress cycle, $\varepsilon_{\perp}$ shows some nonlinear increase at about 60% of the fracture stress. The change in $V_{p\parallel}$ is strikingly reversible and increases with increasing stress, while $V_{p\perp}$ shows a 5% decrease in the second unloading cycle. Then these anisotropic changes in V_p are due to the anisotropic distribution of newly formed cracks as discussed by *Hadley* [1975]. However, quantitative interpretation for this observation needs more studies about the shape and configuration of the cracks related to dilatancy. The changes in resistivity seem to be a simple continuation from the first cycle. Observed anisotropic change in resistivity is rather small relative to the changes with stress. These observations may be explained by the effect of the crack networks with partially water saturation under uniaxial compression.

We have, however, no quantitative explanation for the amount of decreases and the differences of decreasing rates between two directions in resistivity, and the amount of changes in velocity in the second stress cycle. The subject needs further investigation.

Acknowledgement. The authors wish to express their appreciation to W. F. Brace for his helpful suggestions.

REFERENCES

Birch, F., The velocity of compressional waves to 10 kbar, Part I, *J. Geophys. Res., 65,* 1083-1102, 1960.

Bonner, B., Shear wave birefringence in dilating granite, *Geophys. Res. Lett., 1,* 217-220, 1974.

Brace, W. F., Dilatancy-related electrical resistivity changes in rocks, *Pure Appl. Geophys., 113,* 207-217, 1975.

Brace, W. F., A. S. Orange, and T. R. Madden, The effect of pressure on the electrical resistivity of water-saturated crystalline rocks, *J. Geophys. Res., 70,* 5669-5678, 1965.

Brace, W. F., and A. S. Orange, Electrical resistivity changes in saturated rock under stress, *Science, 153,* 1525-1526, 1966.

Brace, W. F., and A. S. Orange, Electrical registivity changes in saturated rocks during fracture and frictional sliding, *J. Geophys. Res., 73,* 1433-1445, 1968*a*.

Brace, W. F., and A. S. Orange, Further studies of the effects of pressure on electrical resistivity of rocks, *J. Geophys. Res., 73,* 5407-5420, 1968*b*.

Gupta, I. N., Seismic velocities in rock subjected to axial loading up to shear fracture, *J. Geophys. Res., 78,* 6936-6942, 1973.

Hadley, K., V_P/V_S anomalies in dilatant rock samples, *Pure Appl. Geophys., 113,* 1-23, 1975.

Keller, G. V., The effect of wettability on the electrical resistivity of sands, *Oil Gas J., 51,* 62-65, 1953.

Matsushima, S., Variation of elastic wave velocities of rocks in the process of deformation and fracture under high pressure, *Dis. Prev. Res. Inst. Bull., 32,* 2-8, 1960.

Mogi, K., Deformation and fracture of rocks under confining pressure. (1), Compression test on dry rock sample, *Bull. Earthq. Res. Inst., 42,* 491-514, 1964.

Nur, A., Dilatancy, pore fluids, and premonitory variations of t_S/t_P travel times, *Bull. Seismol. Soc. Amer., 62,* 1217-1222, 1972.

Nur, A., and G. Simmons, Stress-induced velocity anisotropy in rock: an experimental study, *J. Geophys. Res., 74,* 6667-6674, 1969.

Parkhomenko, E. I., in *Electrical Properties of Rocks,* 314 pp., Plenum Press, New York, 1967.

Scholz, C. H., L. R. Sykes, and Y. P. Aggarwal, Earthquake prediction: a physical basis, *Science, 181,* 803-810, 1973.

Spetzler, H., and R. J. Martin III, Correlation of strain and velocity during dilatancy, *Nature, 252,* 30-31, 1974.

Spetzler, H., R. J. Martin III, N. Soga, and H. Mizutani, Strain fields associated with fracture under high pressure, viewed with holographic interferometry, in *High-Pressure Research: Applications to Geophysics,* edited by M. H. Manghnani and S. Akimoto, Academic Press, New York, 623-633, 1977.

Thill, R. E., Acoustic methods for monitoring failure in rock, in *Proc. 14th Sympos. Rock Mech.,* edited by H. R. Hardy and R. Stefanko, *Am. Soc. Civil Eng.,* New York, 1973.

Tocher, D., Anisotropy in rocks under simple compression, *AGU Trans., 38,* 89, 1957.

Whitcomb, J. H., J. D. Garmany, and D. L. Anderson, Earthquake prediction: Variation of seismic velocities before the San Fernando earthquake, *Science, 180,* 632-635, 1973.

Yamazaki, Y., Electrical conductivity of strained rocks, 1. *Bull. Earthq. Res. Inst., 43,* 783-802, 1965.

Yamazaki, Y., Electrical conductivity of strained rocks, 2, Further experiments on sedimentary rocks, *Bull. Earthq. Res. Inst., 44,* 1553-1570, 1966.

ELASTIC WAVE VELOCITIES IN THE ICHINOME-GATA ULTRAMAFIC NODULES: COMPOSITION OF THE UPPERMOST MANTLE

S. MATSUSHIMA
Institute of Earth Science, School of Liberal Arts
Kyoto University, Kyoto, Japan

K. AKENI
Institute of Geophysics, Faculty of Science
Kyoto University, Kyoto, Japan

Abstract

Ultrasonic wave velocities, $\bar{V}_p$, have been measured to 20 kbar at 800°C in a lherzolite nodule from Ichinome-gata, Japan. The value of $\partial V_p/\partial T$ (-0.49 m/sec·deg) agrees well with that of an ideal olivine aggregate, but the value of $\partial V_p/\partial P$ (13.6 m/sec kbar) is somewhat higher.

The $\bar{V}_p$ and V_p(VRH) calculated from the chemical and petrographic analyses are in fairly good agreement with the observed values. From the comparison of V_p in the nodules with the P_n velocities under Ichinome-gata, it is pointed out that the lherzolite model is consistent with P_n velocity on the Japan Sea side, and the websterite model fits well with P_n velocity on the Honshu Island side.

I. INTRODUCTION

Theoretical and practical methods for determining the elastic wave velocities in rocks from the elastic properties of the constituent minerals have been presented [*Crosson and Lin*,1971; *Birch*,1972;*Baker and Carter*,1972;*Mackenzie*,1972;*Peselnick et al.*, 1974]. Velocities in rocks depend not only on the species and volume fractions of the constituent minerals, but also on their chemical compositions. At present, it is not possible to calculate the velocities in rocks with complicated chemical compositions and uncommon constituent minerals, because the elastic constants of these minerals are not available.

Fortunately, many kinds of nodules, thought to originate from the upper mantle, have relatively simple chemical and mineral compositions, and the elastic constants of their constituent minerals have been measured and reported.

In this paper, we report the measured and calculated ultrasonic velocity of block of lherzolite nodule from Ichinome-gata, in the northeastern part of Honshu Island, Japan. On the basis of the good agreement between the measured and calculated velocities for this lherzolite nodule, the velocities in websterite nodules from the same region are calculated and a conclusion as to the nature of the upper mantle under Ichinome-gata is drawn.

II. EXPERIMENTAL METHODS

A diagram of the piston-cylinder type apparatus, 20 mm in inner diameter and 70 mm in thickness, used in the present study, is shown in Figure 1. Frictional pressure corrections at various temperatures were determined from the hysteresis curves of the load-velocity relations for calcite undergoing polymorphic change at high pressures and temperatures [*Kondo et al.*,1972;*Matsushima et al.*,1974]. The accuracy of the reported pressures is estimated to be within 0.5 kbar.

Cylindrical specimens, 8 mm in diameter and 6 mm long, were placed at the center of a carbon tube heater, 10 mm in inner diameter and 50 mm long. Both ends of the specimen were covered with nickel foils, 30 μm thick, connected with the carbon heater. With this method, the temperature gradient was reduced remarkably. Temperature differences between points A-B and A-C as shown in Figure 2, for nickel foils, 10, and 50 μm in thickness are plotted in Figure 3.

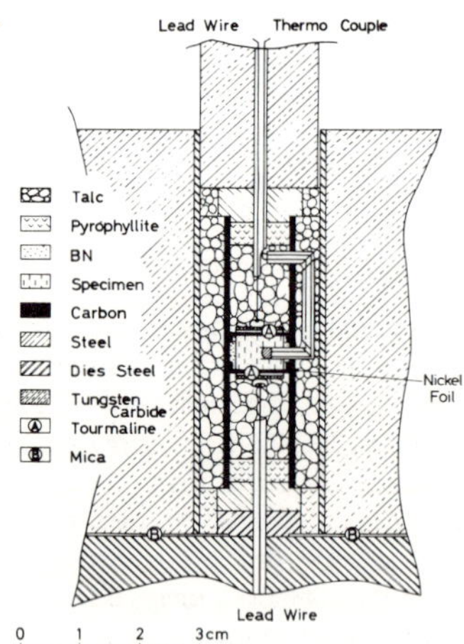

Fig. 1. Schematic representation of high-pressure chamber and sample assembly of piston-cylinder apparatus.

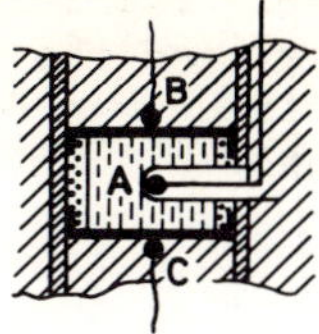

Fig. 2. Points of calibration of the temperature distribution in a specimen.

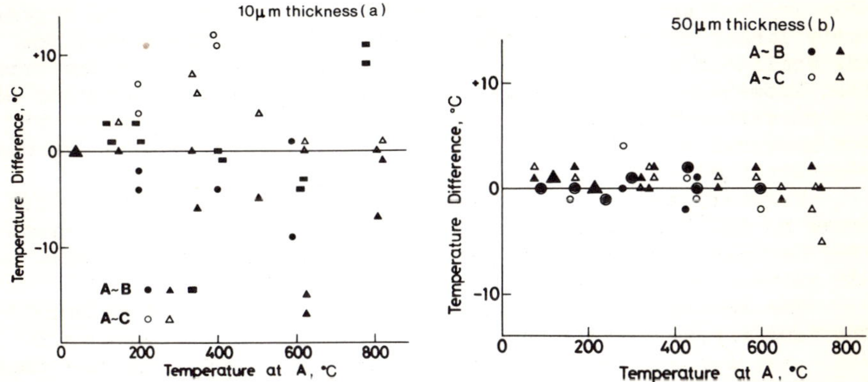

Fig. 3. Temperature differences between A-B and A-C, as shown in Figure 2, for (a) 10 μm, (b) 50 μm thick nickel foil, respectively.

The pulse transmission method was used in the present study. Tourmaline transducers, about 0.25 mm thick, were used to simultaneously generate both compressional and shear waves. The frequencies of these transducers were about 15 MHz for compressional waves and about 10 MHz for shear waves. The wave lengths for both compressional and shear waves in the present specimens were comparable with the mineral grain size (ca 0.5-1 mm), so that the observed compressional velocities correspond to the "short-wave" $\bar{V}_p$ as defined by *Birch*[1972]. Due to polarization of shear waves in two directions in single crystals, it is difficult to correlate the shear waves for the three specimens studied here. Therefore, shear wave velocities are not reported here. The velocity values are corrected for changes in length due to pressure and temperature by using the compressibility and thermal expansion data of olivine [*Anderson et al.*,1968].

A quartz ultrasonic delay line (exactly 5.0 mm long along the c-axis) was used for an accurate time standard. The initial electric pulse signal was simultaneously applied to the specimen and to the quartz delay line. The output signals were displayed on a dual-trace synchroscope, and the transmission time of the ultrasonic wave in the specimens was determined by comparison with the standard time signal.

III. EXPERIMENTAL RESULTS

Three cylindrical specimens were cored from a block of lherzolite nodule, in mutually perpendicular directions. Figure 4 shows compressional wave velocities at various pressures to 20 kbar in each of the three specimens. The effect of pores on the velocity was still slightly present above 10 kbar, and an approximate linear relation was found above 15 kbar. The reason for the nonlinear behavior of velocity with pressure may be the original looseness of the grains and the relatively high hardness of the constituent minerals. However, at higher temperatures, it was observed that the linear variations of velocity with pressure were attained at relatively lower pressures.

Figure 5 shows the V_p measurements on the three lherzolite cores to 800°C at about 20 kbar. Above about 200°C, the velocities decrease linearly with temperature, but the relationship is not quite linear below about 200°C.

The average velocity for the three cores is 8.34 km/sec. The temperature derivative for the present sample agrees well with that for olivine [*Kumazawa and Anderson*,1969], but the pressure derivative is somewhat larger.

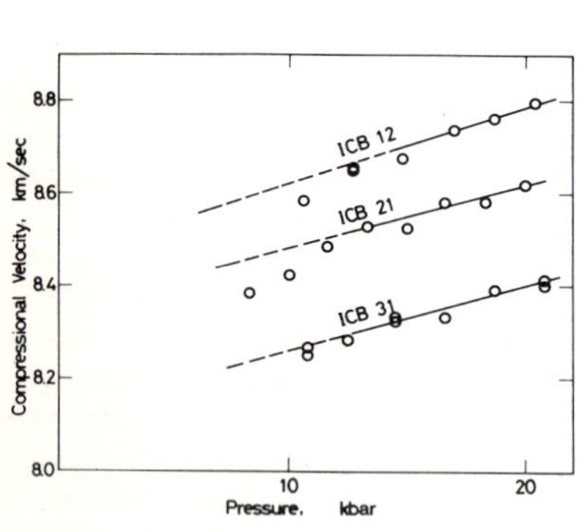

Fig. 4. Plot of V_p versus pressure at room temperature for three mutually perpendicular specimens.

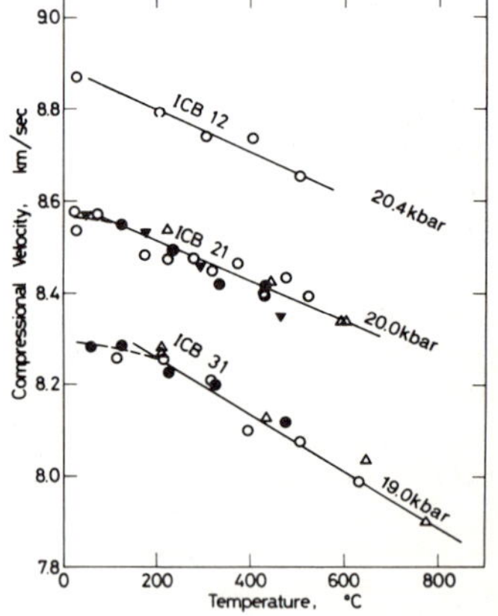

Fig. 5. Plot of V_p versus temperature at about 20 kbar for specimens.

A. Chemical Analyses and Mineral Composition

The chemical analyses of the lherzolite from Ichinome-gata studied here along with the average chemical analyses of other lherzolites and websterites from the same region as reported by *Aoki* [1973] are listed in Table 1.

TABLE 1. Chemical Composition of Ultramafic and Mafic Inclusions from Ichinome-gata (wt%)

	Lherzolite		Websterite	Olivine	Orthopyroxene
Oxide	Present Sample,	Average[a]	Average[a]	Present Sample	Present Sample
SiO_2	42.82	44.54	50.96	41.27	56.93
TiO_2	0.04	0.17	0.39		
Al_2O_3	2.14	2.29	4.48	0.18	2.29
Cr_2O_3	0.15	0.40	0.64		
Fe_2O_3		1.71	1.89		
FeO	8.28	7.25	5.20	9.88	6.29
MnO	0.12	0.14	0.16		
MgO	42.21	40.80	20.27	49.67	34.04
CaO	2.97	2.50	15.45	0.22	0.85
Na_2O	0.06	0.17	0.52		
K_2O	0.01	0.03	0.03		
Spinel[b]	1.20				

a. From *K. Aoki* [1973]
b. undissolved residue

TABLE 2. Estimated Mineral Composition (wt%) and MgO-FeO ratio (mol%) of Lherzolite and Websterite from Ichinome-gata

	Lherzolite		Websterite
	Present Sample	Average	Average
Feldspar	0.06	0.17	4.57
Clinopyroxene	10.85	12.68	65.39
Orthopyroxene	18.16	23.02	21.60
Olivine	69.73	62.93	7.24
Spinel	1.20	1.20	1.20
MgO/MgO + FeO	90.1	90.8	87.1

The mineral composition of the present lherzolite was determined from the chemical, X-ray, and microscopic analyses and is given in Table 2. No trace of serpentine or other hydrated minerals was found. Feldspar was separated by using heavy liquid (ρ = 2.8 g/cm^3) and analysed by X-ray diffraction. Further description of the feldspar is not possible due to the small amount of feldspar powder separated. Olivine and orthopyroxene grains were picked up from the crushed sample with a pincette and analyzed. Their chemical analyses are listed in Table 1. Spinel was separated from the powdered sample by quick dissolution of the silicate minerals in fluoric acid. Also given in Table 2 are the mineral compositions for the average lherzolite and websterite from Ichinome-gata which were estimated, assuming the same MgO-FeO ratio for olivine and pyroxenes and the absence of Al_2O_3 and CaO in olivine and orthopyroxene.

B. Compressional Velocities of Constituent Minerals

It is assumed that the velocity, V, in rock is given by

$$\frac{1}{V} = \sum \frac{f_i}{V_i} \tag{1}$$

where V_i is the velocity in i-th constituent mineral and f_i is its volume fraction. This approximation is valid for a large sample containing a large number of mineral grains. Since this is not quite the case in our lherzolite sample, some discrepancy is expected between our observed and calculated velocities.

Velocities in the minerals depend on their density and chemical composition. For minerals in ultramafic rocks, there is a relationship between the MgO-FeO ratio expressed as "MgO mol%", i. e., MgO/(MgO + FeO), and velocity. The approximate linear relations between MgO mol%-velocity and MgO mol%-density have been presented by *Chung*[1970] for olivine. Density-compressional velocity-MgO mol% diagrams were drawn for each of the constituent minerals (olivine, spinel, clinopyroxene, and orthopyroxene) with the assumption of the linear relations for density-MgO mol% and compressional velocity-MgO mol%. The diagram of velocity and anorthite-albite-orthoclase composition is also drawn for feldspar. Data used for drawing these diagrams are listed in Table 3.

Figure 6 shows the density-velocity-MgO mol% diagram for olivine. Straight lines were drawn through the data points as a first approximation. The assumed compressional velocity of the olivine of the present lherzolite sample is indicated in Figure 6.

The elasticity of the spinel-hercynite system has been summarized by *Wang and Simmons*[1972]. The linear relations are assumed for density, compressional velocity, and MgO mol% for this system. The density, 3.964 g/cm^3, of the spinel of the present

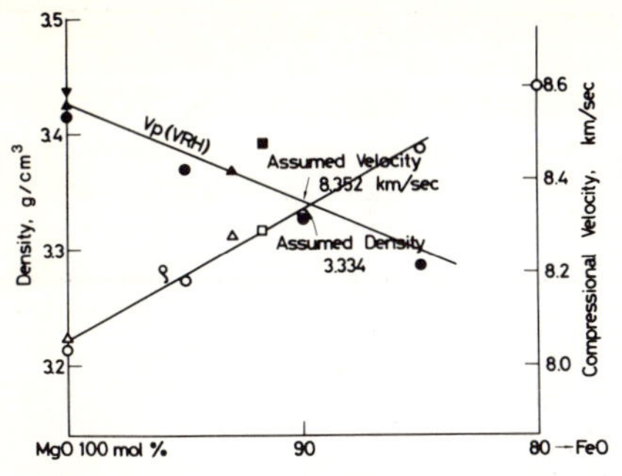

Fig. 6. Density-V_p-MgO molar fraction diagram for olivine, Mg_2SiO_4-Fe_2SiO_4 system.

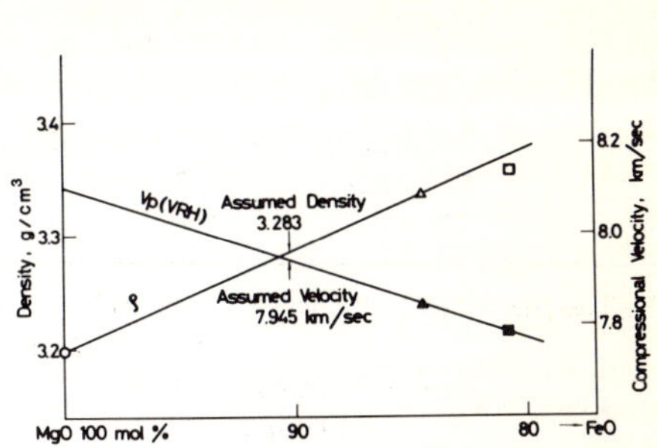

Fig. 7. Density-V_p-MgO molar fraction diagram for orthopyroxene, $MgSiO_3$-$FeSiO_3$ system.

TABLE 3. Density and Compressional Velocity Data, and $\partial V_p/\partial \rho$ for Rock-forming Minerals

	Composition	Density ρ, g/cm^3	Velocity V_p, km/sec	Reference[a]	$\partial V_p/\partial \rho$
Olivine	Fo(Mg_2SiO_4)	3.214		1	-1.88
		3.222	8.593	6	(in the
		3.224	8.564	3	range
	Fo 93.0	3.311	8.419	3	Fo100-
	Fo 91.7	3.324	8.481	4	Fo 85)
	Fo100	3.217	8.534	5	
	Fo 95	3.273	8.422	5	
	Fo 90	3.330	8.317	5	
	Fo 85	3.386	8.216	5	
	Fa(Fe_2SiO_4)	4.393		1	
Spinel	$MgAl_2O_4$	3.582		1	-1.54
			9.712	8, 2	(in the
	$FeAl_2O_4$	4.258		1	range
			8.674	7	$MgAl_2O_4$-$FeAl_2O_4$)
Clino-pyroxene	$CaMg(SiO_3)_2$	3.277		1	-1.98
	Diopside	3.31	7.698	9, 2	(in the
	Augite	3.32	7.216	9, 2	range
	Aegirite-Augite	3.42	6.845	9, 2	Diopside-Aegirite)
	Aegirite	3.50	7.322	9, 2	
	$CaFe(SiO_3)_2$	3.55 ± 0.01		1	

TABLE 3 (continued)

Ortho-pyroxene	$MgSiO_3$	3.198 ± 0.007		1
	Bronzite (En84.1)	3.335 ± 0.001	7.85	10
	Bronzite (En80.7)	3.354	7.78	11
Feldspar	$CaAl_2Si_2O_8$	2.762 ± 0.004		1
	An-Ab (9%An)	2.61	6.066	12, 2
	(24%An)	2.64	6.222	12, 2
	(29%An)	2.64	6.252	12, 2
	(53%An)	2.68	6.567	12, 2
	(56%An)	2.69	6.616	12, 2
	$NaAlSi_3O_8$	2.617 ± 0.005		1
	Or-Ab			
	(53.5%Or)	2.58	5.880	13, 2
	(60.7%Or)	2.57	5.648	13, 2
	(64.9%Or)	2.57	5.743	13, 2
	(66.6%Or)	2.54	5.582	13, 2
	(74 %Or)	2.57	5.786	13, 2
	(75 %Or)	2.54	5.556	13, 2
	(78.5%Or)	2.56	5.912	13, 2
	$KAlSi_3O_8$	2.551 ± 0.008		1

a. 1-*Robie et al.*[1966]; 2-*Simmons and Wang*[1971]; 3-*Kumazawa and Anderson*[1969]; 4-*Verma*[1960]; 5-*Chung*[1970]; 6-*Graham and Barsch*[1969]; 7-*Wang and Simmons*[1972]; 8-*Lewis*[1966]; 9-*Aleksandrov et al.*[1964]; 10-*Kumazawa*[1969]; 11-*Frisillo and Barsch*[1972]; 12-*Ryzhova*[1964]; 13-*Ryzhova and Aleksandrov*[1965]

sample was measured accurately, then its velocity, 9.23 km/sec, was derived from the linear relations.

There are no reliable V_p data for a chemically analyzed clinopyroxene. However, density and composition, and density and V_p data have been published for the diopside-hedenbergite system. These were combined to draw the compressional velocity-density-MgO mol% diagram. To a first approximation a straight line was drawn through the upper bound of the data points. For this line the value of $\partial V_p/\partial \rho$ is quite similar to that of olivine and is assumed to be reasonable. The assumed compressional velocity and density for the clinopyroxene of the present rock are 7.735 km/sec and 3.294 g/cm^3, respectively.

Two velocity data for bronzite [*Kumazawa*,1969;*Frisillo and Barsch*,1972] were plotted on the density-velocity-MgO mol% diagram for orthopyroxene (Figure 7). The value of $\partial V_p/\partial \rho$ is assumed to be -1.90 for orthopyroxene. A straight line with this assumed

$\partial V_p/\partial \rho$ value is seen to fit the two velocity data.

The elastic wave velocities for the anorthite-albite-orthoclase system were also plotted. The linear relation between the compressional velocity and anorthite content for the anorthite-albite system was clearly shown. The data points for albite-orthoclase system were more scattered, and thus the straight line drawn through these points was a very rough approximation. However, the effect of the uncertainty in the compressional velocity of feldspar on the following velocity calculation is negligible, due to the very small amount of feldspar in the rock (Table 3).

C. Comparison between Observed and Calculated Velocities

The Hill average of the compressional velocity, V_p(VRH) for the present lherzolite sample was calculated by using equation (1) and the respective velocities of the constituent minerals determined from the velocity-density-MgO mol% diagrams. The same calculations were performed for the average lherzolite and websterite, and all the results are listed in Table 4.

For comparison with the calculated and the observed velocity, the velocity $\bar{V}_p$ in the lherzolite sample studied here was also calculated. $\bar{V}_p$ was calculated as follows. First, $\bar{V}_p$'s for olivine, ortho- and clinopyroxene, spinel and feldspar were obtained according to the formula [*Birch*,1972]:

$$\frac{1}{\bar{V}_p} = \frac{1}{4\pi}\int_0^{2\pi} d\phi \int_0^{\pi} \frac{1}{V_p(\theta,\phi)} \sin\theta d\theta \tag{2}$$

TABLE 4. Observed and Calculated Compressional Velocities in Lherzolite and Websterite from Ichinome-gata and Comparison with P_n Velocities under Ichinome-gata

	$\bar{V}_p$ km/sec		V_p(VRH) km/sec	V_p(Moho) km/sec
	Obs.	Calc.	Calc.	(8kbar, 500-700°C)
Lherzolite, Present Sample	8.34	8.28	8.21	8.03-8.12
Lherzolite, Average			8.18	7.95-8.04
Websterite, Average			7.66	7.43-7.52[a]
P_n, Honshu side				7.5
P_n, Japan Sea side				8.2

a. The values of pressure and temperature derivatives of V_p of present sample is used.

TABLE 5. Calculated Velocities in Aggregates of Constituent Minerals of Ultramafic and Mafic Rocks

	V_p(VRH) km/sec	$\bar{V}_p$ km/sec	References
Plagioclase	6.616	6.810	*Ryzhova* [1964]
Clinopyroxene	7.698	7.868	*Aleksandrov et al.* [1964]
Orthopyroxene	7.845	7.884	*Kumazawa* [1969]
Olivine	8.419	8.486	*Kumazawa and Anderson* [1969]
Spinel	9.712	9.943	*Lewis* [1966]

where $V_p(\theta,\phi)$ is the highest velocity in the crystal along the direction of polar angle θ, and longitudinal angle ϕ referred to the principal axes. For each mineral, the velocity $V_p(\theta,\phi)$ was computed for directions spaced at intervals of $10°$ of θ and ϕ by solving Christoffel's equation [*Musgrave*, 1970] and using published elastic stiffness. The calculated values $\bar{V}_p$ for each mineral are listed in Table 5, along with the V_p(VRH) values. The chemical compositions of the constituent minerals of the present rock sample are only slightly different from those of the correspondent minerals listed in Table 5. The velocity differences between $\bar{V}_p$ and V_p(VRH) for the minerals listed in Table 5 were used for the calculation of the rock $\bar{V}_p$. The result is given in Table 4 in the column $\bar{V}_p$, calc.

IV. CONCLUSION AND DISCUSSION

Despite the scantiness of data for establishing the velocity-density diagrams and all the approximations in the velocity calculations, there is a good agreement between the observed and calculated velocities for the lherzolite sample studied here. On the basis of this observation, the calculated V_p(VRH) values for the average lherzolite and websterite listed in Table 4 are considered fairly representative.

Using the observed values of the pressure and temperature derivatives of $\bar{V}_p$ for lherzolite, the compressional velocity of the lherzolite and the websterite at the upper mantle conditions under Ichinome-gata was calculated (Table 4). The pressure and temperature in the upper mantle at a depth of about 27 km were estimated to be 8 kbar and 500-700 °C, respectively. These values are compared with those obtained from crustal studies of the same region by explosion and earthquake seismology. The cross section of the Honshu Island through Ichinome-gata after *Yoshii and Asano* [1972] is reproduced in Figure 8, and the P_n velocities are listed in Table 4. It is seen that on the Japan Sea side the lherzolite

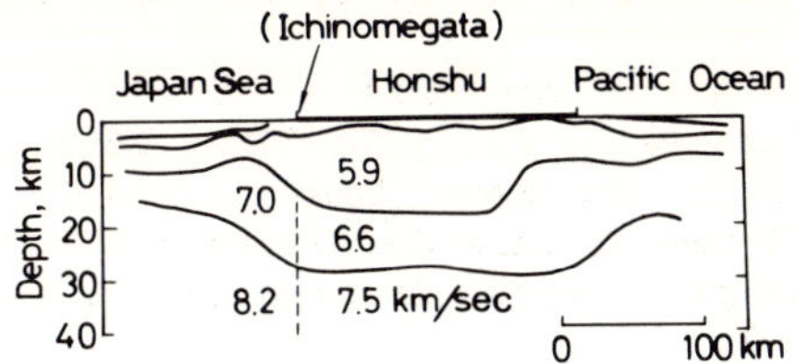

Fig. 8. Crustal structure along Kesennuma-Oga Peninsula (Ichinome-gata) profile, N-E region of Honshu Island, Japan [after Yoshii and Asano, 1972].

model fits the seismological model, while on the Honshu Island side, the websterite model is closer to the seismological model.

Acknowledgments. The authors express their thanks to R. Ramananantoandro and M. H. Manghnani for critically reviewing the paper.

REFERENCES

Aleksandrov, K. S., T. V. Ryzhova, and B. P. Belikov, The elastic properties of pyroxenes, *Sov. Phys. Cryst., 8,* 589-591, 1964.

Anderson, O. L., E. Schreiber, R. C. Liebermann, and N. Soga, Some elastic constant data on minerals relevant to geophysics, *Rev. Geophys., 6,* 491-524, 1968.

Aoki, K., Ejected accidental fragments of Itinome-gata volcano--materials derived from upper mantle and lower crust, (in Japanese) *J. Min. Soc. Jap., 11,* 100-111, 1973.

Baker, D. W., and N. L. Carter, Seismic velocity anisotropy calculated for ultramafic minerals and aggregates, in *Flow and Fracture of Rocks,* Geophys. Mon., Vol. 16, edited by H. C. Heard, I. Y. Borg, N. L. Carter, and C. B. Raleigh, pp. 157-166, Amer. Geophys. Union, Washington, D. C., 1972.

Birch, F., Numerical experiments on the velocities in aggregates of olivine, *J. Geophys. Res., 77,* 6385-6391, 1972.

Chung, D. H., Effects of iron/magnesium ratio on P- and S-wave velocities in olivine, *J. Geophys. Res., 76,* 570-578, 1971.

Frisillo, A. L., and G. R. Barsch, Measurement of single-crystal elastic constants of bronzite as a function of pressure and temperature, *J. Geophys. Res., 77,* 6360-6384, 1972.

Graham, E. K. Jr., and G. R. Barsch, Elastic constants of single-crystal forsterite as a function of temperature and pressure, *J. Geophys. Res., 74,* 5949-5960, 1969.

Kondo, S., K. Suito, and S. Matsushima, Ultrasonic observation of calcite I-II inversion to 700°C, *J. Phys. Earth, 20,* 245-250, 1972.

Kumazawa, M., The elastic constants of single-crystal orthopyroxene, *J. Geophys. Res., 74,* 5973-5980, 1969.

Kumazawa, M., and O. L. Anderson, Elastic moduli, pressure derivatives, and temperature derivatives of single-crystal olivine and single-crystal forsterite, *J. Geophys. Res., 74,* 5961-5972, 1969.

Lewis, M. F., Elastic constants of magnesium aluminate spinel, *J. Acoust. Soc. Amer., 40,* 728-729, 1966.

MacKenzie, D. B., Peridotite fabrics and velocity anisotropy in the earth's mantle, in *Studies in Earth and Space Sciences,* Geol. Soc. Amer., Mem., Vol. 132, edited by R. Shagam, pp. 593-604, Geol. Soc. Amer., Boulder, Colorado, 1972.

Matsushima, S., K. Suito, and S. Kondo, Variation of the elastic properties accompanying the calcite 1-2 transition at high pressure and temperature, pp. 383-388, *Proc. 4th Internat. Conf. High Pressure,* Physico-Chem. Soc. Jap., Kyoto, 1974.

Musgrave, M. J. P., *Crystal Acoustics,* Holden-Day, San Fransisco, 1970.

Peselnick, L., A. Nicolas, and P. R. Stevenson, Velocity anisotropy in a mantle peridotite from the Ivrea Zone: Application to upper mantle anisotropy, *J. Geophys. Res., 79,*1175-1182, 1974.

Robie, R. A., P. M. Bethke, M. S. Toulmin, and J. L. Edwards, X-ray crystallographic data, densities, and molar volumes of minerals, in *Handbook of Physical Constant,* Geol. Soc. Amer., Mem., Vol. 97, edited by S. P. Clark, Jr., pp. 27-73, Geol. Soc. Amer., New York, 1966.

Ryzhova, T. V., Elastic properties of plagioclase, (English Transl.), pp. 633-635, *Izv. Acad. Sci. USSR,* Geophys. Ser., 1964.

Ryzhova, T. V., and K. S. Aleksandrov, The elastic properties of potassium-sodium feldspars, (English Transl.), pp. 53-56, *Izv. Acad. Sci. USSR,* Geophys. Ser., 1965.

Simmons, G., and H. Wang, *Single Crystal Elastic Constants and Calculated Aggregate Properties,* M. I. T. Press, Cambridge, Massachusetts, 1971.

Verma, R. K., Elasticity of some high-density crystals, *J. Geophys. Res., 65,* 757-766, 1960.

Wang, H., and G. Simmons, Elasticity of some mantle crystal structures 1. Pleonaste and hercynite spinel, *J. Geophys. Res., 77,* 4379-4392, 1972.

Yoshii, T., and S. Asano, Time-term analyses of explosion seismic data, *J. Phys. Earth, 20,* 47-57, 1972.

EXPERIMENTAL STUDIES ON THE INFLUENCE OF CO_2 AND H_2O IN THE UPPER MANTLE

P. J. WYLLIE
Department of the Geophysical Sciences
University of Chicago
Chicago, Illinois 60637

Abstract

The system CaO-MgO-SiO_2 includes the minerals olivine (Fo), orthopyroxene (Opx), and clinopyroxene (Cpx). In the presence of excess CO_2, with increasing pressure peridotite undergoes a series of carbonation reactions. At 1100°C, model mantle assemblage Fo + Opx + Cpx + CO_2 is progressively transformed into Fo + Opx + Cd (calcic dolomite) + CO_2, and then Opx + Cd + Cm (magnesite solid solution) + CO_2 near 20 kbar; and into Cm + Cd + Qz + CO_2 just above 30 kbar. For realistic mantle conditions, with only trace amounts of CO_2, the starting assemblage at 1100°C is changed to Fo + Opx + Cpx + Cd near 20 kbar, and this changes to Fo + Opx + Cpx + Cm near 45 kbar. Each of the univariant carbonation reaction boundaries between these assemblages terminates at an invariant point at which carbonate and silicates melt together with CO_2 (1250°-1450°C with pressures increasing from about 25 kbar to 55 kbar), producing a carbonate-rich liquid. In the system CaO-MgO-SiO_2-CO_2-H_2O, the carbonation reactions and melting reactions occur at lower temperatures as H_2O content increases. In the presence of CO_2-H_2O fluids, the carbonate-bearing peridotite acts as a buffer, controlling the CO_2/H_2O ratio as a function of pressure and temperature. This in turn controls the composition of the liquid at temperatures just above the solidus as a function of depth. Along normal geotherms in the mantle, CO_2 is distributed between calcic dolomite and CO_2-H_2O fluid, with carbonate, CO_2 and H_2O dissolving in liquid at the top of the seismic low-velocity zone. Amphibole and phlogopite may become stable in peridotite containing H_2O, Al_2O_3, and K_2O. The compositions of near-solidus mantle magmas at various depths are strongly influenced by the presence or absence of carbonate, amphibole, and phlogopite in the peridotite. There is now enough experimental data from various laboratories to permit an estimate of the maximum areas of stability of carbonate, amphibole and phlogopite on the peridotite-CO_2-H_2O solidus. For high CO_2/H_2O ratios the interstitial liquid at the top of the low-velocity zone is carbonatitic, becoming more silicic (but SiO_2-undersaturated) with increasing depth. Upward migration of this liquid could conceivably lead to the local generation of carbonate-rich peridotites within the lower lithosphere.

I. INTRODUCTION

Exploratory experiments have demonstrated that the presence of H_2O and CO_2 in the upper mantle would exert significant influence on magmatic processes in and above the seismic low-velocity zone, and that the proportion CO_2/H_2O would influence the compositions of liquids generated. Recent experimental studies were reviewed by *Boettcher* [1975] in the quadrenniel U. S. National Report to the International Union of Geodesy and Geophysics. The presence of these volatile components, even in small amounts, could change the physical properties of upper mantle materials.

There is now convincing evidence that both H_2O and CO_2 do exist in the upper mantle. For evidence and reviews see *Moore* [1965, 1970], *Dawson* and *Powell* [1969], *Mason* [1968], *Irving* and *Wyllie* [1973, 1975], *McGetchin* and *Besancon* [1973], *Roedder* [1965], *Green* and *Radcliffe* [1975], *Melton et al.*, [1972], *Dawson et al.*, [1970], and *Dawson* and *Hawthorne* [1973].

There are several experimental approaches to the problem of the influence of volatile components on magma genesis. One is to determine the melting relationships of whole rock compositions (natural or synthetic), either source rocks such as peridotite-H_2O-CO_2 [*Mysen* and *Boettcher,* 1975*a*; 1975*b*] or possible primary magmas derived from the sources such as olivine-melilitite-H_2O-CO_2 [*Brey* and *Green,* 1975, 1976*a*]. Another is to measure the solubilities of H_2O or CO_2 in simple or complex silicate liquids [*Eggler,* 1974; *Brey* and *Green,* 1976*a*, *Mysen et al.*, 1976]. Mineral assemblages representing rock systems, such as forsterite-enstatite-diopside-CO_2 [*Eggler,* 1974, 1976] or forsterite-enstatite-CO_2-H_2O [*Eggler,* 1975*a*], may prove easier to interpret than the complex whole-rock compositions. Melting relationships in systems such as CaO-MgO-SiO_2-CO_2-H_2O provide the framework for identification of the reactions among mineral assemblages, and for comprehension of the relations among them that may be obscured in more complex rock systems [*Kushiro,* 1972; *Huang* and *Wyllie,* 1974; *Wyllie* and *Huang,* 1976*a*].

In this review, I first outline the phase relationships in CaO-MgO-SiO_2-CO_2 to serve as a guide for reactions occurring in peridotite-CO_2. I then examine the effect of H_2O-CO_2 mixtures on the subsolidus carbonation reactions, and construct a schematic solidus surface for the system peridotite-CO_2-H_2O. Combination of the subsolidus and solidus reactions produces diagrams which demonstrate how carbonates in peridotite can buffer the H_2O/CO_2 of the vapor phase. Addition of H_2O to peridotite-CO_2 can stabilize amphibole and phlogopite, and these too may become involved in buffering the subsolidus vapor phase composition.

II. THE SYSTEM $CaO\text{-}MgO\text{-}SiO_2\text{-}CO_2$

A. CO_2-Saturated Liquidus Surface

Figures 1, 2 and 3 illustrate some of the 30 kb reactions in the systems $MgO\text{-}SiO_2\text{-}CO_2$, $CaO\text{-}SiO_2\text{-}CO_2$, and $MgO\text{-}CaO\text{-}CO_2$ that are combined in the opened-up tetrahedron of Figure 4.

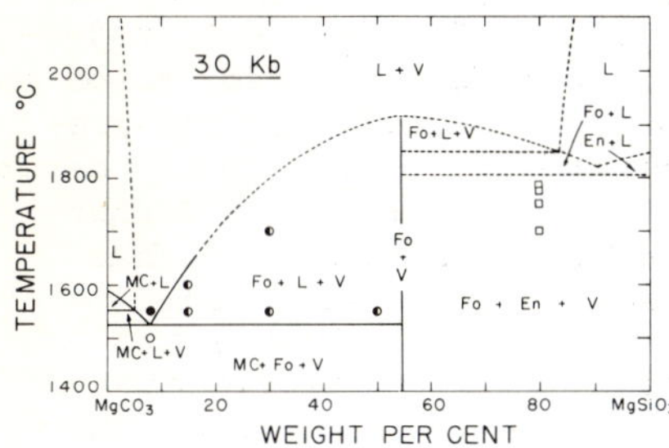

Fig. 1. Isobaric diagram at 30 kbar, largely schematic, showing some of the phase fields in the system $MgO\text{-}SiO_2\text{-}CO_2$ intersected by the join $MgCO_3$-$MgSiO_3$ [Huang and Wyllie, 1964]. Dashed lines are estimated field boundaries. Ternary phase fields intersected near the liquidus piercing point (about 90% $MgSiO_3$) are not represented. Abbreviations: MC = magnesite; Fo = forsterite; En = enstatite; L = liquid; V = vapor (essentially CO_2).

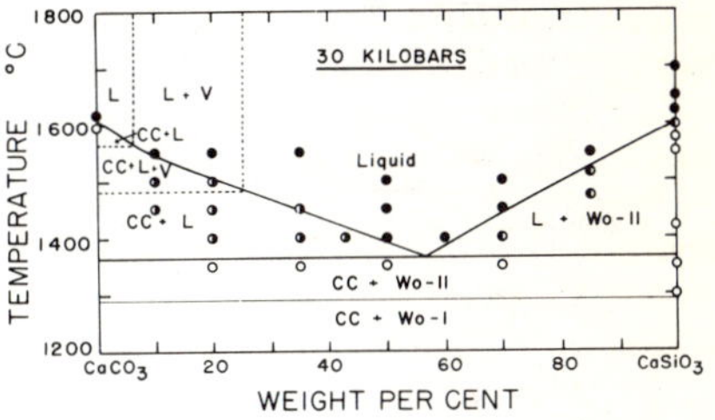

Fig. 2. Isobaric diagram at 30 kbar showing the phase fields in the system $CaO\text{-}SiO_2\text{-}CO_2$ intersected by the join $CaCO_3$-$CaSiO_3$, which is almost binary [Huang and Wyllie, 1974]. Dashed lines indicate uncertain boundaries. Abbreviations: see Fig. 1; CC = calcite; Wo-I = wollastonite I; Wo-II = wollastonite II.

In the system $MgO\text{-}SiO_2\text{-}CO_2$ at this pressure, the field for the primary crystallization of carbonate is separated from other silicate fields by the field for forsterite, with a high temperature maximum (thermal barrier) on its surface (Figures 1 and 4). The carbonate field is small. In contrast, there is a binary eutectic liquid containing almost 45% carbonate in the join $CaSiO_3$-$CaCO_3$ (Figure 2), and the field for the primary crystallization of calcite is quite extensive (Figure 4). These results indicate that the solubility of CO_2 in Mg-silicate liquids is low, as reported by *Eggler* [1974] and *Mysen et al.*, [1976], but the binary eutectic liquid in Figure 2 corresponds to a CO_2 solubility in Ca_2SiO_4 liquid of about 20% (Figure 4), a value subsequently confirmed by *Eggler et al.*, [1976]. The reason for the difference between the two systems $MgO-SiO_2-CO_2$ and $CaO-SiO_2-CO_2$ is that wollastonite and calcite coexist subsolidus at this pressure, whereas enstatite and magnesite react to produce forsterite and CO_2. The latter reaction was determined to 40 kbar by *Newton* and *Sharp* [1975].

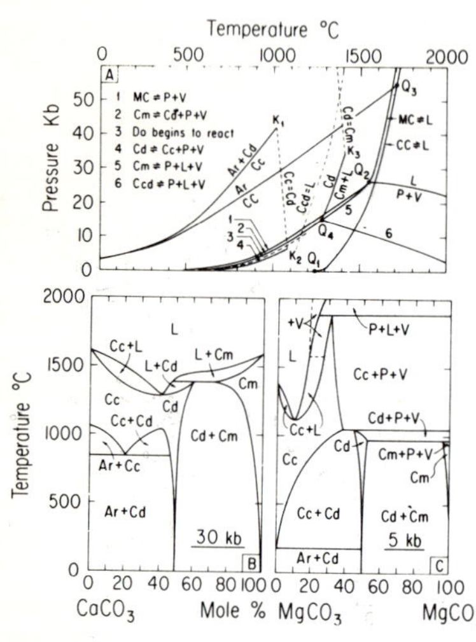

Fig. 3. Reactions involving the join $CaCO_3$-$MgCO_3$ through the system CaO-MgO-CO_2 [Irving and Wyllie, 1975]. A. P-T projection, partly schematic, based on new experimental data and previous results. B. Isobar determined at 30 kbar. C. Isobar based on reactions at 5 kbar shown in Fig. A, and other results cited by Irving and Wyllie [1975]. Abbreviations: see Fig. 2; Do = dolomite; Ar = aragonite; P = periclase; Cc, Cd, Cm = carbonate solid solutions related to calcite, dolomite, or magnesite, respectively, at lower pressures and temperatures; Ccd = intermediate solid solutions.

Figure 3A is based on the 30 kbar results shown in Figure 3B, together with all previously published lower-pressure results. The liquidus from Figure 3B is transferred to the join $CaCO_3$-$MgCO_3$ in Figure 4, and the field boundaries added in Figure 4 for excess CO_2 are based on results of *Huang* and *Wyllie* [1976] on the solubility of CO_2 in $CaCO_3$ and $MgCO_3$. The rest of the system $CaO-MgO-CO_2$ in Figure 5 is schematic. Note the deep thermal trough on the liquidus near dolomite (Figure 4, compare Figure 3B).

Of particular interest in Figure 4 for this paper are the CO_2-saturated liquidus field boundaries that limit the fields of primary crystallization in the ternary systems which have CO_2 as one component. Within the quaternary system, these are connected through the tetrahedron by a CO_2-saturated liquidus surface. On this surface, there is a field boundary separating fields for the primary crystallization of silicates from the crystallization of carbonates. The two ends of this boundary can be located in Figure 4, between Fo and MC, and between Wo and CC; the first point has a CO_2 solubility approaching 40%, and the second a CO_2 solubility approaching 20%.

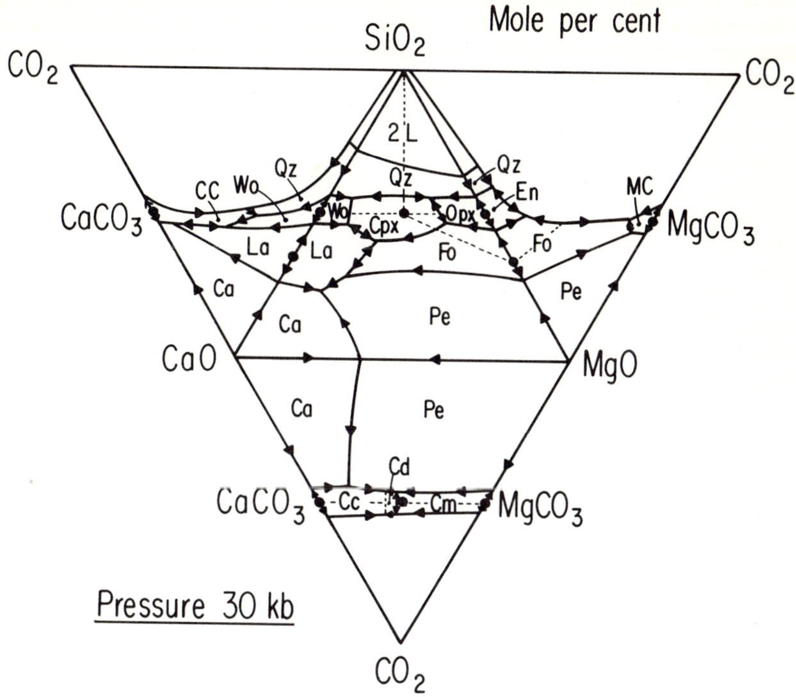

Fig. 4. Liquidus phase relationships at 30 kbar for the systems CaO-MgO-SiO_2, MgO-SiO_2-CO_2, CaO-SiO_2-CO_2, and CaO-MgO-CO_2, partly schematic, based on Figs. 1, 2, and 3 and other data sources cited by Wyllie and Huang [1976a]. Note the phase boundaries limiting the liquidus fields for the primary crystallization of minerals; these give the compositions of liquids saturated with CO_2. The diagram is converted into a tetrahedron by taking the CO_2 corners of the three ternary systems, and folding them up to the same point. The three CO_2-saturated liquidus field boundaries are connected through the quaternary tetrahedron by a CO_2-saturated liquidus surface. See Wyllie and Huang [1975, 1976a] for examples. The pyroxene-carbonate plane through the tetrahedron, $MgSiO_3$-$CaSiO_3$-$CaCO_3$-$MgCO_3$, is shown in Fig. 9. Abbreviations: see Fig. 3; Ca = lime; Pe = periclase; La = larnite (or polymorphs of Ca_2SiO_4); Cpx = clinopyroxene; Opx = orthopyroxene; Qz = quartz (polymorphs of SiO_2).

B. Decarbonation Reactions

There is a whole series of decarbonation reactions in the

system $CaO-MgO-SiO_2-CO_2$, with starting assemblages involving calcite, dolomite, magnesite and quartz reacting to produce free CO_2 and silicate minerals enstatite, forsterite, diopside, and wollastonite, among others. *Bowen* [1940] examined the reactions involved in the progressive metamorphism of siliceous dolomites, and they formed the basis of his petrogenetic grid. The sequence of reactions and subsequent theoretical and experimental work were reviewed by *Turner* [1968, p. 131-151].

Extension of these decarbonation reactions to mantle pressures produces a family of curves involving mantle silicates, CO_2, and carbonates [*Wyllie* and *Huang*, 1975, 1976*a*]. One of these, reaction (6) illustrated in Figure 5, involves all three major mantle minerals, Fo+Opx+Cpx, together with a calcic dolomite (Cd). At constant pressure, the 5-phase assemblage is invariant, and the compositions of the reacting phases are given by the 5-phase tie-figure (with tie-lines to CO_2 omitted for clarity).

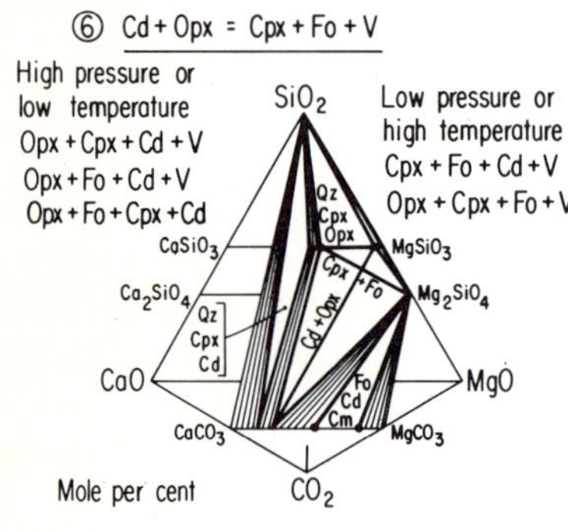

Fig. 5. Isobaric isothermal diagram showing the 5-phase tie-figure for the univariant carbonation reaction (6) which involves model mantle peridotite minerals olivine (Fo), orthopyroxene (Opx), and clinopyroxene (Cpx), together with calcic dolomite (Cd) and vapor (CO_2). All phase fields shown are for excess CO_2, but tie-lines to CO_2 have been omitted for clarity [Wyllie and Huang, 1975]. The 5-phase reaction involves five 4-phase assemblages. The univariant curve for the reaction is shown in Figs. 6, 7, and 8, along with the pairs of 4-phase assemblages involved with excess CO_2 (Fig. 7), and with insufficient CO_2 to carbonate the peridotite completely (Fig. 8). Thus, four of the five 4-phase assemblages can be identified in Figs. 7 and 8. The fifth assemblage is Opx+Cpx+Cd+CO_2, which does not include the Fo-rich model mantle composition.

There are five 4-phase assemblages associated with the reaction, only two of which include all three mantle silicate minerals, Fo+Opx+Cpx. The following discussion is limited to a model mantle assemblage dominated by Fo. The assemblage Fo+Opx+Cpx+Cd dissociates to produce Fo+Opx+Cpx+CO_2. For the reverse reaction of carbonation, with increasing pressure, the assemblage Fo+Opx+Cpx+CO_2 reacts to produce either the vapor-absent assemblage Fo+Opx+Cpx+Cd, or, if there is sufficient CO_2 present, the clinopyroxene becomes completely carbonated to yield the assemblage Fo+Opx+Cd+CO_2.

C. P-T Projection

Reaction boundaries along which carbonates and silicates coexist with CO_2-saturated liquids are generated at intersections between (1) reaction curves for the effect of CO_2 on the melting of silicates, and (2) subsolidus reaction curves for the carbonation of silicate assemblages by CO_2. These intersections provide the key to the silicate-carbonate liquidus relationships through a range of pressures.

Figure 6 compares two published versions of the reactions involving Fo+Opx+Cpx+CO_2. The subsolidus carbonation reaction (6), terminating at the invariant point Q_6 where liquid appears, is plotted according to the data reported by *Wyllie* and *Huang* [*1975*] rather than being slightly displaced for clarity of representation in complex diagrams as in *Wyllie* and *Huang* [1976*a*]. The cause of the difference between the curves may be the recurrent problem of friction correction in piston-cylinder apparatus [*Huang* and *Wyllie,* 1975]. *Eggler* [1976] used the results of *Newton* and *Sharp* [1975] for the forsterite carbonation reaction, but found it necessary to correct them by "1.5 kbar to be compatible with runs in Figure 2". The results of *Wyllie* and *Huang* agree precisely with those of *Newton* and *Sharp,* and therefore they are about 1.5 kbar lower than those of *Eggler*.

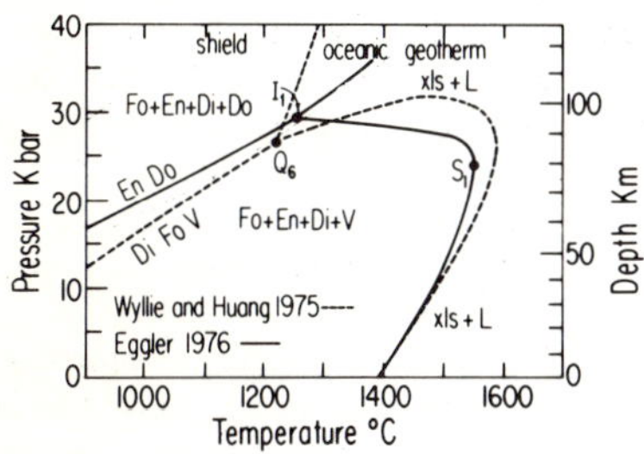

Fig. 6. Comparison of two published versions of reactions for model mantle peridotite (Fo+En+Di, or Fo+Opx+Cpx in Figs. 4 and 5) in the system CaO-MgO-SiO_2-CO_2. There are three univariant curves meeting at an invariant point. The subsolidus carbonation reaction (6) illustrated in Fig. 5 terminates where liquid appears. The solidus is divided into two portions, the low-pressure curve below the invariant point, and the high-pressure curve above the invariant point. The pressure maximum on the solidus for Fo+En+Di+CO_2 shown by Wyllie and Huang [1975, 1976a] was required by the constraints imposed by a point for the assemblage Fo+En+Di+L+CO_2 reported by Eggler [1974] at 30 kbar and 1565°C. Eggler [1976] does not use this point in more recent versions of his silicate-CO_2 work, and if the point is not valid then no maximum is required. See Figs. 7 and 8. The geotherms are from Ringwood [1966].

The solidus curve in Figure 6 is divided into two portions, the low-pressure curve below the invariant point where CO_2 exists as vapor and the CO_2 solubility in the liquid remains low, and the high-pressure curve above the invariant point, where the CO_2 is combined with the peridotite to produce calcic dolomite (Cd in Figure 5). *Eggler* [1976] has recently revised his earlier results [*Eggler,* 1974] on the low-pressure solidus curve, and the pressure maximum shown by *Wyllie* and *Huang* [1975] and reproduced in Figure 6 is no longer required (see figure legend for details, and Figures 7 and 8 for values adopted). *Eggler's* [1976] estimate of dP/dT for the silicate-carbonate solidus rising from I_1 appears to be too low in comparison with the slopes of other silicate, carbonate, and silicate-carbonate melting curves.

Wyllie and *Huang* [1975, 1976*a*, 1976*b*] combined their experimental data in silicate-carbonate systems with other published data, leaning heavily on the silicate-CO_2 results of *Eggler* and others from the Geophysical Laboratory for lower pressures. They constructed a complete, partly schematic P-T-X framework for decarbonation and melting reactions. *Eggler's* recent experiments have overlapped those of our silicate-carbonate studies [*Eggler,* 1976]; at present there are some differences in data and interpretations (Figures 6 and 9).

Univariant reactions for a specific model mantle composition in the system CaO-MgO-SiO_2-CO_2, are shown in Figure 7 with excess CO_2, and in Figure 8 with only 0.2% CO_2 (an arbitrary choice, probably higher than normal mantle CO_2 content). The fixed composition corresponds to a peridotite composed of 65% forsterite, 23% orthopyroxene, and 12% clinopyroxene. Reaction (6) and the solidus curve at pressures below Q_6 are our preferred results (compare Figure 6). Carbonation reactions (4) and (2) and the estimated solidus Q_6-Q_4-Q_2 in Figure 7 are taken from *Wyllie* and *Huang* [1976*a*]. The estimated solidus above Q_6 in Figure 8 is from *Wyllie* and *Huang* [1976*a*], and the subsolidus exchange reaction intersecting the solidus at Q_0 is from *Kushiro et al.*, [1975]. The amount of CO_2 required to complete reaction (6) as depicted in Figure 7, and to carbonate the peridotite even further, as indicated by reactions (4) and (2), is unreasonably high for the mantle. Therefore, Figure 8 gives a more realistic picture of the mineralogy of mantle peridotite containing CO_2. Fo+Opx+Cpx coexists with CO_2 vapor on the low pressure side of reaction (6), and with carbonate and no vapor on the high pressure side.

D. Solubility of CO_2

Until recently, the available data indicated that the solubility of CO_2 in normal silicate magmas at crustal pressures was very small [*Wyllie* and *Tuttle,* 1959]. *Hill* and *Boettcher* [1970] presented evidence for higher CO_2 solubilities in basalt-H_2O-CO_2

at pressures above about 15 kbar. Recent experiments show that solubilities increase to 5-8% in the pressure range 10-30 kbar, for individual minerals such as diopside, and for magmas such as olivine-melilitite-nephelinite [*Eggler*, 1974; *Eggler* and *Mysen*, 1976; *Mysen et al.*, 1976]. *Brey* and *Green* [1976*a*, 1976*b*] revised previous higher estimates [*Brey* and *Green*, 1975] to comparable values.

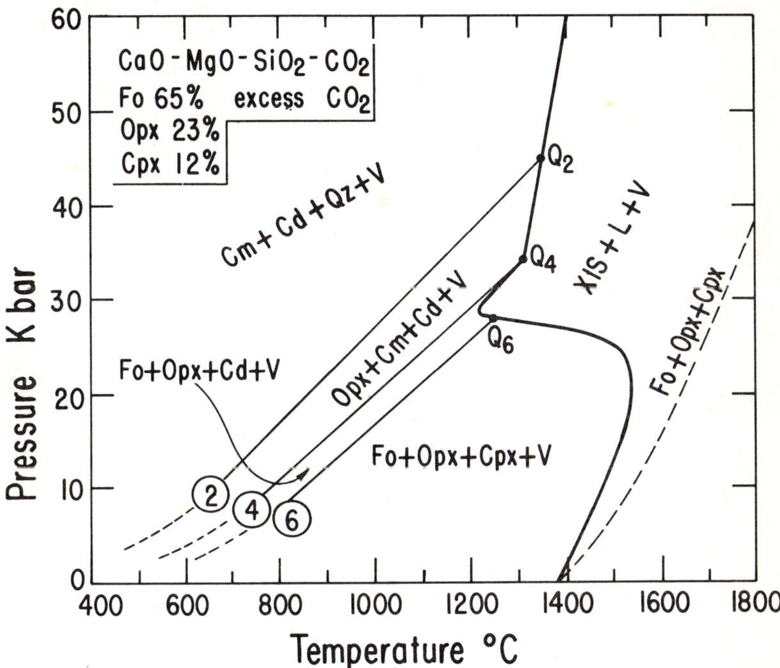

Fig. 7. Partly schematic diagram for a model mantle peridotite (Fo+Opx+Cpx) with excess CO_2, from results and analysis of Wyllie and Huang [1975, 1976a], incorporating revisions of Eggler [1976]. Compare Fig. 6. The dashed melting curve is after Kushiro [1972]. Note that mantle peridotite, Fo+Opx+Cpx, can exist with free CO_2 only at lower pressures, higher temperatures than reaction (6). On the other side of the reaction, one of the mantle minerals, Cpx, is replaced completely by the carbonate, Cd.

The composition of the liquid produced by melting of the silicate-carbonate assemblage along the high-pressure vapor-absent solidus above Q_6 in Figure 8 is carbonatitic with CO_2 content approaching 40% according to *Wyllie* and *Huang* [1975, 1976*a*], and kimberlitic with CO_2 content about 20% according to *Eggler* [1976].

Wyllie and *Huang* [1976*a*, 1976*b*] explained the large increase in CO_2 solubility near Q_6, for near-solidus liquids coexisting with Fo+Opx+Cpx, in terms of the change in liquidus phase relationships caused by intersection of carbonation reaction (6) with the solidus, at point Q_6. *Eggler et al.*, [1976] disagreed with the interpretation as well as the high CO_2 solubilities, but *Wyllie* and *Huang* [1976*c*] re-affirmed their conclusion.

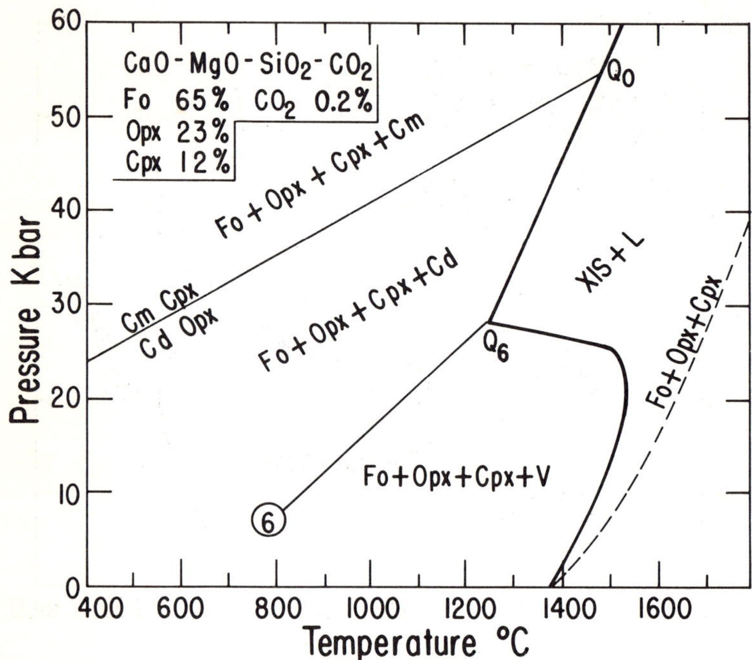

Fig. 8. Partly schematic diagram for the model mantle peridotite depicted in Fig. 7, but with only 0.2% CO_2. *There is insufficient* CO_2 *to carbonate the peridotite completely. Reactions (4) and (2) of Fig. 7 are therefore not intersected (see Fig. 5). Mantle peridotite characterized by the assemblage Fo+Opx+Cpx occurs throughout the subsolidus region, together with a small proportion of carbonate on the high pressure side of reaction (6). An exchange reaction occurs between pyroxenes and carbonates along the curve extending from* Q_0 *[Kushiro, et al., 1975]. Note that melting occurs with free* CO_2 *at pressures below* Q_6. *At higher pressures, the silicate-carbonate assemblages melt to produce a liquid with composition near E in Fig. 9A.*

The fundamental difference between the two studies is in the location of the boundary between the fields for primary silicates and carbonates on the CO_2-saturated liquidus surface, as illustrated in Figure 9 (compare Figure 4). *Wyllie* and *Huang* [1975] used mixtures on the pyroxene-carbonate join, shown by dashed lines in Figure 9A. *Eggler* [1976] used mixtures on the joins Di-Fo and La-Fo with excess CO_2, which intersect the pyroxene-carbonate join along the dashed lines in Figure 9B. I feel confident of the position shown for the silicate-carbonate field boundary in Figure 9A. Figure 9B shows that *Eggler's* [1976] location (projected) required extrapolation.

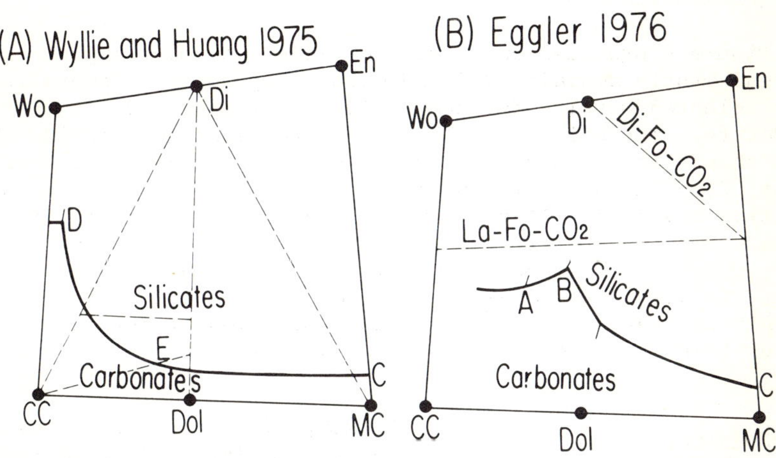

Fig. 9. The silicate-carbonate liquidus field boundary on the CO_2-saturated liquidus surface in the system CaO-MgO-SiO_2-CO_2 at high pressures (weight per cent), projected onto the pyroxene-carbonate join. The two ends of this field boundary can be located in Fig. 4 on the system CaO-SiO_2-CO_2 between fields for Wo and CC, and on the system MgO-SiO_2-CO_2 between the fields for Fo and MC. Dashed lines show composition joins studied by A. Wyllie and Huang [1975], and B. Eggler [1976]. At the invariant point where reaction (6) is joined by a liquid (Q_6 in Figs. 6, 7, and 8), the liquidus field for orthopyroxene just reaches the silicate-carbonate field boundary [see Wyllie and Huang, 1976a, for details] near point E according to Wyllie and Huang [1975], and near A-B according to Eggler [1976]. Liquid E is carbonatitic, with about 40% dissolved CO_2, whereas liquid A-B is kimberlitic with 20% CO_2.

Resolution of the differences shown in Figure 9 is not an academic problem in phase-equilibria. According to *Eggler* [1976], the liquids in the low-velocity zone would be "broadly kimberlitic", with compositions near B in Figure 9B, and CO_2 contents of 20%. I conclude that liquids at the top of the low-velocity zone without H_2O would be carbonatitic, with compositions near E in Figure 9A, and CO_2 contents approaching 40%. These liquids have Ca/Mg ratios greater than 1; they are not magnesite-rich as previously reported by *Eggler* [1975*b*]. With increasing depth and temperature, these would grade into kimberlitic liquids nearer B in composition, but undersaturated with CO_2 [*Wyllie* and *Huang*, 1976*a*, 1976*b*].

III. PERIDOTITE-H_2O

Figure 8 provides one limit for models of mantle processes. The real mantle contains H_2O and other components. Figure 10 shows selected results for the influence of excess H_2O on natural peridotite. The solidus is located at temperatures considerably lower than that in Figure 7 for the synthetic model mantle with excess CO_2. The stability of amphibole is limited by a pressure-sensitive reaction. If the composition of the peridotite is suitable for the stabilization of phlogopite, it reacts in the presence of excess vapor and mantle minerals along the curve shown. At very low and very high pressures, a trace of liquid is probably generated below the solidus for phlogopite-free peridotite-H_2O by the melting of phlogopite.

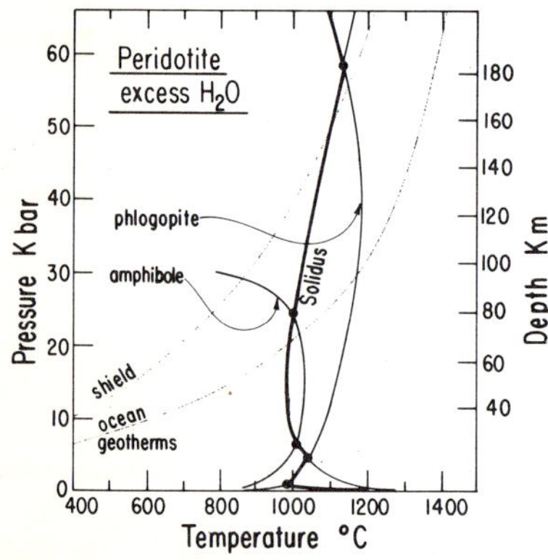

Fig. 10. Peridotite-H_2O, in part schematic. Solidus with excess water [after Kushiro et al., 1968; see Millhollen et al., 1974, for comparison of four sets of experimental results). Amphibole breakdown curve with excess water; see Millhollen et al., [1974] for comparison of four sets of results. Phlogopite breakdown curve with peridotite and excess water, extrapolated from Modreski and Boettcher [1973], see Wyllie [1973]. The solidus is assumed to be effectively coincident with that for peridotite without phlogopite, except for very low pressures and very high pressures, where the phlogopite melting reaction (in association with peridotite minerals) causes melting at lower temperatures. Geotherms are those of Clark and Ringwood [1964].

Note the intersections of the geotherms (lower temperature selections than those in Figure 6) with the three phase boundaries. Amphibole is limited to depths shallower than 100 km in the mantle, whereas phlogopite can exist to depths ranging from 100 to 180 km. The upper boundary of the seismic low-velocity zone in various tectonic environments could be explained by incipient melting of peridotite in the presence of aqueous pore fluid, or by incipient melting of peridotite caused by the melting of amphibole or phlogopite without excess H_2O [*Kushiro et al.*, 1968; *Lambert* and *Wyllie*, 1968, 1970; *Modreski* and *Boettcher*, 1973].

IV. CARBONATION REACTIONS IN THE SYSTEM $CaO-MgO-SiO_2-CO_2-H_2O$

In the presence of excess CO_2-H_2O mixtures, each univariant carbonation reaction of Figure 7 becomes a divariant reaction surface extending down from the univariant curve to lower temperatures and up to higher pressures. *Kerrick* [1974] recently reviewed the theoretical and experimental results for similar reactions with reference to metamorphic processes.

Figure 11A shows a series of schematic isobaric sections through the divariant surface for reaction (6), the carbonation of model mantle peridotite (Figures 5 and 7). The results of *Evans* and *Trommsdorff* [1974] at 7 kbar have been used as a guide for the shape of the intersected surface, but the changing slopes with increasing pressure are just estimated. In Figure 11B, the isobaric section for reaction (6) at 25 kbar is compared with similar sections for the other two carbonation reactions shown in Figure 7.

In Figure 12, points from the estimated isobaric sections in Figure 11A have been used to map the divariant surface for mantle reaction (6) in terms of contours for constant vapor phase composition.

V. PERIDOTITE-CO_2-H_2O

A. Solidus and Carbonation Surfaces

There are now several different experimental results in the literature for reactions in parts of the system peridotite-CO_2-H_2O, as outlined in the legend for Figure 13. The results adopted in this paper are used to illustrate and explore the possible effects of CO_2 and H_2O. The choice of a different set of experimental results would change specific values of pressure and temperature, but the principles and broad conclusions remain unchanged.

In Figure 13A, the solidus for natural peridotite without volatile components, the solidus for peridotite with excess H_2O (Figure 10), and the solidus with excess CO_2 are compared. The latter curve is based on results for the synthetic system in

Figure 7, with temperatures displaced downwards to be consistent with the volatile-free solidus in Figure 13A. It is assumed for graphical convenience that the subsolidus carbonation reactions remain univariant despite the addition and solution of FeO and Al_2O_3; I_2, I_4 and I_6 therefore correspond to points Q_2, Q_4, and Q_6 in Figure 7, but they are displaced to lower temperatures.

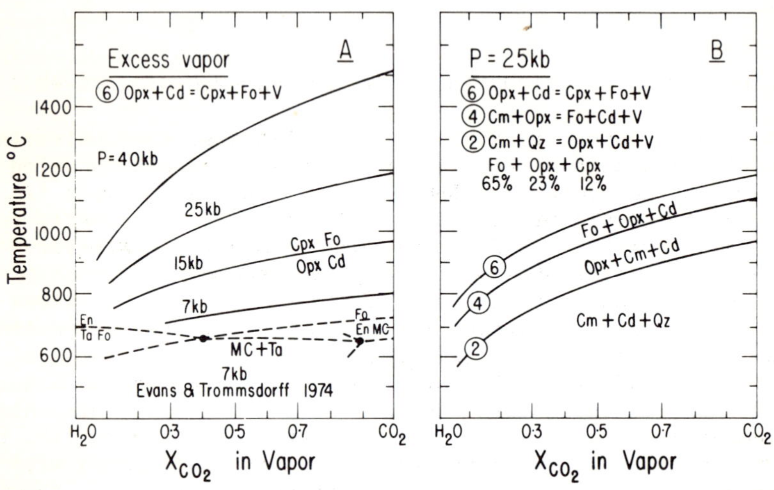

Fig. 11. System CaO-MgO-SiO_2-CO_2-H_2O. Schematic diagrams showing the effect of CO_2-H_2O vapor on the decarbonation reactions (2), (4) and (6), see Figs. 5 and 7. A. The results of Evans and Trommsdorff [1974] for reactions in the system MgO-SiO_2-CO_2-H_2O at 7 kbar illustrate the effect of CO_2 on dehydration reactions, and of H_2O on decarbonation reactions (Abbreviations: Ta = talc, MC = magnesite). These are used to estimate the position of reaction (6) with H_2O at 7 kbar. Experimental data are not available for higher pressures, and the isobaric sections illustrated between 7 kbar and 40 kbar are only schematic. These lines map the divariant surface for reaction (6) with H_2O as an additional component, neglecting the formation of hydrous minerals. B. Schematic 25 kbar section through the system CaO-MgO-SiO_2-CO_2-H_2O showing the effect of H_2O on the temperatures of decarbonation reactions (2), (4) and (6) from Fig. 7, in the presence of excess vapor, neglecting the formation of hydrous minerals.

In Figure 13B, the experimental data of *Mysen* and *Boettcher* [1975*a*] in the system peridotite-CO_2-H_2O have been used as a guide for mapping the solidus surface for different ratios of CO_2/H_2O, although the actual temperatures for specific contours and the shape of the CO_2 solidus differ considerably from their diagrams. *Mysen* and *Boettcher* reported the occurrence of a carbonate phase at total pressures above 20 kbar, and that at 30 kbar, "the charge becomes too blurred for a meaningful optical examination because of this mineral". They noted that the carbonate was probably a quench phase, but the results are consistent with the present interpretation of stable carbonate above 20 kbar, as shown in Figure 14 [*A. L. Boettcher*, personal communication].

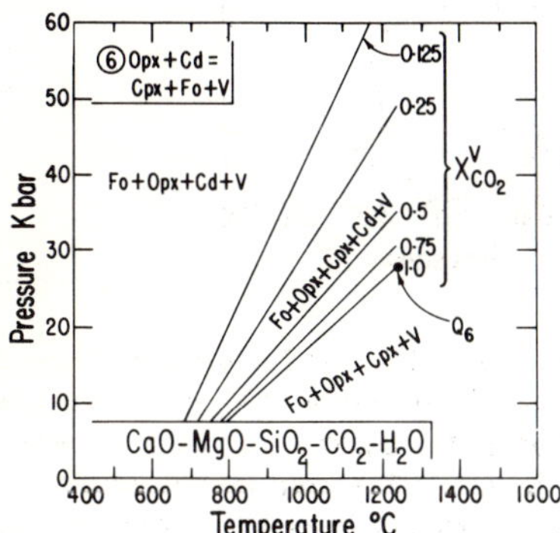

Fig. 12. System $CaO-MgO-SiO_2-CO_2-H_2O$. The divariant surface for the decarbonation reaction (6) with H_2O present, mapped in terms of constant vapor-phase composition, using the schematic values plotted in Fig. 11A. Univariant reaction (6) extends from Q_6 (Figs. 7 and 8); with addition of H_2O the reaction surface extends to lower temperatures, and to higher pressures, from the univariant line.

Figure 14A illustrates the intersection between the divariant solidus surface of Figure 13B and the divariant carbonation surface of Figure 12, assuming that solid solutions in the complex peridotite do not change the carbonation reaction compared with its position in the system $CaO-MgO-SiO_2-CO_2-H_2O$. I also assume at this stage that the peridotite composition is not suitable for the stabilization of amphibole and phlogopite, so that we can examine the influence of the carbonate alone.

The heavy line extending from I_6 through the points of intersections of specific contours on each surface shows the conditions for the beginning of melting of the carbonated assemblage Fo+Opx+Cpx+Cd in the presence of CO_2-H_2O vapors. For each vapor phase composition, both pressure and temperature at which melting begins are fixed. Similarly, at any particular pressure, melting occurs in the presence of vapor of a defined composition only at a specific temperature. Vapor phase compositions are more easily seen in the P-X projection of the solidus surface in Figure 14B. There is little change in P-T for vapors ranging from pure CO_2 to about 50% H_2O, but with further increase of H_2O/CO_2 in the vapor, melting begins at much higher pressures, although at only slightly changed temperature (see Figure 14A).

The divariant surfaces for reactions (4) and (2) of Figure 7 extend similarly from I_4 and I_2 in Figure 14A. They are not shown in this figure, but their lines of intersection with the solidus surface are plotted in the P-X projection of Figure 14B. The shape of the solidus surface can be visualized by the vapor-phase contours in Figure 14A, by the isotherms sketched in Figure 14B, and by the isobaric sections of Figure 15.

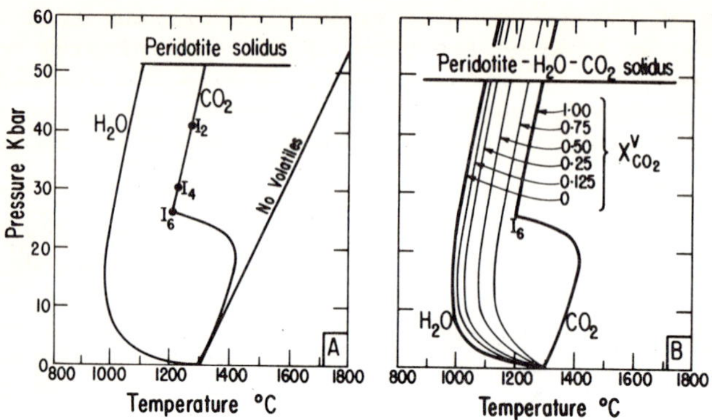

Fig. 13. Schematic diagram illustrating the solidus surface for peridotite-H_2O-CO_2. A. Solidus for peridotite with no volatile components [Wyllie, 1973, for comparison of three different experimental cruves]; with excess H_2O (Fig. 10); and with excess CO_2, based on Fig. 7, but with the sharp curvatures in the solidus at pressures above Q_6 removed for simplicity. The peridotite-CO_2 solidus is located at somewhat lower temperatures than that for the model mantle peridotite in Fig. 7, and this displaces the points Q_6, Q_4, and Q_2 to lower temperatures on reactions (6), (4), and (2), to the points I_6, I_4, and I_2. It is assumed for graphical convenience that reactions (6) (4) and (2) remain univariant in peridotite. B. Schematic contours on the peridotite-vapor solidus surface for vapors of constant composition, using the data of Mysen and Boettcher [1975a] as a guide. Mysen and Boettcher reported a slope of approximately 20°C per 0.1 $X^V_{H_2O}$ unit between 15 and 30 kbar. Their solidus with H_2O was 150°C lower than the one adopted here, and their estimated solidus for peridotite-CO_2 did not drop in temperature to the invariant point I_6 because they had not discovered the subsolidus carbonation reaction (6). However, their contour distribution fits conveniently between the H_2O- and CO_2-solidus curves from Figure 13A, and this is suitable to illustrate types of processes in the following diagrams and text.

Uncarbonated peridotite, Fo+Opx+Cpx, coexists with vapors of all compositions at pressures below I_6 (Figures 15A and B). At higher pressures the point d, the intersection of the carbonation surface (6), migrates to higher H_2O/CO_2 ratios, and the range of uncarbonated peridotite becomes more restricted; it can exist only with H_2O-rich vapors. Carbonated peridotite, Fo+Opx+Cpx+Cd, exists along the phase boundary for reaction (6)--see Figure 14.

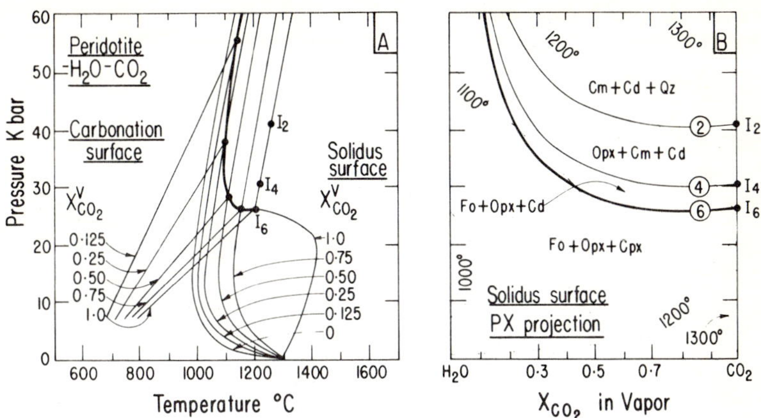

Fig. 14. Intersection of divariant carbonation reactions with the solidus surface for peridotite-H_2O-CO_2, excess vapor, schematic. A. The solidus surface from Fig. 13B and the carbonation surface from Fig. 12. For graphical convenience, it is assumed that the carbonation reactions in peridotite remain univariant, despite the solid solution of FeO and Al_2O_3 in the minerals. The contours on each surface intersect the corresponding contours on the other surface in a series of points, and the curve through these points is the projected line of intersection of the two surfaces. B. P-X projection of the solidus surface, showing lines of intersection with divariant carbonation reactions (6), (4), and (2), and isotherms. Vertical lines in this projection correspond to the contours in Fig. 14A. The intersection of reaction (6) extends from I_6 (compare Figs. 14A and 13A). The positions of I_4 and I_2 are taken from Fig. 13A, and intersections of the surface with divariant reactions (4) and (2) are assumed to be similar to that plotted for reaction (6). Compare and contrast with the isobaric sections of Figs. 11B, and 15.

B. Carbonate as a Buffer of Vapor Phase and Liquid Composition

For mantle peridotite with small quantities of CO_2 and H_2O, the composition of the vapor is buffered by reactions involving carbonate and hydrous minerals in ways that have been described in detail for metamorphic systems. Recent papers using isobaric T-$X^V_{CO_2}$ diagrams such as Figure 15 include the review of *Kerrick* [1974], *Greenwood's* [1975] account of buffered pore fluids, and

Thompson's [1975] evaluation of the capacity of mineral assemblages to buffer pore fluid and of the complexities introduced by Fe-Mg crystalline solutions.

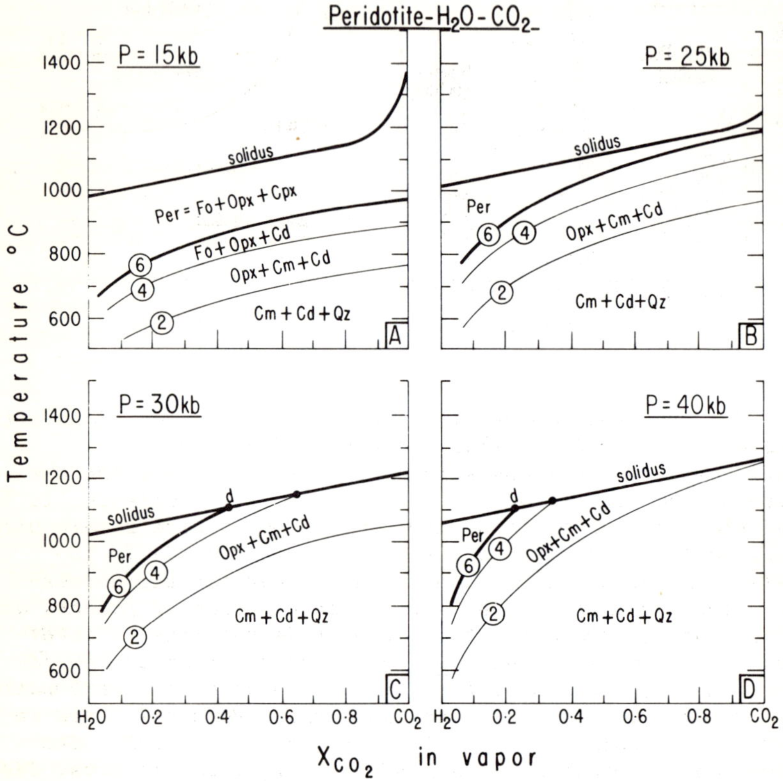

Fig. 15. Schematic isobaric sections through peridotite-H_2O-CO_2 with excess vapor showing the effect of pressure on the solidus surface, and the effect of H_2O on the temperatures of decarbonation reactions (2), (4), and (6), neglecting the formation of hydrous minerals (based on Figs. 11, 13B, and 14A). The decarbonation surfaces intersect the solidus surface at pressures greater than I_6, as shown in Fig. 14. Note migration of the point d in Figs. C and D, corresponding to points on the line (6) in Fig. 14B; with increased pressure, vapor d is enriched in H_2O/CO_2, restricting the stability range of uncarbonated peridotite.

Consider uncarbonated peridotite at any pressure in Figure 15, coexisting with a small proportion of vapor. If the vapor phase is isothermally enriched in CO_2 until it reaches the boundary for reaction (6), then a trace of calcic dolomite, Cd, is formed, and with further addition of CO_2 the system remains on the boundary as more carbonate is produced; the vapor phase composition remains constant. Assuming that there is insufficient CO_2 available to convert Cpx completely to carbonate, the progressive carbonation will cease, and the vapor phase composition remains as defined by reaction (6) for the specific temperature.

Now, if H_2O is permitted to migrate into the system, changing the composition of the vapor phase away from the phase boundary into the field for uncarbonated peridotite, some carbonate will dissociate, changing the vapor phase composition back to that of the phase boundary. The system is forced to remain on the boundary until all of the carbonate has gone.

If the carbonated assemblage is heated or cooled, carbonate will either disassociate or be produced in amounts sufficient to maintain the vapor phase composition corresponding to the boundary for reaction (6).

The carbonated peridotite thus buffers the vapor phase composition. With increasing temperature, as in progressive metamorphism, Figures 15A and B show that the composition of the buffered vapor phase shifts towards high values of CO_2. Figures 15C and D show that at higher pressures, intersection of the carbonation reaction with the solidus at point d restricts the vapor phase composition to low values of CO_2/H_2O.

At pressures below I_6, the composition of the vapor phase at the solidus is given by the total amount of CO_2 and H_2O in the rock. At pressures above I_6, the presence of carbonate in mantle peridotite buffers the vapor phase composition at the solidus to the points d in Figure 15, which correspond to the univariant solidus curve extending from I_6 in Figure 14. This is equivalent to the "polybaric trace of an isobaric invariant point" discussed by *Greenwood* [1975] for metamorphic rocks, but with one phase being a liquid. At any pressure, the vapor phase composition and the temperature of beginning of melting can be read from Figure 14.

It has been established that the composition of the first liquid produced from peridotite in the presence of CO_2-H_2O vapor is strongly dependent on the vapor composition. Near-solidus liquids become more siliceous with increasing H_2O/CO_2 [*Kushiro*, 1972; *Eggler*, 1975*a*; *Mysen* and *Boettcher* 1975*b*]. This is valid at pressures below I_6 for Figures 15A and B. In the presence of *carbonated* peridotite at pressures above I_6, however, I anticipate from limited experimental data in silicate-carbonate-CO_2-H_2O systems that addition of H_2O will be less effective in changing near-solidus liquids away from carbonatitic compositions [*Wyllie* and *Haas*, 1965, 1966; *Boettcher* and *Wyllie*, 1969; *Koster van Groos*, 1975].

C. Carbonate and Vapor Phase in Mantle Sections

Consider mantle peridotite in the presence of excess vapor of fixed composition. Figure 16A shows the solidus and the carbonation reaction (6) for a vapor containing 50% CO_2, reproduced from the contours of Figure 14A. Note the univariant line for other vapor phase compositions connected to I_6 for CO_2. The oceanic geotherm intersects both carbonation reaction and solidus for this vapor phase composition.

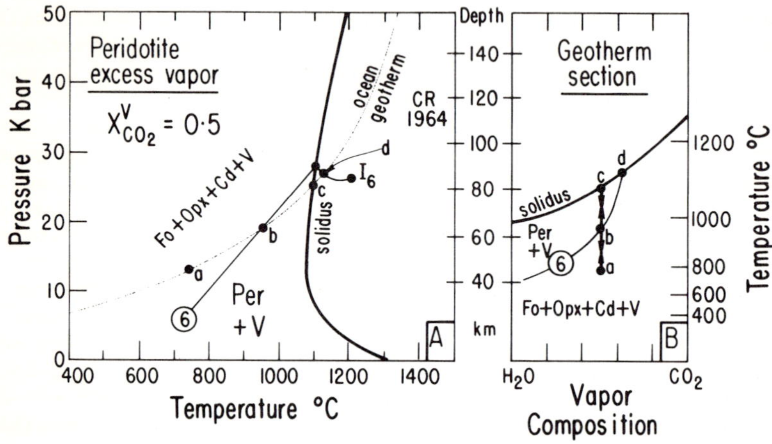

Fig. 16. Peridotite-H_2O-CO_2 with excess vapor, neglecting the formation of hydrous minerals. For peridotite moving up or down the geotherm between a and c (Fig. A) the vapor phase composition and phase fields intersected can be traced on Fig. B. See text for discussion. A. Isopleth with vapor of defined composition $X^V_{CO_2} = 0.5$. The solidus and reaction (6) for these conditions are taken from Fig. 14A--note the connection to I_6, also from Fig. 14A. The geotherm for oceanic regions is that calculated by Clark and Ringwood [1964], somewhat lower in temperature than that of Ringwood [1966]. B. Geotherm section through the phase diagram for P-T-$X^V_{CO_2}$. This is a ruled surface parallel to the axis $X^V_{CO_2}$ following the geotherm of Fig. A. The left-hand axis shows depth (pressure), and the right-hand axis shows the corresponding temperature along the geotherm. The geotherm section intersects the surfaces shown in Fig. 14A. Intersections with the reactions (4) and (2) of Fig. 14B are not plotted.

Figure 16B is a geotherm section through the phase diagram for peridotite-CO_2-H_2O. The vertical axes show the variation in depth (pressure) and temperature along the geotherm, and the two curved lines show the intersection of this P-T trajectory with the two divariant surfaces shown in Figure 14A. Intersections with other carbonation reactions (Figure 14B) have been omitted. The line abc in Figure 16B corresponds to movement of material, with fixed vapor phase composition, up or down the geotherm along abc in Figure 16A.

Consider peridotite with vapor just below the solidus temperature, near c in Figure 16. If this moves slowly upwards through the mantle, remaining in thermal equilibrium, it will react at point b to produce Fo+Opx+Cd. Complete carbonation in this way requires addition to the reacting vapor phase of more CO_2 than we might expect in the upper mantle.

Figure 17A shows the same geotherm section as Figure 16B, and points abc correspond to points on the geotherm in Figure 16A. Assume now that the peridotite with vapor just below the solidus near c in Figures 16 and 17A contains only 0.2% vapor of defined composition, in a closed system. Figure 17A shows that if this is cooled under the same conditions by slow upward movement, carbonation begins as before at point b, but this forces the vapor phase to change composition along the reaction surface ba. Carbonation is not completed, and the assemblage Fo+Opx+Cpx+Cd, plus vapor which becomes progressively enriched in H_2O, persists through a temperature interval.

Consider partially carbonated mantle peridotite containing 0.1% each of H_2O and CO_2, situated at point a on the geotherm section, Figure 17A; the vapor phase is enriched in H_2O because some CO_2 has been incorporated in carbonate. If this material moves to deeper levels with thermal equilibrium being maintained, dissociation of the carbonate will proceed, enriching the vapor in CO_2. The system remains on the carbonation surface as long as any carbonate remains. The last trace of carbonate reacts at point b, where the vapor reaches the composition 50% CO_2, and the system then follows path bc and melting begins at c without carbonate, in the presence of vapor containing 50% CO_2.

Consider the same sequence of events for a mantle containing 0.05% H_2O and 0.15% CO_2. This begins at a in Figure 17B as before, and progressive decarbonation brings the system to point b. At this point, however, with vapor phase containing 50% CO_2, there is still 0.1% CO_2 stored in crystalline carbonate, so the system continues along the reaction surface intersected in bd. The solidus temperature is reached at d before all of the carbonate has dissociated, and the silicate-carbonate assemblage thus begins to melt at d on the univariant line of Figure 14 (reaction 6); see also d in Figure 16A.

The petrology of mantle cross-sections corresponding to the systems depicted in Figures 17A and B are shown in Figures 17C and D. Figure 17C shows carbonated peridotite overlying peridotite without carbonate, overlying a partially melted, vapor-absent layer. This sequence can be traced directly on Figure 16A. The vapor phase composition varies from a to b, but remains constant from b to c. With a higher CO_2/H_2O ratio in the system, as in Figures 17B and 17D, the carbonated peridotite persists all the way down to the partially melted layer, with vapor phase composition varying continuously from a to d. For a peridotite-CO_2 mantle, Figure 17E shows that vapor-free carbonated peridotite overlies the partially melted layer (compare Figures 16A and 8).

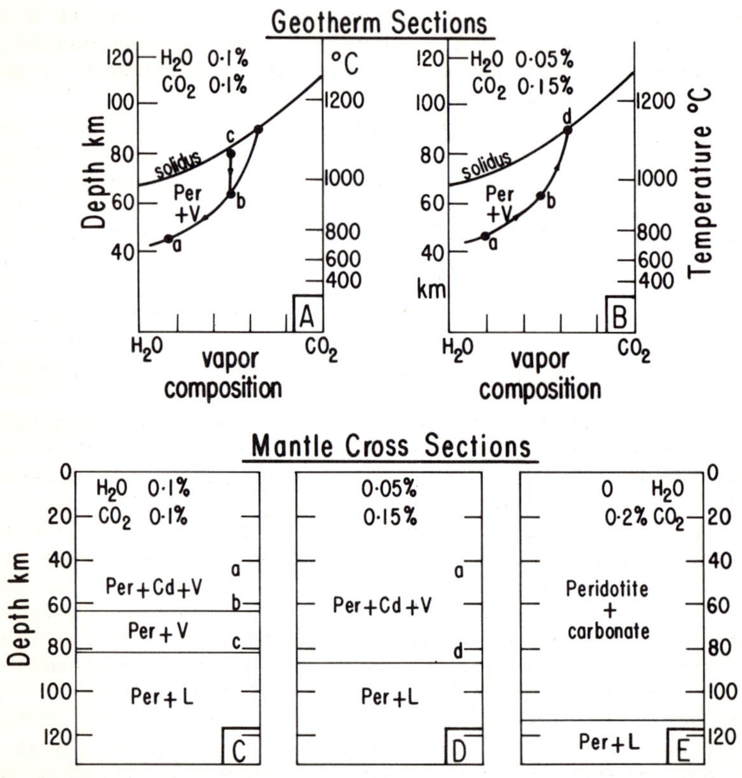

Fig. 17. Geotherm sections and mantle cross-sections for peridotite-H_2O-CO_2 with small quantities of volatile components, neglecting the formation of hydrous minerals. A and B. Geotherm sections showing phase fields intersected, corresponding to Fig. 16B. See text for discussion of paths followed by peridotite-volatile assemblages on these figures and Fig. 16A. C, D, and E. Mantle cross-section with mixed volatiles H_2O-CO_2, as in Figs. A and B, retain excess vapor until melting begins, whereas if no H_2O is present all CO_2 reacts to produce calcic dolomite in the peridotite, as in Figs. 4 and 17E.

The shield geotherm (Figure 6) would intersect the solidus at greater depths than the ocean geotherm, and thus would exclude the peridotite+vapor interval of Figure 17C (compare Figure 16A), except for systems with very high H_2O/CO_2 values. With cooler geotherms, the partially melted layer begins at greater depths, and the vapor phase composition just above the layer, d in Figures 16 and 17, is therefore displaced to higher H_2O/CO_2 ratios, as shown by the univariant solidus curve for reaction (6) in Figure 14. This in turn affects the near-solidus liquid composition.

D. Stability of Hydrous Minerals in Mantle Peridotite

The stabilization of amphibole and phlogopite in the presence of H_2O-CO_2 vapors complicates the simple picture which involves only carbonates in Figures 14 to 17. Estimates for the stability ranges of amphibole and phlogopite in the system peridotite-H_2O with excess H_2O are given in Figure 10. In the presence of CO_2-H_2O mixtures, the subsolidus univariant reaction for each mineral extends to a divariant surface corresponding to the carbonation surface shown in Figures 11A and 12. Each divariant surface intersects the peridotite-CO_2-H_2O solidus surface along a univariant line corresponding to that illustrated for the carbonation reaction (6) in Figure 14. Construction of the carbonate curve was described in some detail. Similar constructions for amphibole and phlogopite require more extrapolation and guess-work. My estimates are shown in Figure 18. There is not much variation possible in the curve for amphibole, but there is considerable uncertainty in the position of the phlogopite curve.

The maximum stability range for amphibole on the solidus surface is limited to a pressure range below 25 kb, and to high ratios of H_2O/CO_2 in the vapor phase. The amphibole stability field does not overlap the carbonate stability field on the solidus (although it does in subsolidus regions, Figure 19), showing that amphibole is not involved in the generation of carbonatitic liquids from mantle peridotite.

Phlogopite has a wide range of stability on the solidus surface, from low pressures up to pressures corresponding to depths of nearly 200 km, and for vapors ranging to high CO_2/H_2O ratios. There is a large area of overlap between the maximum stability fields for carbonate and phlogopite on the solidus, extending between about 25 to 55 kbar, and between vapor phase compositions from 10% to more than 90% CO_2.

Figure 19 reproduces the isobaric sections of Figures 15B and 15C, showing the intersections of the solidus surface and the carbonation surface. In addition, the intersections of the estimated divariant surfaces for the maximum subsolidus stability ranges of amphibole and carbonate are shown, extending down from their intersections with the solidus. The point f lies on the univariant phlogopite intersection in Figure 18, and the point e lies on the univariant amphibole intersection.

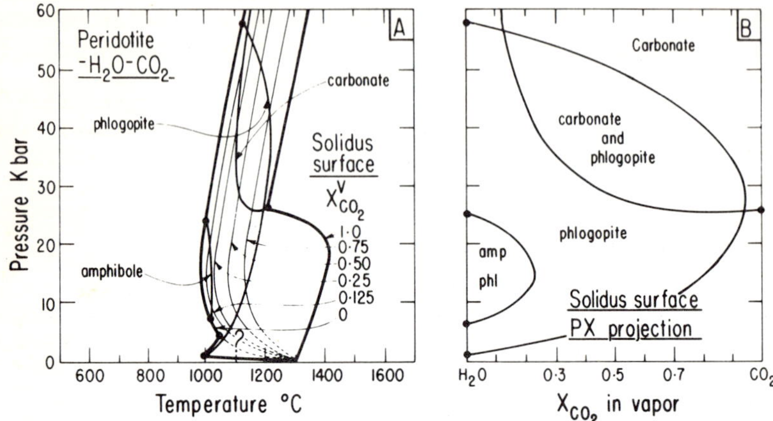

Fig. 18. Peridotite-H_2O-CO_2, in large part schematic, showing the maximum ranges of stability of carbonate and hydrous minerals on the solidus surface. See Fig. 10 for sources. A. Solidus surface from Fig. 13B, with complications at low pressures caused by stability of phlogopite (Fig. 10). There are three curves on the solidus, marking where the subsolidus divariant surfaces for the stability limits of carbonate, amphibole, and phlogopite, in the presence of H_2O-CO_2 vapors, intersect the solidus surface. Construction of the carbonate curve was illustrated in Fig. 14. Estimation of the amphibole and phlogopite curves were even more approximate. B. P-X projection of the solidus surface, showing estimated lines of intersection of the divariant dissociation reactions for carbonate, amphibole and phlogopite, transferred from Fig. 18B, and the maximum ranges of stability of these minerals on the surface. This is directly analogous to Fig. 14B, which shows in addition the temperature isotherms for the surface.

Figure 19 illustrates limits only. At points of intersection of the subsolidus curves, isobaric invariant points are generated, together with arrays of additional reactions, including one involving minerals without vapor. Figure 11A shows part of a similar diagram for the stability of a hydrous mineral (talc) and carbonate (magnesite) at 7 kbar in the system MgO-SiO_2-CO_2-H_2O; anthophyllite is also involved with the reactions in this system [*Evans* and *Trommsdorff*, 1974].

Figures 15 and 19 indicate that persistence of phlogopite to the solidus of mantle peridotite would tend to maintain high ratios of CO_2/H_2O in the vapor phase, and persistence of carbonate to the solidus would tend to maintain high ratios of

H_2O/CO_2. What actually happens at the solidus depends on the relative proportions of carbonate and phlogopite, and the subsolidus reactions occurring between them in the presence of Fo+Opx+Cpx. These remain to be determined.

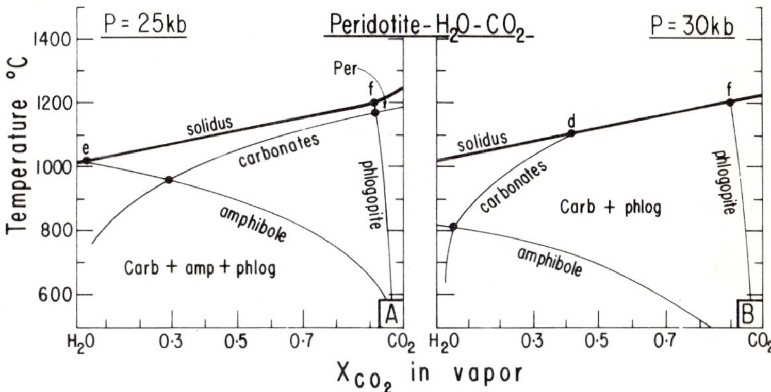

Fig. 19. Schematic isobaric sections through peridotite-H_2O-CO_2 with excess vapor, corresponding to Figs. 15B and 15C, with the maximum stability ranges of the hydrous minerals, amphibole and phlogopite, added to the curves for carbonate stability. At points of intersection of the subsolidus curves, isobaric invariant points are generated, together with arrays of additional reactions. See Fig. 11A.

VI. LIQUID IMMISCIBILITY IN SILICATE-CARBONATE SYSTEMS

Koster van Groos and *Wyllie* [1966, 1968, 1973] examined the effect of excess alkalis on the phase relationships in silicate-carbonate systems at low pressures, and *Koster van Groos* [1975] explored the effect of excess CO_2 on a series of rock compositions at pressures up to 10 kbar. Both studies demonstrated the existence of immiscible silicate and carbonate liquids.

In their experiments with olivine-melilitite-CO_2-H_2O, *Brey* and *Green* [1976*a*] reported carbonate droplets in silicate glass quenched from pressures greater than 20 kb, but concluded that they had too few experimental results to be certain of the inference that this represented liquid immiscibility.

It seems likely that with addition of alkalis and other components to the system CaO-MgO-SiO_2-CO_2, possibly in the form of phlogopite, a field for liquid immiscibility will be generated at mantle pressures. Whether or not this will extend across compositional ranges of petrogenetic relevance remains to be determined.

VII. THE SEISMIC LOW-VELOCITY ZONE

Three schematic mantle cross-sections showing the effect of traces of H_2O+CO_2 with different ratios of H_2O/CO_2 were shown in Figure 17. The diagrams illustrate the partially melted layer commonly equated with the seismic low-velocity zone. Figures 18 and 19 show that only in the presence of vapor with high H_2O/CO_2 would amphibole be involved in melting. If the mantle composition is suitable for the stabilization of phlogopite, Figures 18 and 19 show that this would be present down to the melting zone, except in the presence of vapors with very high CO_2/H_2O.

For peridotite with CO_2 and no H_2O, the CO_2 would be stored as carbonate (calcic dolomite), with the possible exception of oceanic environments with rather high geotherms (Figure 6), where free CO_2 could exist through a limited depth interval. Diapiric uprise of the peridotite under adiabatic conditions would cause dissociation of the carbonate, and release of free CO_2. For peridotite with CO_2 and H_2O, carbonated peridotite coexists with mixed CO_2-H_2O vapors whose composition is buffered by the carbonation reaction. If the peridotite composition is suitable for the stabilization of phlogopite, the vapor phase composition is buffered by reactions involving phlogopite and carbonate.

For a mantle peridotite of fixed composition in thermal equilibrium, Figures 16 and 17 show that the amount of carbonate decreases with depth, as the ratio of CO_2/H_2O increases in the vapor phase (neglecting possible reaction with phlogopite, if present). If a portion of mantle rises from the geotherm under adiabatic conditions, however, the path tracks across the contours on the divariant carbonation surface in a different way (compare Figures 14A and 16A). The amount of carbonate decreases and the ratio of CO_2/H_2O in the vapor phase then increases as the peridotite rises to shallower levels.

For a mantle peridotite with CO_2, and also for CO_2 with H_2O (limits unknown), the liquid at the top of the low-velocity zone is carbonatitic, with 10-15% dissolved silicates, and Ca/Mg ratio greater than 1. The liquid is probably enriched in alkali carbonates, as well, and a host of trace elements. Under static conditions, the interstitial liquid would change in composition from carbonatitic in the upper levels to kimberlitic at deeper levels; the liquid would contain high proportions of CO_2 and H_2O in solution, but the liquid would remain undersaturated with the volatile components. For peridotite with high H_2O/CO_2, the interstitial liquid may be a silicate melt rather than carbonatitic.

Carbonatitic liquids are very fluid, and if they exist they would probably migrate upwards. Local concentrations, and intrusions, or local cooling, could produce layers that are enriched in crystalline carbonates.

This picture is consistent with the proposal of *Boyd* and *Nixon* [1973] that kimberlite magmas may be mixtures of silicate liquids (with kimberlitic affinities) from within the asthenosphere and carbonatitic liquids from near the top of the asthenosphere. It is also consistent with the suggestion of *Franz* and *Wyllie* [1967] that carbonatites could be the fluids involved in the fluidization emplacement of some kimberlites. Eruption of CO_2-undersaturated magmas from the low-velocity layer may become explosive at depths near 80 km where CO_2 must be evolved as the liquid crosses the peridotite-CO_2 solidus boundary dropping down to I_6 (Figures 7 and 13).

Acknowledgments. This research was supported by the Earth Sciences Section, National Science Foundation NSF Grant EAR 74-00157 RES. I have also benefited from support of the Materials Research Laboratory by the National Science Foundation.

REFERENCES

Boettcher, A. L., Experimental igneous petrology, *Rev. Geophys. Space Phys., 13,* 75-79, 1975.

Boettcher, A. L., and P. J. Wyllie, The system $CaO-SiO_2-CO_2-H_2O$: III. A second critical end-point on the melting curve at 32.5 kbars and 515°C, *Geochim. Cosmochim. Acta, 33,* 611-632, 1969.

Bowen, N. L., Progressive metamorphism of siliceous limestone and dolomite, *Jour. Geol., 48,* 225-274, 1940.

Boyd, F. R., and P. H. Nixon, Origin of the ilmenite-silicate nodules in kimberlites from Lesotho and South Africa, in *Lesotho Kimberlites,* edited by P. H. Nixon, Maseru, Lesotho National Development Corporation, 254-268, 1973.

Brey, G., and D. H. Green, The role of CO_2 in the genesis of olivine melilitite, *Contrib. Mineral. Petrol., 49,* 93-103, 1975.

Brey, G. P., and D. H. Green, Solubility of CO_2 in olivine melilitite at high pressures and role of CO_2 in the earth's upper mantle, *Contr. Mineral. Petrol., 55,* 217-230, 1976*a*.

Brey, G. P., and D. H. Green, Reply to Eggler and Mysen, *Contr. Mineral. Petrol., 55,* 237-239, 1976*b*.

Clark, S. P., and A. E. Ringwood, Density distribution and constitution of the mantle, *Review Geophysics, 2,* 35-88, 1964.

Dawson, J. B., and J. B. Hawthorne, Magmatic sedimentation and carbonatitic differentiation in kimberlite sills at Benfontein, South Africa, *Quart. Jour. Geol. Soc. Lond., 129,* 61-85, 1973.

Dawson, J. B., and D. G. Powell, Mica in the upper mantle, *Contr. Mineral. Petrol., 22,* 233-237, 1969.

Dawson, J. B., D. G. Powell, and A. M. Reid, Ultrabasic xenoliths and lava from the Lashaine volcano, Northern Tanzania, *Jour. Petrol., 11,* 519-548, 1970.

Eggler, D. H., Effect of CO_2 on the melting of peridotite, *Carnegie Inst. Washington Year Book, 73,* 215-224, 1974.

Eggler, D. H., Peridotite-carbonate relations in the system CaO-MgO-SiO_2-CO_2, *Carnegie Inst. Washington Year Book, 74,* 468-474, 1975*a*.

Eggler, D. H., Carbonatite generation by a reaction relation in the system CaO-MgO-SiO_2-CO_2 at 30 kbar pressure (abs.), *EOS (Am. Geophys. Union Trans.), 56,* 470, 1975*b*.

Eggler, D. H., Does CO_2 cause partial melting in the low-velocity layer of the mantle? *Geology, 4,* 69-72, 1976.

Eggler, D. H., and B. O. Mysen, The role of CO_2 in the genesis of olivine melilitite: discussion, *Contr. Mineral. Petrol., 55,* 231-236, 1976.

Eggler, D. H., J. R. Holloway, and B. O. Mysen, High CO_2 solubilities in mantle magmas: comment, *Geology, 4,* 198-199, 1976.

Evans, B. W., and V. Trommsdorff, Stability of enstatite + talc, and CO_2-metasomatism of metaperidotite, Val d'Efra, Lepontine Alps, *Amer. Jour. Sci., 274,* 274-296, 1974.

Franz, G. W., and P. J. Wyllie, Experimental studies in the system CaO-MgO-SiO_2-CO_2-H_2O, in *Ultramafic and Related Rocks,* edited by P. J. Wyllie, Wiley and Sons, Inc., New York, 323-326, 1967.

Green, H. W., and S. V. Radcliffe, Fluid precipitates in rocks from the earth's mantle, *Geol. Soc. Am. Bull., 86,* 846-852, 1975.

Greenwood, H. J., Buffering of pore fluids by metamorphic reactions, *Amer. Jour. Sci., 275,* 573-593, 1975.

Hill, R. E. T., and A. L. Boettcher, Water in the earth's mantle: melting curves of basalt-water-carbon dioxide, *Science, 167,* 980-982, 1970.

Huang, W. L., and P. J. Wyllie, Eutectic between wollastonite II and calcite contrasted with thermal barrier in MgO-SiO_2-CO_2 at 30 kilobars, with applications to kimberlite-carbonatite petrogenesis, *Earth and Planet. Sci. Letters, 24,* 305-310, 1974.

Huang, W. L., and P. J. Wyllie, Melting reactions in the system $NaAlSi_3O_8$-$KAlSi_3O_8$-SiO_2 to 35 kilobars, dry and with excess water, *Jour. Geol., 83,* 737-748, 1975.

Huang, W. L., and P. J. Wyllie, Melting relationships in the systems CaO-CO_2 and MgO-CO_2 to 33 kilobars, *Geochim. Cosmochim. Acta, 40,* 129-132, 1976.

Irving, A. J., and P. J. Wyllie, Melting relationships in CaO-CO_2 and MgO-CO_2 to 36 kilobars, with comments on CO_2 in the mantle, *Earth and Planet. Sci. Letters, 20,* 220-225, 1973.

Irving, A. J., and P. J. Wyllie, Subsolidus and melting relationships for calcite, magnesite, and the join $CaCO_3$-$MgCO_3$ to 36 kilobars, *Geochim. Cosmochim. Acta, 39,* 35-53, 1975.

Kerrick, D. M., Review of metamorphic mixed volatile (H_2O-CO_2) equilibria, *Amer. Miner., 59,* 729-762, 1974.

Koster van Groos, A. F., The effect of high CO_2 pressures on alkalic rock and its bearing on the formation of alkalic ultrabasic rocks and the associated carbonatites, *Amer. Jour. Sci., 275,* 163-185, 1975.

Koster van Groos, A. F., and P. J. Wyllie, Liquid immiscibility in the system Na_2O-Al_2O_3-SiO_2-CO_2 at pressures up to 1 kilobar, *Amer. Jour. Sci., 264,* 234-255, 1966.

Koster van Groos, A. F., and P. J. Wyllie, Liquid immiscibility in the join $NaAlSi_3O_8$-Na_2CO_3-H_2O and its bearing on the genesis of carbonatites, *Amer. Jour. Sci., 266,* 932-967, 1968.

Koster van Groos, A. F., and P. J. Wyllie, Liquid immiscibility in the join $NaAlSi_3O_8$-$CaAlSi_2O_8$-Na_2CO_3-H_2O, *Amer. Jour. Sci., 273,* 465-487, 1973.

Kushiro, I., Effect of water on the composition of magmas formed at high pressures, *Jour. Petrol., 13,* 311-334, 1972.

Kushiro, I., Y. Syono, and S. Akimoto, Melting of a peridotite nodule at high pressures and high water pressures, *Jour. Geophys. Res., 73,* 6023-6029, 1968.

Kushiro, I., H. Satake, and S. Akimoto, Carbonate-silicate reactions at high pressures and possible presence of dolomite and magnesite in the upper mantle, *Earth and Planet. Sci. Letters, 28,* 116-120, 1975.

Lambert, I. B., and P. J. Wyllie, Stability of hornblende and a model for the low velocity zone, *Nature, 219,* 1240-1241, 1968.

Lambert, I. B., and P. J. Wyllie, Low-velocity zone of the Earth's mantle: incipient melting caused by water, *Science, 169,* 764-766, 1970.

Mason, B., Eclogitic xenoliths from volcanic breccia at Kakanui, New Zealand, *Contr. Mineral. Petrol., 19,* 316-327, 1968.

McGetchin, T. R., and J. R. Besancon, Carbonate inclusions in mantle-derived pyropes, *Earth and Planet. Sci. Letters, 18,* 408-410, 1973.

Melton, C. E., C. A. Salotti, and A. A. Giardini, The observation of nitrogen, water, carbon dioxide, methane and argon as impurities in natural diamonds, *Amer. Mineral., 57,* 1518-1523, 1972.

Millhollen, G. L., A. J. Irving, and P. J. Wyllie, Melting interval of peridotite with 5.7 per cent water to 30 kilobars, *Jour. Geol., 82,* 575-587, 1974.

Modreski, P. J., and A. L. Boettcher, Phase relationships of phlogopite in the system K_2O-MgO-CaO-Al_2O_3-SiO_2-H_2O to 35 kilobars: a better model for micas in the interior of the earth, *Amer. Jour. Sci., 273,* 385-414, 1973.

Moore, J. G., Petrology of deep-sea basalt near Hawaii, *Amer. Jour. Sci., 263,* 40-52, 1965.

Moore, J. G., Water content of basalt erupted on the ocean floor, *Contr. Mineral. Petrol., 28,* 272-279, 1970.

Mysen, B. O., and A. L. Boettcher, Melting of a hydrous mantle: I. Phase relations of natural peridotite at high pressures and temperatures with controlled activities of water, carbon dioxide, and hydrogen, *Jour. Petrology, 16,* 520-548, 1975*a*.

Mysen, B. O., and A. L. Boettcher, Melting of a hydrous mantle: II. Geochemistry of crystals and liquids formed by anatexis of mantle peridotite at high pressures and high temperatures as a function of controlled activities of water, hydrogen, and carbon dioxide, *Jour. Petrology, 16,* 549-593, 1975*b*.

Mysen, B. O., D. H. Eggler, M. G. Seitz, and J. R. Holloway, Carbon dioxide in silicate melts and crystals. I. Solubility measurements, *Amer. Jour. Sci., 276,* 455-479, 1976.

Newton, R. C., and W. E. Sharp, Stability of forsterite + CO_2 and its bearing on the role of CO_2 in the mantle, *Earth and Planet. Sci. Letters, 26,* 239-244, 1975.

Roedder, E., Liquid CO_2 inclusions in olivine-bearing nodules and phenocrysts from basalts, *Amer. Mineral., 50,* 1746-1782, 1965.

Ringwood, A. E., Mineralogy of the mantle, in *Advances in Earth Sciences,* edited by P. M. Hurley, M.I.T. Press, Cambridge, Massachusetts, 357-399, 1966.

Thompson, A. B., Mineral reactions in a calc-mica schist from Gassetts, Vermont, U.S.A., *Contr. Mineral. Petrol., 53,* 105-127, 1975.

Turner, F. J., *Metamorphic Petrology.* McGraw-Hill, New York, 403 pp.

Wyllie, P. J., Experimental petrology and global tectonics: a preview, in *Experimental Petrology and Global Tectonics,* edited by P. J. Wyllie, *Tectonophysics, 17,* 189-209, 1973.

Wyllie, P. J., and J. L. Haas, The system $CaO-SiO_2-CO_2-H_2O$: I. Melting relationships with excess vapor at 1 kilobar pressure, *Geochim. Cosmochim. Acta, 29,* 871-892, 1965.

Wyllie, P. J., and J. L. Haas, The system $CaO-SiO_2-CO_2-H_2O$: II. The petrogenetic model, *Geochim. Cosmochim. Acta, 30,* 525-544, 1966.

Wyllie, P. J., and W. L. Huang, Peridotite, kimberlite, and carbonatite explained in the system $CaO-MgO-SiO_2-CO_2$, *Geology, 3,* 621-624, 1975.

Wyllie, P. J., and W. L. Huang, Carbonation and melting reactions in the system $CaO-MgO-SiO_2-CO_2$ at mantle pressures with geophysical and petrological applications, *Contr. Mineral. Petrol., 54,* 79-107, 1976*a*.

Wyllie, P. J., and W. L. Huang, High CO_2 solubilities in mantle magmas, *Geology, 4,* 21-24, 1976*b*.

Wyllie, P. J., and W. L. Huang, High CO_2 solubilities in mantle magmas: reply, *Geology, 4,* 199-200, 1976*c*.

Wyllie, P. J., and O. F. Tuttle, Effect of carbon dioxide on the melting of granite and feldspars, *Amer. Jour. Sci., 257,* 648-655, 1959.

THE ROLE OF AMPHIBOLES AND WATER IN CIRCUM-PACIFIC VOLCANISM

A. L. BOETTCHER
Institute of Geophysics and Planetary Physics and The Department of Earth and Space Sciences, University of California at Los Angeles, Los Angeles, California 90024

"All the rivers flow to the sea; yet the
sea is not full; unto the place from whence
the rivers come, thither they return again."
Ecclesiastes 1:8.

Abstract

It is generally acknowledged that water plays a major role in the origin and evolution of calc-alkaline magmas at plate boundaries, such as those of the circum-Pacific zone. Recent petrographic, chemical, and high-pressure experimental evidence strongly suggests that amphiboles are a major carrier of H_2O in subducted oceanic slabs and that crystal-liquid equilibria involving amphiboles are fundamental to mechanisms involved in the genesis of these calc-alkaline magmas.

High-pressure experiments on basaltic and andesitic compositions reveal that amphiboles are important liquidus or near-liquidus phases over a wide range of pressures, bulk compositions, f_{O_2}, and f_{H_2O}. Fractionation of these silica-poor amphiboles (40-46% SiO_2) or partial melting of amphibolite in the basalt-crust component of a subducting slab can produce the voluminous basalt-andesite clan of the orogens. Chemical and petrographic evidence reveals that polymineralic crystal clots that commonly occur in orogenic andesites are products from pre-existing amphiboles with compositions similar to those produced experimentally. The distribution patterns of the rare-earth elements are generally consonant with the amphibole-fractionation scheme, but they are not supportive of the involvement of eclogite in magma genesis in subduction zones.

Amphiboles are not stable at depths greater than about 80 km; at greater depths, serpentine, humites, and other hydrous silicates may be important sources of H_2O for melting, but phlogopite probably does not occur in this geologic setting. Nevertheless, the basaltic-andesitic magmas of the circum-Pacific zone probably originate at depths less than 80 km, within or near the limits of the stability of amphibole, at the top of the subducting slab. Thus the depths of magma generation of these plate boundaries are less than those of the seismic Benioff zones.

I. INTRODUCTION

Numerous research papers dealing with the genesis and evolution of these volcanic chains have been published in the last half decade; reviews of the veracity of proposed modes of origin in light of this plethora of observational, chemical, petrographic, and experimental data have also appeared [e.g., *Wyllie*, 1973; *Ringwood*, 1974; *Boettcher*, 1973].

Most petrologists, it appears, are disposed to believe that volcanism at the distal regions of lithospheric plates, including the circum-Pacific zone, is chemically and physically distinct because of the prominent role played by volatiles, principally H_2O. Evidence of H_2O in these magmas has been published elsewhere [e.g. *Yoder*, 1969; *Anderson*, 1973, 1974].

This paper reasserts my prejudice, and that of others [e.g., *Cawthorn and O'Hara*, 1976], that amphiboles are the major carrier of H_2O in the crustal component of subducted slabs in the regions under question and that melting, reaction, or fractionation of amphiboles is a major mechanism in the development of these circum-Pacific lavas. This paper is not a review; rather, it presents new data, mostly supportive of these concepts, and is biased to that extent.

Three mechanisms have been proposed whereby amphiboles could play a significant role in the genesis of these orogenic zone volcanics: (1) partial melting of subducted oceanic crust in the amphibolite facies, resulting in an amphibole-rich residuum and a SiO_2-saturated basaltic or andesitic anatectic liquid [e.g., *Green and Ringwood*, 1968; *Boettcher*, 1973], (2) crystallization of SiO_2-poor amphiboles from a hydrous basaltic liquid, leaving a quartz-normative (e.g., andesitic) melt [e.g., *Bowen*, 1956; *Helz*, 1976; *Boettcher*, 1973], and (3) subsolidus, high-pressure dehydration of amphiboles in a subducted slab, releasing H_2O for anatexis of overlying mantle peridotite to produce SiO_2-saturated liquids [e.g., *Ringwood*, 1974]. Let us examine recent evidence to evaluate these concepts.

II. AMPHIBOLE REACTION AND FRACTIONATION

A. Experiments at High Pressures

For references to all previous experimental work in hydrous peridotitic, basaltic, and andesitic systems, the reader is referred to the review papers cited above. *Allen et al.* [1975] have recently published a detailed study of hydrous basaltic, andesitic, and nephelinitic systems as a function of vapor pressure, temperature, and f_{O_2}.

As shown in Figure 1, amphiboles are liquidus[1] or near-liquidus phases to pressures as high as 25 kbar and to temperatures in excess of 1000°C for basaltic compositions in the presence of an H_2O-rich vapor where the mole fraction of H_2O in the vapor ($X^V_{H_2O}$) ∿ 1.0 .

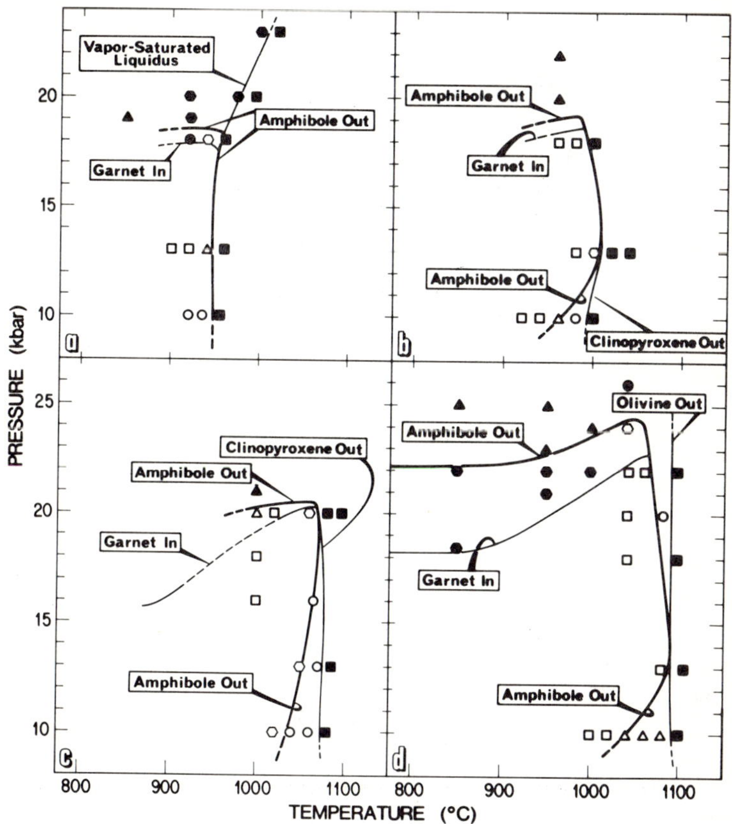

Fig. 1. Pressure-temperature projections of melting relations in the presence of H_2O-rich vapor ($X^V_{H_2O}$ ∿ 1.0) with T-f_{O_2} ∿ N-NO for (a) Mt. Hood andesite, (b) Picture Gorge quartz tholeiite, (c) 1921 Kilauea olivine tholeiite, and (d) Hualalai alkali basalt (after Allen et al., 1975).

---

[1] I refer to the vapor-saturated liquidus as the "liquidus."

The effect of various conditions of f_{O_2} is illustrated in Figure 2. The salient feature of these results is that amphiboles coexist with silicate liquids over a wide range of temperatures and f_{O_2} to depths of about 70 km (for the olivine tholeiite, which most nearly approximates ocean-floor basalt).

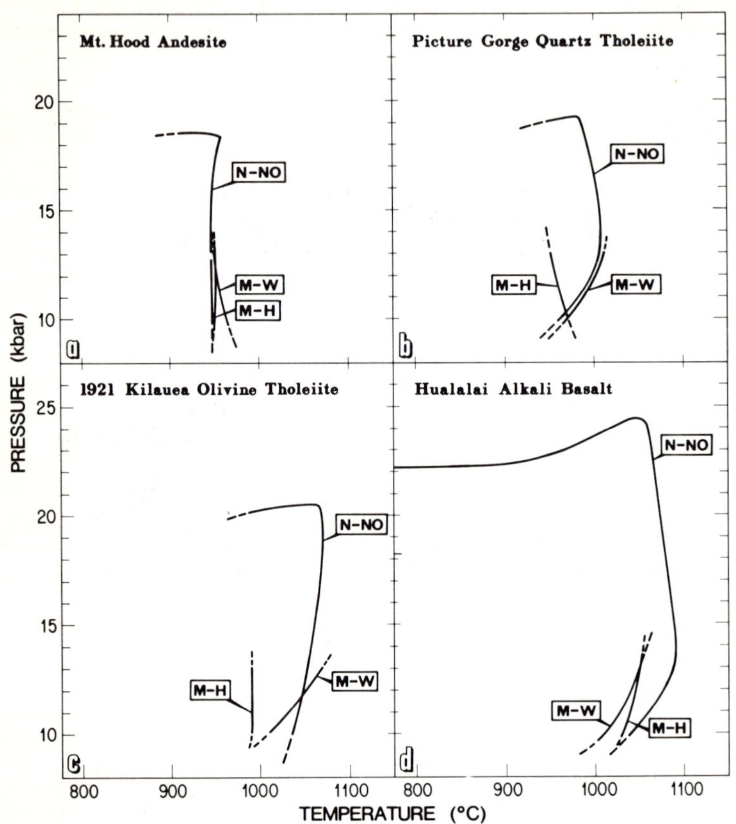

Fig. 2. Pressure-temperature projections of the amphibole-out curves for three basalts and an andesite in the presence of an H_2O-rich vapor ($X^V_{H_2O}$ ∿ 1.0) for T-f_{O_2} conditions for the M-W, N-NO, and M-H buffers [after Allen et al., 1975].

Allen and Boettcher [in preparation; 1975] have expanded these results by adding CO_2 to these systems, primarily to assess the effects of reduced f_{H_2O} on our previous results. The results are shown in Figures 3 and 4 for the olivine tholeiite and the andesite, respectively. Note that the curves in these figures differ from those in Figure 1 because $T\text{-}f_{O_2}$ conditions differ. In the systems with $H_2O\text{-}CO_2$ vapors, it was necessary to buffer f_{H_2} with magnetite-hematite (M-H) to prevent precipitation of carbon from the vapor. This produces f_{O_2} conditions during the run that equal those of M-H for $X^V_{H_2O} \sim 1.0$ but are less for $X_{H_2O} < 1.0$ [*Boettcher et al.*, 1973]. These values of f_{O_2} probably are high compared to those in magma chambers, but they produce the desirable effect that loss of Fe to the noble-metal sample capsules is almost eliminated [see *Allen et al.*, 1975, p. 1070]. This latter result is particularly significant when determining the upper pressure limits on the stability of amphibole. That is, loss of Fe to the capsule, as occurs at the lower values of f_{O_2}, results in a more pyrope-rich garnet and pushes the pressure at which the garnet becomes stable relative to amphibole to much higher values. It was for this reason that we did not determine the stability of amphiboles to pressures above 13 kbar for $X^V_{H_2O} \sim 1.0$ at the M-W buffer (see Figure 2).

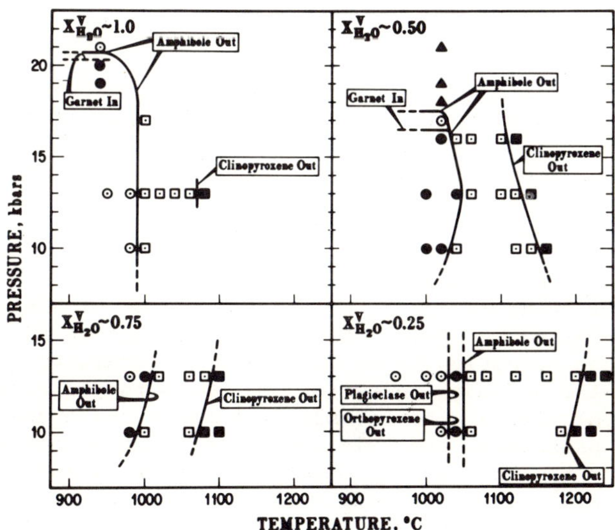

Fig 3. Pressure-temperature projections of an olivine tholeiite [see Allen et al., 1975] under conditions of $X^V_{H_2O}$ ranging from ∿ 1.0 to 0.25 and f_{H_2} buffered at M-H [Allen and Boettcher, in preparation]

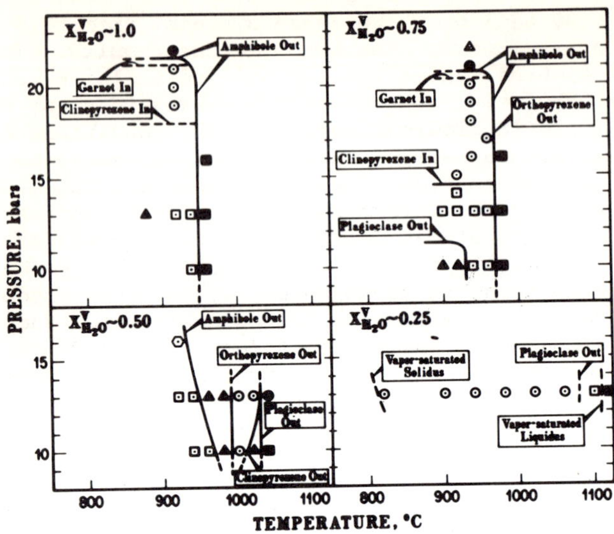

Fig. 4. Pressure-temperature projections of an andesite [see Allen et al., 1975] under conditions of $X^V_{H_2O} \sim 1.0$ to 0.25 and f_{H_2} buffered at M-H [Allen and Boettcher, in preparation].

It was interesting to discover, in the H_2O-CO_2 experiments on the olivine tholeiite (Figure 3), that the overall phase relationships remain similar for values of $X^V_{H_2O}$ as low as $\sim$ 0.5 or less, with amphibole stable to temperatures as high as 1050°C. These data add credence to the suggested mechanisms of magma genesis involving amphiboles.

B. Petrographic Evidence

Additional evidence upholding the role of amphiboles has recently been aired by *Stewart* [1975] and *Stewart et al.* [1976]. Previously, one objection to my model was the scarcity of amphiboles in basalts and many andesites, although they are common in dacites. Amphiboles are not stable in basalts and andesites at near-surface pressures, and this led me to believe that we should be looking for the breakdown products of amphibole rather than amphibole itself. Conversations with Professor C. P. Thornton of The Pennsylvania State University revealed that clots of crystals of plagioclase + orthopyroxene + clinopyroxene + magnetite are common in calc-alkaline andesites the world over (Figures 5 and 6). This is the assemblage that should occur when amphiboles break down attendant upon rapid rise and relief of pressure at magmatic temperatures (e.g., see the experimental results of

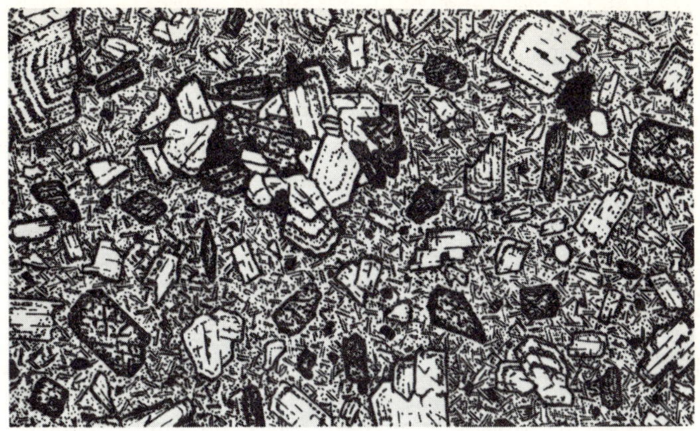

Fig. 5. Drawing of photomicrograph of Mt. Jefferson andesite showing clot consisting of magnetite, pyroxenes, and plagioclase, from cover of McBirney [1969a]. Field of view approximately 10 x 7 mm.

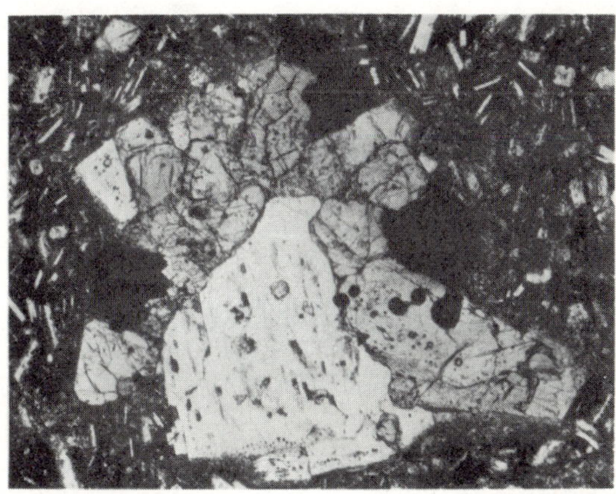

Fig. 6. Photomicrograph of a crystal clot in Crater Lake andesite, plane light, from Stewart [1975]. Field of view approximately 4 x 3 mm.

Eggler [1972] and *Yoder and Tilley* [1962]. Rarely, rapidly quenched examples of incomplete breakdown are revealed by cores of amphiboles with rims of plag + opx + cpx + mt [*Nicholls*, 1971, Plate 1-D; *Jakeš and White*, 1972; *Stewart*, 1975] (Figure 7).

These crystal clots have received little attention from petrologists; they apparently were perceived as glomeroporphyritic accumulations of phenocrysts. However, the petrographic work of *Stewart* [1975] on the Crater Lake andesites reveals that there are significant differences between the constituent minerals of the clots and the phenocrystal phases. For examples, ilmenite and hornblende are common as phenocrysts but are absent in the clots. Manganiferous apatite is a common accessory in the clots but is very scarce as a phenocryst in the andesite. Plagioclase in the clots is "clean," whereas that in phenocrysts and xenoliths contains abundant glass; the clot plagioclase is also generally unzoned except for a reaction zone where in contact with the matrix andesite. Finally, the modal mineralogical composition of the clots differs significantly from the relative abundance of the phenocrystal phases [*Stewart*, 1975, p. 199].

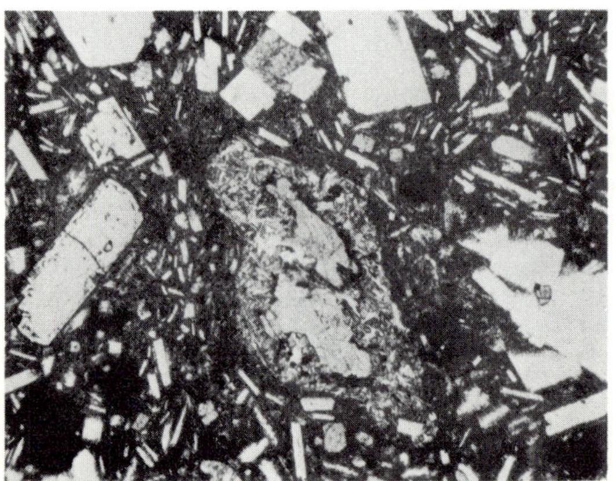

Fig. 7. Photomicrograph of a crystal clot with an amphibole core (see Table 1) in Crater Lake andesite, plane light, from Stewart [1975]. Field of view approximately 4 x 3 mm.

On the basis of chemical analyses of the constituent minerals and modal analyses of the clots, Stewart calculated the average composition of the clots, which is compared in Table 1 with that of the unreacted amphibole in Figure 7 and with amphiboles synthesized at high pressures in basaltic compositions

TABLE 1. Chemical Composition of Clots Compared to Amphiboles

Oxide	A	B	C	D	E
SiO_2	46.6	43.8	45.6	42.0	45.0
TiO_2	1.8	1.2	1.3	2.2	2.0
Al_2O_3	10.3	14.7	11.3	12.7	11.4
FcO[b]	18.8	14.4	14.0	16.0	9.8
MgO	10.8	11.7	14.0	11.8	16.9
CaO	10.1	9.3	10.3	9.0	11.7
Na_2O	1.2	2.5	2.2	2.5	2.3
K_2O	0.4	0.6	0.3	0.5	0.2
Total[c]	99.5	99.7	100.5	98.2	100.6

a. A. Average composition of clots [*Stewart*, 1975].

B. Amphibole synthesized near vapor-saturated liquidus (18 kbar / 920°C) of Mt. Hood andesite + H_2O (N-NO buffer) [*Allen et al.*, 1975].

C. Similar to (B) except 13 kbars/900°C.

D. Similar to (B) except Picture Gorge quartz tholeiite at 10 kbar / 920°C.

E. Microprobe analysis of amphibole in core of clot in Figure 7. [*Boettcher*, unpublished data].

b. Total iron as FeO.

c. Totals include 1.5% H_2O (estimated).

The similarity between the bulk composition of the clots, which constitute as much as 5 volume percent and locally as much as 10 percent of the Crater Lake andesites [*Stewart*, 1975], and that of the amphiboles strongly supports the proposal that amphibole fractionation may have played an important role in the derivation of the andesitic liquids from a basaltic or basaltic-andesitic parent magma. For example, *Stewart et al.* [1976] have demonstrated that removal of about 45 percent of amphibole equivalent to the Crater Lake clots from a quartz tholeiite magma (51% SiO_2) produces a derivative liquid equivalent to the Mt. Hood andesite (59% SiO_2). Because additional phases will also separate with the amphibole, this must be considered as a simple working model--a zeroith approximation. Nevertheless, a careful search for clots in basalts and andesites elsewhere must be undertaken before it can be concluded that amphiboles were not present.

C. Chemical Evidence

Perhaps the strongest evidence pertinent to the genesis of andesites will be forthcoming from the study of the partitioning of trace elements and isotopes. Although an abundance of data is on hand, our ability to interpret it continues to improve with advances in analytical technology and with the analyses of more rock suites.

Perhaps the most important contributions from isotope geochemistry center around the questions of the extent to which pelagic sediments are involved in the genesis of andesites in subduction zones. A recent reevaluation of this question by *Church* [1976] on the basis of lead isotope data suggests that a small but significant crustal component is involved; this probably results from assimilation as the magmas rise and stope through the crust, which is in concert with the conclusions of *Karig and Sharman* [1975] and others that most of the pelagic sediments are plastered onto the continents in imbricate sheets rather than being subducted with the basaltic crust.

Although uncertainties exist [see *Boettcher*, 1973, p. 225], $^{87}Sr/^{86}Sr$ values of orogenic andesites and basalts are not supportive of the concept of the involvement of sediments or continental sialic rocks having high contents of radiogenic strontium [e.g. *Gledhill and Baker*, 1973].

Similarly, stable-isotope geochemistry lends no support to large-scale involvement of sediments in the production of andesites. For example, *Taylor* [1968] reports values of $^{18}O/^{16}O$ of most andesites that are indistinguishable from those of basalt and which are too low to permit significant assimilation of sediments or old sedimentary rocks. *Matsuhisa et al.* [1973] presented oxygen isotope data for the basalt-andesite clan in Japan that are consistent with the concepts of primary, mantle-derived magmas or differentiation of basaltic magmas and not those involving assimilation of ^{18}O-rich sediments.

A number of recent publications have suggested that anatexis of wet or dry subducted oceanic crust in the eclogite facies is the mechanism by which most orogenic andesites originate [e.g., *Green and Ringwood,* 1968; *Green,* 1972; *Ringwood,* 1974; *Condie and Hayslip,* 1975; *Marsh,* 1976]. However, such a mechanism is dissonant with the partitioning of rare-earth elements (REE) between liquid and eclogite [*Gill,* 1974] and with the REE content of orogenic andesites [*Gill,* 1974; *Nicholls and Whitford,* in press; *Thorpe et al.* 1976]. *Lopez-Escobar et al.,*[in press] find for the andesites of central Chile that many of the rocks have heavy rare earth element (HREE) depletion greater than that of the calculated pattern resulting from anatexis of amphibolite or separation (crystallization) of amphibole. Nevertheless, the depletion is not as large as that predicted for eclogite or garnet fractionation. These results point out that models, such as mine, involving separation of only one phase are oversimplified. I will consider this later in this paper, but for the present I prefer to seek alternate processes and to discount models involving only the fractionation of eclogite (or garnet).

The contents of other trace elements in orogenic basalts and andesites have been cited as evidence in favor of a variety of models, but nearly all of the data also are permissive of amphibole-liquid equilibria [e.g., *Thorpe, Potts, and Francis,* 1976; *Nicholls and Whitford,* in press; *Condie and Hayslip,* 1975; *Anderson and Gottfried,* 1971].

More chemical data from rock suites together with laboratory investigations of the partitioning of trace elements between liquids and crystals (e.g. amphiboles) will shed light on the formation of andesites. Nevertheless, most of the extant data support or permit the conclusion that anatexis of amphiboles or fractionation of amphiboles from basaltic liquids are likely mechanisms.

III. Conclusions

Consider now a lithospheric slab of hydrated sea-floor basalt (i.e., amphibolite) descending beneath continental lithosphere and asthenosphere. To determine the behavior of the basalts, we must have knowledge of the distribution of temperature. This problem is far from being solved, but there have been several attempts to calculate temperatures in and surrounding subducting lithospheric plates [*Oxburgh and Turcotte,* 1970; *Toksöz et al.,* 1971; *Griggs,* 1972]. *Wyllie* [1973] plotted these calculated isotherms for a subducting slab and overlying mantle and superimposed upon them the experimentally determined stability limits of various hydrous minerals. With this approach, he was able to assess the potential of hornblende, micas, serpentine, and other minerals as possible sources of H_2O for the genesis of magmas. Along these same lines, I have plotted calculated isotherms for the hanging wall of a down-going slab to evaluate the potential role of amphiboles, based on our recent experimental data.

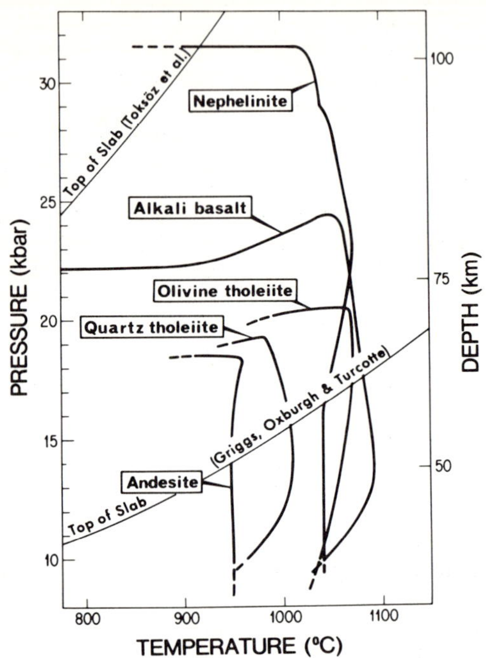

Fig. 8. Pressure-temperature projection showing curves limiting the stabilities of amphiboles in various rocks in the presence of an aqueous vapor ($X^V_{H_2O}$ ∿ 1.0) and with f_{O_2} buffered at N-NO [Allen et al., 1975] compared to calculated geotherms at the top of a descending slab.

Figure 8 shows that as the slab descends, a depth will be reached where amphiboles become unstable relative to a high-temperature, liquid-rich assemblage or a lower temperature, garnet-rich assemblage, depending upon which geotherm is selected. In the former case, anatexis of the amphibolite could produce an amphibole-rich residuum in equilibrium with magmas ranging from basaltic to andesitic, depending upon the degree of melting and the f_{H_2O}. On the other hand, if temperatures at the top of the slab are low, as calculated by *Toksöz, et al.*, [1971], amphiboles at depths less than 75 km will become unstable relative to the denser, garnet-bearing assemblage (eclogite). Figure 9 demonstrates that under these pressure-depth conditions, temperatures will be too low to melt peridotite or even olivine tholeiite,even in the presence of an aqueous vapor phase. Consequently, H_2O liberated with the breakdown of

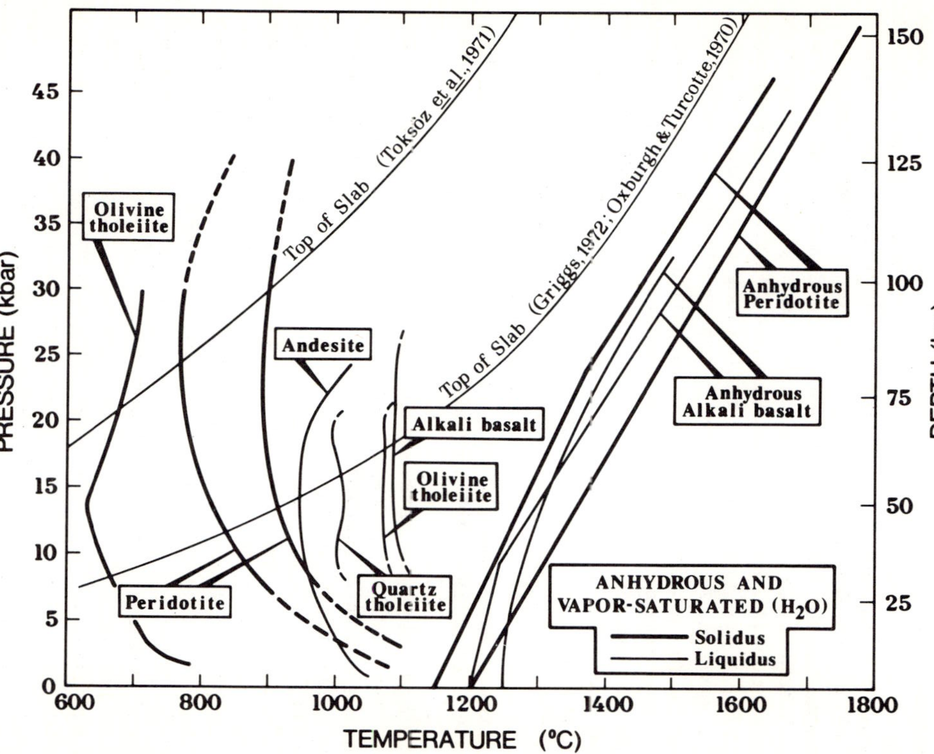

Fig. 9. Pressure-temperature projection showing vapor saturated ($X^V_{H_2O}$ ∿ 1.0) and anhydrous solidi and liquidi for rocks compared to calculated geotherms at the top of a descending slab.

amphiboles may rise into the overlying mantle peridotite [*McBirney*, 1969*b*]. *Ringwood* [1974], among others, has proposed that this released H_2O would cause melting of the peridotite to produce tholeiitic magmas. However, it is conceivable that at least some of this H_2O may become incorporated into phases such as talc or serpentine [*Kitahara et al.*, 1966] or more refractory hydrous pyroxenes [*Sclar et al.*, 1966; *Ringwood and Major*, 1967]. If temperatures in the peridotite are above those at which these hydrous phases exist and melting does occur, there is considerable controversy as to the nature of the magmas formed. *Mysen and Boettcher* [1975] and *Kushiro* [1972] would argue that quartz-normative magmas could be generated and could differentiate, via separation of amphibole, pyroxenes, and other phases, to produce the tholeiite-andesite series. On balance, *Nicholls* [1974] and *Ringwood* [1974] argue that olivine tholeiites would be the product of such anatexis. For reasons discussed below, I believe that the depths of generation of the andesitic magmas are less than 100 km, and under these shallow conditions it is more likely that the hydrous magmas will be quartz normative. However, considering the problems of diffusion of H_2O through the mantle and the possibility of the reduction in f_{H_2O} by formation of hydrous minerals in the overlying mantle, I prefer an origin for the calc-alkaline magmas by the method discussed above, i.e., by anatexis of the hydrous basalt of the subducted oceanic crust.

One aspect of this model in which amphiboles are the major carriers of H_2O in subducting slabs that is objectionable to many petrologists and geophysicists is that amphiboles are stable to depths less than 100 km, whereas magma genesis is thought to occur along the seismic (Benioff) zone beneath the active volcanic belts. The depth to this zone, based on seismology [e.g., *Hamilton*, 1974] or on the distance of active volcanoes from the continental margins (assuming a dip of 45° for the Benioff zone), is 150 to 250 km. As a result, petrologists have proposed that other hydrous phases fractionate H_2O at depths below those where amphiboles are stable. For example, *Jakeš and White* [1970], *Fitton* [1971], *Allen et al.*, [1972], and *Fyfe and McBirney* [1975] have developed modes in which phlogopite contains most of the H_2O at depths between 100 and 250 km. However, there is no experimental evidence to indicate that tholeiites, either quartz or olivine normative, contain phlogopite under any conditions. Based on our high-pressure experiments, only alkali basalts have the necessary compositions to yield phlogopite, at least to pressures of 35 kbars.

Along similar lines, *Ringwood* [1974] proposed that serpentine and other hydrated magnesian silicates would maintain high values of f_{H_2O} in the subducted oceanic crust to depths of about 150 km. I am opposed to this model for two major reasons.

First, there is evidence that serpentine minerals are not abundant in oceanic crust, occurring principally along fracture zones as demonstrated by seismology and laboratory determinations of the acoustical properties of serpentine [*Christensen*, 1972], by dredge hauls [*Barrett and Aumento*, 1970], and by results from the DSDP. I must admit, however, that *Aumento and Loubat* [1971] and *Melson and Thompson* [1970] have dredged samples from the Mid-Atlantic Ridge at 45°N and the Romanche Fracture, respectively, that appear to be parts of layered basic complexes in the lower oceanic crust, some of which are serpentinized. Nevertheless, crustal serpentinites are apparently scarce.

Second, Ringwood's model, in which serpentinite is a major source of H_2O at depth in subducted crust, centers around his conclusion that partial melting of the eclogite in the slab under conditions of high f_{H_2O} imposed by the serpentine produces rhyodacite magmas that rise and react with overlying peridotite to produce pyroxenites that in turn rise, differentiate by eclogite and amphibole fractionation and produce calc-alkaline magmas. As pointed out above, the calc-alkaline magmas of the circum-Pacific zone do not bear the REE patterns calculated by *Gill* [1974] for eclogite fractionation. More work on this important aspect of petrology is warranted.

Other hydrous phases, including zoisite and other members of the epidote group [*Boettcher*, 1970], humites [*McGetchin et al.* 1970; *Merrill et al.*, 1972], or nominally anhydrous phases, such as olivines and pyroxenes that contain some OH [*Martin and Donnay*, 1972], have also been suggested as sources of H_2O at depths in the earth and may play major roles, particularly at depths greater than 60 km.

On the other hand, amphiboles may be stable to depths greater than those discussed earlier in this paper. For example, *Holloway and Ford* [1975] have demonstrated experimentally that F-bearing amphiboles are stable to greater pressures than their OH analogs. We have experimental data revealing that higher values of $Mg/(Mg+Fe^{+2})$ tend to stabilize amphiboles to higher pressures with respect to garnet-bearing assemblages because of the increased pyrope content of the garnets. For example, we [*Allen and Boettcher*, unpublished data] have found that amphiboles are stable to pressures well in excess of 30 kbars for a Mg-rich composition otherwise similar to the Mt. Hood andesite (compare with Figures 1 to 4). Such compositions could result as residues following anatexis at shallower levels. It is not beyond the realm of reasonableness to expect amphiboles to exist to depths of 100 km in subducted crust.

There is no difficulty in applying this amphibole model to regions where the dip of the subduction zone is shallow; for example it is as shallow as $< 15°$ in the central parts of Chile [*Lopez-Escobar et al.*, in press]. The apparent problem occurs in regions where the volcanoes are 150 to 200 km or more above the seismic Benioff zone, such as in Indonesia [*Hamilton*, 1974]. Nevertheless, I propose that the seismic zone coincides with the cooler, central zone of the subducting slab where brittle, not plastic, conditions prevail. The zone of magma genesis is at shallower depths--at the top of the slab. Consequently, the models involving anatexis of amphibolite or settling of amphiboles from magma are consonant with spatial relationships of the circum-Pacific andesite-basalt volcanoes and do not require depths in excess of 100 km.

As a final word, the subduction of hydrated oceanic crust is in concert with the H_2O budget and geochemical cycle of the Earth. Making what appear to be reasonable assumptions about the present rate of outgassing of the Earth through volcanic and hot-spring activity and assuming that this rate has remained constant through the last 3 to 4 billion years, I have calculated that the amount of H_2O released exceeds that in the entire hydrosphere and continental crust by as much as an order of magnitude. *Anderson* [1974] arrived at a factor of 0.3 to 4. Thus, based on our crude estimates, H_2O has been recycled many times, largely through the agencies of subduction and volcanism.

Acknowledgments. Jack Allen of The Pennsylvania State University and Bucknell University, and Dion Stewart of The Pennsylvania State University performed much of the work described herein. Discussions with C. P. Thornton of The Pennsylvania State University regarding calc-alkaline volcanism and volcanic rocks are appreciated. Thanks are also due R. H. Jahns of Stanford University for his review. The research was supported by the Earth Sciences Section, National Science Foundation, NSF Grant EAR73-00220 A02. This paper constitutes contribution no. 1661 of the Institute of Geophysics and Planetary Physics.

REFERENCES

Allen, J. C., and A. L. Boettcher, The stability of amphiboles in andesite-H_2O-CO_2 and basalt-H_2O-CO_2 at high pressures (abstract), *Geol. Soc. Amer. Programs*, 973, 1975.

Allen, J. C., A. L. Boettcher, and G. Marland, Amphiboles in andesite and basalt: I. Stability as a function of P-T-f_{O_2}, *Amer. Mineral.*, 60, 1069-1085, 1975.

Allen, J. C., P. J. Modreski, C. Haygood, and A. L. Boettcher, The role of water in the mantle of the Earth: the stability of amphiboles and micas, *Intl. Geol. Congress*, Section 2 (Petrology), Montreal, 231-240, 1972.

Anderson, A. T., Jr., The before-eruption water content of some high-alumina magmas, *Bull. Volcanol.* 37, 530-552, 1973.

Anderson, A. T., Jr., Chlorine, sulfur, and water in magmas and oceans, *Geol. Soc. Amer. Bull.*, 85, 1485-1492, 1974.

Anderson, A. T., and D. Gottfried, Contrasting behavior of P, Ti, and Nb in a differentiated high-alumina olivine tholeiite and a calc-alkaline andesite suite, *Geol. Soc. Amer. Bull.*, 82, 1929-1942, 1971.

Aumento, F., and H. Loubat, The Mid-Atlantic Ridge near 45°N: XVI. Serpentinized ultramafic intrusions, *Can. J. Earth Sc.*, 8, 631-663, 1971.

Barrett, D. L., and F. Aumento, The Mid-Atlantic Ridge near 45°N: XI. Seismic velocity, density, and layering of the crust, *Can. J. Earth Sci.*, 7, 1117-1124, 1970.

Boettcher, A. L., The system $CaO-Al_2O_3-SiO_2-H_2O$ at high pressures and temperatures, *J. Petrol.*, 11, 337-379, 1970.

Boettcher, A. L., Volcanism and orogenic belts--the origin of andesites, *Tectonophysics*, 17, 223-240, 1973.

Boettcher, A. L., B. O. Mysen, and J. C. Allen, Techniques for the control of water fugacity and oxygen fugacity for experimentation in solid-media high-pressure apparatus, *J. Geophys. Res.*, 78, 5898-5901, 1973.

Bowen, N. L., *The Evolution of the Igneous Rocks*, pp. 334, Dover Publications, Inc., New York, 1956 .

Cawthorn, R. G., and M. J. O'Hara, Amphibole fractionation in calc-alkaline magma genesis, *Amer. J. Sci.*, 276, 309-329, 1976.

Christensen, N. I., The abundance of serpentinites in the oceanic crust, *J. Geol.*, 80, 709-719, 1972.

Church, S. E., The Cascade Mountains revisited: a re-evaluation in light of new lead isotope data, *Earth Planet Sci. Lett.*, 29, 175-188, 1976.

Condie, K. C., and D. L. Hayslip, Young bimodal volcanism at Medicine Lake volcanic center, Northern California, *Geochim. Cosmochim. Acta*, 39, 1165-1178, 1975.

Eggler, D. H., Water-saturated and undersaturated melting relations in a paricutin andesite and an estimate of water content in the natural magma, *Contr. Mineral. Petrol.*, 34, 261-271, 1972.

Fitton, J. G., The generation of magmas in island arcs, *Earth Planet. Sci. Lett.*, 11, 63-67, 1971.

Fyfe, W. S., and A. R. McBirney, Subduction and the structure of andesitic volcanic belts, *Amer. J. Sci.*, 275A. 285-297, 1975.

Gill, J. B., Role of underthrusting oceanic crust in the genesis of a Fijian calc-alkaline suite, *Contrib. Mineral. Petrol.*, 43, 29-45, 1974.

Gledhill, F. and P. E. Baker, Strontium isotope ratios in volcanic rocks from the South Sandwich Islands, *Earth Planet Sci. Lett.*, 19, 369-372, 1973.

Green, T. H., Crystallization of calc-alkaline andesite under controlled high-pressure hydrous conditions, *Contr. Mineral. Petrol.*, 34, 150-166, 1972.

Green, T. H., and A. E. Ringwood, Genesis of the calc-alkaline igneous rock suite, *Contr. Mineral. Petrol.*, 18, 105-162, 1968.

Griggs, D. T., The sinking lithosphere and the focal mechanism of deep earthquakes, in *The Nature of the Solid Earth*, edited by E. C. Robertson, pp. 361-384, McGraw-Hill, New York, 1972.

Hamilton, W., Earthquake map of the Indonesian region, *U. S. Geol. Survey Map* I-875-C, 1974.

Helz, R. T., Phase relations of basalts in their melting ranges at P_{H_2O} = 5 kb. Part II. Melt compositions, *J. Petrol.*, 17, 139-193, 1976.

Holloway, J. R., and C. E. Ford, Fluid-absent melting of the fluorohydroxy amphibole pargasite to 35 kilobars, *Earth. Planet. Sci. Lett.*, 25, 44-48, 1975.

Jakeš, P., and A. J. R. White, K/Rb ratios of rocks from island arcs, *Geochim. Cosmochim. Acta*, 34, 849-856, 1970.

Jakeš, P., and A. J. R. White, Hornblendes from calc-alkaline volcanic rocks of island arcs and continental margins, *Amer. Mineral.*, 57, 887-902, 1972.

Karig, D. E., and G. F. Sharman III, Subduction and accretion in trenches, *Geol. Soc. Amer. Bull.*, 86, 377-389, 1975.

Kitahara, S., S. Takenouchi, and G. C. Kennedy, Phase relations in the system $MgO-SiO_2-H_2O$ at high temperatures and pressures, *Amer. J. Sci.*, 264, 223-233, 1966.

Kushiro, I., Effect of water on the composition of magmas formed at high pressures, *J. Petrol.*, 13, 311-334, 1972.

Marsh, B. D., Some Aleutian andesites: their nature and source, *J. Beol.*, 84, 27-45, 1976.

Martin, R. F., and G. Donnay, Hydroxyl in the mantle, *Amer. Mineral.*, 57, 554-570, 1972.

Matsuhisa, Y., O. Matsubaya, and H. Sakai, Oxygen isotope variations in magmatic differentiation processes of the volcanic rocks in Japan, *Contr. Mineral. Petrol.*, 39. 277-288, 1973.

McBirney, A. R. (editor), Proceedings of the Andesite Conference, *Oreg. Dept. Geol. Mineral. Ind. Bull.*, 65. 1969a.

McBirney, A. R., Compositional variations in Cenozoic calc-alkaline suites, *Oreg. Dept. Geol. Mineral. Ind. Bull.*, 65, 185-189, 1969b.

McGetchin, T. R., L. T. Silver, and A. A. Chodos, Titanoclinohumite: a possible mineralogical site for water in the upper mantle, *J. Geophys. Res.*, 75, 255-259, 1970.

Melson, W. G., and G. Thompson, Layered basic complex in oceanic crust, Romanche Fracture, equatorial Atlantic Ocean, *Science*, 168, 817-820, 1970.

Merrill, R. B., J. K. Robertson, and P. J. Wyllie, Dehydration reaction of titanoclinohumite: reconnaissance to 30 kilobars, *Earth Planet. Sci. Lett.*, 14, 259-262, 1972.

Mysen, B. O., A. L. Boettcher, Melting of a hydrous mantle: II. Geochemistry of crystals and liquids formed by anatexis of mantle peridotite at high pressure and high temperature as a function of controlled activities of water, hydrogen, and carbon dioxide, *J. Petrol.*, 16, 549-593, 1975.

Nicholls, I. A., Petrology of Santorini Volcano, Cyclades, Greece, *J. Petrol.*, 12, 67-119, 1971.

Nicholls, I. A., Liquids in equilibrium with periodotitic mineral assemblages at high water pressures, *Contr. Mineral. Petrol.*, 45, 289-316, 1974.

Oxburgh, E. R., and D. L. Turcotte, Thermal structure of island arcs, *Geol. Soc. Amer. Bull.*, 81, 1665-1688, 1970.

Ringwood, A. E., The petrological evolution of island arc systems, *J. Geol. Soc. London*, 130, 183-294, 1974.

Ringwood, A. E., and A. Major, High pressure reconnaissance investigations in the system Mg_2SiO_4-MgO-H_2O, *Earth Planet. Sci. Lett.*, 2, 130-133, 1967.

Sclar, C. B., L. C. Carrison, and O. M. Stewart, High pressure synthesis of a new hydroxylated pyroxene in the system MgO-SiO_2-H_2O (abstract), *Trans. Amer. Geophys. Union*, 48, 226, 1966.

Stewart, D. C., Crystal clots in calc-alkaline andesites as breakdown products of high-Al amphiboles, *Contrib. Mineral. Petrol.*, 53, 195-204, 1975.

Stewart, D. C., J. C. Allen, and A. L. Boettcher, Genesis of andesitic magma by amphibole fractionation: petrographic and experimental evidence (abstract), *EOS*, 57, 344, 1976.

Taylor, H. P. The oxygen isotope geochemistry of igneous rocks, *Contrib. Mineral. Petrol.*, 19, 1-71, 1968.

Thorpe, R. S., P. J. Potts, and P. W. Francis, Rare earth data and petrogenesis of andesite from the north Chilean Andes, *Contrib. Mineral. Petrol.*, 54, 65-78, 1976.

Toksöz, M. N., J. W. Minear, and B. R. Julian, Temperature field and geophysical effects of a downgoing slab, *J. Geophys. Res.*, 76, 1113-1138, 1971.

Wyllie, P. J., Experimental petrology and global tectonics--a preview, *Tectonophysics*, 17, 189-210, 1973.

Yoder, H. S., Jr., Calc-alkaline andesites, experimental data bearing on the origin of their assumed characteristics, *Oreg. Dept. Geol. Mineral. Ind. Bull.*, 65, 77-89, 1969.

Yoder, H. S., Jr., and C. E. Tilley, Origin of basalt magmas: an experimental study of natural and synthetic rock systems, *J. Petrol.*, 3, 342-532, 1962.

DUAL PLATE TECTONICS MODEL

M. KUMAZAWA and Y. FUKAO
Department of Earth Science
Nagoya University
Nagoya, Japan

Abstract

The physical properties of materials in the vicinity of the 650-km seismic discontinuity are studied based on (1) phase relations (solid-solid, and solid-liquid) inferred from recent high-pressure experiments; (2) possible distribution of mineralogical and chemical compositions; and (3) shear softening in γ-$(Mg,Fe)_2SiO_4$ and its effect on melting temperatures and flow properties. It is predicted that the horizon just above the 650-km discontinuity is a low-velocity zone, in particular in shear wave velocity, a layer of local low melting temperature, and a layer of the active chemical migration and fractionation. At the same time this horizon is also identified as a low-viscosity layer (lower LVZ).

These features lead to a concept of dual plate tectonics models. The layer between 200- and 550-km depths is sandwiched between two relatively soft layers (upper and lower LVZs) and is expected to behave as a rigid plate (mesoplate). The interaction of two groups of plates (lithoplate and mesoplate) is discussed as a particular type of convective motion in the mantle. The upper LVZ (100 - 200-km depth) is not a zone of counter flow of lithoplate motion, but is a shear zone decoupling the lithoplate from the mesoplate. The mesoplates are spreading away from the trench, where the lithospheric materials diffuse into the lower LVZ and the mesoplates. The diapirism for generating hot spots is supposed to originate from the lower LVZ just above the 650-km depth.

PHYSICAL AND CHEMICAL NATURE OF THE LITHOSPHERE AND ASTHENOSPHERE, ESPECIALLY AT THEIR BOUNDARY

K. ITO
Department of Earth Sciences
Faculty of Science, Kobe University
Nada, Kobe 657, Japan

Abstract

Thermal models of the oceanic and continental lithosphere are presented. Variation of heat flow as a function of crustal age is used to compute the temperature distribution in the lithosphere. The lithosphere thickness is determined as a depth where the temperature reaches the melting temperature, which is estimated on the basis of high-pressure experiments under water-deficient conditions. The lithosphere-asthenosphere boundary is thermodynamically an open system in which H_2O and CO_2 behave as mobile components. The melting temperature is roughly 1100°C. The oceanic lithosphere thickens rapidly during the first 30-50 m.y. and grows little after 80-100 m.y. The continental lithosphere grows much more slowly but becomes thicker than does the oceanic one. The density of the asthenosphere under the ocean is calculated on an assumption that isostasy is maintained at the lithosphere-asthenosphere boundary. The computed results require that the lower part of the lithosphere must be denser than peridotite. A mechanism to produce a dense layer in the lower part of the lithosphere is proposed. The dense layer is considered to consist of garnet-rich eclogite. The heat flow from the asthenosphere to the lithosphere is twice larger, and the temperature of the asthenosphere at a same depth is significantly higher under the oceans than under the continents. Geophysical implications of this contrast are discussed.

I. INTRODUCTION

Physical and chemical properties of the lithosphere and its evolution help explain various processes in the earth's interior. In currently accepted plate models, the lithosphere is a solid slab overlying the partially molten asthenosphere. The oceanic lithosphere is created at a mid-ocean ridge, and it cools and thickens as it spreads away, and subducts at a trench through the asthenosphere. The lithosphere-asthenosphere boundary is delineated by the melting temperature of the lithosphere.

The nature of the continental lithosphere is less clear. It is as thin as 30 km in young orogenic regions but thicker than 100 km in old shield areas. It is likely, therefore, that the continental lithosphere also thickens as does the oceanic one, though its evolution seems to be much more complicated.

In this paper, the thermal and mineralogical structures of the oceanic and continental lithosphere are estimated, the conditions of the lithosphere-asthenosphere boundary are proposed, the mechanism of growth of the lithosphere is outlined, and the nature of the asthenosphere is deduced from observations of the lithosphere.

II. A THERMAL MODEL OF THE OCEANIC LITHOSPHERE

The oceanic lithosphere is generated from the asthenosphere when the liquid in the asthenosphere freezes. *Parker and Oldenburg*[1973] and *Yoshii*[1973] independently proposed similar thermal models of the oceanic lithosphere in which the latent heat of freezing was taken into account for the thermal balance at the lithosphere-asthenosphere boundary. Their original models predicted linear relationships (1) between the lithosphere thickness l and the square root of crustal age of the ocean floor $t^{1/2}$, (2) between the water depth h and the square root of crustal age $t^{1/2}$, and (3) between the heat flow Q_S and the inverse of square root of crustal age $t^{-1/2}$. *Oldenburg*[1975] and *Yoshii et al.* [1976] later improved their original models.

Regional variations of Rayleigh wave dispersion in the oceanic region indicate that the lithosphere is thicker in regions of greater crustal ages, though it is not clear whether or not the predicted linear l-$t^{1/2}$ relationship holds [*Forsyth*, 1973; *Leeds*, 1975; *Yoshii*, 1975].

The ocean floor deepens as the crustal age increases. The linear h-$t^{1/2}$ relationship holds well up to the crustal age of 80 m.y. [*Davis and Lister*, 1974], and not for the ocean floor older than 80 m.y. The water depth does not increase as rapidly as predicted but tends to be nearly constant. This deviation suggests that the oceanic lithosphere older than at least 100 m.y. has a finite thickness [*Sclater et al.*, 1975].

The linear Q_S-$t^{-1/2}$ relationship does not hold either. The heat flow tends to be steady at old ocean floors [*Sclater and Francheteau*, 1970], indicating that the oceanic lithosphere thickens insignificantly after about 100 m.y. [*Ito*, 1976].

Ito [1976], improving the thickening lithosphere model, used the observed Q_S-t relationship to calculate the temperature distribution in the lithosphere and determined the depth to the lithosphere-asthenosphere boundary, delineated by the melting temperature of the lithosphere. The equation for the temperature T in the lithosphere is

$$\frac{\partial T}{\partial t} = k\left(\frac{\partial^2 T}{\partial z^2}\right) \tag{1}$$

where k is the thermal diffusivity and z the depth from the surface. The coordinate z is fixed in the lithosphere, moving horizontally with a speed of the half spreading rate. The horizontal component of heat flow and the heat production in the lithosphere are neglected.

The solution for equation (1) is given according to Carslaw and Jaeger's solution 2.1.14 [*Carslaw and Jaeger*, 1951],

$$T = \phi + \frac{z^2}{k\ 2!}\dot{\phi} + \frac{z^4}{k^2\ 4!}\ddot{\phi} + \cdots\cdots$$

$$+ z\psi + \frac{z^3}{k\ 3!}\dot{\psi} + \frac{z^5}{k^2\ 5!}\ddot{\psi} + \cdots\cdots \tag{2}$$

where ϕ and ψ are any functions of the time t and dots denote differentiation with respect to t.

The surface boundary conditions are given as

$$T = 0 \qquad \text{at} \qquad z = 0 \qquad \text{for all } t, \tag{3}$$

and

$$\frac{\partial T}{\partial z} = \frac{Q_S(t)}{K} \qquad \text{at} \qquad z = 0, \tag{4}$$

where $Q_S(t)$ is the surface heat flow at the crustal age t and K is the thermal conductivity. Putting (3) and (4) into (2), we have $\phi = 0$ and $\psi = Q_S/K$. Thus, equation (2) can be rewritten as

$$T = \frac{1}{K}\left(zQ_S + \frac{z^3}{k\ 3!}\dot{Q}_S + \frac{z^5}{k^2\ 5!}\ddot{Q}_S + \cdots\right) \tag{5}$$

Q_S was formulated in the power series of t up to 6th order [*Ito*, 1976]; in this paper, Q_S is fitted to *Sclater and Francheteau's* data [1970] as a simple and explicit function:

$$Q_S = A + B\exp(-Ct) \tag{6}$$

A represents the steady state heat flow, $A + B$ is the heat flow at $t = 0$, and C gives the decreasing rate of heat flow with age. For A,B and C, the values used are 1.10, 2.0, and 0.033, respectively. Though Q_S is formulated slightly different from the formulation used in the previous paper [*Ito*, 1976], the conclusions derived in this paper do not differ.

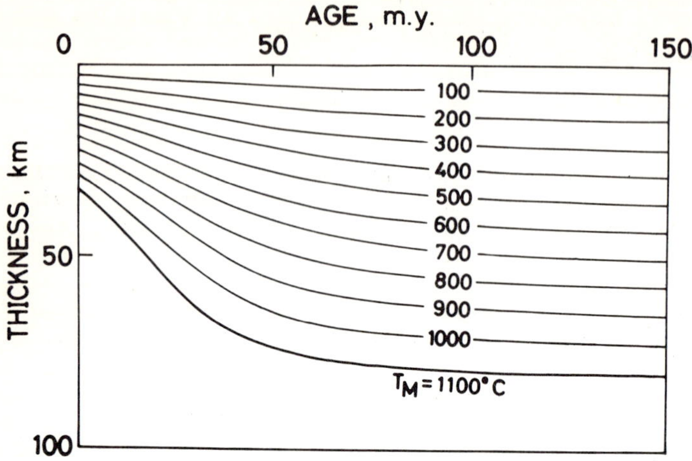

Fig. 1. Temperature distribution and thickness of the oceanic lithosphere.

Since the average thickness of the oceanic lithosphere is estimated to be 70 km [*Kanamori and Press*, 1970], and the melting temperature T_M is 1100°C (see section III), the values selected for average physical parameters of the lithosphere are: $K = 0.008$ cal/cm sec °C, specific heat $c_p = 0.25$ cal/g °C, density $\rho = 3.4$ g/cm^3, and hence $k = 0.00094$ cm^2/sec. The temperature distribution in the lithosphere and the thickness of the lithospheric plate are shown in Figure 1. One of the most important features indicated is that the lithosphere thickens very slowly after about 70 km. The implication will be discussed later.

III. TEMPERATURE AT THE LITHOSPHERE-ASTHENOSPHERE BOUNDARY

The temperature at the lithosphere-asthenosphere boundary is assumed to be the melting temperature of the lithosphere, which strongly depends on the existence of volatiles in the asthenosphere and, at 20 kbar, ranges from 900°C under H_2O-saturated conditions to 1350°C under anhydrous conditions.

Mid-ocean ridge basalts, which are products of the partial melting of the asthenosphere beneath the ridges, contain 0.2% or less H_2O [*Moore*, 1970]. Oceanic island basalts, however, which represent the liquid fraction existing beneath the oceanic lithosphere far away from the ridges, contain more H_2O. The difference suggests that the H_2O content of liquid, $X^L(H_2O)$, at the top of the asthenosphere increases with the crustal age of the overlying lithosphere. There are chemical and physical evidences to suggest that a liquid layer exists at the lithosphere-asthenosphere boundary, which is enriched in volatiles and incompatible elements such as alkali, U, Th, and rare earth elements [*Ito*, in press]. These elements are accumulated at the top by convection, diffusion, and vapor transfer in the entire asthenosphere.

The boundary between the liquid layer and the other parts of the asthenosphere may not be clear, but it is convenient in the following discussion to separate the liquid layer as a small system distinguished from the other parts of the asthenosphere. This lithosphere-asthenosphere boundary (LAB) system is thermodynamically open. Liquid ascending by convection enters into the system. Solid crystallizing by cooling of the lithosphere is taken out of the system. Assuming that the volume of the LAB system is constant, the masses of ascending liquid and crystallizing solid are equal, but the volatiles and incompatible elements vary. The thermodynamic state of such a system is defined not by H_2O content but by its fugacity. It is controlled by the flow rate of H_2O from the underlying asthenosphere into the system and the crystallization rate of hydrous minerals taken out of the system.

We assume that the H_2O concentration in the LAB system at $t = 0$ is equal to 0.1 wt %, which is the concentration of H_2O in basaltic magma at mid-ocean ridges. The $X_L(H_2O)$ increases with age because H_2O flows into the system by ascending from the underlying asthenosphere. As $X_L(H_2O)$ increases, the fugacity of H_2O increases, and the freezing temperature of the LAB system, that is, the melting temperature of the lithosphere T_M decreases. The effect of $X_L(H_2O)$ on T_M can be estimated from high-pressure experimental data of melting under the H_2O-deficient condition.

Solidus of a peridotite as a function of $X^V(H_2O)$ in vapor is shown in Figure 2, modified from *Mysen and Boettcher* [1975]. Relations between $X^V(H_2O)$ and $X^L(H_2O)$ have not been determined for basaltic melts but are known for albite melt at 20 kbar and 1450°C [*Mysen*, 1975]. We assume that the same relationship as the one between a normalized H_2O concentration in liquid, $X^L(H_2O)/\{X^L(H_2O)$ when $X^V(H_2O) = 1\}$, and $X^V(H_2O)$ for albite melt is applicable also for basaltic melts. The solubility of H_2O in basaltic melts at high pressures is taken from *Hamilton et al.* [1964] and *Green* [1973]. From these data, solidi of the peridotite as a function of $X^L(H_2O)$ are obtained as shown in Figure 3.

Assuming the increasing rate of $X^L(H_2O)$, we can estimate the melting temperature of the LAB system as a function of the crustal age. In Figure 3, the melting temperature for different cases of $X^L(H_2O)$ is shown as a function of the pressure calculated from the overlying lithosphere thickness shown in Figure 1. The temperature at the base of the lithosphere is about 1100°C, but is slightly higher near the mid-ocean ridge. Therefore, the lithosphere thickness shown in Figure 1 is slightly underestimated at young ages.

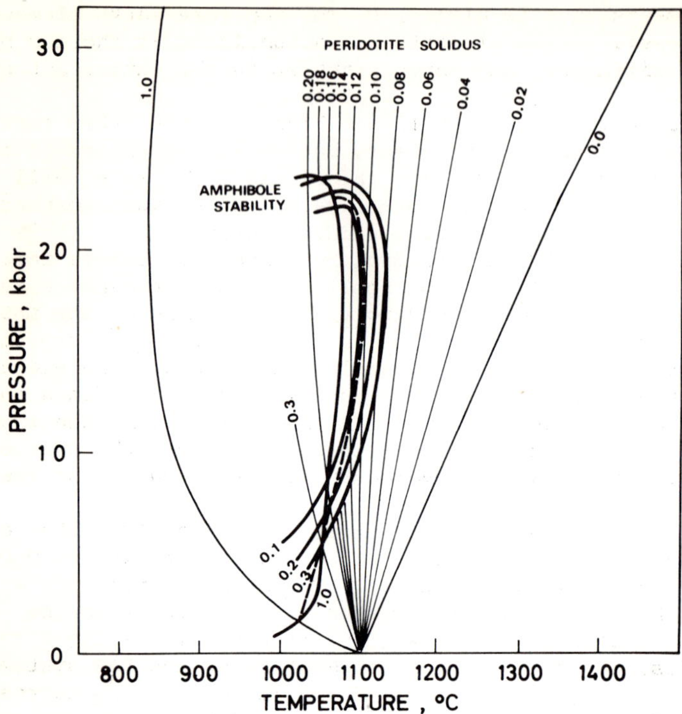

Fig. 2. Solidi of peridotite (thin lines) and stability limits of amphibole (thick lines) as functions of $X^V(H_2O)$ in vapor. Small numbers represent $X^V(H_2O)$ in weight fraction. Anhydrous solidus is taken from Ito and Kennedy [1967], hydrous solidi from Mysen and Boettcher [1975], amphibole stability limits below 10 kbar from Holloway [1973], and above 10 kbar from Mysen and Boettcher [1975]. A dashed line represents the melting curve of amphibole in the lithosphere-asthenosphere boundary system.

As $X^L(H_2O)$ increases, amphibole may become stable. The stability limits of amphibole as a function of $X^V(H_2O)$, extrapolated from experimental data by *Holloway* [1973] and *Mysen and Boettcher* [1975], are superimposed on the peridotite solidi in Figure 2. In spite of some uncertainty of the amphibole stability in molten peridotites, most experiments demonstrate that amphibole is not stable at temperatures higher than 1100°C and pressures greater

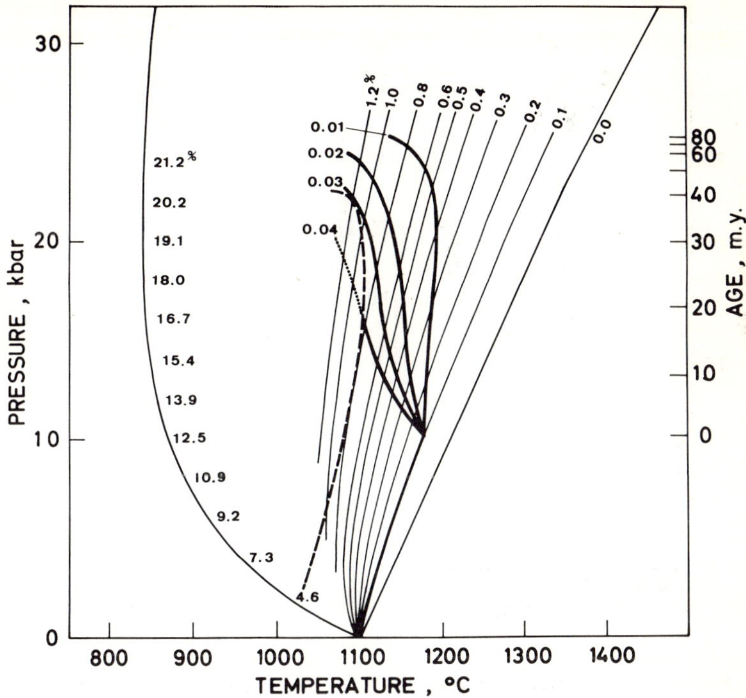

Fig. 3. Solidi of peridotite (thin lines) as a function of $X^L(H_2O)$ in liquid and melting curve of the lithosphere (thick lines) as a function of rate of increase of $X^L(H_2O)$ in the lithosphere-asthenosphere boundary system. Small numbers on the thin lines represent $X^L(H_2O)$ in wt %. Small numbers on the thick lines represent the rate of increase of $X^L(H_2O)$ in wt % per m.y. A dashed broken line is the melting curve of amphibole. The pressure versus age relation is obtained from the thickness versus age relation shown in Figure 1.

than 25 kbar.

The melting curve of amphibole in the open system of LAB, represented by intersections of the peridotite solidi and the amphibole stability curves for the same $X^V(H_2O)$, is superimposed on the melting curve of the lithosphere in Figure 3. If the rate of increase of $X^L(H_2O)$ is less than 0.03 wt % per m.y., amphibole is not stable in the oceanic lithosphere, and the basal

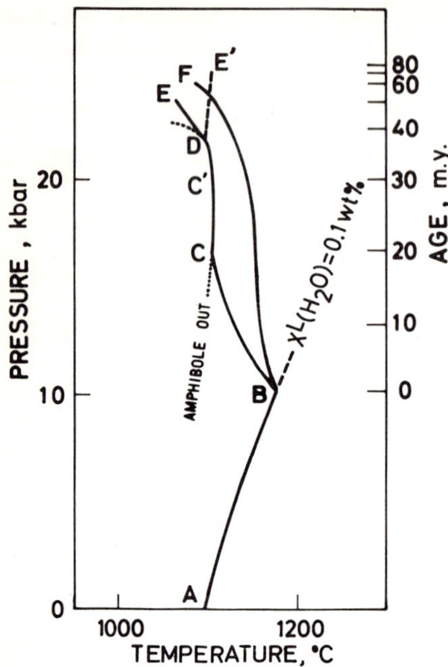

Figure 4. Temperature at the base of the lithosphere as a function of crustal age of the ocean floor. A-B and a dashed extension line represent the solidus of peridotite when $X^L(H_2O)$ = 0.1 wt %, B the basal temperature and pressure condition of the lithosphere beneath the mid-ocean ridge, and B-F an example of the basal temperature change when the rate of increase of $X^L(H_2O)$ is less than 0.03 wt % per m.y. C-C'-D and dotted extension lines represent the breakdown curve of amphibole, and B-C-C'-D an example of the basal temperature change when the rate of increase of $X^L(H_2O)$ is larger. C represents the depth at which amphibole becomes stable, D the depth at which amphibole becomes unstable, D-E the basal temperature change when phlogopite is not stable, and D-E' when phlogopite is stable.

temperature of the lithosphere changes along the curve B-F in Figure 4, for example. If amphibole exists in the oceanic lithosphere, as petrologists suggest, the rate of increase must be greater than 0.03 wt % per m.y. We have little knowledge about the rate of increase of $X^L(H_2O)$ in the LAB system, which depends on not only the flow rate of H_2O into but also the size of the

LAB system.

Though the presence of amphibole may be geochemically important, whether or not amphibole exists in the oceanic lithosphere does not affect much the estimation of temperature at the base of the lithosphere. In case the rate of increase of $X^L(H_2O)$ is greater than 0.03 wt % per m.y. and amphibole becomes stable, the basal temperature changes along the curve B-C-D-E, for instance, in Figure 4. At point C in Figure 4, the melting curve of the lithosphere intersects the stability limit of amphibole. Therefore, amphibole is crystallized from the LAB system, buffering the $X^L(H_2O)$. The flow of H_2O into the LAB system continues, but a part of H_2O is taken out of the system by amphibole crystallization. Amphibole contains about 1 wt % H_2O.

While it takes about 10 m.y. from C to C' (Figure 4), the lithosphere thickens by 10 km and $X^L(H_2O)$ increases by 0.1 wt %. The rate of increase of $X^L(H_2O)$, if amphibole does not crystallize, is 0.4 wt % over this 10 m.y.; therefore, 0.3 wt % of H_2O is taken out by amphibole. The amount of amphibole in the lithosphere depends on the size of the LAB system: if the LAB system is 1 km thick, for instance, the amount of amphibole is 3 wt % of the lithosphere. The size of the LAB system cannot be large, however, and 3 wt % may be overestimated by one order of magnitude. This small content may explain why amphibole is not found from xenocrysts and xenoliths in oceanic island basalts.

As the T_M approaches the knick point D, $X^L(H_2O)$ in the LAB system must increase to stabilize amphibole. When the required rate of increase exceeds 0.04 wt % per m.y., amphibole becomes unstable, and the T_M changes along line D-E. At the depths where amphibole tends to become unstable, the thickening rate of the lithosphere is very slow. Though it is shown in Figure 4 that amphibole first stops crystallizing before the lithosphere stops thickening, in nature the opposite might be the case.

After amphibole becomes unstable, phlogopite may start crystallizing and buffer $X^L(H_2O)$ in the LAB system. In that case, the T_M does not decrease along line D-E, but follows the phlogopite breakdown curve D-E'. We do not know the stability relation of phlogopite in the H_2O-deficient condition. The curve D-E' is only hypothetical.

Along line C-D-E', $X^L(H_2O)$ in the LAB system is buffered by amphibole and/or phlogopite. Thus, $X^L(H_2O)$ is kept nearly constant at about 1 wt % H_2O, while flows of CO_2 and alkali elements into the LAB system continue. This makes the chemical composition of the LAB system more alkali-rich and more strongly silica-undersaturated due to the increased CO_2/H_2O ratio. Hawaii alkali basalts contain 1-2 wt % H_2O^+ [*Moore*, 1970], thus it is reasonable to assume that the Hawaii alkali basalt magma represents the liquid in the LAB system beneath the Hawaiian Islands overlying the oceanic crust about 100 m.y. old.

IV. DENSITY OF THE OCEANIC LITHOSPHERE AND ASTHENOSPHERE

Over the oceanic basins, the free-air gravity anomaly is nearly zero [*Talwani et al.*, 1965], suggesting that isostasy is maintained for the oceanic lithosphere. For the lithospheric plate shown in Figure 5, the equation of isostasy is

$$\rho_W h(t) + \int_0^{l(t)} \rho_L(t,z)dz + \rho_A \Delta z$$
$$= \rho_W h(t + \Delta t) + \int_0^{l(t+\Delta t)} \rho_L(t + \Delta t, z)dz \quad (7)$$

Here, ρ_W, ρ_L and ρ_A are the density of the water, lithosphere and asthenosphere, respectively, and h, l and z are the water depth, lithosphere thickness, and depth from the ocean floor, respectively. Δz is defined as $h(t + \Delta t) + l(t + \Delta t) - h(t) + l(t)$.

The water depth as a function of the crustal age $h(t)$ is given by *Sclater et al.* [1971]. The lithosphere thickness as a function of the crustal age $l(t)$ and the density of the lithosphere as a function of the age and depth $\rho_L(t,z)$ can be obtained from the present thermal model of the lithosphere.

Assuming that the lithosphere consists of peridotite, using the peridotite density as a function of pressure and temperature given by *Ito* [1974], and taking the thermal expansivity as 0.00003 per deg. and the bulk modulus as 1260 kbar, we obtain the density distribution in the lithosphere. The density of the asthenosphere, thus obtained from equation (7), is shown in Figure 6. For comparison, the density of the peridotitic lithosphere is also shown.

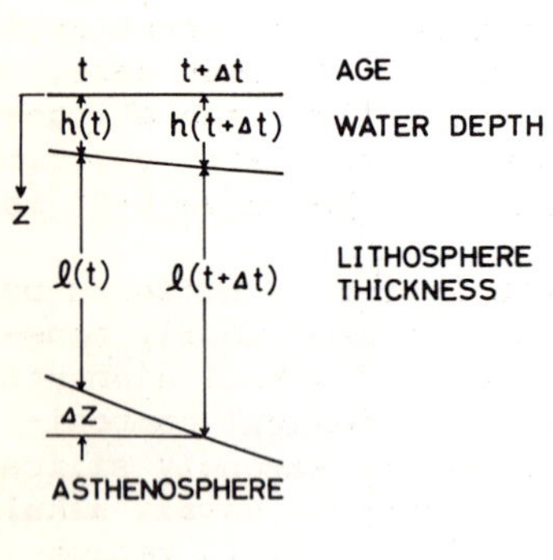

Fig. 5. Lithosphere-asthenosphere model.

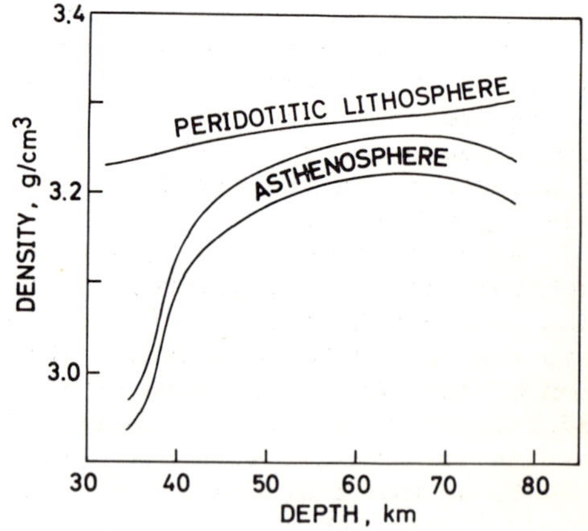

Fig. 6. Densities of the asthenosphere and peridotitic lithosphere.

Because the asthenosphere is partially molten, it should be lighter than the basal part of the lithosphere. Because the degree of melting in the asthenosphere is expected to be larger at shallower depths, it is reasonable that the density difference shown in Figure 6 between the lithosphere and asthenosphere is large at shallow depth and decreases with depth. However, it is unlikely that the density of the asthenosphere decreases at depths greater than 60 km (Figure 6). This contradiction rules out the homogeneous peridotite model of the lithosphere. In order that the computed density of the asthenosphere does not decrease with depth, the lower part (greater than at least 50 km) of the lithosphere must be denser than peridotite. A possible candidate for the lower part of the lithosphere is garnet-rich eclogite [*Forsyth and Press*, 1971; *Ito*, 1974].

V. PETROLOGICAL MODEL OF THE OCEANIC LITHOSPHERE

The isostasy of the lithosphere requires that the lower part of the lithosphere be denser than peridotite. Seismic wave velocity profiles obtained by long-range explosion experiments in the eastern Atlantic Ocean [*Steinmetz and Hirn*, 1974] and the western Pacific Ocean [*Asada and Shimamura*, 1976] revealed the existence of a high, compressional velocity layer (>8.6 km/sec) in the lower part of the lithosphere.

Yoshii et al. [1976] presented a petrological model of the oceanic lithosphere consisting of, from top to bottom, dunite or harzburgite, garnet pyroxenite, and eclogite. Their model was based on the assumption that the latent heat of the asthenosphere-to-lithosphere transition is large. In the present thermal model, however, the latent heat is required to be less than 20 cal/g to the depth of about 60 km [*Ito*, 1976]. If this is the case, the petrological model must be revised. It is more likely that the asthenosphere solidifies as a whole to become the lithosphere; thus, the chemical composition of the lithosphere is that of the asthenosphere down to about 60 km. Below 60 km, however, the latent heat can be larger and the fractional accretion process proposed by *Yoshii et al.* [1976] probably takes place. At pressures higher than about 20 kbar, garnet in peridotite is stable. Because of a reaction relationship olivine + orthopyroxene + liquid = garnet + clinopyroxene at 20-30 kbar [*Ohara and Yoder*, 1967], the liquid in the asthenosphere fractionally crystallizes to garnet and clinopyroxene or to eclogite [*Yoshii et al.* 1976].

When the lithosphere thickens, the pressure and partial pressure of H_2O of the LAB system increase. The volume of the primary phase garnet increases at the expense of clinopyroxene with increased pressure [*Cohen et al.*, 1967] and with increased P_{H_2O} [*Kushiro*, personal communication]. Therefore, the composition of the LAB system which is the cotectic composition coexisting with garnet

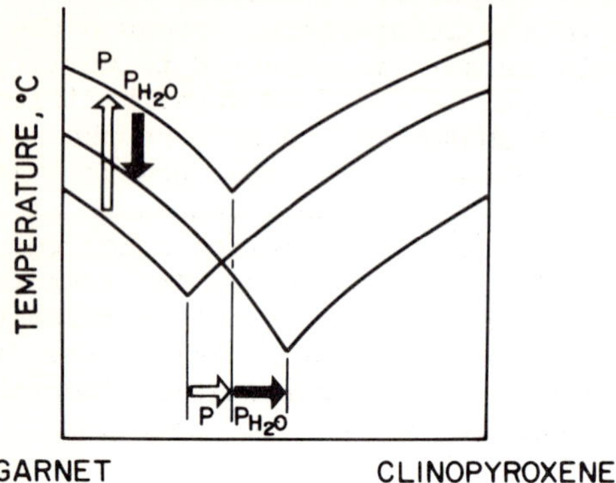

Figure 7. Changes of the cotectic relation between garnet and clinopyroxene. White arrows indicate the changes due to increased pressure, and black arrows the changes due to increased P_{H_2O} in liquid.

and clinopyroxene as well as olivine and orthopyroxene in the asthenosphere, shifts from the garnet side toward the clinopyroxene side (Figure 7), as the lithosphere thickens. Consequently, the garnet/clinopyroxene ratio in the crystallizing solid must be greater than that of the cotectic composition under a fixed condition. In an extreme case, if the thickening rate of the lithosphere is very slow, the mineral assemblage crystallizing from the asthenosphere should be only garnet.

As this fractional accretion process becomes dominant at the base of the lithosphere below about 60 km, garnet content increases with increased depths, and, thus, the basal part of the lithosphere consists of nearly 100% garnet. The compressional wave velocity of 8.6 km/sec found at the base of the lithosphere near the Mariana rough [*Asada and Shimamura*, 1976] can be explained best by the existence of a garnet-rich layer at the base of the lithosphere.

VI. THERMAL MODEL OF THE CONTINENTAL LITHOSPHERE

The continental and oceanic lithospheres are distinguished by radioactive heat sources within the crust (Figure 8). The measured heat flow Q_S in many parts of North America is found to be related to the measured heat production in near-surface plutonic rock $A(0)$ by

$$Q_S = q^* + DA(0) \quad (8)$$

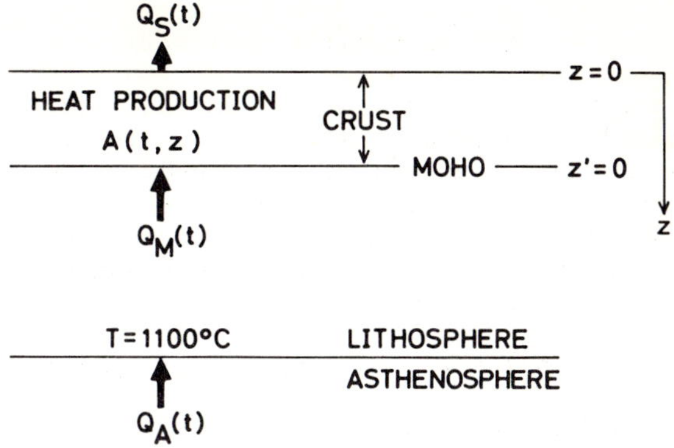

Figure 8. Thermal balance in the continental lithosphere and asthenosphere.

where q^* and D are relatively constant within each province [*Roy et al.*, 1968]. The parameter q^* represents the fraction of heat flow from the mantle Q_M. The parameter D relates to the distribution of heat source in the crust. *Lachenbruch* [1970] interpreted the relation that the heat production in the continental crust decreases with depth z in an exponential form

$$A(z) = A(0)\ \exp(-z/D) \tag{9}$$

D varies little from one province to another, ranging between 7.5 and 10 km. $A(0)$ depends on the degree of erosion of the plutonic rock and is a function of the age of orogeny in the province. Then, we may generalize equation (9) as a function of the age and

TABLE 1. Heat flow from the mantle q^* in various province

Province	Age, m.y.	q^*, HFU	Reference[a]
Australian Shield	>2500	0.64	(1)
Eastern U.S.	250-1100	0.80	(2)
Basin and Range	20	1.40	(2)
Sierra Nevada	80	0.40	(2)

a. (1) *Jaeger* [1970]; (2) *Roy et al.* [1968].

depth as

$$A(t,z) = A_0(t)\ \exp(-z/D) \tag{9A}$$

The parameter q^* varies from 0.4 in the Sierra Nevada to 1.4 in the Basin and Range [*Roy et al.*, 1968]. The low q^* in the Sierra Nevada is anomalous and is considered to be due to subduction of the cold lithosphere under the Sierra Nevada. Excepting q^* for the Sierra Nevada, other data given in Table 1 show that q^* decreases with the age of orogeny. We roughly fit the data to a simple exponential function

$$q^* = Q_M(t) + 0.60 + 0.85\ \exp(-0.003t) \tag{10}$$

The surface heat flow Q_S in the continents decreases with the age of orogeny [*Polyak and Smirnov*, 1968; *Sclater and Francheteau*, 1970]. Though the data by *Roy et al.*[1968] indicate that the surface heat flow depends on not only the age but also the local convective pattern in the mantle, we assume that *Polyak and Smirnov*'s and *Sclater and Francheteau*'s data represent a general tendency (Figure 9). We roughly fit the data to the simple exponential function

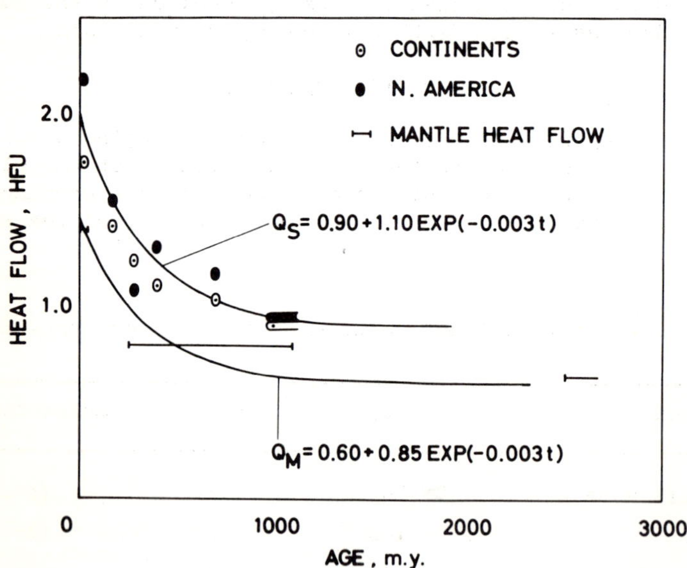

Fig. 9. Surface heat flow Q_S [Polyak and Smirnov, 1968; Sclater and Francheteau, 1970] and mantle heat flow Q_M (see Table 1) against the age of orogenic province for continents.

$$Q_S(t) = 0.90 + 1.10 \exp(-0.003t) \tag{11}$$

If the thermal state within the crust can be approximated by the steady state, the heat source distributed in the form of equation (9A) contributes to the surface heat flow by $DA_0(t)$. Thus,

$$Q_S(t) = Q_M(t) + DA_0(t) \tag{12}$$

From (10), (11), and (12), we have the fraction of heat flow contributed by the heat generated within the crust,

$$DA_0(t) = 0.30 + 0.25 \exp(-0.003t) \tag{13}$$

The temperature distribution in the crust can be approximated by the steady state solution of the thermal equation

$$\frac{\partial^2 T}{\partial z^2} = -\frac{A(t,z)}{K} \tag{14}$$

The solution is

$$T = \frac{Q_M}{K} z + \frac{D}{K} A_0(t)\{1 - \exp(-z/D)\} \tag{15}$$

From (10), (13), and (15), we have the temperature in the crust $T(t,z)$ as a function of the depth and the age of orogeny. The temperature at the Moho discontinuity T_{MOHO} is given by T for $z = z_{MOHO}$ in equation (15). The constants D and z_{MOHO} are taken as 10 km and 40 km, respectively.

The equation for the temperature in the continental lithosphere below the Moho discontinuity is the same with equation (1) for the oceanic lithosphere. Equation (1) can be solved in the same manner.

Taking z' as zero at the Moho discontinuity, we have the boundary conditions

$$T = T_{MOHO}(t) \qquad \text{at } z' = 0 \tag{16}$$

$$\frac{\partial T}{\partial z'} = \frac{Q_M(t)}{K} \qquad \text{at } z' = 0 \tag{17}$$

Therefore, in *Carslaw and Jaeger*'s [1951] solution given by equation (2) of this paper, $\phi = T_{MOHO}(t)$ and $\psi = Q_M(t)/K$. The two functions can be obtained from equations (15) and (10), respectively.

The same physical parameters used for the thermal model of the oceanic lithosphere are used. The lithosphere thickness is determined at the depth where the temperature reaches the melting temperature, which is assumed to be 1100°C. The change of

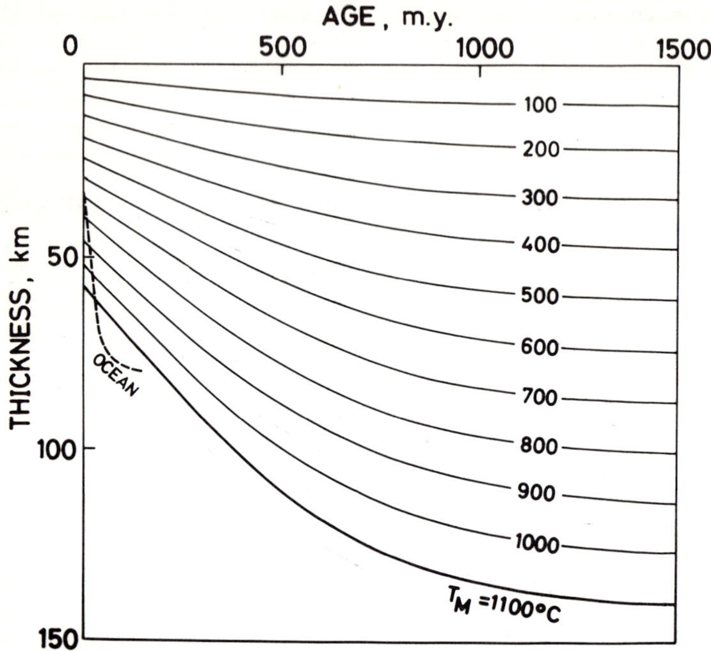

Figure 10. Temperature distribution and thickness of the continental lithosphere. Thickness of the oceanic lithosphere is superimposed by a dashed line.

temperature distribution and thickness of the continental lithosphere with increasing age of orogeny is shown in Figure 10.

Compared with the oceanic lithosphere, the continental lithosphere thickens very slowly [*Kono and Amano*, in press] but grows thicker. The heat flow from the asthenosphere to the lithosphere is much smaller under the continents than under the ocean.

The lithosphere-asthenosphere boundary under the continent is less clear than under the ocean, but it may be defined by either a decrease in the seismic wave velocity or an increase of Q^{-1}. The lithosphere thickness estimated by such definition is generally greater under the continent than under the ocean. Under the stable continents, the lithosphere thickness is estimated to be 110-150 km [*Lehman*, 1961; *Brune and Dorman*, 1963; *Wickens and Pec*, 1968]. Average thickness is estimated to be 120 km under the continent, whereas it is 70 km under the ocean [*Burton and Bennell*, 1976]. The Andes area, which overlies the subducting lithosphere, seems to have anomalous lithosphere, as thick as 400 km [*Sacks and Okada*, 1974].

The thickness of the continental lithosphere and the heat flow depend on not only the age of orogeny but also the local pattern of convection in the asthenosphere. Though there are a few anomalous areas such as the Andes and the Sierra Nevada, the continental lithosphere in general grows from thin plate under the young orogenic areas to thick plate under the old shield areas. The present thermal model of the continental lithosphere, as shown in Figure 10, is compatible with this observed evolution of the continental lithosphere.

VII. CONVECTION IN THE ASTHENOSPHERE

In the present thermal model, we have the following boundary condition

$$K \frac{\partial T}{\partial z} = Q_A + \rho L \frac{dl}{dt} \tag{18}$$

where L is the latent heat of crystallization and Q_A is the heat flow from the asthenosphere to the lithosphere. In the present thermal models, Q_A is higher under the oceans than under the continents. The high Q_A in the old ocean basins requires effective heat transfer by active convection. Under the ocean, the primary convection cell involves the creation and subduction of the lithosphere plate. The secondary convective cell may be the roll-shaped one with the axis of roll perpendicular to the spreading axis [*Richter and Parsons*, 1975]. Small scale convective cells are inferred from the high Q_A in the old ocean basins, the low Q_A near the mid-ocean ridges, and the accumulation of liquid and volatiles in the top of the asthenosphere.

The present thermal models indicate large temperature differences in the asthenosphere under the ocean and under the continent. The temperature of basaltic magmas erupting at the mid-ocean ridges is about 1200°C or higher [*Scheidegger*, 1973]. When the temperature drop due to the latent heat of fusion and adiabatic expansion is taken into account, the temperature of the asthenosphere beneath the mid-ocean ridges must be higher than 1400°C at 200 km [see *Cawthorn*, 1975, for estimation]. The thermal state of the asthenosphere under the ocean and continent is schematically shown in Figure 11.

Why the heat flow from the asthenosphere to the lithosphere Q_A is by a factor of two lower under the continent than under the ocean is a puzzle. The contrast may imply that the temperature difference persists down to the lower mantle as suggested by *Jordan* [1975].

At the same depths, the temperature of the asthenosphere is much higher beneath the ocean than under the continent. When a

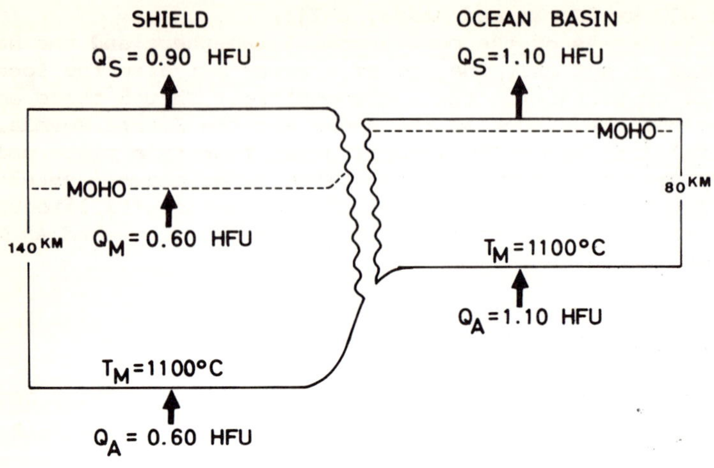

Figure 11. Thermal model of the oceanic and continental lithosphere and asthenosphere.

continent starts drifting over the asthenosphere previously overlain by the oceanic lithosphere, the base of the continental lithosphere must be heated from 1100°C to a temperature of nearly 1400°C. This heating effect is seen beneath the African continent. *Boyd* [1973] has found that the geotherm estimated from pyroxene-geothermometer applied to nodules in African kimberlites shows a knick at 1100°C at about 160 km. At depths greater than 160 km, the geotherm is steeper and reaches 1400°C at 200 km. *Boyd* considered that the steepening of the geotherm was due to frictional heating. In the present thermal model, no frictional heating is needed. The base of the continental lithosphere will be heated if the continent drifts toward the ocean. This transient heating causes dehydration, decarbonatization, and partial melting of the lithosphere, resulting in kimberlite and carbonatite eruption.

The present thermal model of the lithosphere neglects this possible reheating at the base of the continental lithosphere. The thermal state of the continental lithosphere is easily disturbed when the continent drifts. Neglecting this disturbance in the present thermal model is clearly an oversimplification. To deduce realistic models of the continental lithosphere, it is important to compare the thermal data for the lithosphere of a rapidly drifting continent with those of a continent that has been stationary for a long geological time.

VIII. CONCLUSIONS

1. The oceanic lithosphere thickens rapidly during the first 30-50 m.y. but grows little after 80-100 m.y.
2. The temperature at the base of the lithosphere is nearly 1100°C but is slightly higher near the spreading center.
3. The accumulation rate of H_2O to the top of the asthenosphere is important to determine the thermal state at the lithosphere-asthenosphere boundary.
4. Amphibole may or may not be present in the oceanic lithosphere. If present, it buffers the concentration of H_2O to about 1 wt % at the lithosphere-asthenosphere boundary.
5. If the average thickness of the oceanic lithosphere is 70 km and that of the continental one 120 km, the average thermal conductivity of the lithosphere is required to be 0.008 cal/cm^2 °C sec.
6. The lower part of the oceanic lithosphere must be denser than peridotite. Garnet content increases below 60 km, and the basal part of the oceanic lithosphere may be composed of nearly pure garnet.
7. The continental lithosphere thickens as the oceanic lithosphere, though the growth rate is about one order of magnitude smaller.
8. The heat flow from the asthenosphere to the lithosphere is 1.1 HFU under the ocean basins but 0.6 HFU under the stable shields. The temperature of the asthenosphere at the same depths is much higher under the oceans than under the continents.
9. When the continent starts drifting, the base of the continental lithosphere is heated by the hot asthenosphere previously overlain by the oceanic lithosphere. This transient heating of the continental lithosphere causes extrusion of kimberlites.

Acknowledgements. I thank C. L. Drake for reviewing the paper. This work was supported by Scientific Research Grant No. 043022 from the Ministry of Education.

REFERENCES

Asada, T., and H. Shimamura, Observation of earthquakes and explosions at the bottom of the western Pacific: Structure of oceanic lithosphere revealed by longshot experiment, in *The Geophysics of the Pacific Ocean Basin and Its Margin*, *Geophys. Monogr. Ser.*, vol. 19, edited by G. H. Sutton, M. H. Manghnani, and R. Moberly, pp. 135-153, Amer. Geophys. Union, Washington, D. C., 1976.

Boyd, F. R., A pyroxene geotherm, *Geochim. Cosmochim. Acta*, *37*,

2533, 1973.

Brune, J., and J. Dorman, Seismic waves and earth structure in the Canadian shield, *Bull. Seismol. Soc. Am., 53*, 167, 1963.

Burton, P. W., and J. D. Bennell, Q^{-1} and lithospheric thickness, *Earth Planet. Sci. Lett., 30*, 151, 1976.

Carslaw, H. S., and J. C. Jaeger, *Conduction of Heat in Solids*, Oxford Univ. Press, Cambridge, England, 1951.

Cawthorn, R. G., Degrees of melting in mantle diapirs and the origin of ultrabasic liquids, *Earth Planet. Sci. Lett., 27*, 113, 1975.

Cohen, L. H., K. Ito, and G. C. Kennedy, Melting and phase relations in an anhydrous basalt to 40 kilobars, *Am. Jour. Sci., 265*, 475, 1967.

Davis, E. E., and C. R. B. Lister, Fundamentals of ridge crest topography, *Earth Planet. Sci. Lett., 21*, 405, 1974.

Forsyth, D. W., The early structural evolution and anisotropy of the oceanic upper mantle, Ph.D. Thesis, Mass. Inst. Tech., Cambridge, Mass., 1973.

Forsyth, D. W., and F. Press, Geophysical tests of petrological models of the spreading lithosphere, *J. Geophys. Res., 76*, 7963, 1971.

Green, D. H., Contrasted melting relations in a pyrolite upper mantle under mid-oceanic ridge, stable crust and island arc environments, *Tectonophys., 17*, 285, 1973.

Hamilton, D. L., C. W. Burnham, E. F. Osborn, The solubility of water and effects of oxygen fugacity and water content on crystallization in the mafic magmas, *J. Petrol., 5*, 21, 1964.

Holloway, J. R., The system pargasite-H_2O-CO_2: A model for melting of a hydrous mineral with a mixed-volatile fluid. I. Experimental results to 8 kbar, *Geochim. Cosmochim. Acta, 37*, 651, 1973.

Ito, K., Petrological models of the oceanic lithosphere: geophysical and geochemical tests, *Earth Planet. Sci. Lett., 21*, 169, 1974.

Ito, K., Heat flow and thickness of the oceanic lithosphere, *Earth Planet. Sci. Lett., 30*, 65, 1976.

Ito, K., Magma genesis in a dynamic mantle, in *Genesis of the Metalliferous Ore Deposits in Japan and East Asia*, edited by H. Imai, Tokyo Univ. Press, Tokyo, 1977.

Ito, K., and G. C. Kennedy, Melting and phase relations in a natural peridotite to 40 kbar, *Am. Jour. Sci., 265*, 519, 1967.

Jaeger, J. C., Heat flow and radioactivity in Australia, *Earth Planet. Sci. Lett., 8*, 285, 1970.

Jordan, T. H., The continental tectosphere, *Rev. Geophys. Space Phys., 13*, 1, 1975.

Kanamori, H., and F. Press, How thick is the lithosphere?, *Nature, 226*, 330, 1970.

Kono, Y., and M. Amano, Thickening plate model of continental plate, *J. Phys. Earth*, in press.

Lachenbruch, A. H., Crustal temperature and heat production: Implications of the linear heat flow relation, *J. Geophys. Res., 75*, 3291, 1970.

Leeds, A. R., Lithospheric thickness in the Western Pacific, *Phys. Earth Planet. Interiors, 11*, 61, 1975.

Lehman, I., S and the structure of the upper mantle, *Geophys. J. Roy. Astron. Soc., 4*, 124, 1961.

Moore, J. G., Water content of basalt erupted on the ocean floor, *Contrib. Mineral. Petrol., 28*, 272, 1970.

Mysen, B. O., Solubility of volatiles in silicate melts at high pressure and temperature: The role of carbon dioxide and water in feldspar, pyroxene, and feldspathoid melts, *Carneg. Inst. Yearb. 74*, 454, 1975.

Mysen, B. O., and A. L. Boettcher, Melting of a hydrous mantle: I. Phase relations of natural peridotite at high pressures and temperatures with controlled activities of water, carbon dioxide, and hydrogen, *J. Petrol., 16*, 520, 1975.

O'Hara, M. J., and Y. S. Yoder, Jr., Formation and fractionation of basic magmas at high pressures, *Scottish J. Geol., 3*, 67, 1967.

Oldenburg, D. W., A physical model for the creation of the lithosphere, *Geophys. J. Roy. Astron. Soc., 43*, 425, 1975.

Parker, R. L., and D. W. Oldenburg, Thermal model of ocean ridges, *Nature, Phys. Sci., 242*, 137, 1973.

Polyak, F. M., and Ya. B. Smirnov, Relationship between terrestrial heat flow and the tectonics of continents, *Geotectonics*, 205, 1968.

Richter, F. M., and B. Parson, On the interaction of two scales of convection in the mantle, *J. Geophys. Res., 80*, 2529, 1975.

Roy, R. F., D. D. Blackwell, and F. Birch, Heat generation of plutonic rocks and continental heat flow provinces, *Earth Planet. Sci. Lett., 5*, 1, 1968.

Sacks, I. S., and H. Okada, A comparison of the anelasticity structure beneath western South America and Japan, *Phys. Earth Planet. Interiors, 9*, 211, 1974.

Scheidegger, K. F., Temperatures and compositions of magmas ascending along mid-ocean ridges, *J. Geophys. Res., 78*, 3340, 1973.

Sclater, J. G., and J. Francheteau, The implications of heat flow observations on current tectonic and geochemical models of the crust and upper mantle of the Earth, *Geophys. J. Roy. Astron. Soc., 20*, 509, 1970.

Sclater, J. G., R. A. Anderson, and M. L. Bell, Elevation of ridges and evolution of the central Eastern Pacific, *J. Geophys. Res., 76*, 7888, 1971.

Sclater, J. G., L. A. Lawver, and B. Parson, Composition of long-wavelength residual elevation and free air gravity anomalies in the North Atlantic and possible implications for the thickness of the lithospheric plate, *J. Geophys. Res., 80*,

1031, 1975.
Steinmetz, L., and A. Hirn, Sub-Moho stratification from an oceanic long-range profile, *EOS, Trans, AGU, 55*, 360, 1974.
Talwani, M., X. LePichon, and M. Ewing, Crustal structure of mid-ocean ridges, *J. Geophys. Res., 70*, 341, 1965.
Wickens, A. J., and K. Pec, A crust-mantle profile from Mould Bay, Canada, to Tucson, Arizona, *Bull. Seism. Soc. Am., 58*, 1821, 1968.
Yoshii, T., Upper mantle structure beneath the North Pacific and the marginal seas, *J. Phys. Earth, 21*, 313, 1973.
Yoshii, T., Regioanlity of group velocities of Rayleigh waves in the Pacific and thickening of the plate, *Earth Planet. Sci. Lett., 25*, 305, 1975.
Yoshii, T., Y. Kono, and K. Ito, Thickening of the oceanic lithosphere, in *The Geophysics of the Pacific Ocean Basin and Its Margin, Geophys. Monogr. Ser.*, vol. 19, edited by G. H. Sutton, M. H. Manghnani, and R. Moberly, pp. 423-430, Amer. Geophys. Union, Washington, D. C., 1976.

UPPER MANTLE P-STRUCTURE AND THE 650-KM DISCONTINUITY

Y. FUKAO
Department of Earth Sciences
Nagoya University
Chikusa, Nagoya (464)
Japan

Abstract

Two kinds of studies were made on the *P* wave velocity structure in the upper mantle. One is a $dT/d\Delta$ measurement of *P* waves at epicentral distances of less than 30°. The results were inverted to obtain a velocity model for the upper 800 km mantle on the oceanic side of the Japan-Kurile arc. The other is a study of precursors to seismic core phase *P'P'* on the basis of the world-wide set of seismograms. A precursor was interpreted in terms of the reflection from the 650-km transition zone. This zone spreads out over the thickness of 40-50 km but the bottom part was suggested to be very sharp. A thickness of 10 km was suggested for the 400-km transition zone.

I. $dT/d\Delta$ Measurement

The $dT/d\Delta$ measurements were made at the Wakayama Micro-Earthquake Observatory, Japan, for *P* waves from earthquakes in the Kurile-Kamchatka region [Fukao, in preparation]. As may be inferred from Figure 1, the travel paths are largely confined to the oceanic side (trench-side) of the deep seismic zone of the Japan-Kurile arc. The epicentral distances range from 13 to 30°. Figure 2 shows the results of the analysis, in which the measured values of $dT/d\Delta$ were plotted against the epicentral distances, Δ, corrected to surface focus. The open circles show the first arrival data and the filled circles show the later arrival data. The bar attached to each data point indicates the standard deviation, which is less than 3% for every measurement. This standard deviation is consistent with the scatter of the data, which are fitted with a continuous curve having branches from C to G. The first arrival emerges on the C branch in the distance range 14-20°, on the E branch in the distance range 20-24° and on the G branch at distances greater than 24°. There are two large cusps near 13.5° and 19.5° and two sharp drops in $dT/d\Delta$

near 21.5° and 26°. Another important feature of this curve is the near flatness of the C branch, which is consistent with our observation that the amplitude of the C arrival is always very small. The weakness of the C arrival was first found by *Kishimoto* [1956], who described it as the "abnormally strong dim-

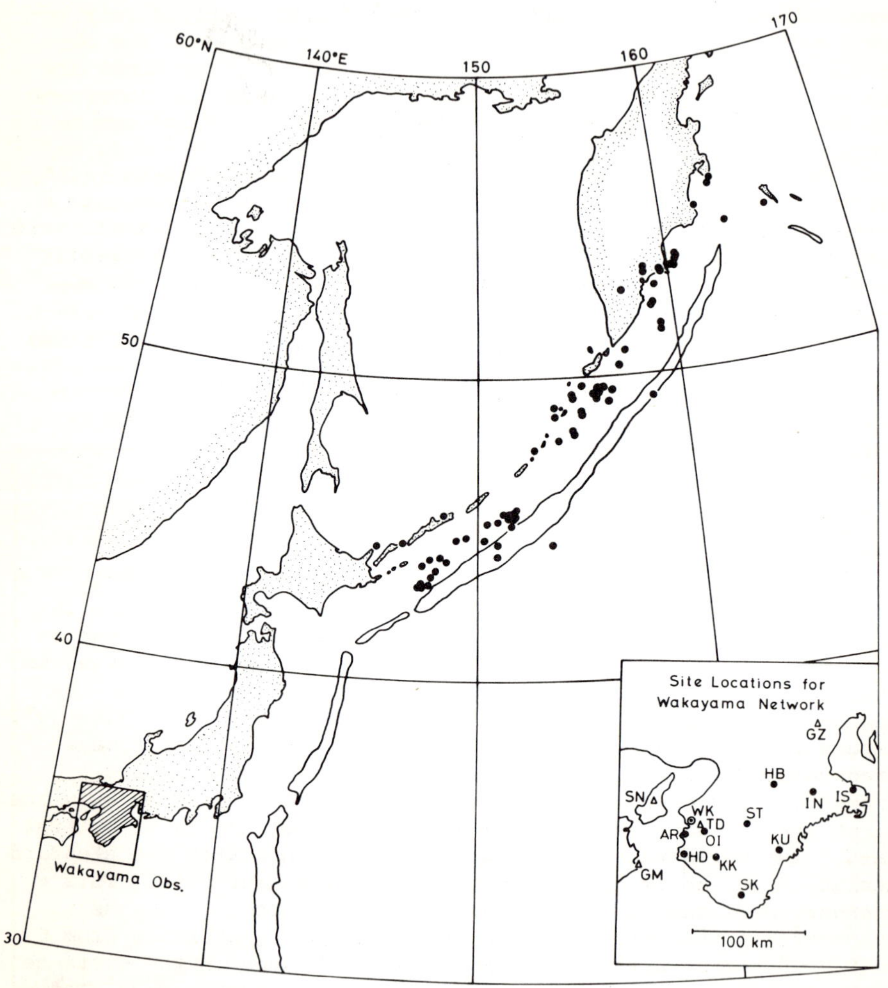

Fig. 1. Epicenters of the earthquakes for which dT/dΔ measurements were made. Site locations for the Wakayama Micro-Earthquake Observatory are given in the insert. Station code is given by two letters.

inution" of *Pd* waves (= C waves in our notation).

The results can be inverted if a model of the *P* velocity were established from surface down to some 200 km. To construct a model to such depths, we collected copies of the seismograms of the Kurile earthquake of December 02, 1971, from about twenty stations on the Pacific coast (Figure 3). This event was used because of its unique location and its sharp onset. The *P* waves travel through the uppermost mantle almost exclusively beneath the deep-sea trench and its landward seabottom. The observed travel time residuals from the Jeffreys-Bullen times are plotted against the epicentral distances in Figure 4. Important are the following: (1) the first arrival data up to 14° can be fitted by a curve with the apparent velocity of 8.30 ± 0.01 km/sec; (2) this first arrival branch does not extend beyond 14.5°; (3) a new later phase emerges near 13° about 4 sec after the first arrival, which becomes the initial phase beyond 14.5° without intersecting the forerunning phase; and (4) the data at larger distances (not shown in Figure 4) indicate that the prograde branch for the above new phase corresponds to the C branch as defined in Figure 2. By a trial-and-error method we searched

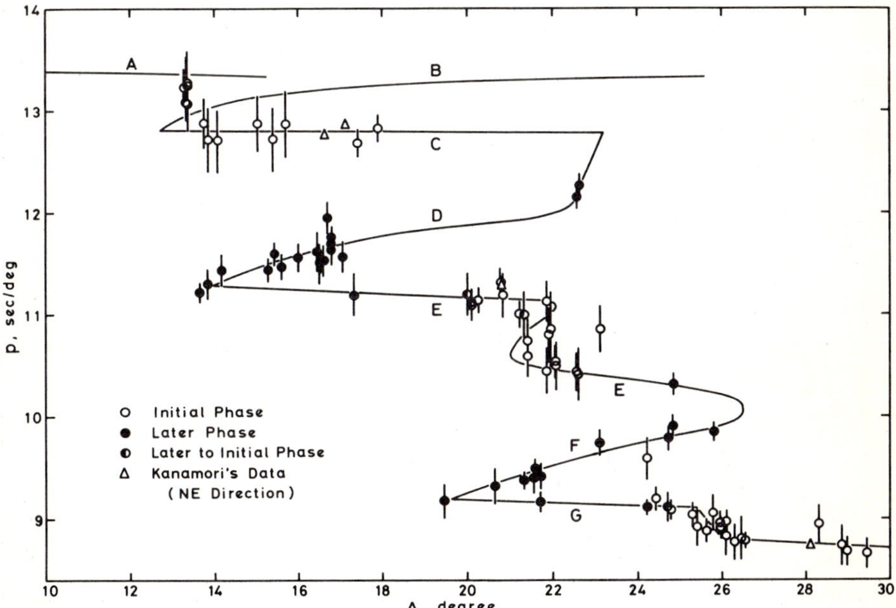

Fig. 2. Measured values of dT/dΔ. The C to G branches were determined by fitting a curve to the dT/dΔ curve. The A and B branches were determined from the travel time study.

for a velocity model to explain these observations and the upper 200 km of the earth shown in Figure 5 is the result. The model includes a high-velocity lid extending down to 85 km depth with a low-velocity zone (LVZ) below it. The base of the LVZ consists of a rapid velocity increase, starting at 165 km and terminating at 200 km. The travel times calculated for the model are compared with the observations (Figure 4), and the agreement is very good except for the constant base line shift required presumably by the hypocentral determination error.

In Figure 2 the $dT/d\Delta$ calculated for the above model defines the whole A and B branches and the reverse end of the C branch. Starting from this point the observed $dT/d\Delta$ curve was successively integrated by the Herglotz-Wiechert method to obtain a velocity-depth profile below 200 km. The result is shown

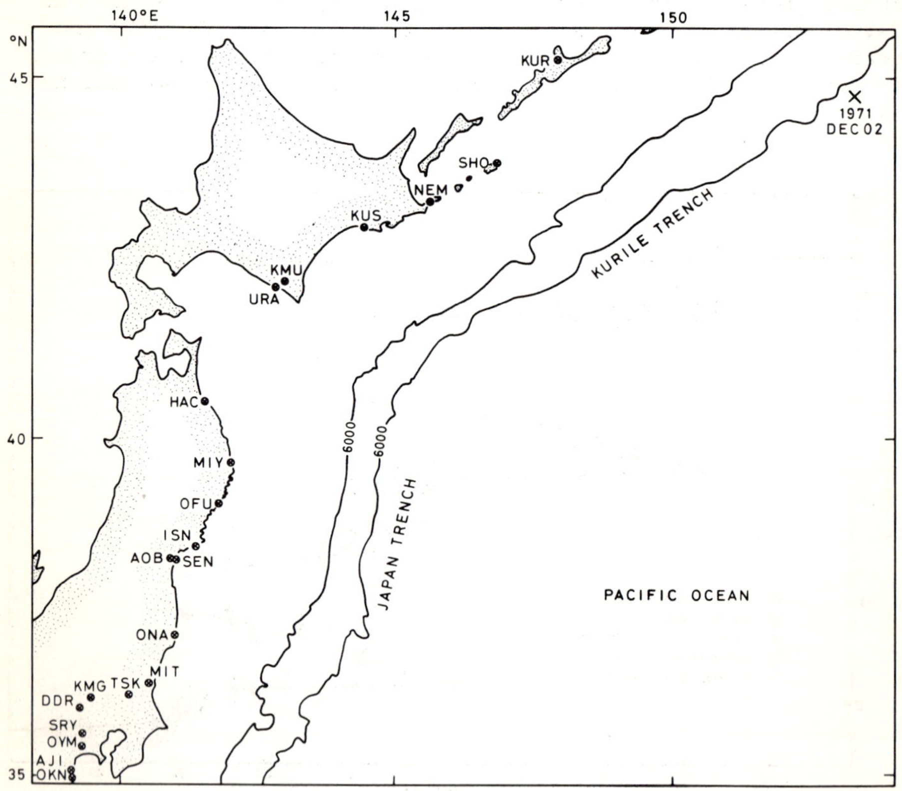

Fig. 3. Epicenter of the Kurile earthquake of December 02, 1971, and location of the observatories. Station code is given by three letters.

in Figure 5. Velocity does not increase with depth or the velocity gradient is very slightly negative just below the base of the LVZ, which is a consequence of the near flatness of the C branch. Some depth interval above the 400-km transition is characterized by a gradual increase in velocity gradient, which causes a focusing of energy at 22.5-23° as a large later arrival (see Figure 2). The 400-km transition consists of about a 6% increase in velocity over the thickness of less than 15 km. The near critical reflection from this sharp transition zone produces the D later-arrival extending back to 13.5-14°. At depths around 500 km the velocity increases rapidly, but not so rapidly as to produce a large triplication of the E branch near 21.5°. A much more rapid increase in velocity occurs at depths around 650 km. This major transition consists of a 6-7% increase in velocity and is sharp, especially near the base. Such sharpness near the bottom is related to the substantial backward extension of the F branch up to 19.5°. The model also possesses a minor transition zone at depths around 740 km, which is the result of the $dT/d\Delta$ behavior near 26°.

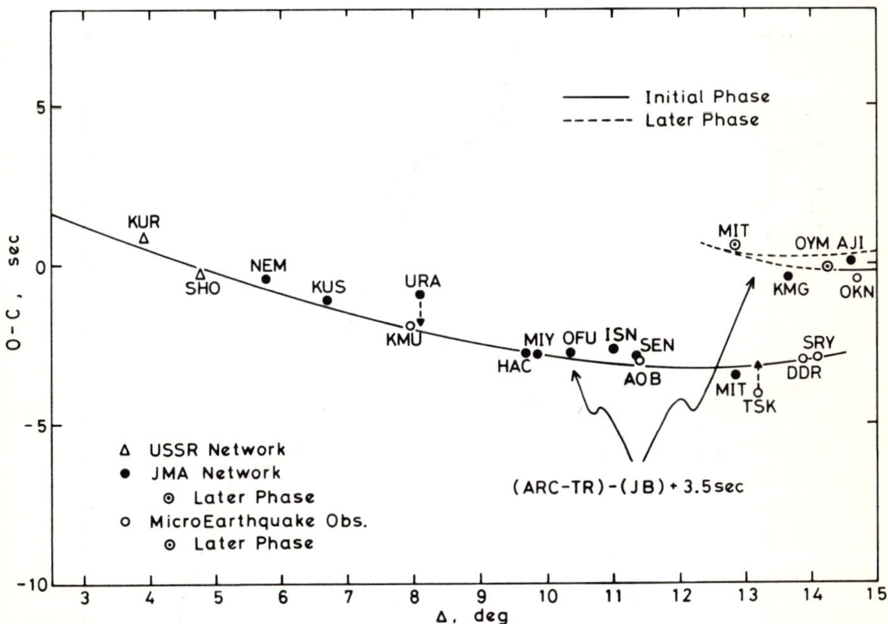

Fig. 4. Observed travel time residuals from Jeffreys-Bullen times compared with the ARC-TR model for which a base line shift of 3.5 sec is made. Station code is given by three letters.

The model labeled ARC-TR represents the upper mantle on the oceanic side of the Japan-Kurile arc. In the LVZ, the average velocity is 8.1 km/sec, which is 1.5% lower than the lithospheric velocity (8.23 km/sec) but is considerably higher than the LVZ velocities in other regions. Just below the base of the LVZ the velocity does not increase with depth, a feature found in

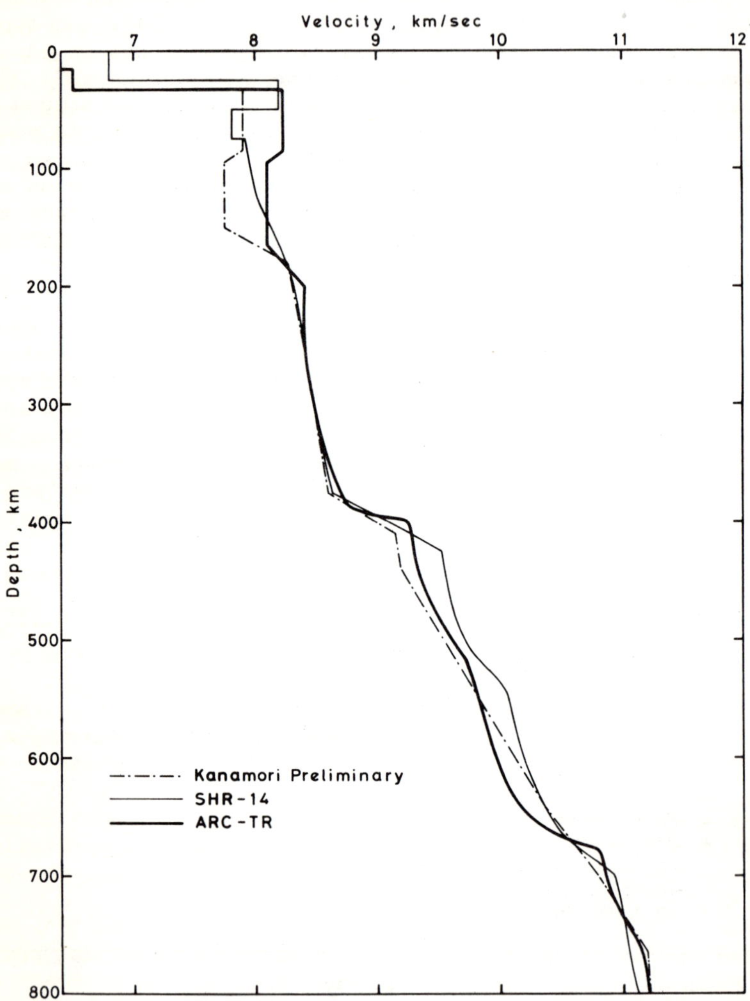

Fig. 5. Upper mantle velocity models: Preliminary [Kanamori, 1967] and SHR-4 [Helmberger and Engen, 1974] compared with ARC-TR.

some recent *P* velocity models [e.g., *Wiggins and Helmberger*, 1973]. The 400-km transition is perhaps too small and too sharp to be explained by the 100% olivine mantle. In general, an olivine-spinel transition involves more than a 12% increase in velocity [*Liebermann*, 1975], but here the velocity increase is only 6%. The solid solution of pyroxene in the garnet structure may play an important role for the smallness and the sharpness of the 400-km transition [*Ringwood*, 1975]. As far as this study is concerned, the 650-km transition is less sharp than the 400-km transition. If the 6-7% increase in velocity occurred near 650 km within the thickness of 10 km, for example, then the forward extension of the F branch must go far beyond 28°. There is no positive evidence for this, however. The above conclusion seems to be quite contradictory to the conclusion obtained from the *P'dP'* studies [*Whitcomb and Anderson*, 1970; *Adams*, 1971; *Nakanishi and Fukao*, in preparation].

II. *P'dP'* Observation

The *P'dP'* phase is a reflected wave of the main *P'P'* phase (*PKPPKP*) from the underside of the discontinuity at a depth of *d* km. Figure 6 shows the composite seismograms for the three Novaya Zemlya explosions of October 27, 1966, October 14, 1970, and October 27, 1973. The epicentral distances range from 55 to 85°. The most conspicuous waves seen on the right are the main *P'P'* waves, for which the theoretical travel time curves are drawn. Because of the structure of the earth's core, the travel time curves are divided into several branches. The *P'dP'* can be recognized 2.5 min before each *P'P'* arrival and the observed arrivals are best fitted by the travel time curves calculated for *d* = 660 km. The curve fitness indicates that the uncertainty in the depth determination is less than 15 km for a given velocity model, in this case, Herrin-Qamar's mantle-core model [*Herrin*, 1968 *Qamar*, 1973]. We also studied the seismograms for the Amchitka Island explosion of November 06, 1971, in the distance range 60-85° and for the deep Okhotsk Sea earthquake of August 30, 1970, (h = 645 km) in the distance range 45-85°. In both cases, the observed arrivals can be best explained again, by the travel times calculated for *d* = 660 km. The reflection points are spread over a wide area beneath Antarctica, the South Atlantic Ocean, and the South Indian Ocean.

For each set of the clear arrivals of *P'dP'* and *P'P'*, we measured the peak-to-peak amplitudes and the predominant periods. Figure 7 shows the amplitude ratio histogram of *P'dP'* to *P'P'*, which is not very different among the three events studied. The average amplitude ratio is 0.18. The histograms for the predominant periods of *P'dP'* and *P'P'* show both a relatively sharp maximum at about 1.2 sec. The results can be used to estimate

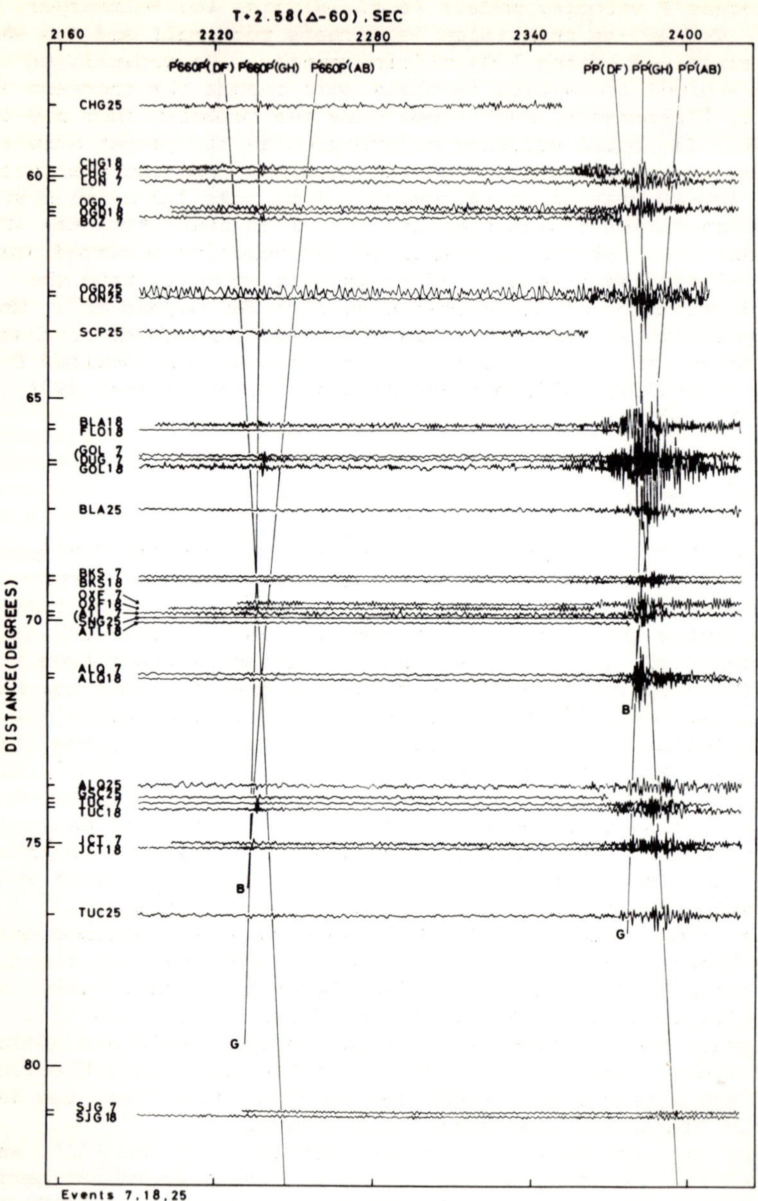

Fig. 6. Composite seismograms for the Novaya Zemlya explosions of October 27, 1966, October 14, 1970 and October 27, 1973. Both P'P' and P'dP' (d = 660 km) arrivals have three branches (DF, GH and AB, see Qamar [1973]) in this distance range.

the thickness and the elastic properties of the 660-km discontinuity. We calculated the synthetic seismograms of *P'dP'* and *P'P'* for various second-order discontinuities placed at a depth of 660 km with two models of the earth's crust. The matrix methods described by *Haskell* [1962] and *Richards* [1971] were used to generate complex reflection spectra, which were then Fourier synthesized. Figure 8 shows the relation between the transition thickness and the elastic property change across it, which is consistent with the observation. The result somewhat depends on the value of *Q*, quality factor, for attenuation in the upper mantle. But regardless of the *Q* value, the thickness of the transition zone is nevertheless smaller than 4 km [*Richards*, 1972]. If the thickness is more than 5 km, it is practically impossible to observe the *P'dP'* phase. In Figure 8, the relative change across the transition is assumed to be the same for the *P* and *S* wave velocities and the density. Note, however, that the number attached to the horizontal axis can be read approximately as $(\Delta Vp/Vp + \Delta\rho/\rho)/2$ because of the nature of the near vertical reflection of *P* waves. If the upper mantle *Q* is 300, for example, the sum of the *P* velocity jump and the density jump is then 18-22% for the transition thickness of 4 km and 10-14% for the

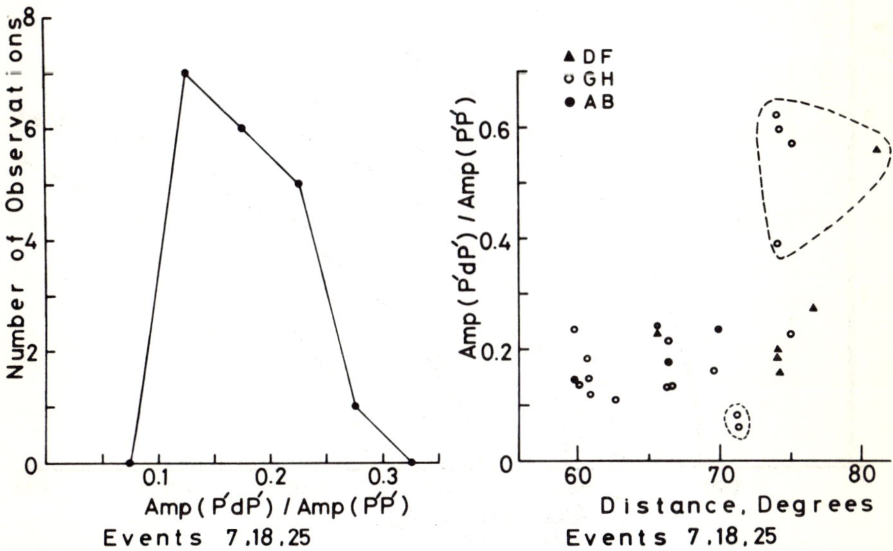

Fig. 7. Amplitude ratio histogram of P'dP' to P'P' for the three Novaya Zemlya explosions. The ratio is also plotted as a function of epicentral distance on the right. Dotted lines encircle the data for which two arrivals of different branches are overlapped. These data are excluded from the histogram.

thickness of 2 km. These values are required to explain the observed amplitude ratio of *P'dP'* to *P'P'*, which is not corrected for the possible threshold of detection capability of *P'dP'*. The asymmetry of the histogram (Figure 7) suggests that such a threshold may exist.

III. Conclusions

The ARC-TR model represents the upper mantle on the oceanic side of the Japan-Kurile arc. The observed *P'dP'* phase is a reflection from the discontinuity beneath the Antarctic Continent and the surrounding oceans. The possible regional difference in mantle structure makes it somewhat difficult to reconcile the results of our two studies. The following conclusions can be derived if the lateral heterogeneity did not extend to greater depths in the upper mantle. (1) The 400-km transition is very sharp but the thickness is not less than 5 km because of the

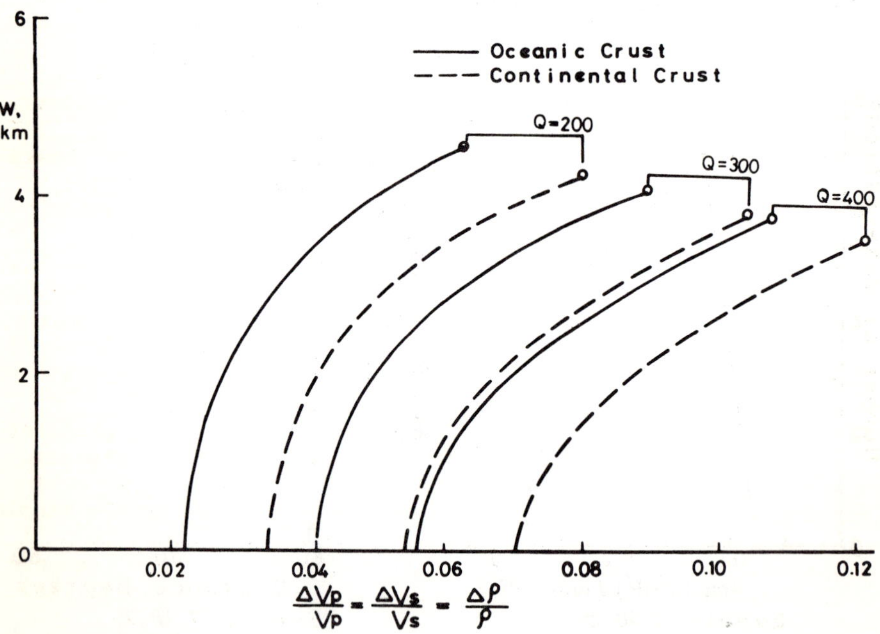

Fig. 8. Transition thickness and the elastic property change across it. The ranges consistent with observation, are shown by the solid and dotted curves. The range depends on the average Q in the upper mantle and the crustal structure relevant to the P'P' reflection.

failure to detect the clear vertical reflection from it. A thickness of 10 km may be appropriate. (2) In the ARC-TR model the 650-km transition is not as sharp as suggested from the *P'dP'* studies. The bottom part of this transition is, however, quite sharp and can be still sharper than indicated in the model. This part may be responsible for the clear arrival of the *P'dP'* phase.

Acknowledgment. The final version of this paper was written while I was a Geophysical Research Fellow at Seismological Laboratory, California Institute of Technology.

REFERENCES

Adams, R. D., Reflections from discontinuities beneath Antarctica, *Bull. Seism. Soc. Am.*, *61*,1441-1451, 1971.

Haskell, N. A., Crustal reflection of plane P and SV waves, *J. Geophys. Res.*, *67*, 4751-4767, 1962.

Helmberger, D. V., and G. R. Engen, Upper mantle shear structure, *J. Geophys. Res.*, *79*, 4017-4028, 1974.

Herrin, E., Introduction to the 1968 seismological tables for P phases, *Bull. Seism. Soc. Am.*, *58*, 1193-1195, 1968.

Kanamori, H., Upper mantle structure from apparent velocities of P waves recorded at Wakayama Micro-Earthquake Observatory, *Bull. Earthquake Res. Inst. Tokyo Univ.*, *45*, 657-678, 1967.

Kishimoto, Y., Seismometric investigation of the earth's interior, Part 3: on the structure of the earth's mantle (1), *Mem. Coll. Sci. Kyoto Univ., Ser. A*, *28*, 117-142, 1956.

Liebermann, R. C., Elasticity of olivine (α), beta (β), and spinel (γ) polymorphs of germanates and silicates, *Geophys. J. Roy. Astro. Soc.*, *42*, 899-929, 1975.

Qamar, A., Revised velocities in the earth's core, *Bull. Seism. Soc. Am.*, *63*, 1073-1105, 1973.

Richards, P. G., Elastic wave solutions in stratified media, *Geophysics*, *36*, 798-809, 1971.

Richards, P. G., Seismic waves reflected from velocity gradient anomalies within the earth's upper mantle, *Z. Geophysik*, *38*, 517-527, 1972.

Ringwood, A. E., Composition and petrology of the earth's mantle, McGraw-Hill, New York, 1975.

Whitcomb, J. H., and D. L. Anderson, Reflection of P'P' seismic waves from discontinuities in the mantle, *J. Geophys. Res.*, *75*, 5713-5728, 1970.

Wiggins, R. A., and D. V. Helmberger, Upper mantle structure of the western United States, *J. Geophys. Res.*, *78*, 1870-1880, 1973.

HYDROXYL-CLINOHUMITE AND HYDROXYL-CHONDRODITE: POSSIBLE H_2O-BEARING MINERALS IN THE UPPER MANTLE

S. AKIMOTO
Institute for Solid State Physics, University of Tokyo
Roppongi, Minato-ku, Tokyo 106, Japan

K. YAMAMOTO
Water Research Institute, Nagoya University
Chikusa-ku, Nagoya 464, Japan

K. AOKI
Institute of Mineralogy, Petrology and Economic Geology
Tohoku University
Aoba, Sendai 980, Japan

Abstract

The system Mg_2SiO_4-MgO-H_2O was investigated at pressures between 29 and 77 kbar and at temperatures between 470 and 1225°C. Hydroxyl-clinohumite, $Mg(OH)_2 \cdot 4Mg_2SiO_4$, and hydroxyl-chondrodite, $Mg(OH)_2 \cdot 2Mg_2SiO_4$, are found to be stable for nearly the same pressure-temperature conditions, ranging from approximately 700 to 1100°C, and from 29 to 77 kbar. Other members of humite group minerals such as hydroxyl-norbergite, $Mg(OH)_2 \cdot Mg_2SiO_4$, and hydroxyl-humite, $Mg(OH)_2 \cdot 3Mg_2SiO_4$, have no stability field at pressures treated in this work. Phase A, $3Mg(OH)_2 \cdot 2Mg_2SiO_4$, is found to be stable at pressures above approximately 50 kbar and at temperatures relatively lower than the stability fields for hydroxyl-clinohumite and hydroxyl-chondrodite. Present laboratory data suggest that clinohumite, chondrodite, and phase A are possible H_2O-bearing minerals in the upper mantle, especially in the subduction zone and in the overlying peridotite wedge.

I. INTRODUCTION

Humite group minerals consist of four minerals, chemical formulae of which are generally expressed by $Mg(OH,F)_2 \cdot nMg_2SiO_4$ where n = 1,2,3, and 4. The mineral with n = 1 is norbergite, n = 2 is chondrodite, n = 3 is humite, and n = 4 is clinohumite. *McGetchin et al.*[1970] suggested that humite group minerals may be an important mineralogical site for bound H_2O in the upper mantle and may be more abundant than commonly recognized because of their similarity to olivine in structural, optical, and chemical properties [*Ribbe et al.*, 1968].

Ringwood and Major [1967] investigated the system Mg_2SiO_4-MgO-H_2O at pressures above 100 kbar at temperatures between 600 and 1100°C and discovered three new hydrated magnesium silicate phases denoted A, B, and C. In a previous paper on the system MgO-SiO_2-H_2O, *Yamamoto and Akimoto* [1974] determined the chemical formula of phase A as $3Mg(OH)_2 \cdot 2Mg_2SiO_4$ and further found a new hydrated magnesium silicate phase, phase D, whose cell parameters are close to those of natural chondrodite. Later, they concluded that phase D is hydroxyl-chondrodite [*Yamamoto and Akimoto*, 1977].

In view of the possible importance of these hydrated magnesium silicates in the hydration and dehydration process in the upper mantle, further detailed investigation on the high pressure-temperature stability relations of the hydroxyl-humite group minerals was made in this work.

A careful examination of Buell Park, Arizona, kimberlite was also made with the objective of finding these dense hydrated magnesium silicates in natural rocks derived from the upper mantle. Occurrence of titanochondrodite in the kimberlite is briefly described.

II. EXPERIMENTAL METHODS

Mixtures of guaranteed grades of the reagents $Mg(OH)_2$ and silicic acid ($SiO_2 \cdot yH_2O$) were used as starting materials. The compositions of mixtures, $xMg(OH)_2 + SiO_2 \cdot yH_2O = xMgO + SiO_2 + (x + y)H_2O$, were determined from mixing ratios of $Mg(OH)_2$ to $SiO_2 \cdot yH_2O$, where x denotes the mol ratio of MgO to SiO_2. Ten mixtures (x = 2, 17/8, 9/4, 7/3, 5/2, 11/4, 3, 7/2, 4, and 5) were prepared for investigating the stability of humite group minerals. The compositions of these starting mixtures are represented on the $MgSiO_3$-MgO-H_2O triangular diagram (Figure 1).

To prevent water loss during the run, powder samples of the starting materials were placed in gold capsules and sealed by dc carbon arc welding. The capsules were subjected to high pressures between 29 and 77 kbar in a tetrahedral-anvil-type apparatus. The pressure values were calibrated at room temperature on the basis of the pressure-fixed points recommended at the U.S. National Bureau of Standards Symposium, 1968, using resistance transitions of Bi I-II (25.5 kbar), Ba I-II (55 kbar), and Bi III-V (77 kbar). Reaction temperatures were between 470 and 1225°C. Temperature was measured with Pt/Pt-13%Rh thermocouple in contact with the gold capsules. No correction was made for the effect of pressure on the emf of the thermocouple. Lengths of run time were varied from 0.5 to 12 hours, depending upon the reaction temperature. 122 runs in total were carried out in the present study.

Samples were quenched in the conventional method under work-

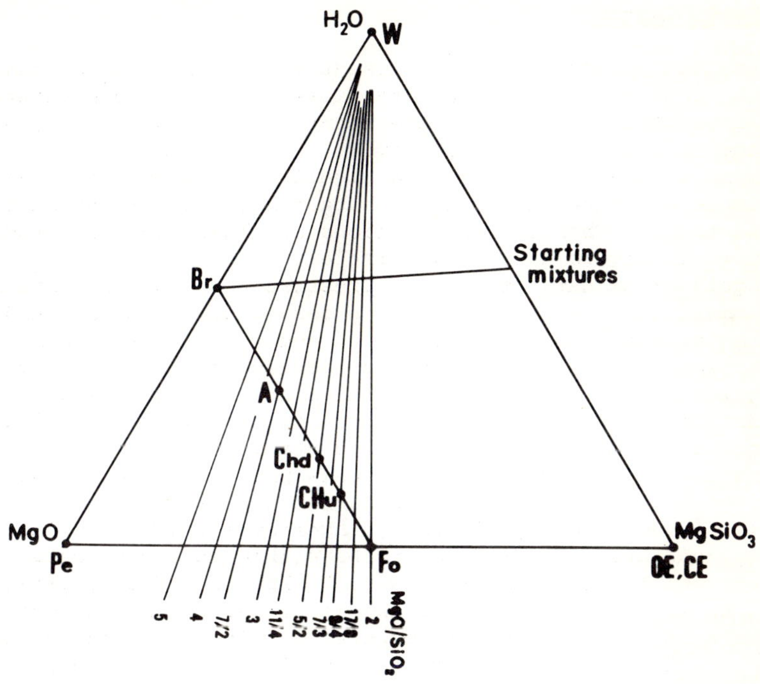

Fig. 1. The system $MgSiO_3$-MgO-H_2O. Compositions of all the starting mixtures are represented by cross points between the line labeled "starting mixtures" and the lines with constant MgO/SiO_2 mol ratios. The compositions of phases that appeared in the present experiments are plotted by solid circles. Abbreviations: Pe, periclase; Br, brucite; A, phase A; Chd, hydroxyl-chondrodite; CHu, hydroxyl-clinohumite; Fo, forsterite; OE, orthoenstatite; CE, clinoenstatite; W, water.

ing pressure. The quenched products were examined by powder X-ray diffraction technique. Structure determination of the samples was carried out with the aid of a Weissenberg camera and an automatic four-circle diffractometer, when single crystals were available.

More detailed description on the experimental procedure is given in a separate publication [*Yamamoto and Akimoto*, 1977].

III. RESULTS AND DISCUSSION

A. Phase Diagram

Reaction products of the present experiments are brucite, phase A, chondrodite, clinohumite, forsterite, orthoenstatite, and clinoenstatite. These were identified using an X-ray diffractometer and a Weissenberg camera. Effusion of water from the capsule when it was cut or a significant degree of wetness of the sample was taken as indication that H_2O had been produced by the reaction. Chemical compositions of these reaction products are shown in Figure 1 by solid circles.

Equilibrium data at various pressure-temperature conditions are summarized in Figure 2, where a few invariant points and their associated lines of univariant equilibrium are shown. Composition-paragenesis diagrams projected on the $MgSiO_3$-MgO-H_2O triangular system are also illustrated in the figure around significant invariant points. Data points are omitted in Figure 2 to avoid the complexity. Detailed tables showing pressure-temperature-time conditions and phases present in individual runs are given in a separate paper [*Yamamoto and Akimoto*, 1977].

Among humite group minerals, hydroxyl-clinohumite and hydroxyl-chondrodite are found to be stable in a wide range of pressure and temperature from 29 to 77 kbar and from approximately 700 to 1100°C. This work is the first report on the successful synthesis of the hydroxyl endmembers of clinohumite and chondrodite. Cell parameters and refractive indices of hydroxyl-clinohumite and hydroxyl-chondrodite are given in Table 1. Hydroxyl-norbergite[$Mg(OH)_2 \cdot Mg_2SiO_4$] and hydroxyl-humite[$Mg(OH)_2 \cdot 3Mg_2SiO_4$] were not obtained in the present investigation. It is likely that these phases have no stability field at pressures ranging from 29 to 77 kbar. The synthesis of the higher members of the humite group [$Mg(OH)_2 \cdot nMg_2SiO_4$ with $n \geq 5$], whose existence has been suggested by *Bragg and Claringbull* [1965] was also unsuccessful.

Hydroxyl-clinohumite is formed from brucite + forsterite and phase A + forsterite by solid-solid reaction, and from forsterite + magnesium-rich water. Hydroxyl-chondrodite is formed from brucite + hydroxyl-clinohumite and phase A + hydroxyl-clinohumite by solid-solid reaction, from the decomposition of phase A, and from hydroxyl-clinohumite + magnesium-rich water. The upper stability limit of hydroxyl-clinohumite and hydroxyl-chondrodite depends upon the bulk composition of the reaction system. In Figure 2 the upper stability limit for $2.5MgO + SiO_2 + 2.67H_2O$ ($x = 5/2$, hydroxyl-chondrodite + exess water) is shown.

B. Crystal Structure of Phase A

In Figure 2, phase A is found to be stable above approximate-

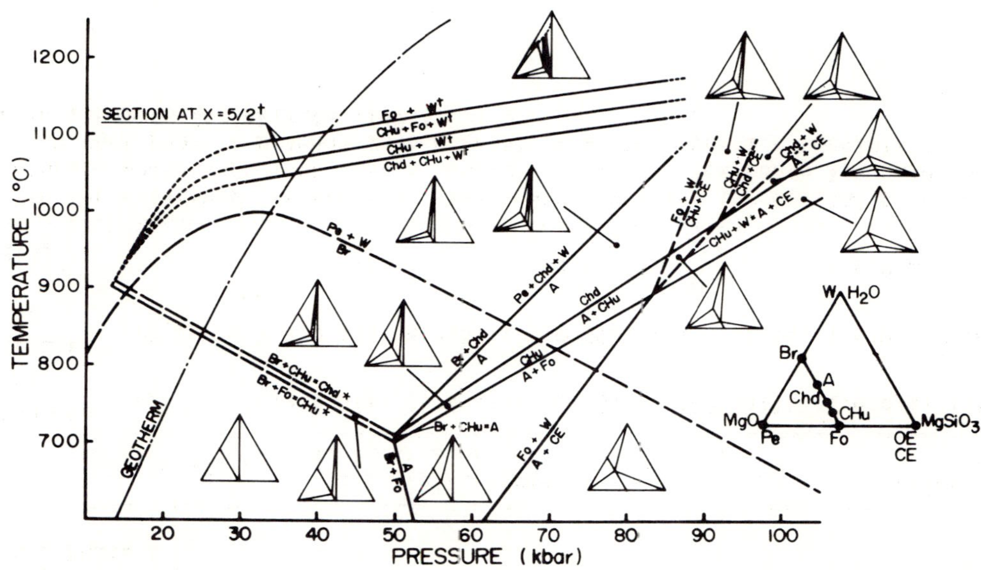

Fig. 2. High-pressure and high-temperature synthesis diagram in the system $MgSiO_3$-MgO-H_2O. Abbreviations are the same as in Figure 1.
† The upper limit of the synthesis fields of hydroxyl-chondrodite and hydroxyl-clinohumite is represented for the composition $2/5Mg(OH)_2 + SiO_2 \cdot 0.166H_2O$.

TABLE 1. Crystallographic and Optical Data for Hydroxyl-clinohumite, Hydroxyl-chondrodite and Phase A

Phase	Space Group	Cell Parameters				Refractive Indices		
		a, Å	*b*, Å	*c*, Å	β, °	α	β	γ
Hydroxyl-clinohumite $Mg(OH)_2 \cdot 4Mg_2SiO_4$	$P2_1/c$	13.695(1)	4.7474(1)	10.284(1)	100.64(1)	1.638(2)	1.641(2)	1.669(1)
Hydroxyl-chondrodite $Mg(OH)_2 \cdot 2Mg_2SiO_4$	$P2_1/c$	7.914(2)	4.752(1)	10.350(2)	108.71(5)	1.630(2)	1.642(2)	1.658(1)
Phase A[a] $3Mg(OH)_2 \cdot 2Mg_2SiO_4$	$P6_3$	7.866(2)		9.600(3)		1.638(1)	1.640(2)	1.649(1)

a. There is an inconsistency between the X-ray measurement and the optical observation: the Weissenberg photograph indicates a hexagonal symmetry for phase A, while conoscopic figures show that phase A belongs to a biaxial positive crystal. Possible explanation is given in *Yamamoto and Akimoto* [1974].

ly 50 kbar. Its stability field is generally in the higher pressure and lower temperature side than that of clinohumite and chondrodite. Cell parameters and refractive indices of phase A are also listed in Table 1.

As shown in Figure 1, chemical composition of phase A is on the join between $Mg(OH)_2$ and Mg_2SiO_4 in common with humite group minerals. The structure analysis now going on with the aid of the automatic four-circle diffractometer, however, suggests that phase A should not be termed as a member of humite group minerals [*Morimoto and Yamamoto*, in preparation]. Particularities of the structure of phase A are seen in the following points.

1. In the structure of phase A no shared edge is found between Si-tetrahedra and Mg-octahedra. This is a striking contrast to the structure of humite group minerals in which Si-tetrahedra always share three edges with Mg-octahedra.
2. The structure of humite group minerals, like olivine, is based on a slightly distorted hexagonal close-packed array of anions, whose stacking is AB····· sequence. On the other hand, in the structure of phase A, stacking sequence of the anion-array is found to be ABAC····· along the c-axis.
3. There is no brucite layer in the structure of phase A. This is a great difference from the structure of humite group minerals, which consist of alternate slabs of the forsterite structure and of brucite.

C. Geophysical Implications

Recently, petrologic models on the dehydration mechanism of oceanic lithosphere in subduction zones and the formation of orogenic type magmatism caused by migration of released water, have been proposed by a number of investigators. *Ringwood* [1975, 1976] suggested that in the descending lithosphere slabs and overlying peridotite wedge, serpentine, talc and brucite transform over a wide range of pressure and temperature above approximately 50 kbar and 600°C into the dense hydrated magnesium silicates mentioned in Section IIIA. Figure 2 demonstrates that phase A can coexist with forsterite and enstatite up to 800°C at 77 kbar. This strongly supports Ringwood's suggestion that phase A is a possible H_2O-bearing mineral phase in the subduction zone in the upper mantle.

It is noted that both clinohumite and chondrodite are stable even along the geotherm at pressures between 25 and 40 kbar (corresponding to ca. 70 and 120 km at depths). The geotherm in Figure 2 is reproduced from a paper by *Herrin* [1972]. Although these humite group minerals are stable only in rocks much richer in MgO than the accepted mantle rocks, they are still of considerable interest as possible sites for H_2O in the upper mantle based on the following lines of reasoning.

The phase diagram shown in Figure 2 suggests the reaction

forsterite + water ⇄ clinohumite + enstatite in pressure-temperature conditions ranging from 83 to 90 kbar and 900 to 1100°C (thick, dashed line in Figure 2). It is very likely that this type of reaction would occur actually in the upper mantle. Above the stability fields of serpentine, talc, or brucite, water liberated gradually from the cold descending slab would react with olivine in anhydrous peridotite to produce clinohumite.

Furthermore, *Nakamura and Kushiro* [1974] suggested in their recent experimental investigation on the Mg_2SiO_4-SiO_2-H_2O system that water passing upward through the upper mantle peridotite will carry large amounts of SiO_2 in solution, leaving the mantle as a whole relatively depleted in SiO_2. If this is the case, as *Ringwood* [1975, 1976] pointed out, the possible existence of clinohumite and chondrodite in the subduction zone and in the overlying peridotite wedge would be enhanced substantially.

D. Occurrence of Titanoclinohumite and Titanochondrodite in Kimberlites

Clinohumite with high content of titanium has already been found in kimberlite from three localities: Buell Park, Arizona [*Allen and Balk*, 1954], Ruslovaya pipe in Yakutia, Siberia [*Voskresenskaya et al.*, 1965; *Shchelchkova and Brovkin*, 1969] and Moses Rock dike, Utah [*McGetchin and Silver*, 1970; *McGetchin et al.*, 1970]. The last authors concluded from their petrological observation that titanoclinohumite in the Moses Rock kimberlite had equilibrated in the mantle at depths ranging from approximately 50 to 150 km at moderately high temperatures generally less than 1000°C before transport to the surface during the emplacement of the kimberlite. Present laboratory data are consistent with their petrological inference.

Recently, one of the present authors (K. Aoki) also made a careful examination of Buell Park, Arizona, Kimberlite diatreme and first found titanochondrodite in addition to titanoclinohumite. On the basis of kimberlite mineralogy and petrography, *Aoki et al*, [1976] suggested that these titanochondrodite and titanoclinohumite are crystallized from kimberlite magma at a depth of ca. 100 km and at 1000°C. Present experimental data on the Mg_2SiO_4-MgO-H_2O system also support their suggestion.

In Table 2 the results of microprobe analyses on these natural titanochondrodite and titanoclinohumite are summarized. More detailed description of the petrographic, chemical and crystallographic properties is given in a previous publication [*Aoki et al.*, 1976].

Acknowledgments. The authors thank N. Morimoto, Osaka University for his kind help in the structure analysis of phase A crystals synthesized in this work. They are heartily grateful

TABLE 2. Microprobe Analyses of Titanochondrodites and Titanoclinohumites from Buell Park Kimberlite, Arizona

	Titanochondrodite		Titanoclinohumite	
	Average	Range	Average	Range
SiO_2	32.32	(31.87 - 32.86)	35.61	(34.44 - 36.19)
TiO_2	9.30	(7.32 - 10.19)	4.83	(3.60 - 5.28)
Al_2O_3	0.01	(0.00 - 0.03)	0.01	(0.00 - 0.01)
Cr_2O_3	0.08	(0.03 - 0.12)	0.03	(0.00 - 0.12)
Fe_2O_3				
FeO	10.49	(9.97 - 11.04)	10.43	(8.40 - 12.05)
MnO	0.19	(0.15 - 0.24)	0.18	(0.11 - 0.23)
MgO	43.61	(42.40 - 44.64)	46.04	(44.60 - 48.01)
NiO	0.17	(0.15 - 0.23)	0.19	(0.09 - 0.28)
CaO	0.00	(0.00 - 0.01)	0.00	(0.00 - 0.01)
H_2O+	2.83[a]		1.92[a]	
Total	99.00		99.24	
Si	2.000		4.000	
Ti	0.433		0.408	
Al	0.000		0.001	
Cr	0.004		0.002	
Fe	0.543		0.980	
Mn	0.010		0.017	
Mg	4.020		7.704	
Ni	0.009		0.017	
Ca	0.000		0.000	
Mg/(Mg+Fe)	0.879	(0.873 - 0.887)	0.885	(0.866 - 0.906)
M/Si[b]	2.51	(2.47 - 2.54)	2.28	(2.22 - 2.35)
No. of grains analysed	12		38	

a. Calculated from theoretical formula.
b. M = Mg + Fe + Ni + Mn + Ca + Cr + Al + Ti.

to K. Fujino and M. Akaogi for their laborious work in finding titanochondrodite from Buell Park kimberlite. A part of expense of this study was defrayed by a grant from the Geodynamics Project of Japanese Government Expenditure.

REFERENCES

Allen, J. E., and R. Balk, Mineral resources of Fort Defiance and Tohachi quadrangles, Arizona and New Mexico, *N. Mex. Bur. Mines Miner. Resour. Bull.*, *36*, 192p., 1954.

Aoki, K., K. Fujino, and M. Akaogi, Titanochondrodite and titano-clinohumite derived from the upper mantle in the Buell Park kimberlite, Arizona, USA, *Contrib. Mineral. Petrol.*, *56*, 243-253, 1976.

Bragg, L., and G. F. Claringbull, *Crystal Structures of Minerals*, 409p., G. Bell and Sons, London, 1965.

Herrin, E., A comparative study of upper mantle models: Canadian shield and basin and range provinces, in *The Nature of the Solid Earth*, edited by E. C. Robertson, pp. 216-231, McGraw Hill, New York, 1972.

McGetchin, T. R., and L. T. Silver, Compositional relations in minerals from kimberlite and related rocks in the Moses Rock dike, San Juan County, Utah, *Am. Mineral.*, *55*, 1738-1771, 1970.

McGetchin, T. R., L. T. Silver, and A. A. Chodos, Titanoclino-humite: a possible mineralogical site for water in upper mantle, *J. Geophys. Res.*, *75*, 255-259, 1970.

Nakamura, Y., and I. Kushiro, Composition of the gas phase in Mg_2SiO_4-SiO_2-H_2O at 15 kbar, *Carnegie Inst. Wash. Yearb.*, *73*, 255-258, 1974.

Ribbe, P. H., G. V. Gibbs, and N. V. Jones, Cation and anion substitutions in the humite minerals, *Mineral. Mag.*, *36*, 966-975, 1968.

Ringwood, A. E., *Composition and Petrology of the Earth's Mantle*, 618p., McGraw Hill, New York, 1975.

Ringwood, A. E., Phase transformations in descending plates and implications for mantle dynamics, *Tectonophys.*, *32*, 129-143, 1976.

Ringwood, A. E., and A. Major, High pressure reconnaissance investigations in the system Mg_2SiO_4-MgO-H_2O, *Earth Planet. Sci. Lett.*, *2*, 130-133, 1967.

Shchelchkova, L. G., and A. A. Brovkin, Titanolivine from the Siberian kimberlites, *Zap. Vses. Miner. Obshch.*, *98*, 246-247, 1969.

Voskresenskaya, V. B., V. V. Koval'skii, K. N. Nikishov, and Z. F. Parinova, Discovery of titanolivine in Siberian kimberlites, *Zap. Vses. Miner. Obshch.*, *94*, 500-503, 1965.

Yamamoto, K., and S. Akimoto, High pressure and high temperature investigations in the system MgO-SiO_2-H_2O, *J. Solid State Chem.*, *9*, 187-195, 1974.

Yamamoto, K., and S. Akimoto, The system MgO-SiO_2-H_2O at high pressures and temperatures: stability field for hydroxyl-chondrodite, hydroxyl-clinohumite and 10 Å-phase, *Amer. J. Sci.*, in press, 1977.

A PRESSURE-INDUCED PHASE TRANSITION IN As_2Se_3 GLASS

R. OTA[1] and O. L. ANDERSON
Institute of Geophysics and Planetary Physics
University of California
Los Angeles, California 90024

Abstract

Measurements of the elastic modulus of arsenic triselenide glass under pressure up to 20 kbar show that this glass undergoes a phase transition at 10 kbar. Both phases are apparently amorphous, and the reaction is reversible. The phase change is defined in terms of the bulk modulus and its isothermal pressure derivative as a function of pressure. The change in phase is described by a structural factor in the equation of state, and the variation of this factor represents the mechanical property of glass across the transition pressure.

I. EXPERIMENTAL METHODS

Sound velocities were measured using an ultrasonic interferometry technique (Fig. 1). A glass sample, 5 mm thick, was compressed in a Teflon container filled with a pressure liquid in a conventional piston-cylinder pressure apparatus [*Ota and Anderson,* in preparation].

The true pressure was determined from the ram pressure from a pressure calibration using a manganin coil and the transition of Bismuth (1) to Bismuth (2) as the pressure standard (Fig. 2). At 20 kbar of piston pressure, pressure loss was 1.7 kbar due to the friction between the walls of the pressure vessel and the Teflon container.

From the pulse repetition frequency for longitudinal and shear waves, we determined the sound velocities as a function of pressure, and we observed curvature near 10 kbar for both longitudinal and shear wave velocities (Fig. 3).

The density under pressure was calculated by the well-known *Cook's* [1957] method

---

[1]Now at Department of Industrial Chemistry, Faculty of Engineering, Kyoto University, Yosida, Kyoto, Japan.

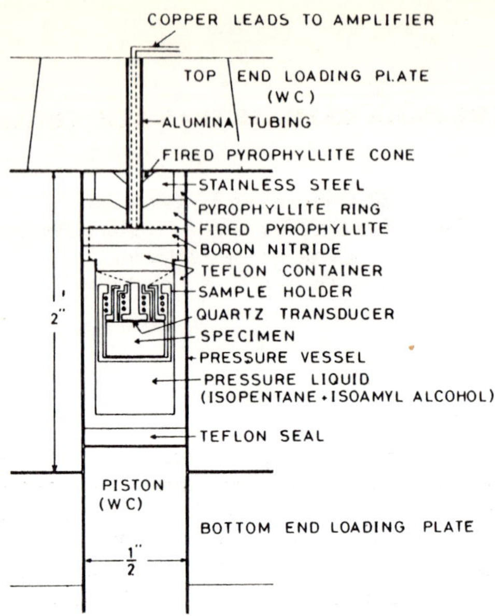

Fig. 1. A schematic drawing of the pressure vessel and the accessories for sound velocity measurement.

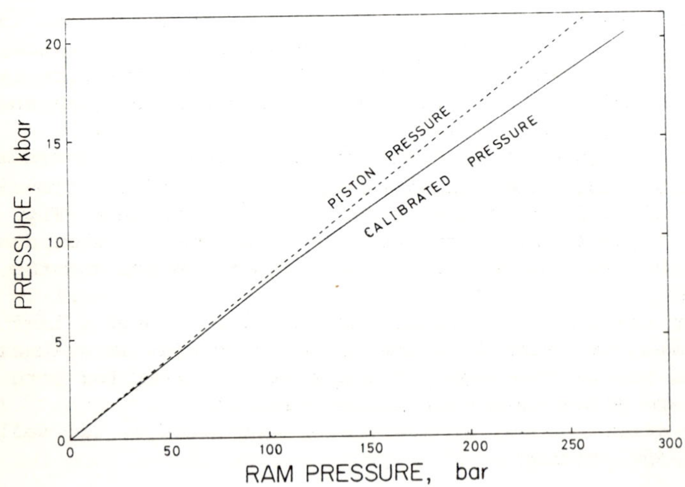

Fig. 2. The calibration curve for the true pressure versus ram pressure determined by the manganin coil.

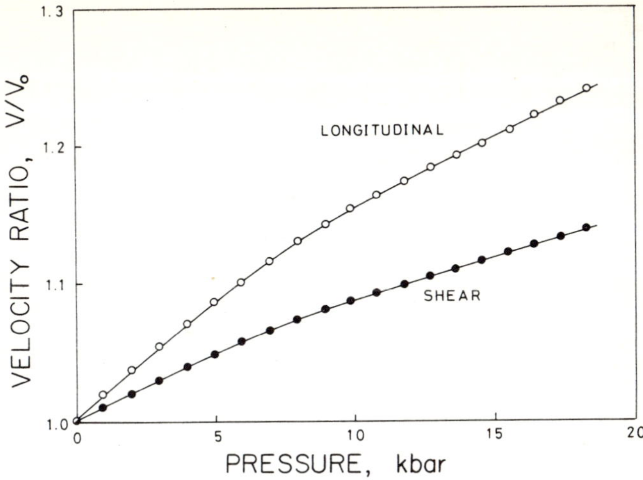

Fig. 3. Pressure dependence of the sound velocities of longitudinal and shear waves in As_2Se_3 glass at 25°C.

$$\left(\frac{\rho}{\rho_o}\right)^{1/3} = 1 + \frac{1 + 3\alpha\gamma T}{12\, L_o^2\, \rho_o^2} \int_o^P \frac{dP}{f_\ell^2 - \frac{4}{3} f_t^2} \qquad (1)$$

In Figure 4 the solid line is the experimental measurement of density by equation (1). Line A is the Murnaghan equation, and line B is the lattice theory equation. The experimental values begin to deviate from A and B at about 10 kbar.

The bulk modulus calculated from the sound velocities and the density increases almost linearly up to 10 kbar, at which pressure the slope apparently becomes less positive (Fig. 5).

The change of the slope of the bulk modulus-pressure data is most clearly demonstrated in Figure 6, in which we plot the pressure derivative K' versus pressure. Here again, A is the constant, found to be 7.86 at small pressures, and B is the bulk modulus from the data according to the lattice theory equation. According to either the lattice theory or the finite strain theory, the pressure derivative K' decreases by 1.3 to 1.7 from zero to infinite pressure. In the pressure range of 0-20 kbar, the theoretical drop of K' must be less than 0.6. On the other hand, we find that K' for As_2Se_3 glass drops as much as 2.2 in this pressure range, so the discrepancy 1.6 is real and not accountable by the behavior of a single homogeneous phase. Thus, we can assume that this sharp drop of K' is due to some kind of

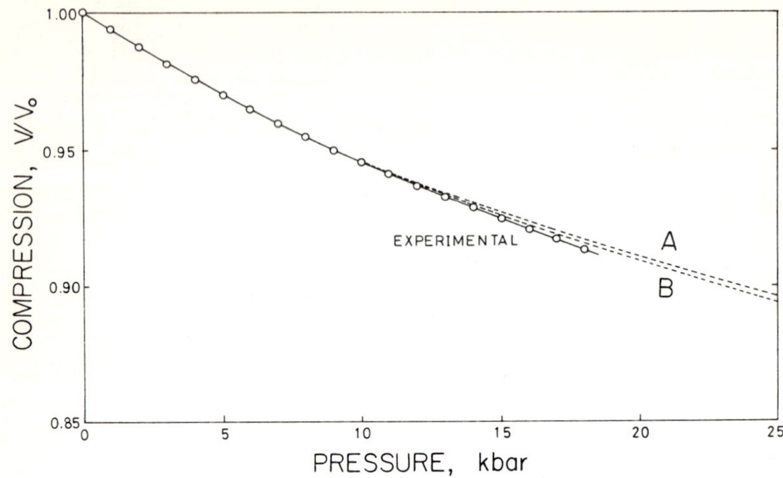

Fig. 4. The compression of As_2Se_3 glass. Solid line is experimental. Dashed line A is the Murnaghan equation [1949] and dashed line B is the lattice theory equation for central forces with a boundary condition K_T = 140.38 kbar and K_T' = 7.86 [Anderson, 1970].

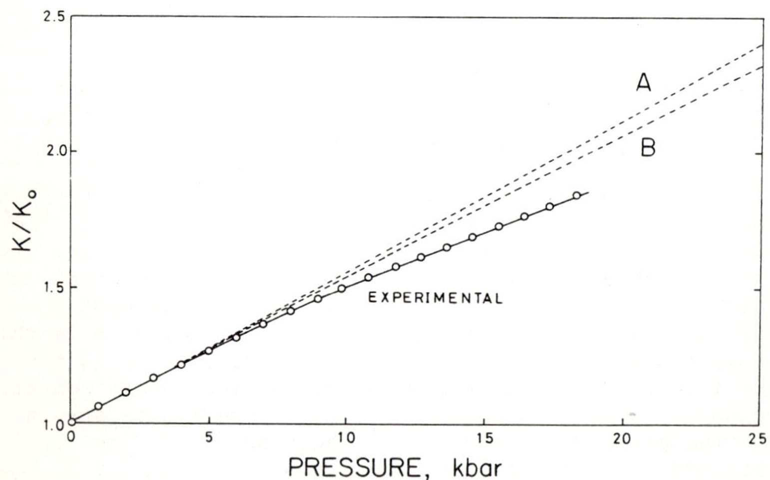

Fig. 5. The isothermal bulk modulus of As_2Se_3 glass as a function of pressure at 25°C. Dashed lines A and B are the theoretical values calculated using a boundary condition K_T = 140.38 kbar and K_T' = 7.86.

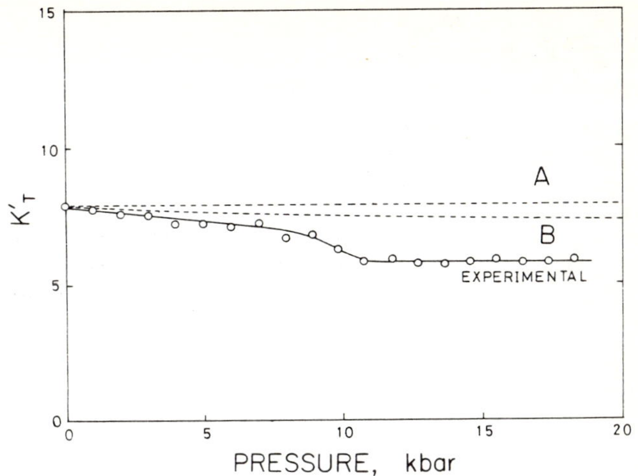

Fig. 6. The pressure derivative of the isothermal bulk modulus of As_2Se_3 glass as a function of pressure. Dashed lines A and B are the computed values using a boundary condition K_T = 140.38 kbar and K_T' = 7.86.

structural transformation in the glass from a low-pressure to a high-pressure phase.

II. PROPERTIES OF THE HIGH-PRESSURE PHASE

Using the law of corresponding states, all the mechanical properties of the high-pressure phase can be computed. Taking the low-pressure state as 1 and the high-pressure state as 2, at the transition pressure, the bulk moduli are equal, so that using the *Murnaghan* [1949] equation of state

$$K_{1o} + K_{1o}' P^* = K_{2o} + K_{2o}' P^* \tag{2}$$

where P^* is the transition pressure. Solving for K_{2o} we have

$$K_{2o} = 140.38 + 1.86 P^* \tag{3}$$

Taking P^* as 10 kbar as shown in Figure 6, we then have K_{2o} = 159 kbar. Applying the law of corresponding states [*Anderson*, 1966] to determine density, we have

$$\left(\frac{K_{2o}}{K_{1o}}\right) = \left(\frac{\rho_{2o}}{\rho_{1o}}\right)^X \tag{4}$$

where $X = 4$ or 5. The density of state 2 is 4.72 to 4.75, and the density of the crystalline As_2Se_3 is 4.75. This indicates that the high-pressure phase is very close, if not identical, to the crystalline phase (Fig. 7).

The decrease of K' from phase 1 to phase 2 is consistent with the change of density. If we take an empirical law [*Anderson et al.*, 1968] K' is inversely proportional to the density to some factor Y

$$\left(\frac{K'_{2o}}{K'_{1o}}\right) = \left(\frac{\rho_{1o}}{\rho_{2o}}\right)^Y \tag{5}$$

Using the density value of the high-pressure phase, we find $Y = 9$. This is shown in Figure 8 where all the data except As_2Se_3 were taken from *Anderson et al.* [1968]. The value of K' is shown to decrease with density in a sequence of oxides where the mean atomic weight stays constant. In a phase change, the mass does not change. As_2Se_3 glass shows a rapid decrease of K' with density similar to the behavior for oxides compounds. From the law of corresponding states, it is clear that in any high-pressure phase change, the zero pressure bulk modulus and the zero pressure density will increase while passing to the high-pressure phase, but, at the same time, the value of K' will decrease.

III. STRUCTURAL TRANSFORMATION IN GLASS

Assuming the lattice potential function of an atom pair in a glass is approximately the same before and after the transition, and the total energy of the system depends only on the sum of all energy pairs, then the volume, V, is connected by a structural factor, C, to the average atomic separation, r. (In a NaCl structure $C = 2$, and is larger for the CsCl structure and smaller for the ZnS structure.)

$$V = C\,N\,r^3 \tag{6}$$

Assuming C_o is the constant for the crystalline phase and C is a changing variable appropriate to the glassy phase, the volume ratio is related to C as follows:

$$C/C_o = \frac{V/V_o}{(V/V_o)_C} \tag{7}$$

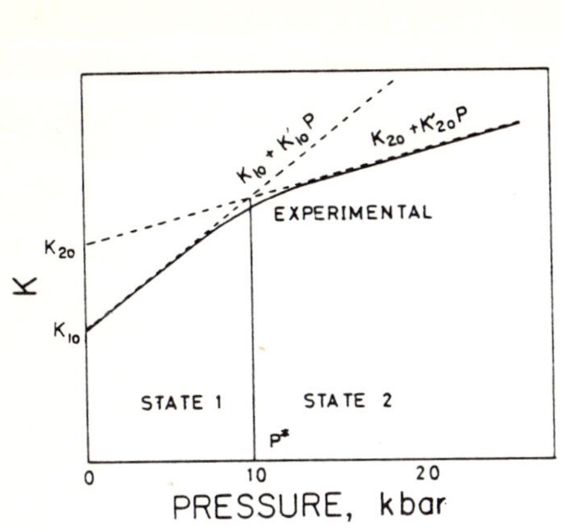

Fig. 7. An illustration of the isothermal bulk modulus of As_2Se_3 glass across the phase-transition pressure and the bulk modulus of the high-pressure phase.

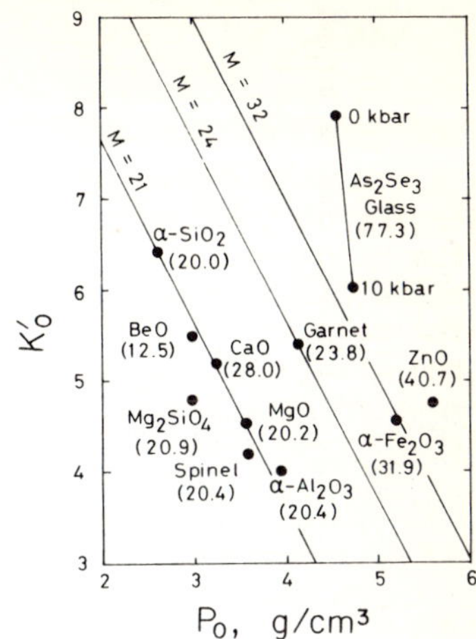

Fig. 8. The pressure derivative of the adiabatic bulk modulus at zero pressure K_0' versus density ρ_0 with mean atomic weight $\overline{M}$ in parentheses. The figure for oxides is reproduced from Anderson et al. [1968].

where $(V/V_0)_C$ is the volume ratio when C stays constant and C/C_0 corresponds to the volume ratio of the glassy to the crystalline phase. Using the lattice theory equation by *Anderson* [1970], C/C_0 can be considered to be a function of pressure. Its value as a function of pressure decreases with pressure as shown in Figure 9. Based on the random network theory for the glass structure, the variation of C can be understood quantitatively if C is interpreted to be the bending of atomic bonds in glass.

IV. APPLICATION OF PHASE TRANSITION

Variation of viscosity of a melt of glass forming compounds may be explained by applying the generalized phase transition concept developed above. We can assume that under pressure a melt as well as a glass may transform gradually from a low-pressure to be a high-pressure phase. From glass research, we know that viscosity can increase or decrease with pressure depending on the composition. Some melts--for example, B_2O_3 [*Sperry and Mackenzie*, 1968]--exhibit a normal increase of viscosity with pressure. However, some silicate melts--such as jadeite [*Kushiro*, 1977]-- show a decreasing viscosity with

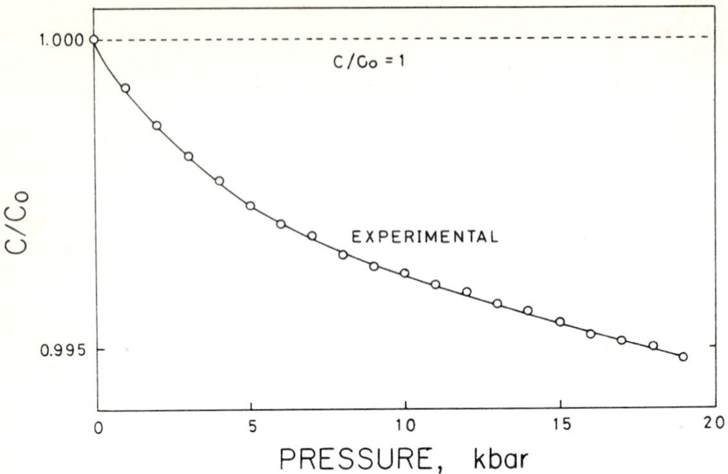

Fig. 9. Pressure dependence of C/C_o for As_2Se_3 glass. The solid line is calculated from equation (7) and the Murnaghan equation with a boundary condition K_{2o} = 159 kbar and K_{2o}' = 6. The dashed line is a constant C/C_o = 1.

increasing pressure. A normal glass melt must increase its viscosity with pressure if there are no phase transitions according to standard theories of melts [*Cohen and Turnbull*, 1959; *Macedo and Litovitz*, 1965]. However, if a phase transition occurs in a melt as a result of pressurization, the viscosity may rise or fall across the phase transition pressure. It is possible to speculate that in a melt of jadeite, the low-pressure phase has a higher viscosity than the high-pressure phase. The possibilities of this occurring in jadeite can be demonstrated by examination of the parameters in the viscosity equation appropriate to glassy melts. Identifying viscosity with the hybrid equation by *Macedo and Litovitz* [1965], we have

$$\log \eta = \log \eta_o + \frac{\gamma V_o}{V_f} + \frac{E}{RT} \tag{8}$$

where V_f is the free volume, V_o the molecular volume, γ a constant, and E the activation energy for viscous flow.

It is possible to assume that the activation energy E is smaller for the high-pressure phase than for the low-pressure phase because the average inter-molecular distance will be reduced in the high-pressure phase, thus also reducing the viscosity of the high-pressure phase. Another possible explanation for jadeite glass is that the activation volume V_o for the high-pressure configuration of glass is less than that for the low-pressure phase of glass, while the reverse is true for the configurational volume V_f .

Acknowledgments. The authors wish to thank M. H. Manghnani and R. Liebermann for suggestions on an earlier version of the manuscript. This work was supported by the National Science Foundation, grant GH-43469. This paper is Contribution No. 1647 of the Institute of Geophysics and Planetary Physics, University of California, Los Angeles, CA 90024.

REFERENCES

Anderson, O. L., A proposed law of corresponding states for oxide compounds, *J. Geophys. Res., 71,* 4963-4971, 1966.

Anderson, O. L., Elastic constants of the central force model for three cubic structures, pressure derivatives, and equations of state, *J. Geophys. Res., 75,* 2719-2740, 1970.

Anderson, O. L., E. Schreiber, F. C. Liebermann, and N. Soga, Some elastic constant data on minerals relevant to geophysics, *Rev. Geophys., 6,* 491-524, 1968.

Cohen, M. H., and D. Turnbull, Molecular transport in liquid, *J. Chem. Phys., 31,* 1164-1169, 1959.

Cook, R. C., Variation of elastic constants and static strains with hydrostatic pressure: method for calculation from ultrasonic measurements, *J. Acous. Soc. Am., 29,* 445-449, 1957.

Kushiro, I., Changes in viscosity and structure of silicate melt at high pressures, in *High-Pressure Research: Applications to Geophysics,* edited by M. H. Manghnani and S. Akimoto, Academic Press, New York, 25-37, 1977.

Macedo, P. B., and T. A. Litovitz, On the relative roles of free volume and activation energy in the viscosity of liquids, *J. Chem. Phys., 42,* 245-256, 1965.

Murnaghan, F. D., Foundations of the theory of elasticity in *Proceedings of the Symposium on Applied Mathematics,* Volume 1 (American Mathematicl Society), 219, 1949.

Sperry, L. L., and J. D. Mackenzie, Pressure dependence of viscosity of B_2O_3, *Phys. Chem. Glasses, 9,* 91-95, 1968.

CRYSTAL STRUCTURES OF PHASES PRODUCED BY DISPROPORTIONATION OF K-FELDSPAR UNDER PRESSURE

N. KINOMURA, M. KOIZUMI
Institute of Scientific and Industrial Research, Osaka University
Yamadakami, Suita, Osaka 565 Japan

S. KUME
College of General Education, Osaka University
Machikaneyama-cho, Toyonaka, Osaka 560 Japan

Abstract

Potassium feldspar disproportionates into a mineral assemblage of kyanite, coesity, and $K_2SiSi_3O_9$ under pressures up to about 100 kbar. An X-ray powder diffraction showed that $K_2SiSi_3O_9$ is isostructural with wadeite, $K_2ZrSi_3O_9$. The use of single crystals of this substance made it possible to refine the structure and to determine the atomic coordinates. When the interatomic distances of $K_2SiSi_3O_9$ were compared to those of $K_2ZrSi_3O_9$, little difference was observed in the size and shape of SiO_4 tetrahedra, but a noticeable change was seen between SiO_6 and ZrO_6 octahedra. The distance between Si and O in SiO_6 octahedra of $K_2SiSi_3O_9$ is approximately the same as that of compounds whose structure is composed of SiO_6 octahedra.

I. INTRODUCTION

Experiments on both static and dynamic compressions have shown an existence of pressure-induced phase transformation of K-feldspar to a hollandite-type phase [*Kume et al.*, 1966; *Ringwood et al.*, 1967; *Ahrens*, 1969]. When applied pressures are below 100 kbar, a disproportionation takes place instead of the phase transformation [*Kinomura et al.*, 1975]. In this report, the results of two kinds of experiment will be presented; the first being on the disproportionation of K-feldspar, and the second on the structural analysis of $K_2SiSi_3O_9$ produced by the disproportionation.

II. EXPERIMENTAL METHODS AND RESULTS

A. Disproportionation of K-feldspar

Two kinds of solutions of KOH and $Al(NO_3)_3$ were mixed with silica gel in the molar ratio of K-feldspar, $KAlSi_3O_8$. The mixture was fired at 600°C after evaporating the solution. The obtained material was charged in a cylindrical capsule, 3 mm in diameter and 5 mm long, of either gold or platinum. The capsule was inserted to a high-pressure cell [*Yanagisawa and Kume*, 1973; *Endo et al.*, 1976] and was subjected to various combinations of temperature and pressure up to 1000°C and 100 kbar for 1 hr, using a cubic anvil-type apparatus. The pressures worked on the sample were standardized by electrical changes of metals (Bi at 25.5 and 77, Ba at 55, and Sn at 100 kbar).

When a run was over, the sample was quenched to room temperature and atmospheric pressure. Phases produced by these treatments were investigated by X-ray powder diffraction. At the applied pressures below 40 kbar, K-feldspar was made, and this result is consistent with those reported by *Seki and Kennedy* [1964]. When pressures exceeded 100 kbar, the sample was found to have the hollandite-type structure, and this is in agreement with the results obtained by *Ringwood et al*. [1967]. When the applied pressures were approximately in the range 40-100 kbar, the X-ray patterns of the products were interpreted to be those of the mineral assemblage shown in Table 1. In this assemblage, $K_2SiSi_3O_9$ was identified to be isostructural with wadeite, $K_2ZrSi_3O_9$ [*Kinomura et al.*, 1974]. Since Zr in wadeite is in an octahedral and Si in a tetrahedral site, the isostructural relation between $K_2ZrSi_3O_9$ and $K_2SiSi_3O_9$ indicates that a quarter of the Si in $K_2SiSi_3O_9$ is situated in octahedral and the remaining three-quarters in tetrahedral sites. Since the coexistence of 4- and 6-coordinated Si ions in a silicate has not been previously reported, more precise analysis is required. A result of analysis using a single crystal of $K_2SiSi_3O_9$ grown under pressure is presented in the following section.

B. Structural Analysis of $K_2SiSi_3O_9$

Starting material was obtained by mixing silica gel and KNO_3 powder in a molar ratio of $K_2Si_4O_9$ and fired at 500°C in air to decompose the nitrate. After a few drops of 0.3*N* KOH solution was added, the starting material was treated in the same way as stated in the previous section. The experimental condition was at 1000°C and 60 kbar for 6 hr. Potassium hydroxide worked as a mineralizer on the reaction, and, when the run was over, hexagonal dipyramidal crystals of 10 to 70 μm in size were produced. A small amount of coesite was also found

TABLE 1. Example of Mineral Assemblage Disproportionated from K-feldspar at 900°C and 90 kbar

d (Å)	*I*	Identification[a]
4.92	w	W
4.33	w	K
3.66	w	W
3.43	w	C
3.35	w	K
3.31	w	W
3.18	m	K
3.13	m	W
3.09	s	C
2.77	vs	W,C
2.74	vs	W
2.71	vs	W
2.70	w	K,C
2.52	w	K
2.16	vw	W,K
2.11	m	W
1.96	w	K,W
1.93	m	K
1.79	w	W
1.77	m	W,K

a. C-phase identified as coesite; K-phase identified as kyanite; W-phase identified as wadeite-type $K_2SiSi_3O_9$. X-ray data for coesite and kyanite are referred from the ASTM card file.

in the products.

Some of the dipyramidal crystals were confirmed to be the wadeite-type phase by X-ray powder diffraction. One of the crystals largest in size was selected, and its space group was determined by the Weissenberg and Burger precession method utilizing Cu Kα radiation.

Three-dimensional intensity data were collected on the same sample by a Rigaku four-circle automatic diffractometer using the $2\theta-\omega$ scan technique in Ni-filtered Cu Kα radiation. Counts for each reflection were integrated over $1.0 + 0.175\tan\theta$ on both sides of a peak maximum with a scanning rate of 2°/min. The c axis was the rotation axis, and 840 reflections within the hemisphere $2\theta \leq 110°$ were collected.

The integrated intensities were converted to structure factors after applying the Lorenz-polarization corrections. Neither absorption nor extinction correction was undertaken in the present analysis. The atomic scattering factors were referred from the *International Tables for X-ray Crystallography*

[1974], and the analysis was continued using a program of "Block Diagonal Matrix Least-square Refinements" described by *Ashida* [1973].

The Weissenberg and precession photographs showed that systematic presences of reflections were 00l with $l = 2n$. Accordingly, the possible space groups are $P6_3/m$ and $P6_3$. The structure determination was started on the assumption that the structure has a center of symmetry, and the space group is the same as wadeite, $P6_3/m$. The lattice parameters were given to be $a = 6.6136$ and $c = 9.5118$ Å, respectively.

When the atomic parameters proposed by *Henshaw* [1955] for wadeite were adopted, the results gave an $R = 0.34$, and further refinements using the least-square method with isotropic temperature factors reduced to $R = 0.057$ and $Rw = 0.052$. The parameters obtained are listed in Table 2, and the observed and calculated structure factors are shown in Table 3. Table 4 illustrates the interatomic distances and angles.

III. DISCUSSION

The experimental results on the disproportionation of K-feldspar are represented as follows.

$$\underset{\text{K-feldspar}}{KAlSi_3O_8} \longrightarrow \underset{\text{kyanite}}{Al_2SiO_5} + \underset{\text{coesite}}{SiO_2} + \underset{\text{wadeite-type}}{K_2SiSi_3O_9}$$

$$\longrightarrow \underset{\text{hollandite-type}}{KAlSi_3O_9}$$

These phase changes accompany increases in densities. The increment after the disproportionation is 21%, reaching 35% with the direct conversion from feldspar to the hollandite-type phase.

As far as the structural analysis is concerned, an agreement

TABLE 2. Fractional Atomic Coordinates (x, y, and z) and Isotropic Temperature Factors (B)[a]

	x	y	z	B
K	.6666(0)	.3333(0)	.5575(4)	.624
Si(1)	.3652(8)	.2337(8)	.2500(0)	.009
Si(2)	.0 (0)	.0 (0)	.0 (0)	.126
O(1)	.2323(11)	.2093(12)	.1073(5)	.433
O(2)	.4942(17)	.0798(16)	.2500(0)	.226

$R = 0.057$ $Rw = 0.052$

$a = 6.6136 \pm 0.0004$Å, $c = 9.5118 \pm 0.0005$Å

a. The R and Rw factors are also shown.

TABLE 3. Observed (Fo) and Calculated (Fc) Structure Factors

hkl	Fo	Fc	hkl	Fo	Fc	hkl	Fo	Fc
010	217	190	122	308	288	334	147	149
020	302	294	132	286	294	414	340	336
030	500	482	142	711	708	015	139	125
040	130	107	152	58	38	025	684	673
050	239	213	212	90	60	035	48	33
110	381	375	222	121	128	045	366	357
120	36	29	232	90	78	115	75	65
130	473	471	242	37	37	125	411	407
140	500	484	312	278	250	135	362	347
150	80	78	322	252	234	145	226	222
210	382	349	332	451	422	215	362	364
220	730	728	412	170	151	225	83	87
230	61	58	422	12	10	235	340	338
240	26	12	512	325	308	315	179	162
310	361	337	013	710	715	325	95	91
320	428	410	023	101	71	415	78	69
330	194	193	033	105	88	006	998	1055
410	765	780	043	156	142	016	208	198
420	80	129	053	110	67	026	387	367
510	313	288	113	203	168	036	122	53
011	203	193	123	436	406	046	212	214
021	689	696	133	149	142	116	227	185
031	90	84	143	238	228	126	86	86
041	224	209	213	49	26	136	144	127
051	289	267	223	231	226	216	372	345
111	393	376	233	109	90	226	258	224
121	605	586	243	93	80	236	253	240
131	161	146	313	819	838	316	456	438
141	249	240	323	714	715	326	474	460
151	42	46	333	428	429	017	387	390
211	197	167	413	15	19	027	228	214
221	72	55	423	27	43	037	26	35
231	138	133	004	564	567	117	302	294
241	270	255	014	195	174	127	138	139
311	307	278	024	294	253	137	132	132
321	172	151	034	137	111	217	422	425
331	125	126	044	248	252	227	94	74
411	158	151	054	46	60	317	419	422
421	26	35	114	108	97	008	580	615
511	105	77	124	543	515	018	338	335
002	231	227	134	323	316	028	168	159
012	203	172	144	322	309	038	219	221
022	79	44	214	343	325	118	345	357
032	623	608	224	589	571	128	419	408
042	246	234	234	58	40	218	193	170
052	64	58	314	86	76	019	185	194
112	785	853	324	29	12	119	70	79

TABLE 4. Interatomic Distances and Angles

	Distances (Å)		
K-O(1)	3.00(1)		
K-O(2)	3.28(1)		
Si(1)-O(1)	1.58(1)		
Si(1)-O(2)	1.62(1)		
Si(1)-O(2)'	1.66(1)		
Si(2)-O(1)	1.79(1)		
O(1)-O(1)	2.71(1)		tetra[a]
O(1)-O(2)	2.66(1)		tetra[a]
O(1)-O(2)'	2.56(1)		tetra[a]
O(2)-O(2)'	2.57(1)		tetra[a]
O(1)-O(1)	2.54(1)	same plane	octa[b]
O(1)-O(1)	2.51(1)	different plane	octa[b]
	Angles (degree)		
O(1)-Si(1)-O(1)	118.3(6)		tetra[a]
O(1)-Si(1)-O(2)	112.1(6)		tetra[a]
O(1)-Si(1)-O(2)'	104.7(6)		tetra[a]
O(2)-Si(1)-O(2)'	103.2(6)		tetra[a]
Si(1)-O(2)-Si(1)	136.8(6)		tetra[a]
Si(1)-O(2)'-Si(1)	136.8(6)		tetra[a]
O(1)-Si(2)-O(1)	90.6(6)	same plane	octa[b]
O(1)-Si(2)-O(1)	89.4(6)	different plane	octa[b]

a. tetra-oxygens forming tetrahedrons.
b. octa-oxygens forming octahedrons.

between the observed and calculated structure factors is evidence that $K_2SiSi_3O_9$ is isostructural with $K_2ZrSi_3O_9$, and three-quarters of the Si ions are situated in tetrahedral and a quarter in octahedral sites. When the Si-O distances of $K_2SiSi_3O_9$ are compared to those of Zr-O and Si-O of $K_2ZrSi_3O_9$ [*Henshaw* 1955], the distances in the SiO_6 tetrahedrons are almost the same, but those in SiO_6 and ZrO_6 octahedrons differ by a noticeable amount. When the ratio of ionic radius, $r_{Zr}(VI)/r_{Si}(VI)$ is calculated, it comes to 1.8 [*Shannon and Prewitt*, 1969], which is large enough to explain the above difference in the size of octahedrons. *Stishov and Belov* [1962] have calculated the Si-O and the O-O distances in the rutile-type SiO_2, and *Ringwood et al.* [1967] report an average bond length of octahedron of the hollandite-type $KAlSi_3O_8$. Comparison of these distances to the present results show that the corresponding values are nearly the same, indicating that the shape of octahedrons in $K_2SiSi_3O_9$ is similar to those of silicates containing 6-coordinated Si ions. Two compounds, SiP_2O_7 AIII and AIV, are also known to possess SiO_6-octahedrons [*Bissert and Liebau*, 1970; *Liebau and Hesse*, 1971]. Similarities in Si-O and O-O separations are again observed in these phosphates, and the Si-O distances are, in general, in the range from 1.716 to 1.872 Å [*Stishov and Belov*, 1962] regardless of the nature of compounds.

REFERENCES

Ahrens, T. J., Shock compression of feldspars, *J. Geophys. Res.*, *74*, 2727, 1969.

Ashida, T., *The Universal Crystallographic Computing System-Osaka*, The Computation Center, Osaka University, Osaka, 55, 1973.

Bissert, G., and F. Liebau, Die Kristallstruktur von monoklinem Siliziumphosphat SiP_2O_7 AIII: Eine Phase mit [SiO_6] - Oktaedern, *Acta Cryst.*, *B26*, 233-240, 1970.

Endo, T., S. Kume, N. Kinomura, and M. Koizumi, A new compound $K_2Cr_8O_{16}$ with Hollandite type structure, *Mat. Res. Bull.*, *11*, 609-614, 1976.

Henshaw, D. E., The structure of wadeite, *Mineral. Mag.*, *30*, 585-595, 1955.

International Tables for X-ray Crystallography, edited by J. A. Ibers, and W. C. Hamilton, Intern. Union of Cryst., Kynoch Press, Birmingham, *IV*, 71, 1974.

Kinomura, N., S. Kume, and M. Koizumi, Stability of $K_2Si_4O_9$ with wadeite-type structure, *Proc. 4th Internat. Conf. High Pressure*, Kyoto, 211-214, 1974.

Kinomura, N., S. Kume, and M. Koizumi, Synthesis of $K_2SiSi_3O_9$ with silicon in 4- and 6-coordination, *Mineral. Mag.*, *40*, 401-404, 1975.

Kume, S., T. Matsumoto, and M. Koizumi, Dense form of germanate orthoclase ($KAlGe_3O_8$), *J. Geophys. Res.*, *71*, 4999-5000, 1966.

Liebau, F., and K. R. Hesse, Die Kristallstruktur einer zweiten monoklinen Siliciumdiphosphatphase, SiP_2O_7 AIV, mit oktaedrisch koordiniertem Silicium, *Zeit. Krist.*, *133*, 213-224, 1971.

Ringwood, A. E., A. F. Reid, and A. D. Wadsley, High-pressure transformation of alkali aluminosilicates and aluminogermanates, *Earth Planet. Sci. Lett.*, *3*, 38-40, 1967.

Seki, Y., and G. C. Kennedy, The breakdown of potassium feldspar, $KAlSi_3O_8$, at high temperatures and high pressures, *Amer. Mineral.*, *49*, 1688-1706, 1964.

Shannon, R. D., and C. T. Prewitt, Effective Ionic Radii in Oxides and Fluorides, *Acta Cryst.*, *B25*, 925-946, 1969.

Stishov, S. M., and N. B. Belov, Crystallographic structure of new dense modification of silica SiO_2, *Geokhimiya*, *143*, 951-954, 1962.

Yanagisawa, Y., and S. Kume, A new polymorph of cerium polysulfide CeS_2, *Mat. Res. Bull.*, *8*, 1241-1246, 1973.

SESSION II

PHASE TRANSITIONS RELATED TO EARTH'S DEEP INTERIOR

SILICATE ILMENITES AND THE POST-SPINEL TRANSFORMATIONS

E. Ito and Y. Matsui
Institute for Thermal Spring Research
Okayama University
Misasa, Tottori-ken 682-02, Japan

Abstract

High-pressure phase relations of several ferromagnesian silicates were investigated, using a double-stage split-sphere type apparatus.

Clinoenstatite disproportionates at about 170 kbar and 1000°C into stishovite plus β-Mg_2SiO_4, which is replaced by γ-Mg_2SiO_4 at 190 kbar. The γ-Mg_2SiO_4 reacts with stishovite to form the $MgSiO_3$ ilmenite at about 260 kbar. The hexagonal unit cell dimensions of $MgSiO_3$ ilmenite were determined to be a = 4.7284 (4) Å and c = 13.5591 (16) Å.

At 1000°C the $ZnSiO_3$ clinopyroxene transforms directly into ilmenite at 120 kbar, and β-Zn_2SiO_4 disproportionates into $ZnSiO_3$ (ilmenite) plus ZnO (rocksalt) at 130 kbar. The hexagonal unit cell dimensions of $ZnSiO_3$ ilmenite were determined to be a = 4.7462 (3) Å and c = 13.7569 (10) Å. $ZnSiO_3$ (ilmenite) was found to decompose into ZnO (rocksalt) plus stishovite at about 200 kbar.

It is interesting to note that silicate ilmenite is nonexistent in transition-metal compounds. The γ phases of Ni_2SiO_4, Co_2SiO_4, and Fe_2SiO_4 decompose directly into their oxide mixture (rocksalt plus stishovite).

Neither β- nor γ-Mn_2SiO_4 exists, and α-Mn_2SiO_4 disproportionates into MnO (rocksalt) plus $MnSiO_3$ (garnet), which decomposes into MnO (rocksalt) plus stishovite at about 220 kbar.

These experimental results indicate the possibility that ilmenite and related minerals are significant constituents in the lower-half part of the transition zone, provided the Mg/Fe value is high.

I. INTRODUCTION

In the last 20 years, high-pressure phase equilibria of a large number of silicates and related compounds have been investigated in order to reveal the mineralogical constitution of the deep mantle.

The high-pressure phase relations in the system Mg_2SiO_4-Fe_2SiO_4 have attracted special attention, as olivine has been believed to be the most dominant mineral in the upper mantle. It has been experimentally confirmed that the olivine (α) - modified spinel (β) transformation is chiefly responsible for the seismic discontinuity found near the depth of 400 km, and the modified spinel is considered to be dominant mineral in the upper-half part of the transition zone [e.g., *Akimoto et al.*, 1976].

On the other hand, *Ringwood and Major* [1966] observed that pyroxenes react with garnet at about 100 kbar, forming a garnet with complex composition, and *Ringwood* [1970] claimed that this "pyroxene-garnet" transformation would also contribute to the 400 km discontinuity.

These features of the formation of the transition zone have stimulated us to investigate the successive transformation of silicate spinels, the so-called post-spinel transformation, and to search for denser phases of metasilicates and aluminosilicates. This high-pressure polymorphism is expected to be of great significance for the interpretation of the next first-order discontinuity at around the depth of 650 km and should be informative for the constitution of the lower mantle. Along this line, we have studied the high-pressure phase relations for several ferromagnesian silicates, using a double-stage split-sphere-type apparatus. In this report, we summarize these results.

II. EXPERIMENTAL METHODS

The split-sphere-type apparatus was originally designed by *Kawai* [1966] at Osaka University. The description of the apparatus was given by *Kawai and Endo* [1970], and its application to high-pressure phase-equilibrium studies has been reported by *Kawai et al.*[1970] and *Suito* [1972]. The significant characters of the apparatus are as follow. First, the capability to compress a charge of large volume up to very high pressures makes it possible to obtain pressures higher than 200 kbar over a volume of about 5 mm^3. Therefore, samples of several milligrams are easily compressed up to pressures higher than 200 kbar and heated up to temperatures higher than 1000°C by utilizing an internal heating system. Second, a pressure environment in the specimen is maintained in a highly hydrostatic state by multi-anvils, which are equivalent to each other in shape and in movement, in spite of the use of a solid medium for the pressure transmission.

The pressure medium is usually pyrophyllite dehydrated in air at 800°C for 30-40 min. Pressure values are calibrated against the oil pressure at room temperature, using several fixed points. It has been pointed out that the use of pyrophyllite causes a definite reduction of the pressure value at high temperatures, especially when a small specimen is heated with the internal heating system [*Inoue*, 1975]. Although it is reported that this disadvantage could be avoided by using the alternative materials, such as sintered MgO, as the pressure medium [*Suito*, 1977], we report here the experimental results using the conventional pyrophyllite medium, with the pressure values thoroughly re-calibrated according to the new pressure scale proposed by *Bundy* [1975].

Sample temperature is monitored with Pt/Pt-13%Rh thermocouple without any correction for the pressure effect on emf. In the runs at pressures higher than 200 kbar, however, temperature is in most cases estimated from the electric power supply to the heater. Though the latter method posesses rather large uncertainties and is not suitable for the precise determination of phase equilibria, it is convenient for use at extremely high pressures.

Finely pulverized starting materials are charged into the cylindrical platinum or carbon heater embedded in a small pyrophyllite octahedron. After being held at the desired pressure and temperature conditions for an appropriate time, the sample is quenched by shutting off the electric power supply. The pressure is released slowly, and the product is recovered in the ambient condition.

Identification of the phases in the central portion of the quenched sample is made by the powder X-ray diffraction technique. The use of a graphite crystal monochromator plus a single channel pulse height analyzer makes it possible to obtain the powder diffraction pattern with very high S/N ratio for all the silicates studied utilizing Cu Kα radiation.

III. RESULTS AND DISCUSSION

A. $MgSiO_3$

On the basis of the currently available information, *Ringwood* [1970] inferred that $MgSiO_3$(clinopyroxene) might transform to either the assemblage of β-Mg_2SiO_4 plus stishovite or to the garnet structure, and, further, to the ilmenite structure at higher pressures.

Ito et al. [1972] observed that $MgSiO_3$(clinopyroxene) disproportionates into stishovite plus β-Mg_2SiO_4 at 170 kbar and 1000°C, and this assemblage is replaced by stishovite plus γ-Mg_2SiO_4 at 190 kbar. In Figure 1(a) the powder X-ray tracing of the product quenched at 240 kbar is reproduced.

Taking these results into consideration, *Kawai et al.* [1974] compressed $MgSiO_3$(clinopyroxene) up to pressures higher than the

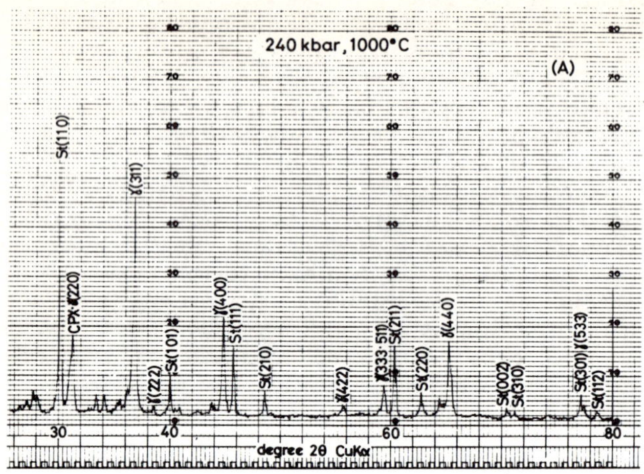

Fig. 1 (a). The powder X-ray diffraction chart of $MgSiO_3$ run at about 240 kbar and 1000°C.

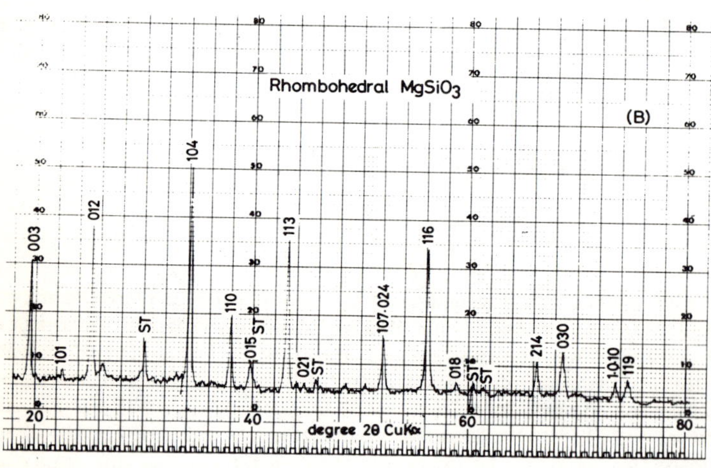

Fig. 1 (b). The powder X-ray diffraction chart of $MgSiO_3$ (ilmenite). Presence of stishovite is due to the excess SiO_2 contained in the starting material.

TABLE 1. Powder X-ray Data for $MgSiO_3$ Ilmenite[a]

h	*k*	*l*	d_{obs}, Å	d_{calc}, Å	Intensity
0	0	3	4.522	4.520	55
1	0	1	3.922	3.920	5
0	1	2	3.507	3.505	70
1	0	4	2.611	2.611	100
1	1	0	2.364	2.364	30
0	1	5	2.263	2.261	15
1	1	3	2.094	2.095	70
0	2	1	2.026	2.025	<5
1	0	7	1.7524	1.7510	25
0	2	4	1.7524	1.7526	
1	1	6	1.6335	1.6336	65
0	1	8	1.5661	1.5661	5
2	1	4	1.4078	1.4079	15
0	3	0	1.3651	1.3650	20
1	0	10	1.2869	1.2872	10
1	1	9	1.2705	1.2705	10
2	2	0	1.1820	1.1821	5

Unit-cell dimensions

Hexagonal cell (Z = 6):	Rhombohedral cell (Z = 2)
a = 4.7284 ± 0.0004 Å	a = 5.2802 ± 0.0004 Å
c = 13.5591 ± 0.0016 Å	α = 53°11'56" ± 25"
V = 262.54 ± 0.05 $Å^3$	V = 87.45 ± 0.02 $Å^3$
ρ_{calc} = 3.795 ± 0.002 g/cm^3	

a. Previous data on $MgSiO_3$ [*Kawai et al.*, 1974] contained considerable errors. In this article, the newly redetermined data are presented.

GaP point and heated it at about 1000°C. More than ten extra peaks appeared in the powder X-ray chart of the quenched product, as shown in Figure 1 (b). All the extra peaks could be successfully indexed on the basis of a hexagonal lattice, just as in the cases of the $MgGeO_3$ and $MnGeO_3$ ilmenites [*Ringwood and Seabrook*, 1962, 1963]. The presence of 00(3*n*) reflections, for example 003, leads to the conclusion that the high-pressure polymorph obtained belongs to the rhombohedral system with the ilmenite structure (space group $R\bar{3}$). Detailed X-ray data are given in Table 1. The calculated density is 3.795(2) g/cm^3, which is 2% denser than the isochemical assemblage of γ-Mg_2SiO_4 plus stishovite. The pressure value required for the formation of $MgSiO_3$(ilmenite) is estimated to be about 260 kbar at 1000°C.

$MgSiO_3$(garnet) was not observed in these studies. The stabi-

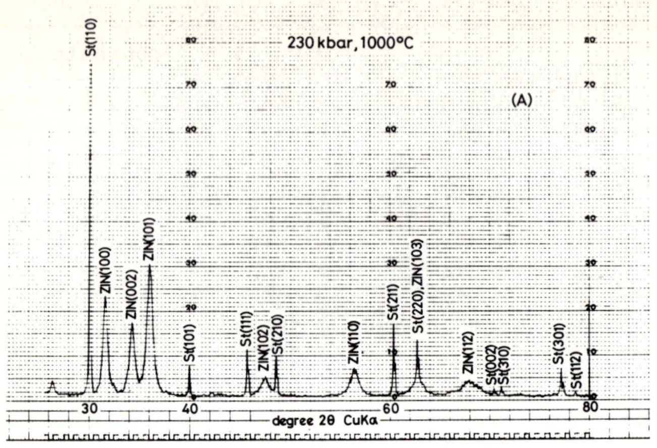

Fig. 2 (a). The powder X-ray diffraction chart of $ZnSiO_3$ run at about 230 kbar and 1000°C.

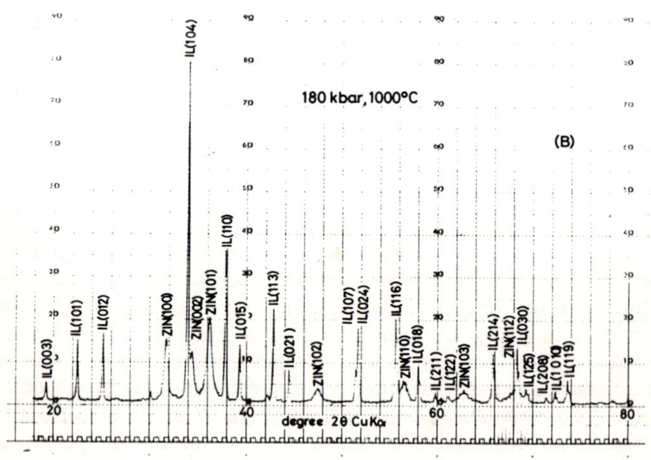

Fig. 2 (b). The powder X-ray diffraction chart of Zn_2SiO_4 run at 180 kbar and 1000°C.

lity of the garnet cannot be ruled out completely, because the experimental temperature was fixed at approximately 1000°C. However, because the assemblage of β-Mg_2SiO_4 plus stishovite is estimated to be about 4% denser than $MgSiO_3$(garnet), the stability field of $MgSiO_3$ (garnet) would be limited perhaps at very high temperature if it does exist. Nevertheless, it is possible that the substitution of a small amount of larger cations, viz., Mn(II) for Mg(II) or Al(III) for Si(IV), would stabilize the the garnet structure. This view is supported by the synthesis of $(Mg_{0.6}Mn_{0.4})SiO_3$ (garnet)[*Tachimori*, 1974] and of the mineral majorite [*Ringwood and Major*, 1971].

In Mg_2SiO_4, we observed that the γ phase persists up to 300 kbar. The stablity of $MgSiO_3$ (ilmenite), however, suggests the possibility of disproportionation of γ-Mg_2SiO_4 into $MgSiO_3$(ilmenite) plus periclase.

B. Zinc Silicates

Ito and Matsui [1974] confirmed that $ZnSiO_3$ (clinopyroxene) directly transforms into the ilmenite structure at 120 kbar and 1000°C. Detailed powder X-ray data on $ZnSiO_3$ (ilmenite) are given in Table 2. Calculated density of 5.253 g/cm^3 is believed to be the highest value among all the ferromagnesian silicates known so far. It was found that $ZnSiO_3$ (ilmenite) decomposes at about 200 kbar into ZnO (rocksalt) plus stishovite as shown in Figure 2 (a). In this chart, the crystal structure of zinc oxide is not the rocksalt but the zincite-type. The extreme broadness of the diffraction lines of the zincite, however, suggests strongly that this phase is the product of reversion from the rocksalt during the course of quenching. The existence of ZnO (rocksalt) has been demonstrated by *Inoue* [1975] by the *in situ* X-ray diffraction experiment under high pressures.

Ito and Matsui [1974] discovered the disproportionation of β-Zn_2SiO_4 (designated as Zn_2SiO_4-V by *Syono et al.*[1971*a*]) into $ZnSiO_3$ (ilmenite) plus ZnO (rocksalt) as shown in Figure 2 (b). In this X-ray diffraction tracing, the crystal structure of ZnO is again that of zincite with very broad lines.

Syono et al.[1971*a*] suggested that the well-known preference for tetrahedral coordination of Zn(II) ion makes the crystal structures of high-pressure polymorphs and phase relations different from those in other ferromagnesian silicates up to moderate pressures. *Ito and Matsui* [1974], however, pointed out that the chemical characteristics of Zn(II) ion tends to diminish with increasing pressure, and the transformation scheme in zinc silicates at very high pressure might serve as a useful guide for the study of magnesium silicates.

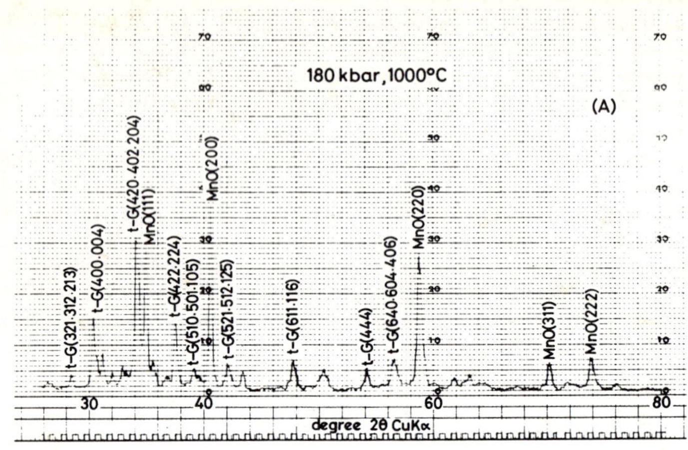

Fig. 3 (a). The powder X-ray diffraction chart of Mn_2SiO_4 run at 180 kbar and 1000°C.

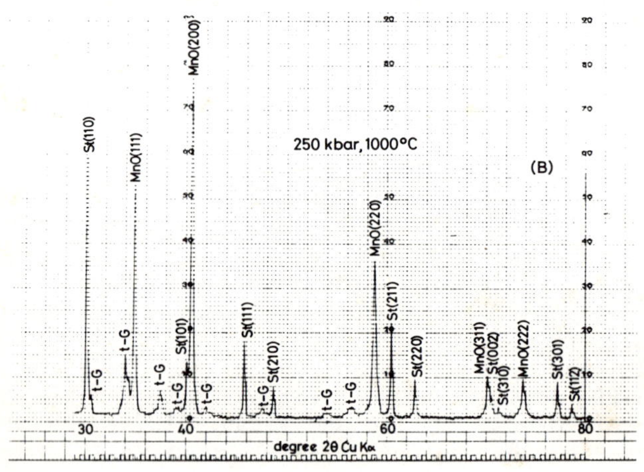

Fig. 3 (b). The powder X-ray diffraction chart of $MnSiO_3$ run at about 250 kbar and 1000°C.

TABLE 2. Powder X-ray Data for $ZnSiO_3$ Ilmenite[a,b]

h	k	l	d_{obs}, Å	d_{calc}, Å	Intensity
0	0	3	4.579	4.586	10
1	0	1	3.933	3.938	15
0	1	2	3.524	3.528	15
1	0	4	2.635	2.638	100
1	1	0	2.373	2.373	35
0	1	5	2.287	2.286	15
1	1	3	2.107	2.108	30
0	2	1	2.032	2.033	10
1	0	7	1.7730	1.7730	15
0	2	4	1.7644	1.7642	20
1	1	6	1.6488	1.6489	20
0	1	8	1.5863	1.5864	15
2	1	1	1.5438	1.5438	<5
0	0	9	1.5283	1.5285	5
1	2	2	1.5147	1.5154	<5
2	1	4	1.4158	1.4158	10
0	3	0	1.3702	1.3701	10
1	2	5	1.3529	1.3528	<5
2	0	8	1.3189	1.3188	<5
1	0	10	1.3047	1.3046	5
1	1	9	1.2853	1.2850	10

Unit-cell dimensions

Hexagonal cell ($Z = 6$):
$a = 4.7462 \pm 0.0003$ Å
$c = 13.7569 \pm 0.0010$ Å
$V = 268.38 \pm 0.03$ Å^3
$\rho_{calc} = 5.253 \pm 0.002$ g/cm^3

Rhombohedral cell ($Z = 2$):
$a = 5.3420 \pm 0.0003$ Å
$\alpha = 52°44'56" \pm 17"$
$V = 89.46 \pm 0.01$ Å^3

a. Previous data on $ZnSiO_3$ [*Ito and Matsui*, 1974] contained some errors. In this article, the newly re-determined data are presented.

b. The (1 2 2) reflection is not used for the unit-cell calculation.

C. Manganese Silicates

Ito et al.[1974] established that tephroite, Mn_2SiO_4 (olivine), disproportionates into $MnSiO_3$ (garnet) [cf. *Akimoto and Syono*, 1972] plus MnO (rocksalt) at 140 kbar and 1000°C (Figure 3 (a)). The $MnSiO_3$ (garnet) decomposes into MnO (rocksalt) plus stishovite at about 220 kbar (Figure 3 (b)). In their systematic investigation, neither the β nor the γ phase was observed. This behavior is evidently the reflection of the characters of Mn(II) ion, which has a rather large ionic radius and is free from the crystal field stabilization in its high-spin state [*Syono et al.*,

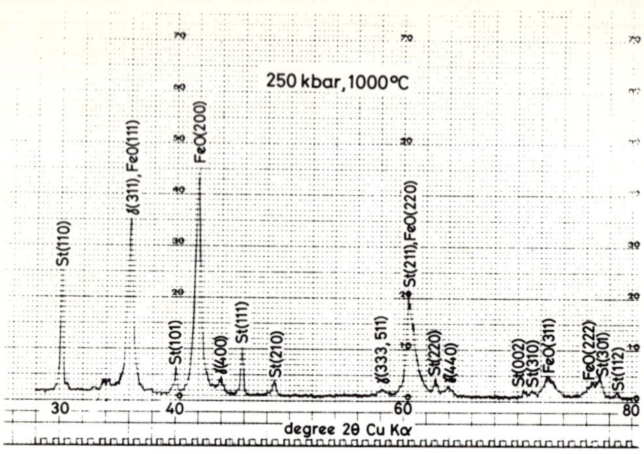

Fig. 4. The powder X-ray diffraction chart of Fe_2SiO_4 run at about 250 kbar and 1000°C.

1971*b*].

D. Iron, Cobalt, and Nickel Silicates

These three transition metal silicates show quite similar transformation scheme. The γ phases of Fe_2SiO_4, Co_2SiO_4, and Ni_2SiO_4 decompose into their oxide mixtures (rocksalt plus stishovite) at 1000°C and 230, 175, and 220 kbar, respectively. The X-ray diffraction chart reproduced in Figure 4 evidently demonstrates the decomposition of γ-Fe_2SiO_4. In Co_2SiO_4, the boundary between the γ phase and the oxide mixture was found to have a definitely positive slope, which can be roughly approximated by the relation: P (kbar) = 0.065 T (°C) + 110 [*Ito*, 1975]. The decomposition of these γ phases has also been recognized by several authors [*Bassett and Ming*, 1972; *Liu*, 1975*a*,*b*].

In the metasilicates, on the other hand, all the products were either the assemblages of the γ phase plus stishovite or those of oxide mixtures. In other words, no evidence was obtained for the existence of high-pressure phases of $FeSiO_3$, $CoSiO_3$, and $NiSiO_3$. Therefore, it is interesting to note that the silicate ilmenite is non-existent in the cases of transition-metal compounds, in spite of the fact that ionic radii of Fe(II), Co(II), and Ni(II) are comparable to those of Mg(II) and Zn(II). It is highly likely that the low symmetry in the metal position in the ilmenite (point symmetry 3) may be less favorable for the ligand field stabilization of *d*-electrons than in the cases of the rocksalt structure where the symmetry of the metal position is the highest (*m3m*).

IV. A POSSIBLE CONSTITUTION OF THE LOWER-HALF PART OF THE TRANSITION ZONE

On the basis of the above experimental results, it is possible to construct a tentative mineralogical model for the lower-half part of the transition zone. Two major transitions may be responsible for the 650-km discontinuity: (1) the disproportionation of the β or γ phase into silicate ilmenite plus rocksalt, and (2) transformation of the complex garnet to an ilmenite-like solid solution. In the latter transition, the $CaSiO_3$ component may exsolve as a perovskite form. Therefore, the lower-half part of the transition zone (depth interval from 700 to 1000 km) may be composed of complex ilmenite (75%), ferropericlase (20%), and $CaSiO_3$ perovskite (5%), assuming the pyrolite composition throughout the transition zone. The zero pressure density ρ_o is then expected to increase from 3.68 to 3.96 g/cm^3 across the discontinuity. Since the bulk modulus of $MgSiO_3$ (ilmenite) has recently been estimated to be about 2.3 Mbar [*Y. Sato*, 1976, personal communication], the seismic parameter Φ_o of this assem-

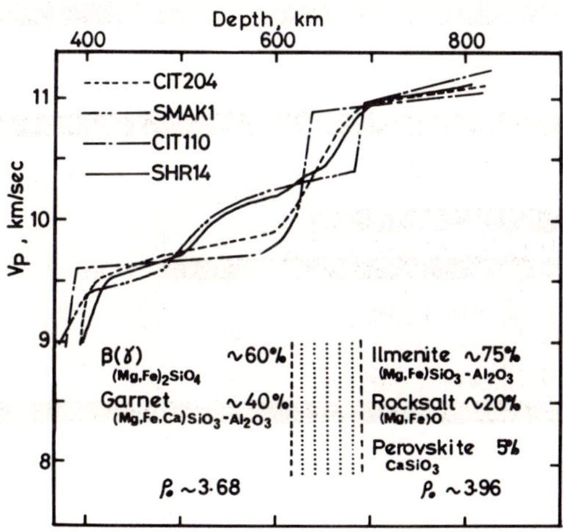

Fig. 5. A tentative model for the transition zone assuming the pyrolite composition.

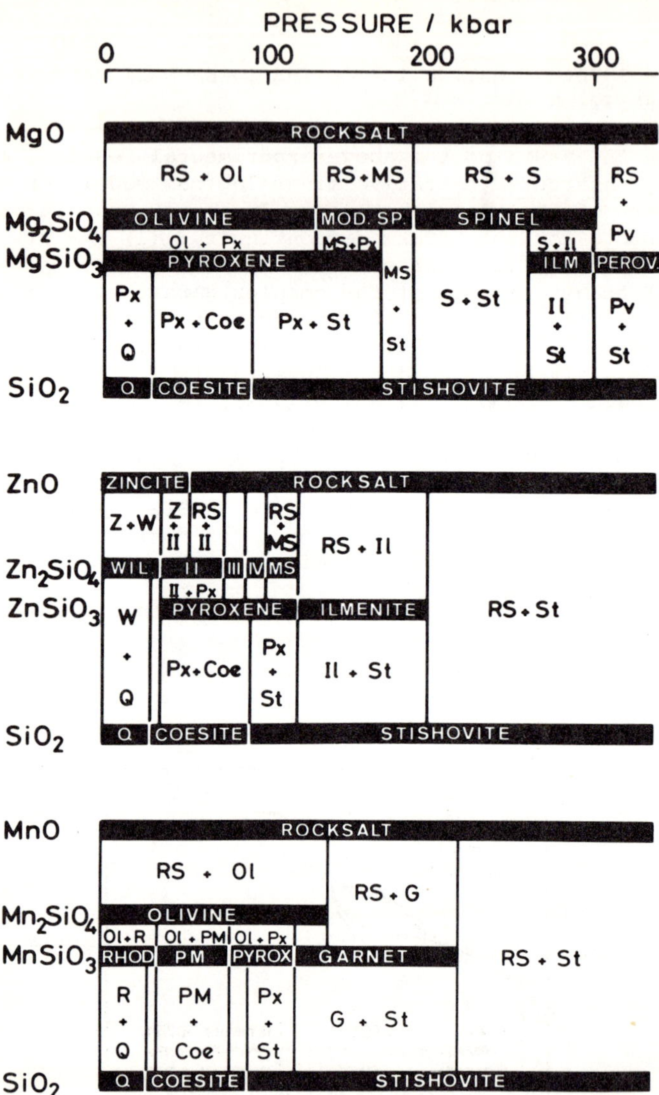

Fig. 6. The revised Akimoto diagram showing the isothermal phase relations in the system $MO-SiO_2$ (M = Mg, Zn, Mn, Ni, Co, and Fe) at 1000°C [cf. Akimoto et al., 1976, Fig. 13]. (continued)

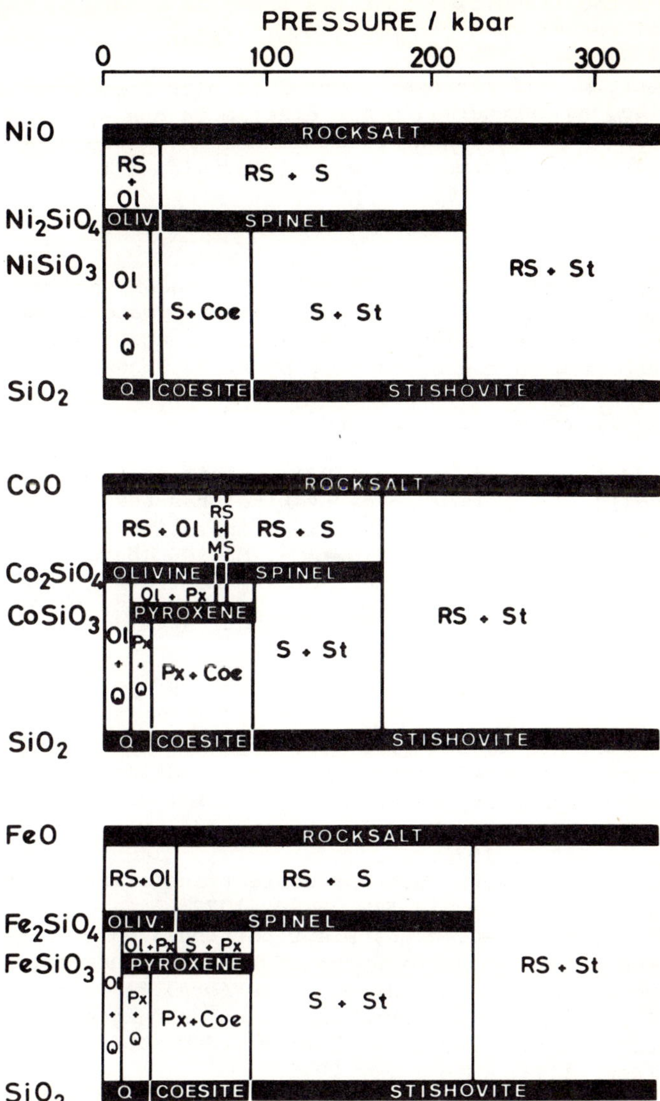

Fig. 6 (continued)
Note: Phase relations in the $MgO-SiO_2$ system above ca. 300 kbar are only schematical.

blage would have the value of around 60 $(km/sec)^2$. These characteristics of the transition zone are shown in Figure 5, together with several recent V_P-depth profiles. A similar model has been proposed by *Ringwood* [1975].

In Figure 6, all the experimental results discussed above are summarized as isothermal phase diagram in the system $MO-SiO_2$ (M = Mg, Zn, Mn, Ni, Co, and Fe) at 1000°C in the form of the *Akimoto* diagram [cf. *Akimoto et al.*, 1976, Fig. 13].

Acknowledgments. We thank Kaichi Suito and Masaharu Tachimori for their stimulating discussion. We are grateful to the late Takashi Matsumoto and to Naoto Kawai for their support. The careful review on the manuscript by A. E. Ringwood is much appreciated. This work was financially supported by the research fund for the Japanese Geodynamic Project.

Note added in proof: Liu's [1975*c*] claim on the stabilization of $MgSiO_3$ (orthorhombic perovskite) has been confirmed in our laboratory [*Ito and Matsui*, in preparation]. The necessary modification is made in Figure 6. $MgSiO_3$ (perovskite) should be an important constituent of the mantle below the depth of about 800 km.

REFERENCES

Akimoto, S., Y. Matsui, and Y. Syono, High-pressure crystal chemistry of orthosilicates and the formation of the mantle transition zone, in *The Physics and Chemistry of Minerals and Rocks*, edited by R. G. J. Strens, Wiley Interscience, London, 327-363, 1976.

Akimoto, S., and Y. Syono, High pressure transformation in $MnSiO_3$, *Amer. Mineral., 57*, 76-84, 1972.

Bassett, W. A., and L.-C. Ming, Disproportionation of Fe_2SiO_4 to $2FeO + SiO_2$ at pressures up to 250kbar and temperatures up to 3000°C, *Phys. Earth Planet. Interiors, 6*, 154-160, 1972.

Bundy, F. P., Ultrahigh pressure apparatus using cemented tungsten carbide pistons with sintered diamod tips, *Rev. Sci. Instr., 46*, 1318-1324, 1975.

Inoue, K., Development of high temperature and high pressure X-ray diffraction apparatus with energy dispersive technique and its geophysical application, Ph.D. Thesis, University of Tokyo, Tokyo, 1975.

Ito, E., High pressure decompositions in cobalt and nickel silicates, *Phys. Earth Planet. Interiors, 10*, 88-93, 1975.

Ito, E., and Y. Matsui, High pressure synthesis of $ZnSiO_3$ ilmenite, *Phys. Earth Planet. Interiors, 9*, 344-352, 1974.

Ito, E., T. Matsumoto, and N. Kawai, High pressure decompositions in manganese silicates and its geophysical implications, *Phys. Earth Planet. Interiors, 8,* 241-245, 1974.

Ito, E., T. Matsumoto, K. Suito, and N. Kawai, High pressure breakdown of enstatite, *Proc. Japan Acad., 48,* 412-415, 1972.

Kawai, N., A static high pressure apparatus with tapering multi-piston formed a sphere I, *Proc. Japan Acad., 42,* 385-388, 1966.

Kawai, N., and S. Endo, The generation of ultrahigh hydrostatic pressure by a split sphere apparatus, *Rev. Sci. Instr., 4,* 425-428, 1970.

Kawai, N., S. Endo, and K. Ito, Split sphere high pressure vessel and phase equilibrium relation in the system Mg_2SiO_4-Fe_2SiO_4, *Phys. Earth Planet. Interiors, 3,* 182-185, 1970.

Kawai, N., M. Tachimori, and E. Ito, A high pressure hexagonal form of $MgSiO_3$, *Proc. Japan Acad., 50,* 378-380, 1974.

Liu, L.-G., Disproportionation of Ni_2SiO_4 to stishovite plus bunsenite at high pressures and temperatures, *Earth Planet. Sci. Lett., 24,* 357-362, 1975*a*.

Liu, L.-G., High-pressure disproportionation of Co_2SiO_4 spinel and implications for Mg_2SiO_4 spinel, *Earth Planet. Sci. Lett., 25,* 286-290, 1975*b*.

Liu, L.-G., Post-oxide phases of forsterite and enstatite, *Geophys. Res, Lett., 2,* 417-419, 1975*c*.

Ringwood, A. E., Phase transformations and the constitution of the mantle, *Phys. Earth Planet. Interiors, 3,* 109-155, 1970.

Ringwood, A. E., *Composition and Petrology of the Earth's Mantle,* McGraw-Hill, Inc., New York, 481-514, 1975.

Ringwood, A. E., and A. Major, High pressure transformations in pyroxenes, *Earth Planet. Sci. Lett., 1,* 351-357, 1966.

Ringwood, A. E., and A. Major, Synthesis of majorite and other high pressure garnets and perovskites, *Earth Planet. Sci. Lett., 12,* 411-418, 1971.

Ringwood, A. E., and M. Seabrook, High pressure transformations of $MgGeO_3$ from pyroxene to corundum structure, *J. Geophys. Res., 67,* 1690-1691, 1962.

Ringwood, A. E., and M. Seabrook, High pressure transformations in germanate pyroxenes and related compounds, *J. Geophys. Res., 68,* 4601-4609, 1963.

Suito, K., Phase transformations of pure Mg_2SiO_4 into a spinel structure under high pressures and temperatures, *J. Phys. Earth, 20,* 225-243, 1972.

Suito, K., Phase relations of pure Mg_2SiO_4 up to 200 kbar, in *High-Pressure Research: Applications to Geophysics,* edited by M. H. Manghnani and S. Akimoto, Academic Press, New York, 255-266, 1977.

Syono, Y., S. Akimoto, and Y. Matsui, High pressure transformations in zinc silicates, *J. Solid State Chem., 3,* 369-380,

1971*a*.

Syono, Y., M. Tokonami, and Y. Matsui, Crystal field effect in the olivine-spinel transformation, *Phys. Earth Planet. Interiors, 4,* 347-352, 1971*b*.

Tachimori, M., Phase equilibrium study in the system $MgSiO_3$-$MnSiO_3$ at very high pressures and temperatures (in Japanese), M.S. Thesis, Osaka University, Osaka, 1974.

PHASE TRANSITIONS IN RUTILE-TYPE STRUCTURES

J. C. JAMIESON
Department of the Geophysical Sciences
University of Chicago
Chicago, Illinois 60637

Abstract

Transitions in rutile-type RX_2 compounds are as ubiquitous in these phases as in the better studied silica phases arising from the quartz structure. These transitions may be either reconstructive or displacive. If reconstructive, the basic coordination number of 6 changes to 8 or 9. In SiO_2 (rutile), volume changes of 10%, arising from a change from a 6 to 8 coordination, decrease the Gibbs free energy by 3 kcal/mole for a transition at 100 kbar, or 30 kcal at 1 Mbar. The larger transitions corresponding to a 6 to 9 coordination change have $\Delta V/V$ of 20-30% corresponding to energy changes of 6-9 kcal/mole for a transition at 100 kbar. Displacive transitions may be real or virtual. They seem to be invariably signaled by $dC_S/dP < 0$. They are virtual when they are not observed due to the occurrence of a large volume phase change at pressures lower than that at which $C_S = 0$. Data will be presented on NiF_2 which has a predicted (*A. Wu*) magnetic phase transition at about 45 kbar, which is apparently of second or higher order.

I. INTRODUCTION

In 1952, *Birch* offered his celebrated hypothesis that the mineralogy of the earth below a certain depth, 900-1000 km, was predominantly oxide phases. This hypothesis remained experimentally untestable until the availability of high-pressure/high-temperature apparatus led to the discovery of a new form of SiO_2 (stishovite) [*Stishov* and *Popova*, 1961], some 62% denser than α-quartz. This seemed to confirm Birch's hypothesis since reactions of the type

$$Mg_2SiO_4 \leftrightarrow 2\ MgO + SiO_2 \tag{1}$$

$$MgSiO_3 \leftrightarrow MgO + SiO_2 \qquad (2)$$

would be driven to the right by pressure. However, *Bridgman* [1945, 1951] had already suggested an important role for polymorphism in geological and geophysical processes. *Bernal* [1936] had also suggested polymorphism on the LHS of (1) and (2) from germanate analogues, which would equally apply, of course, to the RHS due to the known polymorphism of GeO_2. Hence, the effect of pressure on (1) and (2) is really unknown until we have located the region of stability in (P,T) space of several pure endmembers, to say nothing of variation with composition [in (1) and (2), Mg stands for any mixture of divalent cations].

II. AVAILABLE PHASES

I now wish to introduce the concept of availability of a given structure, namely that two crystal structures are known (even in two different chemical compositions but identical formula), which, if they exist in any one compound, may be related in an equilibrium, subject to thermodynamic constraints--e.g., the phase existing at the higher temperature has the higher entropy, and the higher pressure phase has the greater density. Na, K, and Rb halides are well known to transform to CsCl structures at high pressures. Hence, an available structure for MgO in (1) and (2) would be that of CsCl. Still another possibility is that of NiAs, since all three are known to exist in AgF [*Jamieson et al.*, 1975]. If we neglect the latter possibility, it might be thought that the possibility of MgO in the CsCl structure should be seriously considered for the earth's mantle. This would have a strong effect on the result of studies on crystal field splitting of Fe in our hypothetical CsCl-type MgO, due to the change in coordination number from 6 to 8. However, a consideration of Figure 1 shows that MgO is an unlikely candidate for the CsCl structure type. The $\Delta V/V_o$, as given by *Liu* and *Bassett* [1973] is plotted for eight alkali halides with our data [*Jamieson et al.*, 1975] on AgF versus radius ratio, r_A/r_X, using the radii of *Prewitt* and *Shannon* [1969]. A systematic relationship appears in which $-\Delta V/V_o$ can vanish for r_A/r_X between 0.5 and 0.55. When this happens, the CsCl type is less dense than the NaCl type and, hence, may not be adopted at high pressures. Whether or not this same relationship holds for oxides is unknown, but if it does, MgO with $r_A/r_X = 0.51$ is completely indeterminate as either $\Delta V/V_o > 0$, or else it is so small as to be undetectable in shock-wave experimentation. Incidentally, the presence of Fe in solid solution leads to a smaller effective r_A/r_X, hence making the transition even less probable. A negative fact bolsters Figure 1, namely the remaining Ag halides with ratios 0.40 to 0.49 transform to different crystal structures at high pressures [*Schock* and *Jamieson*, 1969], nor are any transitions observed in LiBr (0.38) and LiCl (0.41) [*Carter*, 1973]. LiF (0.56) would have a transition predicted by Figure 1, and *Carter* [1973] adds the possibility that one exists at ~900 kbar. He also reports a firm transition in NaF (0.77). The available data on NaBr (0.52) and NaI (0.46) are too sparse to make any statement. Figure 1 combined with *Liu* and *Bassett*'s [1973] relationship that ΔS is a linear function of $-\Delta V/V_o$ means that that quantity may also be related to r_A/r_X. We can neglect any phase changes in MgO here (however, a completely unexpected change to a structure different from CsCl might occur) and discuss SiO_2, assuming that it is already in the stishovite form.

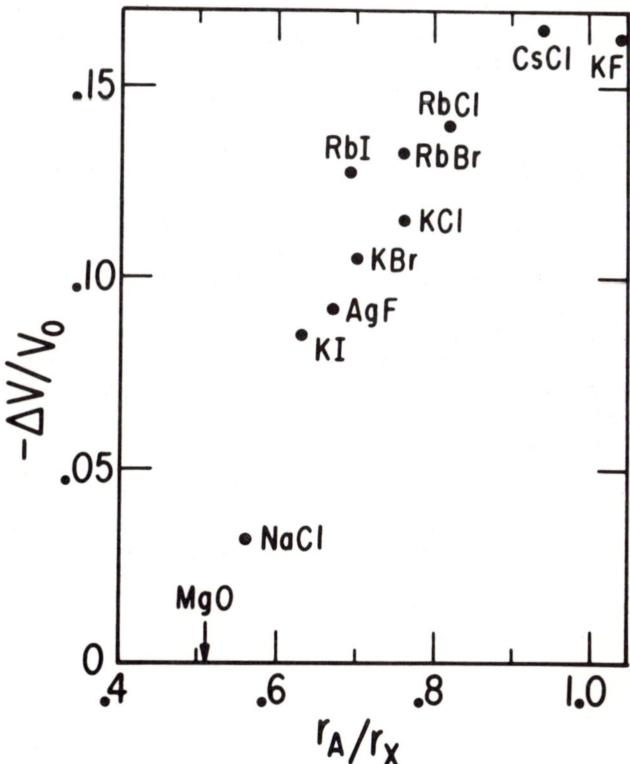

Fig. 1. $\Delta V/V_o$ versus r_A/r_X, the radius ratio for alkali halides showing a transition with pressure of the NaCl-CsCl type. KF is taken from Vaidya and Kennedy [1971], RbF from Weir and Piermarini [1964]. KF and CsF from the Wier and Piermarini reference lie outside the plot in qualitative agreement.

III. FLUORIDE ANALOGUES

The use of fluorides as analogues to oxides and germanates to silicates has been known for some time. An excellent review of the concepts and literature has recently appeared [*Jones*, 1976]. In this report, only fluorides and analogous oxides are discussed.

The only nontransition metal difluorides occurring in the rutile structure are MgF_2 and ZnF_2. In a study on MgF_2, no transitions were found to over 100 kbar [*Kabalkina* and *Popova*, 1964; *Dandekar* and *Jamieson*, 1969]; however, ZnF_2 has supplied a wealth of phenomena. In *Kabalkina* and *Popova's* study [1964], X-ray diffraction was used on ZnF_2 after exposure to pressure. It was reported to be in the α-PbO_2 form and suggested to be, indeed, the high-pressure form. The hazards of this type of study were later confirmed [*Kabalkina et al.*, 1968; *Dandekar* and *Jamieson*, 1969] when with in situ X-ray analysis, the true high-pressure form was found to be a distorted fluorite type, which, on pressure release, transformed to the α-PbO_2 type. *Kabalkina et al.*, [1968] also reported a cubic form of the fluorite type at 300°C and 70-80 kbar with $\Delta V/V_o = 9\%$, where V_o corresponds to the rutile form. Further studies by *Lityagina et al.*, [1974] established a phase diagram for ZnF_2 with a triple point at 250°C and ~85 kbar, $dP/dT < 0$ for rutile → fluorite. They found a similar phase diagram for MnF_2 at lower pressures. In our room-temperature studies on MnF_2 [*Dandekar* and *Jamieson*, 1969], we have verified a distorted CaF_2 form, but more recent hydrostatic in situ X-ray analysis on MnF_2 has revealed [*Yagi* and *Jamieson*, in preparation 1976] still a different fluorite-like form which is also tetragonal. In addition, at circa 80 kbar an orthorhombic form appears indexable as the α-$PbCl_2$ structure, some 10% denser than the distorted fluorite at their equilibrium. This high-pressure form of fluorite has also been found in CaF_2, SrF_2, and BaF_2 [*Seifert*, 1966; *Dandekar* and *Jamieson*, 1969; and *Fursenko* and *Kirkinskii*, 1970] and reversible in CdF_2 [*Dandekar* and *Jamieson*, 1969]. $\Delta V/V$ are all ~10%--i.e., there is no noticeable trend in $\Delta V/V$ with cation contrary to the univariant halides above.

Using MnF_2 as a type for rutile transitions, the following occurs at room temperature:

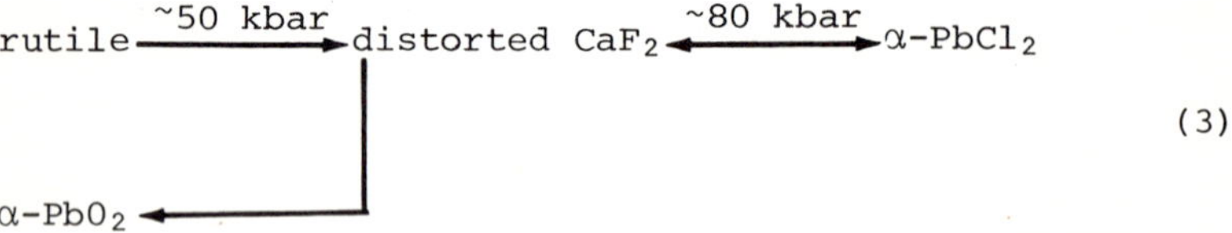

(3)

The latter is some 30% denser at 80 kbar [*Dandekar* and *Jamieson*,

1969]. This is quite reminiscent of occurrences in shock-wave studies on TiO_2 [*McQueen et al.*, 1967], in which a large volume change detectable on the Hugoniot was represented by the recovery of only the slightly denser α-PbO_2 form. In laser heating-diamond cell studies, *Liu* [1974] recovered an α-PbO_2 form of SnO_2 from P = 155 kbar and T = 1000-1400°C. Contrasting its small density difference to rutile with the large difference (30%) found in shock-wave studies, *Liu* too concluded that the recovered α-PbO_2 form was not the large transition, or transitions, detected by shock studies (the techniques commonly used do not detect a second transition on the Hugoniot if they are too close--about 700 kbar in the case of rutile TiO_2). The pressure reported is that of the first transition; ΔV that of the combination.

The above discussion allows one to draw a "rubber" phase diagram for AX_2 compounds as shown in Figure 2. (The rutile-fluorite and fluorite-distorted fluorite lines are adopted from MnF_2 and ZnF_2 in *Lityagina et al.*, [1974]). The slope of the fluorite/α-$PbCl_2$ line stems from the fact that α-$PbCl_2$ is denser in all known cases and in PbF_2 is the low-temperature form relative to the fluorite form [*K.-H. Hollwege*, 1973]. The slope of the α-$PbCl_2$ to distorted-fluorite forms is unknown, except that α-$PbCl_2$ must be the high-pressure form, and its slope is also constrained by the triple-point conditions with fluorite. A phase field for the α-PbO_2 structure has been drawn to be consistent with the data for α-PbO_2 and fluorite of *Syono* and *Akimoto* [1968]. A very recent hydrostatic in situ X-ray diffraction study [*T. Yagi*, private communication 1976] suggests that this slope is very likely negative rather than positive but still steep. The dashed metastable extension of this is indicated at lower pressures. A second dashed line immediately to the left of the rutile-fluorite-distorted fluorite represents a metastable extension from the metastable triple point α-PbO_2-rutile-distorted fluorite. I believe these metastable extensions are the key to understanding phase behavior at P and T of rutile-type AX_2 compounds--in particular, the α-PbO_2 to distorted fluorite. It explains the ubiquitous appearance of α-PbO_2 types in quench studies from elevated pressures. This also suggests that the finding of α-PbO_2 types in post-mortum analyses is a telltale that says one or more large-volume transitions have occurred. If we accept this postulate, then fluorite, distorted fluorite, and α-$PbCl_2$ are all "available" phases for stishovite and should be considered in reaction (1) and (2), especially in light of the finding by *German et al.*, [1973] of the α-PbO_2-form in SiO_2, which had been shock-loaded to 700-900 kbar. On the other hand, the lack of any reported transition--other than that from quartz to stishovite in the data of *Wackerle* [1962] as analyzed by *McQueen et al.*, [1963]--seems to preclude a further large-volume transition in SiO_2. To settle this matter, a shock-wave experiment with stishovite as starting material is needed so as to drastically lower the temperatures attained along the Hugoniot.[1]

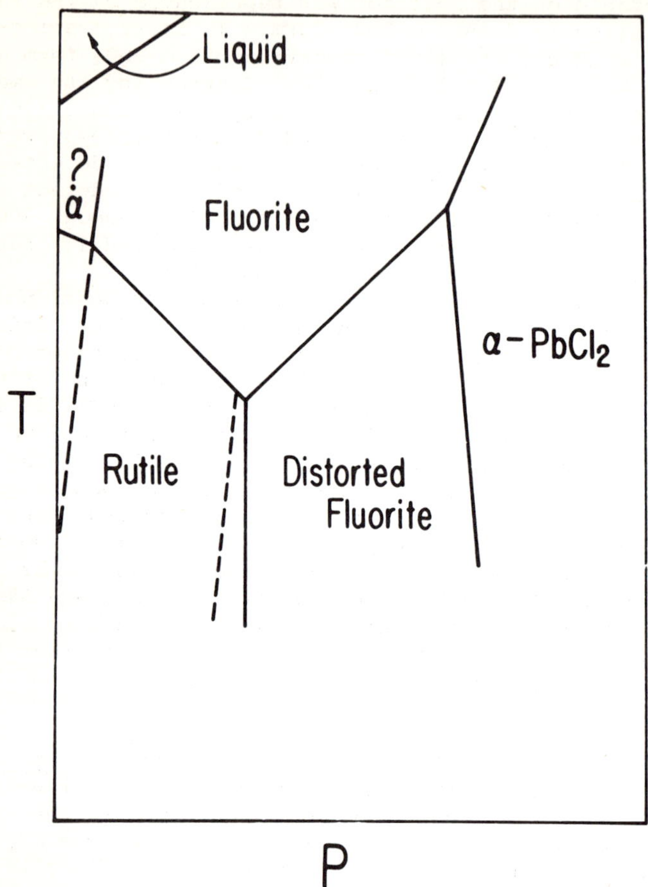

Fig. 2. "Rubber" phase diagram for AX_2 compounds showing availability of large possible $\Delta V/V_o$ for phase transitions from the rutile type.

---

[1] A very recent example of the lack of detection in shock-wave experimentation of a transition known to occur in static experimentation are the experiments of *Dandekar* [Private communication, 1976]. Using a high-velocity gas gun, he studied BaF_2 (fluorite structure) to 100 kbar and found no trace of a transition to the α-$PbCl_2$ form with $\Delta V/V_o$ ~11% known to occur in static experiments at $P < 20$ kbar [*Seifert*, 1966; *Dandekar* and *Jamieson*, 1969]. Similar studies on PbF_2 with a known static transition at 4 kbar, revealed no transition to 40 kbar.

Other known pressure transitions in rutile-type structures are of small ΔV, many magnetic in origin, but they at least form an "available" class for SiO_2. *Austin* [1969] reported quenching a new structure in NiF_2 and CoF_2 from elevated temperature and pressure. Although not rigorously proven, *Nagel* and *O'Keefe* [1971] have identified this as the $CaCl_2$ type, a slightly distorted rutile form. The density differences are exceedingly small, less than 0.5% [*Austin*, 1969]. We have studied NiF_2 with in situ X-ray diffraction and found *Austin's* phase [*Wu* and *Jamieson*, submitted to *Jour. Appl. Phys.*, 1976]. Our study used Cu Kα radiation with the sample immersed in a truly hydrostatic fluid. To the best of our resolving power, the transition was continuous: starting about 20 kbar, appropriate lines, such as 101 and 011, began to separate and became clearly separable at about 40-50 kbar. Here we have a continuous distortion in a rutile structure to an orthorhombic one. Pictures taken after the pressure studies showed the normal rutile form in contrast to *Austin's* results. While this change in NiF_2 may be explained by magnetic ordering, the existence of this apparently same structure in $CaCl_2$ can not. In the case of NiF_2, the elastic constant $C_S = 1/2\ (C_{11} - C_{12})$ was found to dominish rapidly with pressure. Such behavior is said to reflect the occurrence of a transition to another form at high pressure, when $C_S = 0$ and a lattice instability results. This has been rigorously verified [*Peercy* and *Fritz*, 1974; *Worlton* and *Beyerlein*, 1975] for TeO_2, which changes continuously from tetragonal to orthorhombic at about 9 kbar where C_S has vanished. Such transitions may be virtual in that another transition, reconstructive and large ΔV, may occur before $C_S = 0$. In this case, the decreasing C_S is not predictive of the transition that actually occurs.

IV. GEOPHYSICAL CONCLUSIONS

Liu [1974] has discussed (1) and (2) with SiO_2 in the α-PbO_2 and concludes that a mixture of periclase and perovskite $MgSiO_3$, or perovskite $MgSiO_3$ alone, would be the stable assemblage in the lower mantle. This is based on 1-bar density differences and does not take into account possible differences in compressibility of the various phases. This in itself is important as the density differences are small, 1-3%. Transformation of SiO_2 (rutile) to any of its available structures would make the oxides more dense than the perovskite side of the reaction. If one estimates the energy ($P\Delta V$) to be gained by such transition, the high pressures of the earth predominate. For example, at the 900-km depth $P \sim$ 300 kbar, a change of stishovite to a denser (10%) fluorite form (neglecting compressibility) gives 9 kcal, while transition to the α-$PbCl_2$ form (30%) gives 27 kcal to the RHS of (1) and (2).

With such large energy differences and the known availability of much denser AX_2 forms than rutile, it is hard to justify any hypothesis other than the original oxide one of *Birch* [1952] at this time. Figure 3, however, should give one pause in coming to any conclusion. It is a (P,T) diagram for the earth's mantle and is a redrawing of one originally presented in 1966 [*Jamieson*, 1966]. Superimposed is the currently known (P,T) diagram for iron; surrounding that is a field in which the current techniques of laser-heating diamond cells are providing so much valuable information today. In conclusion, I feel there are sufficient alternate possibilities so that the question as to whether the lower mantle is the perovskite structure or oxides can only be given the Scottish verdict of "Unproven."

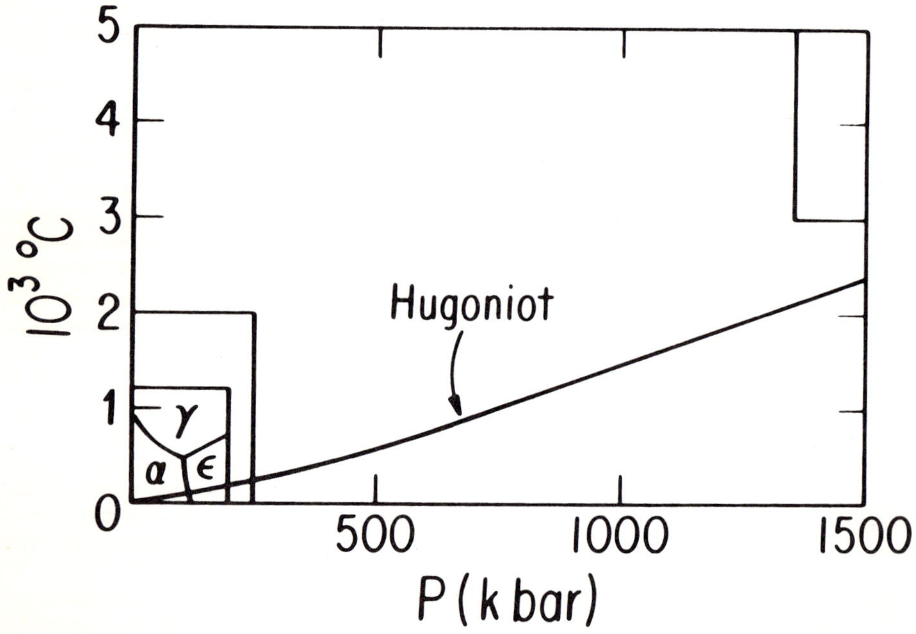

Fig. 3. The range of pressure and temperature in the earth's mantle. Superimposed (lower left), our current knowledge of Fe phases, the current P,T conditions for study of silicates and oxides, and the Fe Hugoniot. The box at upper right is a rough estimate of conditions at the base of the mantle.

Acknowledgments. This research was supported by the Earth Sciences Section, National Science Foundation, NSF Grant EAR 73-00636 A02. We have also benefited from support of the Materials Research Laboratory by the National Science Foundation. Correspondence with T. Yagi was very helpful.

REFERENCES

Austin, A. E., High pressure transformations of transition metal difluorides, *J. Phys. Chem. Solids, 30,* 1282-1285, 1969.

Birch, F., Elasticity and constitution of the earth's interior, *J. Geophys. Res., 57,* 227-286, 1952.

Bernal, J. D., Remark, *Observatory, 59,* 268, 1936.

Bridgman, P. W., Polymorphic transitions and geological phenomena, *Am. J. Sci., 243a,* 90-97, 1945.

Bridgman, P. W., Some implications for geophysics of high-pressure phenomena, *Bull. Geol. Soc. Amer., 62,* 533-535, 1951.

Carter, W. J., Hugoniot equation of state of some alkali halides, *High Temperatures-High Pressures, 5,* 313-318, 1973.

Dandekar, D. P., and J. C. Jamieson, Some high pressure phases of RX_2 fluorides, *Trans. Amer. Cryst. Assoc., 5,* 19-28, 1969.

Fursenko, B. A., and V. A. Kirkinskii, X-ray investigation of polymorphic transition in CaF_2 at high pressure, Akad. Nauk SSSR (Novasibersk) in *Experimental Studies in Mineralogy,* 1970-1971.

German, V. N., M. A. Podurets, and R. F. Trunin, Synthesis of a high density form of silicon dioxide in shock waves, *Sov. Phys.-JETP, 37,* 107, 1973.

Hollwege, K.-H., ed., Landolt-Börnstein New Series, Group III, Vol. 7. Springer-Verlag Berlin 1973.

Jamieson, J. C., Pressure, the new dimension in Inorganic Chemistry, *Trans. Roy. Soc. Canada 4th series Vol. IV Sect. III,* 263, 1966.

Jamieson, J. C., P. M. Halleck, R. B. Roof and C. W. F. T. Pistorius, Additional polymorphism and non-stoichiometry in AgF, *J. Phys. Chem. Solids, 36,* 939-944, 1975.

Jones, Leonie E. A., High temperature elasticity of fluoride and oxide analogues, Ph.D. Thesis, Australian National University, April 1976.

Kabalkina, S. S., and S. V. Popova, Phase transitions in zinc and manganese fluorides at high pressures and temperatures, *Soviet Physics-Doklady, 8,* 1141-1143, 1964.

Kabalkina, S. S., L. F. Vereschagin and L. M. Lityagina, X-ray diffraction study of ZnF_2 at pressures up to 130 kb and at temperatures of 25 and 300°, *Soviet Physics-Doklady, 12,* 946-949, 1968.

Lityagina, L. M., M. F. Kachan, S. S. Kabalkina and L. F. Vereschagin, Fluorite and fluorite-like modification on the phase diagrams of MnF_2, ZnF_2 and CoF_2, *Doklady Akad. Nauk SSSR, 216,* 1066-1069, 1974.

Liu, L., and W. A. Bassett, Compression of Ag and phase transformation of NaCl, *J. Appl. Phys., 44,* 1475-1479, 1973.

Liu, L., Synthesis of a new high pressure phase of tin dioxide and some geophysical implications, *Phys. Earth and Planetary Int., 9,* 338-343, 1974.

McQueen, R. G., J. N. Fritz and S. P. Marsh, On the equation of state of stishovite, *J. Geophys. Res., 68,* 2319-2322, 1963.

McQueen, R. G., J. C. Jamieson and S. P. Marsh, Shock wave compression and X-ray studies of titanium dioxide, *Science, 155,* 1401-1404, 1967.

Nagel, L., and M. O'Keeffe, Pressure and stress induced polymorphism of compounds with the rutile structure, *Mat. Res. Bull., 6,* 1317-1320, 1971.

Peercy, P. S., and I. J. Fritz, Phenomenological theory of the high-pressure structure phase transition in paratellurite (TeO_2), *Sol. State Comm., 16,* 1197-1200, 1975.

Prewitt, C. T., and R. D. Shannon, Use of radii as an aid to understanding the crystal chemistry of high pressure phases, *Trans. Amer. Cryst. Assoc., 5,* 51-60, 1969.

Schock, R. N., and J. C. Jamieson, Pressure-induced phase transformations in the Bl Ag-halides, *J. Phys. Chem. Solids, 30,* 1527-1533, 1969.

Seifert, K.-F., Strukturunwandlungen von Halogeniden des Typus AX_2 unter höheren Drücken, *Ber. Bunsen Gel. Physik. Chem., 70,* 1041-1042, 1966.

Stishov, S. M., and S. V. Popova, A new dense form of silica, *Geokhimiya, 10,* 837, 1961.

Syono, Y., and S. Akimoto, High-pressure synthesis of fluorite-type lead dioxide, *Mater. Sci. Res. Bull., 3,* 153-157, 1968.

Vaidya, S. N., and G. C. Kennedy, Compressibility of 27 halides to 45 kbar, *J. Phys. Chem. Solids, 32,* 951-964, 1971.

Wackerle, J., Shock-wave compression of quartz, *J. Appl. Phys., 33,* 922-937, 1962.

Weir, C. E., and G. I. Piermarini, Lattice parameters and lattice energies of high-pressure polymorphs of some alkali halides, *J. Res. NBS, 68A,* 105-111, 1964.

Worlton, T. G., and R. A. Beyerlein, Structure and order parameters in the pressure-induced continuous transition in TeO_2, *Phys. Rev. B 12,* 1899-1907, 1975.

ORTHORHOMBIC PEROVSKITE (Mg, Fe)SiO_3 AND CONSTITUTION OF THE LOWER MANTLE

H. SAWAMOTO
Department of Earth Science
Nagoya University
Nagoya, Japan

Abstract

The orthorhombic perovskite $(Mg_{1-x}Fe_x)SiO_3$ ($0 \leq x \leq 0.6$) is synthesized at 350 kbar and 1300°C by means of a MASS-318 type high-pressure apparatus. The lattice constants of the perovskite at $x = 0$ are $a = 4.815 \pm 0.004$ Å, $b = 5.012 \pm 0.004$ Å and $c = 6.916 \pm 0.004$ Å, and they vary linearly with x. The extrapolated values at $x = 1$ are $a = 4.833 \pm 0.004$ Å, $b = 5.058 \pm 0.004$ Å, and $c = 6.931 \pm 0.004$ Å. The present values are considerably larger than those presented by *Liu* [1976a]. It is argued that perovskite $(Mg,Fe)SiO_3$ has some imperfections involving crystal-chemical complexity.

The elastic parameters of the high-pressure phases of olivine and pyroxene at ambient condition are estimated by *Birch*'s [1961] and the other similar velocity-density systematics. The velocity-density profiles are computed from the Birch-Murnaghan equation of state and compared with the seismological data. It is noted that the observed jump in *Vp* at the 650-km depth is not explained if uniformity of chemical composition is assumed in the mantle. An overall fit of the experimental data to the seismological data leads to the following conclusion; although the mineralogy of the upper mantle is represented mostly by olivine and its high pressure polymorphs with Fe/(Fe + Mg) = 0.05 - 0.15, the chemical composition of the lower mantle should be close to pyroxene stoichiometry with Fe/(Fe + Mg) = 0.0 - 0.1. In order to account for the significant difference in chemical composition of the upper and lower mantles, there should be some global process of chemical differentiation by which Mg and Fe are extracted from the lower mantle.

I. INTRODUCTION

Recently the layered structure in the mantle has attracted

the attention of solid state geophysicists. It is widely believed that the stepwise increase in seismic wave velocity in the upper mantle is caused by a series of pressure-induced phase transitions in the ferromagnesian silicates.

Previous interpretations of seismic velocity profiles in the mantle have been carried out by *D. Anderson* [1970], *Ringwood* [1972], and others on the basis of recent high-pressure data. These interpretations have suggested that the distinct stepwise increase in seismic waves is attributed to the phase change of magnesium-rich olivine to a β-phase (modified spinel structure) near 400-km depth, further to a γ-phase at about a 550-km depth, and finally to a post-spinel phase at about a 650-km depth. Recent research has focused on the nature of the post-spinel phase and, hence, the mineralogical constitution of the lower mantle below 650-km.

In a previous paper [*Kumazawa et al.*, 1974], we reported the decomposition of pure forsterite to periclase plus stishovite. *Ming and Bassett* [1975b] also reported the same decomposition in $(Mg,Fe)_2SiO_4$. Therefore, the major phase of the lower mantle was thought to be mixed oxides [*Al'tschuler et al.*, 1965]. However, *Ringwood* [1972], *Burdick and Anderson* [1975] and others have reported that the lower mantle has a velocity higher than that predicted by mixed oxides corresponding to olivine stoichiometry, which is a compatible major chemical composition in the upper mantle. Therefore, there are three possibilities: (1) further phase transitions give rise to denser phases appropriate to the lower mantle; (2) chemical composition of the lower mantle is different from that of the upper mantle; and (3) the combination of (1) and (2). Using a diamond anvil device, *Liu* [1975] has recently found an orthorhombic perovskite-like phase of $MgSiO_3$. This polymorph is denser than the corresponding mixed-oxide assemblage, as predicted by *Reid and Ringwood* [1969, 1975]. In view of the existence of the perovskite phase in magnesian silicate, *Liu* [1975] has concluded that oxide mixtures are not stable mineral assemblages under the physical condition of the lower mantle and that $MgSiO_3$ perovskite is the most important phase of the lower mantle.

The purpose of this paper is (1) to report the properties of the orthorhombic perovskite phase in the $MgSiO_3$-$FeSiO_3$ system synthesized at high pressure and high temperature, and (2) to present a consistent mineralogical and chemical model for the lower mantle and discuss the nature of a seismological discontinuity at a 650-km depth.

II. EXPERIMENTAL PROCEDURE

A. High-Pressure Apparatus

The high-pressure instrumentation used was the MASS 318 type apparatus driven by a couple of RH3 guide blocks in a uniaxial press [*Kumazawa et al.*, 1972]. The mechanism and mechanics of the MASS-3I8-type device have been reported in previous paper [*Masaki et al.*, 1975], and are summarized briefly in this paper.

The two N-anvils (truncated cube), six W-anvils (rectangular prism), twelve sheets of compressible pads, and pressure medium are assembled together to constitute a cube. When the assembly is isotropically compressed in a cubic press or between two RH3 guide blocks, the pressure medium is uniaxially compressed by a couple of N-anvils (cemented WC). The W-anvils (cemented WC) displace outward, giving rise to the stroke of N-anvils and supplying laterally supporting stress to each side of the anvils to reduce stress deviator. The laminated mica sheet and stainless steel plates, perforated and unperforated, are used as compressible pads. The effective compressibility of the pads is varied by changing the overall thickness of the pads and the fraction of perforation in the stainless steel. The optimum compressibility of the pads is determined on a trial-and-error basis. The use of perforated stainless steel pads is a new technique, enabling easier control of their compressibility. Elastic deformation of anvils takes place in the region adjacent to the pressurized pressure medium. In order to compensate for anvil deformation, mica sheets are inserted between the anvils as the modified elastic design proposed by *Kumazawa* [1973]. In order to provide the electric insulation and to reduce friction between the anvils, a sheet of mica is placed between contacting anvils. The overall size of the cube-shaped MASS assembly is 68.2 mm on edge.

B. Pressure Cell and Calibration

The shape of the pressurized cavity is that of two interpenetrating trigonal pyramids. In practice, it is prepared by utilizing two truncated trigonal pyramids (largest edge length, 4.2 mm; height, 1.0 mm). The total volume of the pressurized cavity is 11.4 mm^3. The peripheral portion of the pressure medium is made of pyrophyllite, but the central part is made of boron nitride, which surrounds a graphite tube heater.

The starting material is directly packed into a graphite tube without any seal. The volume of the specimen before compression is 0.95 mm^3. The thermocouple connection is brought out through two of the W-anvils from the inside faces of guide blocks.

The generated pressure is calibrated to press load at room temperature by means of the resistance changes associated with the phase change in Bi (I-II, 26 kbar; IV-V, 76 kbar [*Jeffery et al.*, 1966]), GaAs (semiconductor-metal, 180 kbar), and GaP (semiconductor-metal, 220 kbar) [*Piermarini and Block*, 1975]. The actual pressure of high-temperature and high-pressure runs includes an uncertainty of about ± 50 kbars.

C. Temperature Measurement

In the high-pressure and high-temperature runs, temperature is measured with a Pt/Pt-13%Rh thermocouple without any correction for the pressure effect on the thermocouple emf.

We have found in some case that the emf of the thermocouple

decreases significantly at high temperature because of the dissolution of the carbon in the thermocouple wire. Further, the electric current through the carbon heater decreases considerably with time as a result of the reduction of the electrical conductivity due to the direct conversion of carbon to diamond. Therefore, the relation between the input electrical power to the furnace and the temperature in the sample is recorded during the first several minutes of each run. In order to keep the temperature of the sample at the prescribed value, the input power is maintained at constant value given by an interpolation or extrapolation of the relation between the power and temperature obtained during the first several minutes of the high-pressure and high-temperature runs. Because of the above problems, we believe our temperature, at high pressure, has an uncertainty of at least ± 200°C.

D. Sample Preparation and Examination

The starting material $(Mg,Fe)SiO_3$ is prepared by a tetrahedral anvil-type press at 50 kbar and 1200°C from synthetic Mg_2SiO_4, Fe_2SiO_4, and α-SiO_2. Mg_2SiO_4 and Fe_2SiO_4 are of laboratory stock originally synthesized from $MgCO_3$, $Fe(COOH)_2$, and α-SiO_2 of reagent grade using standard techniques.

$MgSiO_3$, $FeSiO_3$, and $(Mg,Fe)SiO_3$, thus synthesized, are found to be a single-phase orthopyroxene and, hence, are used as starting materials for the high-pressure runs.

Pressure is applied to the sample first; then the temperature is brought to the desired value and held constant for on the order of 10^3 sec. The furnace power is then turned off, and the assembly is allowed to be quenched to ambient temperature within a few seconds before pressure is released. In the quenched runs, a considerable amount of contaminant, such as the materials of heater, electric leads, thermocouple, some parts of the pressure medium, and their high-pressure polymorphs, firmly adhere to the specimen. The contaminants are carefully removed under a binocular microscope. The sample is then ground into powder. The amount of sample recovered is usually $\sim 10^{-4}$ gm and is mounted on a quartz plate. The samples are analyzed by X-ray diffractometry.

III. RESULTS

A. Perovskite Phase

The experimental conditions and the products of each run are summarized in Table 1. The data of X-ray diffraction obtained for $MgSiO_3$ are shown in Table 2, in which γ-Mg_2SiO_4 (spinel), stishovite, periclase, gold, graphite and some unknown phase are identified. Gold is undoubtedly a contaminant from the electric leads. Enstatite has been found to first break down into a mixture of γ-Mg_2SiO_4 (spinel) plus SiO_2 (stishovite) [*Ringwood and Major*, 1968; *Ito et al*., 1972], and then to recombine to form the hexagonal $MgSiO_3$ (ilmenite structure), and then again to dispro-

TABLE 1. Results of the P-T Runs in the $MgSiO_3$-$FeSiO_3$ system

Run no.	Starting Materials Composition	Phase[a]	P kbar	T °C	Time min.	Phase[b] Present
01	$MgSiO_3$	en	370	1400	20	p≧st≧m>sp
04	$(Mg_{0.8}Fe_{0.2})SiO_3$	ss	300	1300	50	sp>p≧st>mw
06	$(Mg_{0.6}Fe_{0.4})SiO_3$	en-fs	300	1200	30	sp>st≧p≧mw
07	$(Mg_{0.4}Fe_{0.6})SiO_3$	ss	300	1200	30	sp>st>p≧mw

a. en=enstatite, ss=solid-solution of orthopyroxene, and en-fs=mixture of enstatite and ferrosilite.
b. p=perovskite, st=stishovite, m=periclase, sp=spinel, and mw=magnesiowustite.

portionate into a simple oxide mixture. Therefore, the spinel, stishovite, and ferro-periclase contained in the sample are regarded as intermediate phases of reaction. The presence of periclase is demonstrated by the diffraction line (220), although other diffraction lines are overlapping with those of the other phases present (*e.g.*, periclase (111) with spinel (311), and periclase (200) with perovskite (discussed later) (121) or (031). Ilmenite-structured $MgSiO_3$ reported by *Kawai et al.* [1974], *Ming and Bassett* [1975a] and *Liu* [1976b] was not found in the present X-ray diffraction patterns. By removing all the known diffraction lines in Table 2, 24 X-ray diffraction lines remain to be interpreted. The indexing of these diffraction lines was attempted by assuming a new phase with orthorhombic symmetry similar to the orthorhombic perovskite discovered by *Liu* [1975]. The resulting lattice constants for their fit, of what is assumed to be the new phase, are $a = 4.815 \pm 0.004$ Å, $b = 5.012 \pm 0.004$ Å, and $c = 6.916 \pm 0.004$ Å. The agreement between calculated and observed *d*-spacings is good. The volume of the unit cell is 166.90 ± 0.50 Å^3 with six formula-units per unit cell.

There are six lines in the diffraction pattern obtained in the $MgSiO_3$ experiment(Table 2) which cannot be indexed with those of the new phase. These unidentified lines are also found in the $(Mg_{0.6}Fe_{0.4})SiO_3$ run, but are not found in the $(Mg_{0.8}Fe_{0.2})SiO_3$, and $(Mg_{0.4}Fe_{0.6})SiO_3$ runs. Although these diffraction lines may indicate the presence of another new phase, we are, at present, neglecting their presence in the discussion below. X-ray diffraction data for the specimens from various runs are given in Tables 3, 4, and 5.

TABLE 2. X-ray Diffraction Data for $MgSiO_3$

Observed		Perovskite		Other Phases			
d(obs)	I/I_o[b]	hkl	d(cal)[c]	hkl	d(cal)	I/I_o	Notes
3.47	30	110	3.472				
3.45	20	002	3.458				
3.36	90			002	3.36	100	graphite[e]
2.977[a]	90						
2.955	90			110	2.955	100	stishovite[e]
2.876[a]	100			220	2.855	23	spinel(?)[d]
2.539[a]	25	020	2.506				
2.501	25	112	2.450				
2.452[a]	80						
2.431	40			311	2.435	100	spinel[d]
2.415	20	200	2.408				
2.349	100			111	2.355	100	gold[e]
2.248	25			101	2.247	25	stishovite[e]
2.208[a]	45						
2.116	65	121	2.116	200	2.106	100	periclase[e]
2.100	30	013	2.094				
2.087	25	103	2.079				
2.038	55			200	2.039	52	gold[e]
				101	2.03	3	graphite[e]
2.015	25			400	2.019	50	spinel[d]
1.981	30	202	1.976	111	1.979	40	stishovite[e]
1.922	30	113	1.921				
1.867	25	122	1.870	210	1.869	20	stishovite[e]
1.732	25	220	1.736				
1.723	10	004	1.729				
1.697	15	023	1.697				
1.683	20	221	1.684	004	1.679	8	graphite[e]
1.623	25	031	1.624				
1.605[a]	45						
1.572	15	130	1.579				
1.545	25	131	1.539	333	1.554	25	spinel[d]
		222	1.552	103	1.544	2	graphite[e]
1.529	45			211	1.530	60	stishovite[e]
1.503	20	032	1.504				
1.492	25	311	1.493				
1.485	30			220	1.489	25	periclase[e]
1.477	35			220	1.478	25	stishovite[e]
1.440	40	132	1.436	220	1.442	32	gold[e]
1.427	10			440	1.427	63	spinel[d]

a. unidentified lines.
b. estimated visually.
c. $a_o = 4.815 \pm 0.004$ Å
$b_o = 5.012 \pm 0.004$ Å
$c_o = 6.916 \pm 0.004$ Å
d. $a_o = 8.076$ Å [*Ito et al.*, 1972].
e. ASTM Index cards.

TABLE 3. X-ray Diffraction Data for $(Mg_{0.8}Fe_{0.2})SiO_3$

Observed		Perovskite		Other Phases			
d(obs)	I/I_o [b]	hkl	d(cal) [c]	hkl	d(cal)	I/I_o	Notes
3.48	15	110	3.475				
3.46	10	002	3.459				
3.36	50			002	3.36	100	graphite [e]
2.955	25			110	2.955	100	stishovite [e]
2.868	35			220	2.867	(23)	spinel [d]
2.516 [a]	10	020	2.510				
2.456	40	112	2.452				
2.445	100			311	2.445	(100)	spinel [d]
				111	2.445	(10)	magnesiowustite [f]
2.411	10	200	2.408				
2.247	15			101	2.247	25	stishovite [e]
2.117	10	121	2.119	200	2.118	(100)	magnesiowustite [f]
2.027	35			400	2.027	50	spinel [d]
				101	2.03	3	graphite [e]
2.094	6	013	2.095				
1.979	16	202	1.976	111	1.979	40	stishovite [e]
1.921	7	113	1.922				
1.871	13	122	1.872	210	1.869	20	stishovite [e]
1.746	10	220	1.738				
1.723	12	004	1.729				
1.686	12	221	1.686	004	1.678	8	graphite [e]
1.608	10	300	1.607				
1.587	6	130	1.581				
1.559	20	222	1.553	511	1.560	(25)	spinel [d]
				333	1.560	(25)	spinel [d]
1.530	30			211	1.530	60	stishovite [e]
1.506	6	032	1.506	220	1.497	52	stishovite [e]
1.476	13			220	1.478	(53)	magnesiowustite [f]
1.432	40	132	1.437	440	1.433	(63)	spinel [d]
1.425	7	024	1.424				

a. broad.
b. estimated visually.
c. $a_o = 4.817 \pm 0.004$ Å
$b_o = 5.020 \pm 0.004$ Å
$c_o = 6.918 \pm 0.004$ Å.
d. $a_o = 8.108$ Å.
e. ASTM Index cards.
f. $a_o = 4.233$ Å.

TABLE 4. X-ray Diffraction Data for $(Mg_{0.6}Fe_{0.4})SiO_3$

Observed		Perovskite		Other Phases			
d(obs)	I/I_o [b]	hkl	d(cal) [c]	hkl	d(cal)	I/I_o	Notes
3.49	10	110	3.482				
3.46	8	002	3.462				
3.36	55			002	3.36	100	graphite [e]
2.980 [a]	35						
2.952	55			110	2.955	100	stishovite [e]
2.878	53			220	2.878	(23)	spinel [d]
2.557 [a]	15						
2.518	15	020	2.515				
2.462	100	112	2.455	111	2.457	(10)	magnesiowustite [f]
				311	2.454	(100)	spinel [d]
2.353	15			111	2.355	100	gold [e]
2.252	15			101	2.247	25	stishovite [e]
2.209 [a]	10						
2.167	7	210	2.175				
2.120	10	121	2.120				
				200	2.128	(100)	magnesiowustite [f]
2.044	45	022	2.035	400	2.035	(50)	spinel [d]
				200	2.039	52	gold [e]
				101	2.03	3	graphite [e]
1.979	15	202	1.979				
1.920	10	113	1.924				
1.874	15	122	1.874	210	1.869	20	stishovite [e]
				331	1.867		spinel [d]
1.839	10	212	1.842				
1.741	10	220	1.741				
1.731	10	004	1.730				
1.699	5	023	1.700	004	1.678	8	graphite [e]
1.631	5	031	1.629				
1.606 [a]	10						
1.575	15	130	1.583	130	1.566	(25)	spinel [d]
1.547	6	131	1.544				
		222	1.555	103	1.544	2	graphite [e]
1.530	15			211	1.530	60	stishovite [e]
1.504	10	032	1.508	220	1.505	(52)	magnesiowustite [f]
1.494	10	311	1.495				
1.477	10			220	1.478	25	stishovite [e]
1.446	23	132	1.400	440	1.439	(63)	spine [d]
1.430	6	024	1.426				

a. unidentified lines.
b. estimated visually.
c. $a_o = 4.825 \pm 0.004$ Å
$b_o = 5.030 \pm 0.004$ Å
$c_o = 6.924 \pm 0.004$ Å.
d $a_o = 8.140$ Å.
e. ASTM Index cards.
f. $a_o = 4.257$ Å.

TABLE 5. X-ray Diffraction Data for $(Mg_{0.4}Fe_{0.6})SiO_3$

Observed		Perovskite		Other Phases			
d(obs)	I/I_o[b]	hkl	d(cal)[c]	hkl	d(cal)	I/I_o	Notes
3.48	5	110	3.483				
3.46	5	002	3.459				
3.36	50			002	3.36	100	graphite[f]
2.956	25			110	2.955	100	stishovite[f]
2.891	15			220	2.889	(23)	spinel[d]
2.813	5	102	2.810				
2.714[a]	12						
2.462	100	112	2.455	111	2.470	(10)	magnesiowustite[e]
				311	2.463	(100)	spinel[d]
2.408	2	200	2.409	101	2.247	25	stishovite[f]
2.253	10						
2.174	5	120	2.174				
2.136	2			200	2.139	(100)	magnesiowustite[e]
2.107	4	013	2.098				
2.043	55	022	2.039	400	2.043	(50)	spinel[d]
1.984	10	202	1.977	111	1.979	40	stishovite[f]
1.930	3	113	1.930				
1.873	10	122	1.877	331	1.874		spinel[d]
				210	1.869	20	stishovite[f]
1.733	5	220	1.741				
1.725	5	004	1.729				
1.692	5	023	1.699				
		221	1.689				
1.671	4	203	1.666	004	1.679	80	graphite[f]
1.637	4	031	1.633				
1.609	3	300	1.606				
1.593	7	130	1.589				
1.572	16	301	1.565	333	1.572	(25)	spinel[d]
1.552	3	222	1.555				
1.532	13			210	1.530	60	stishovite[f]
1.504[a]	10						
1.478	5			220	1.478	25	stishovite[f]
1.444	46	132	1.442	440	1.444	(63)	spinel[d]
1.425	2	024	1.427				

a. unidentified lines.
b. estimated visually.
c. $a_o = 4.819 \pm 0.004$ Å
$b_o = 5.042 \pm 0.004$ Å
$c_o = 6.919 \pm 0.004$ Å.
d. $a_o = 8.171$ Å.
e. $a_o = 4.279$ Å.
f. ASTM Index cards.

The new orthorhombic phase has an X-ray diffraction pattern quite similar to the perovskite-structured $(Mg,Fe)SiO_3$ first synthesized by *Liu* [1975] as shown in Figure 1. Therefore, it is interpreted as the same perovskite phase, although the lattice constants are different by 0.2 - 1.4%, which is beyond the margin of experimental error. It is worth noting the difference of the two perovskites on the basis of X-ray diffraction data. The most intense diffraction line of *Liu*'s [1975] $MgSiO_3$ perovskite (112) overlaps the most intense diffraction line of the spinel phase (331). The same situation takes place in the present perovskite, particularly in the Fe-rich case. However, perovskite (112) and

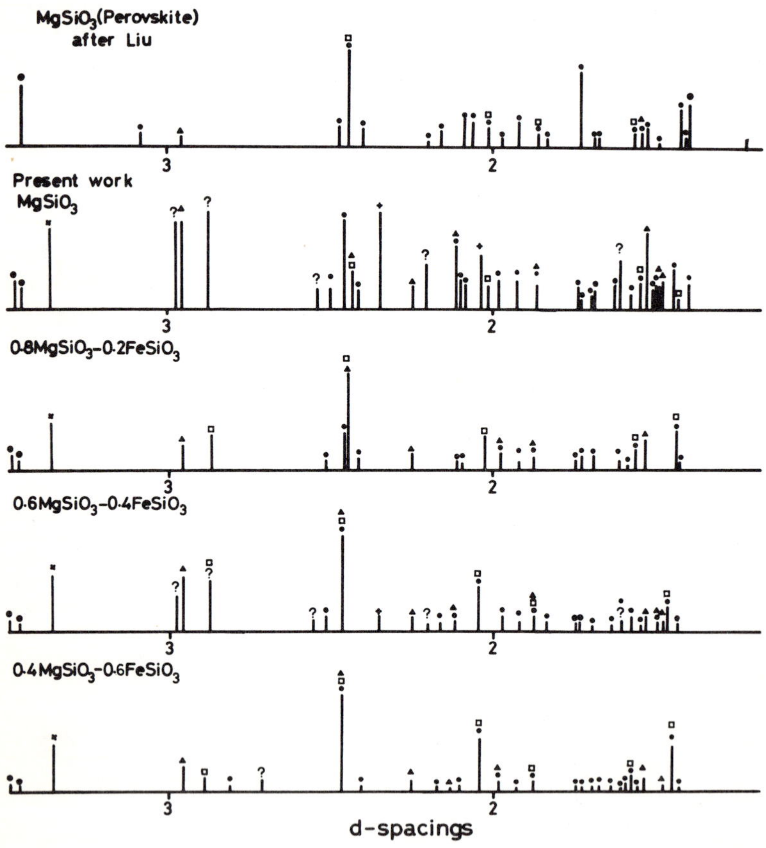

Fig. 1. Graphical representation of X-ray diffraction data for $(1-X)MgSiO_3 + XFeSiO_3$. Top, Liu [1975]; remaining data, present work.

spinel (311) are clearly separated in the $MgSiO_3$ case in the present work as shown in Figure 1. Another important point in comparing the two perovskites is the *d*-spacings of (002) and (110). In *Liu*'s perovskite, two diffraction lines (002) and (110) (thus (004) and (220)) overlap each other. The diffraction lines (220) + (004) and (110) + (002) are respectively the second- and third-most intense lines in *Liu*'s case. On the other hand, in the present case, (110) and (002), and (220) and (004) have different *d*-spacings.

The overall quality of the diffraction patterns and number of detected lines are fewer in the present work, as compared with those of *Liu* [1975]. The present author believes that the similarities and dissimilarities of the perovskites synthesized in different laboratories are significant.

All the lattice constants of the present perovskite increase linearly with the ratio Fe/(Mg + Fe); however, this increase is

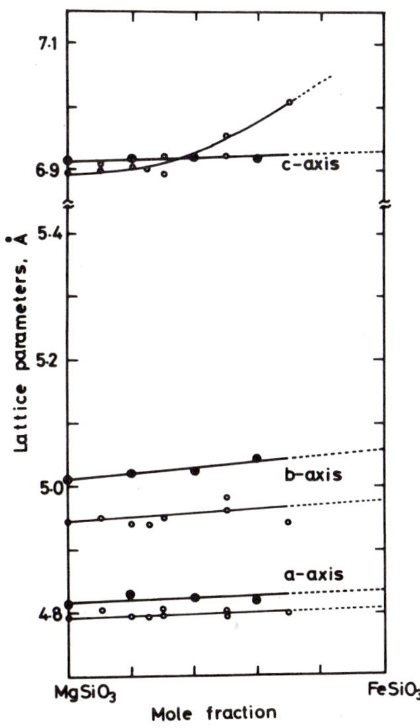

Fig. 2. The lattice parameters in perovskite solid solution (Mg,Fe)SiO_3. Closed circles, present data; open circles, Liu [1976a].

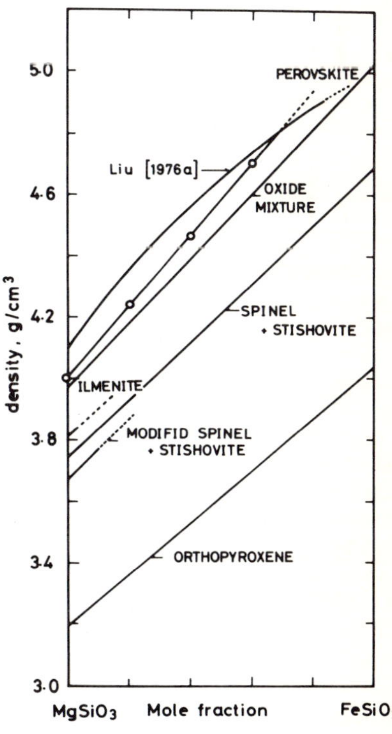

Fig. 3. Comparison of densities in different phases of the $MgSiO_3$-$FeSiO_3$ system. Open circles, present data; squares Liu [1975, 1976a].

very small (Figure 2). The density of the present perovskite at ambient condition (Figure 3) is lower than that obtained by *Liu* and larger than the isochemical mixed phase.

B. Difference in Crystal Structures of Perovskites

The observed difference between the lattice constants of orthorhombic perovskites $(Mg,Fe)SiO_3$ synthesized by *Liu* [1976a] and the present author is quite large, and there should be some reason to account for the difference. One possible interpretation is the flexibility in crystal structure and the resulting imperfection in the perovskite structure.

The possibility of the deviation from ABO_3 stoichiometry with perovskite structure is exemplified in the following:

1. Cation vacancies only at A site - - Cation deficiency at A-site of perovskite lattice can be tolerated over a wide range in some compounds such as $Sr_{1-x}NbO_3$ ($0.05 \leq x \leq 0.3$) [*Ridgley and Ward*, 1955], and $La_{1-x}TiO_3$ ($0 \leq x \leq 0.33$) [*Kestigian and Ward*, 1955].
2. A deficiency of oxygen in the lattice - - $SrFeO_3$ synthesized at low oxygen pressure has the perovskite structure and usually shows the deficiency of oxygen up to the fraction 1/6 [*Watanabe*, 1957].
3. Solid solution with AO or BO_2 component - - A perovskite solid solution takes place in the ABO_3-AO and ABO_3-BO_2 systems (*e.g.*, $BaTiO_{3_{ss}}$ [*Rase and Roy*, 1964]). Although the range of solid solution, *i. e.*, the deviation from stoichiometry, is quite narrow at atmospheric pressure, it has a general tendency to increase with increasing pressure.
4. An excess oxygen in the lattice - - An oxygen deficient pyrochlore with an $A_2B_2O_{7-x}$ formula such as $Pb_2Ru_2O_{7-x}$ ($0.6 \leq x \leq 1$) [*Longo et al.*, 1969], transforms to an orthorhombic perovskite at high pressure.
5. Ordering or disordering of site occupancy - - Garnets with the general formula $A_3B_2C_3O_{12}$, such as YIG[*Shimada et al.*, 1968], transform to perovskite. In such cases, three different ions are accommodated in two different sites. This implies that considerable deviation in site occupancy can occur in the perovskite structure.

The perovskite structure, therefore, has not always the ideal ABO_3; it can tolerate considerable deviation from the ideal stoichiometry (ABO_3) and non-ideal ordering over a wide range of compositions.

The formation of a cation vacancy at an A-site or an oxygen vacancy is expected to contribute to the decrease in lattice parameters. In the case of the perovskite solid-solution, rich in the elements of A-site, the lattice parameter is larger than that of a composition with the ideal stoichiometry, while it is smaller in the case of solid-solutions rich in the B-site elements. If the run product from stoichiometric $(Mg,Fe)SiO_3$ pyroxene at high temperature and pressure is a single-phase perovskite, one can expect that the lattice parameters may reflect values for the ideal perovskite. However, in the present experiment the actual run products included both oxide components and spinels in the intermediate high pressure phases. Almost the same situation occurred in *Liu*'s [1975] experiment. This implies the possibility that both perovskites obtained by *Liu*'s [1975] and by the present author deviate from ideal stoichiometry and that they have different chemical compositions and thus different lattice parameters. When the (Mg,Fe)O component is a solute in perovskite, the lattice parameters are expected to be larger, and when the SiO_2 component is a solute, they are expected to be smaller. Further complication may arise from the relative difference in solubility of Fe^{2+} and Mg^{2+} in a perovskite structure.

Recently, *Reid and Ringwood* [1975] proposed that the unit cell volumes of $A^{2+}B^{4+}O_3$ perovskite can be generally expressed as $V = C(A\text{-}O)^{x}\,(B\text{-}O)^{3-x}$ where C = 24.29, x = 0.970, and (A-O) and (B-O) are, respectively, the average six-coordinate bond lengths for A and B atoms in simple oxide components. The predicted unit-cell volume and density of $MgSiO_3$ perovskite based on the relation are 160.9 ± 0.8 $Å^3$ and 4.14 g/cm^3, respectively. The predicted 4.14 is greater than the 4.00 of the present study, or the 4.08 in *Liu*'s [1976a] work. We infer that the actual value of lattice parameters of the new perovskite $(Mg,Fe)SiO_3$ is not yet determined. This problem of the difference in lattice parameters would be presumably resolved, when the yield of perovskite is sufficient so as to allow a single crystal to be analyzed and detailed and to carry out chemical and X-ray structure analysis. We emphasized the unexpected complication in the perovskite phase $(Mg,Fe)SiO_3$ is taking place, and more detailed experimentation in perovskite synthesis and analysis need to be carried out.

IV. CORRELATION OF LABORATORY AND SEISMOLOGICAL DATA

There are two different approaches to study the correlation between the laboratory and the seismological data for the mantle:

1. The observed seismic wave velocities and deduced densities for the mantle are extrapolated to the ambient values, and then compared with the laboratory data.
2. The acoustic wave velocities and densities of the mantle-candidate materials are extrapolated to high temperature and pressure compatible with the assumed conditions in the earth's interior, and then compared with the seismic velocity and density profiles for the mantle.

In either case, we have to know the elastic wave velocities *Vp*, *Vs*, and density ρ of mantle mineral analogs.

A. Prediction of Elastic Parameters of High-pressure Phases of Olivine and Pyroxene

Birch's empirical relationship between *Vp* and ρ is

$$Vp = a(M) + b\rho \quad (1)$$

where $a(M)$ is a function of mean atomic weight M, and b is a constant. Later *D. Anderson* [1967], *O. Anderson and Soga* [1967], *O. Anderson and Liebermann* [1969], *Shankland* [1972], *Chung* [1972] and *Soga* [1971] modified Birch's law to

$$V = c(M)\rho^{\lambda} \quad (2)$$

where V is Vp or $V\phi$, and $c(M)$ is a constant. On the other hand, when Vp and Vs are normalized by the respective values Vpo and Vso of the isochemical mixed oxide phase, they are linearly related to ρ,

$$Vp/Vpo = c\cdot\rho/\rho o \ , \ Vs/Vso = d\cdot\rho/\rho o \quad (3)$$

where c and d are constants [*Davies*, 1976].

Another useful method for estimating the increases of seismic-wave velocity ΔV, is associated with increases in density $\Delta\rho$ which take place upon phase transitions. By taking a slope of V-ρ curves represented by (1) and (2), we have

$$b = \Delta V/\Delta\rho = \lambda c(M)\rho^{\lambda-1}. \quad (4)$$

This implies that *Birch's b* is dependent on M. In any case,(3) or (4) seems to be quite useful in relating the experimentally determined jump of density associated with phase transition with the seismologically observed jump in seismic waves in the mantle. *Liebermann and Ringwood* [1973] have determined ΔV and $\Delta\rho$ for a number of materials. In Figure 4, the empirical relation of $\Delta V/\Delta\rho$ to M is plotted for the transition from α to γ phases in A_2BO_4-type compounds. Similar type of plot for ABO_3-type compounds is shown in Figure 5. In order to predict the value of $\Delta V/\Delta\rho$, we have utilized a straight line fitting

$$\Delta V \ / \ \Delta\rho = e + fM \quad (5)$$

to the $\Delta V/\Delta\rho$ - M relation for both A_2BO_4- and ABO_3-type compounds. Based on the value of $\Delta\rho$ obtained by X-ray densities, the values of ΔV and, thus, V are calculated.

The Vp values for β-$(Mg,Fe)_2SiO_4$, γ-$(Mg,Fe)_2SiO_4$ and perovskite $(Mg,Fe)SiO_3$ predicted by the methods described above are presented in Table 6.

B. Calculation of Seismic Wave Velosity and Density

The density change at high pressure is calculated by applying the Birch-Murnaghan equation at high temperature

$$P = 3/2K(T)[x^{\frac{7}{3}} - x^{\frac{5}{3}}][1 + \ 3((K'(T)/4 - 1)(x^{\frac{2}{3}} - 1) \quad (6)$$

TABLE 6. Elastic Properties of $(Mg,Fe)_2SiO_4$ and $(Mg,Fe)SiO_3$

Structure	$\frac{Fe}{Fe+Mg}$	ρ_0 g/cm^3	Vp km/sec	K_0 Mbar	$(\frac{\partial K}{\partial P})_T$	$(\frac{\partial \mu}{\partial P})_T$	$(\frac{\partial \mu}{\partial T})_P$	δ	Notes
Olivine	0	3.22	8.59	1.28	4[a]	1.82[f]	-0.134[f]	5[a]	i
	0.1	3.34	8.40						i
	0.2	3.40	8.30						i
Modified	0	3.47	9.76	2.0					d
Spinel	0	3.47	9.35						eq. 1
	0.1	3.60[e]	9.60	2.0	4[a]	0.612[b]	-0.085[b]	5[a]	d
	0.1	3.60	9.19						eq. 1
	0.2	3.73	9.41	2.0					d
	0.2	3.73	9.06						eq. 1
Spinel	0	3.55	10.14	2.0					d
			9.59	-					eq. 1
			10.3	-					g
			10.03	2.01					h
			9.8	2.06					h
			9.66	1.86	4[a]	0.612[b]	-0.085[b]	5[a]	i
			9.73	1.94					j
	0.1	3.68	9.93	2.0					d
	0.1	3.68	9.44	-					eq. 1
	0.2	3.81	9.72	2.0					d
	0.2	3.81	9.30	-					eq. 1
Perovskite	0	4.00	10.66	2.6					d
			10.77	2.54					eq. 1
			10.64	2.6					eq. 3
		4.08	11.01						d
		(*Liu's*)	11.01		4[a]	1.74[c]	-0.223[c]	5[a]	eq. 1
			11.01	2.65					eq. 3
	0.1	4.12	10.06	2.6					d
			10.73						eq. 1
	0.2	4.24	10.55	2.6					d
			10.72						eq. 1

a. Assumed. (see text)
b. The values of $MgAl_2O_4$.
c. The values of Al_2O_3.
d. adopted values. (based on eq. 4)
e. *Akimoto* [unpublished].
f. *Kumazawa and Anderson* [1969].
g. *Mizutani et al.* [1970].
h. *Liebermann* [1970].
i. *Chung* [1971].
j. *Wang and Simmons* [1972].

$$x = \rho(T,P) \; / \; \rho(T,0) = V(T,0) \; / \; V(T,P)$$

where P is pressure, $K(T)$ is the isothermal bulk modulus, $K'(T)$ is its pressure-derivative at ordinary pressure, and V is the specific volume.

The necessary parameters to evaluate the equation of state are $K(T)$, $V(T)$, and $\rho(T)$. They are given by

$$K(T) = K_0 \; exp[-\delta \int_{293}^{T} \alpha(T) \; dT] \tag{7}$$

$$V(T,0) = V_0 \; [1 + \int_{293}^{T} \alpha(T) \; dT] \tag{8}$$

$$\rho(T,0) = \rho_0 \; [1 - \int_{293}^{T} \alpha(T) \; dT] \tag{9}$$

where $\alpha(T)$ is the coefficient of volume thermal expansion at $P = 0$ and δ is an anharmonic parameter (Grüneisen-Anderson parameter) defined by

$$\delta = - \; (\partial lnK/\partial T) \; / \; \alpha$$

The anharmonic properties of the high-pressure forms of olivine and pyroxene have not been measured experimentally. The parameter δ supposed to be independent of temperature, has the value of 4.72 for corundum, 4.64 for periclase, 5.13 for spinel

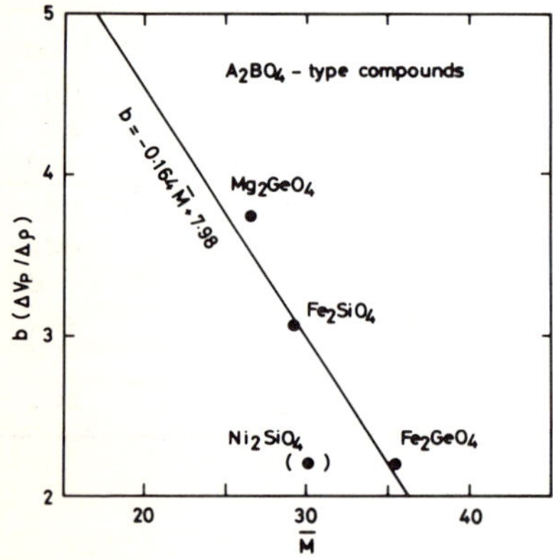

Fig. 4. The relation between the b value in equation (4) and the mean atomic weight $\overline{M}$ for A_2BO_4-type compounds. Data are taken from Liebermann and Ringwood [1973].

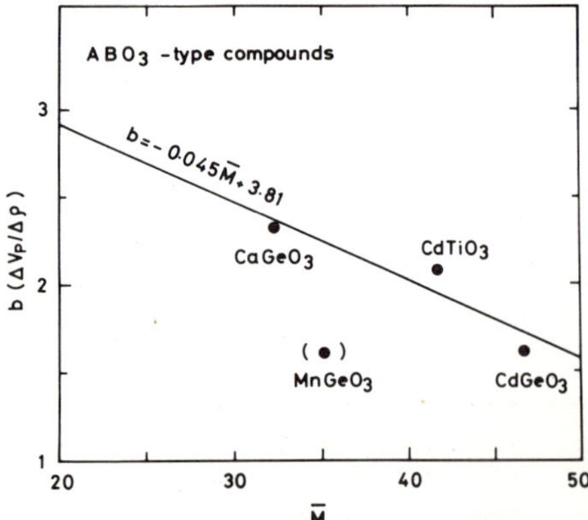

Fig. 5. The relation between the b value in equation (4) and the mean atomic weight M for ABO_3-type compounds. Data are taken from Liebermann and Ringwood [1973].

($MgAl_2O_4$), 5.62 for forsterite, and 6.51 for garnet [*O. Anderson et al.*, 1968]. The value of δ is not much different in different materials, and we adopt $\delta = 5$ irrespective of material to compute $K(T)$ in the present study. $K'(T)$ is assumed to be 4.0 independent of composition and temperature

$$\partial^2 K / \partial P \partial T = 0 \tag{10}$$

For the coefficient of thermal expansion and the pressure and temperature derivatives of rigidity, we used values of $MgAl_2O_4$ for β- and γ-phases of olivine and those of Al_2O_3 for perovskite.

The expression for Vp at high temperature and high pressure is

$$[Vp(T,P)]^2 = [K(T,P) + 4\mu(T,P)/3] / \rho(T,P) \tag{11}$$

where

$$K(T,P) = K(T) + (\frac{\partial K}{\partial P})P \tag{12}$$

$$\mu(T,P) = \mu(0,0) + (\frac{\partial \mu}{\partial T})T + (\frac{\partial \mu}{\partial P})P \tag{13}$$

In the evaluation of seismic wave velocities, the crucial parameters are rigidity μ and its temperature and pressure variations. These parameters are structure-sensitive and are difficult to predict either by solid-state theory or by empirical equations of state. A simple equation (13) is used and, thus, our estimate of Vs profile is expected to be quite uncertain. On the other hand, the seismic Vp is better known than the Vs. Therefore, we first calculate the depth profile of ρ and Vp in the representative mineral phases as a function of depth. The temperature distribution adopted for the calculation is same as the lower estimate of *Clark and Ringwood* [1964]. Approximate representative values are 700°C at 100-km, 1400°C at 400-km, and 1850°C at 500-km depths.

A comparison of the laboratory data of ρ and seismic Vp shows the following features: (1) olivine is appropriate for the depth range above 400 km; (2) modified spinel is appropriate for 400 to 600 km; and (3) high-density materials, such as stishovite, corundum, and perovskite, are appropriate beneath 700 km. Then the seismological data are extrapolated at ambient conditions in accordance with the calculated changes in velocity and density with depth.

C. Constitution of the Mantle Inferred from Seismic Vp and ρ

The possible major constituent metallic elements in the mantle are Mg and Si, which are the most abundant non-volatile elements in the solar system [*Ringwood*, 1970]. Although the solar abundance of Fe follows that of Si, less than 30% of Fe may be in the mantle, since Fe is concentrated into the earth's core.

The amount of Al or Ca in the mantle is expected to be less than 0.1 of Si as inferred from solar abundances. These elements will be concentrated in the uppermost layer of the earth. The remaining elements, except oxygen, have very low concentration in the mantle and their effect on the bulk elastic properties of mantle is ignored. Since the solar abundances show Si:Mg:Fe = 1:1.1:0.9, the chemical composition of the mantle lies between $(Mg,Fe)_2SiO_4$ and $(Mg,Fe)SiO_3$ ($0 < Fe/(Fe + Mg) < 0.3$) as the first approximation. The comparison between the observed and calculated mantle properties is first made for a mantle composition of olivine Mg_2SiO_4 and pyroxene $MgSiO_3$ stoichiometries and then the effects of the other components are considered.

In Figure 6, the seismic *Vp* and density are plotted for mantle minerals. The mineral phases represented are olivine and pyroxene stoichiometries, and oxides; MgO, FeO, CaO (rock salt), SiO_2 (stishovite), and Al_2O_3. The inferred properties of pyrope-almandine garnet are also shown. For perovskite $(Mg,Fe)SiO_3$ phases, there are two different values for *Vp* and ρ from two different lattice parameters reported by *Liu* [1975] and the present paper.

The mode of successive phase transformations in $(Mg,Fe)_2SiO_4$ and in $(Mg,Fe)SiO_3$ at $P > 150$ kbar (below 500 km depth in the mantle) is now being studied extensively in several laboratories. However, there are several controversies in the interpretation of experimental data, particularly with regard to the stability fields of mixed-oxide phase and ilmenite phase [*e.g.*, *Liu*, 1976b]. If each of the expected phase transformations takes place in the mantle, a number of seismic discontinuities should be present in the transition zone in the mantle. However, the resolution of seismological data is not sufficient to permit us to make detailed correlations with the laboratory data. Considering the major seismic discontinuities to be at depth of 370, 550, and 650 km in the mantle, it seems adequate to consider a very simple four-layered model to describe the upper mantle and the uppermost part of the lower mantle.

The four layers to be considered here are B : 200-350 km, C1 : 400-500 km, C2 : 550-600 km, and D : 700-1500 km. A seismological distinction of the C1 and the C2 layers is not always made. On the other hand, several stepwise increases in seismic waves, in the depth range from 700 to 1000 km are reported in some papers. The *Vp* and ρ in each of the four layers at ambient condition are shown by a dotted area in Figure 6.

The simple interpretations of Figure 6 are as follows.

1. In the upper three layers (B, C1 and C2), the chemical composition with an olivine stoichiometry is compatible with the seismological evidences as a whole.
2. If the B layer is composed of olivine, the ratio Fe/(Fe + Mg) is expected to be 0.05 to 0.15. *Ahrens* [1973] reported that the ratio Fe/(Fe + Mg) in olivine in the upper mantle is 0.15 based on the *Vp* profile by *Helmberger and Wiggins* [1971] and the

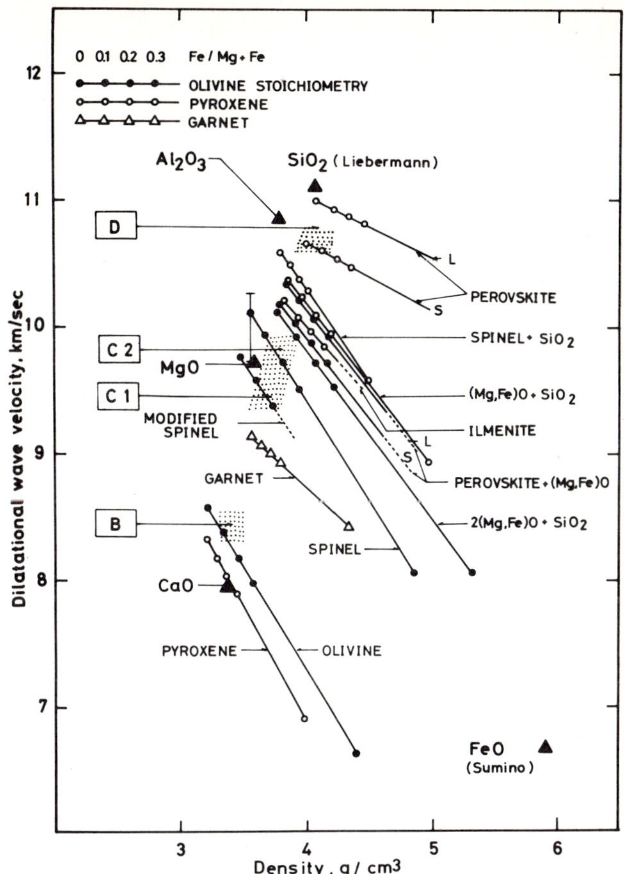

Fig. 6. The Vp-ρ plots for mantle candidate phases and the mantle materials at ambient condition. Perovskite L, Liu [1976a]; Perovskite S, present work; FeO, Sumino [unpublished]; SiO_2, Liebermann [in press 1976]. See text for detail.

density profile by *Dziewonski* [1971], and *Liebermann and Ringwood*'s [1973] diagram. *Burdick and Anderson* [1975] reported the ratio of 0.17 based on the geotherm by *Swift* [1969] and the pressure-depth curve of *Jordan and Anderson* [1974]. However, garnet with Fe/(Fe + Mg) = 0.3 ± 0.1 and orthopyroxene with Fe/(Fe + Mg) = 0.05 ± 0.05 are well accommodated in this layer as the secondary constituent phases, keeping the Fe/(Fe + Mg) of olivine at the acceptable value 0.1.

The C1 and C2 layers are consistent with the chemistry of olivine plus minor garnet; olivine being modified spinel

β-$(Mg,Fe)_2SiO_4$ and spinel γ-$(Mg,Fe)_2SiO_4$, respectively, in the C1 and C2 layers. The phase assembly, such as $(Mg,Fe)SiO_3$ (ilmenite) + (Mg,Fe)O (ferropericlase) or the mixed-oxide phase, gives too high velocity to be compatible with the seismological data. If C1 and C2 layers are composed, respectively, of pure modified spinel and of pure spinel, the expected iron content Fe/(Fe + Mg) is 0.10 - 0.20, which is a little higher than that in the B layer. However, the addition of garnet with high iron content (*e.g.*, Fe/(Fe + Mg) = 0.3 - 0.4) reduces the required iron contents in modified spinel and spinel to 0.10 - 0.15.

The presence of complex garnet [*Ringwood*, 1967] does not affect the Fe content in this layer, since elastic parameter of complex garnet is not much different from that of pyrope.

3. The observed jump in *Vp* from the B layer to the C1 or C2 layer is almost the same as that expected from α- to β- or γ-phase of olivine. This means that the dilution effect by an additional component should be small.

4. In the D layer, most of the compatible constitution is close to pyroxene stoichiometry with perovskite structure. Although a 1:1 mixture of *Liu*'s [1976a] perovskite and ferropericlase gives close values of velocity and density in the D layer, these values are still smaller than the observed data. In order to raise the velocity and density values to fit the seismological data, some fraction of stishovite and/or perovskite should be added to this 1:1 mixture of *Liu*'s[1976a] perovskite and ferropericlase. It is noted that significant content of ferropericlase in this layer, as a consequence of assuming olivine stoichiometry, is not compatible with the seismological data, since the density and velocity of ferropericlase are quite low. This implies that the most consistent chemical composition in the D layer is different from that in the B, C1, and C2 layers. By assuming that the D layer is composed of perovskite and ferropericlase, the ratio of (Mg,Fe)O to SiO_2 in the D layer is expected to be 1.0 - 1.2 based on the comparison in Figure 6. The diversity of the estimated ratio of (Mg,Fe)O to SiO_2 is caused by the differences in density and hence velocity in the two perovskites reported by *Liu* [1976a] and in this paper.

5. Another possible mineralogy in the D layer is spinel plus stishovite. This phase assembly is expected to occur at the pressure below the stability condition of perovskite. Therefore, if this phase assembly is assigned to the material in the D layer, the minor seismic discontinuity in the shallow part of the D layer should be correlated with the recombination reaction of spinel and stishovite to yield perovskite. Although the pressure of this reaction seems too low to be correlated with the physical conditions beneath the 650-km boundary, this possibility is not discarded because of insufficient experimental and seismological data.

6. In the D layer, the expected iron content Fe/(Fe + Mg) is 0.0 - 0.15, which is notably smaller than that in the C1 and C2 layers. Even if a reasonable amount of Al (as Al_2O_3 or as a

component of solid solution in the perovskite phase) were mixed with the materials in the D layer, the required Fe content in this layer would be increased only by a few percent in order to be compatible with the seismological data. The contribution of the CaO component does not seem to affect the Fe content in this layer. If this interpretation is accepted, Fe is depleted in the D layer and is enriched in the shallower layer of the transition zone. This view is different from the previous discussions on the enrichment of Fe in the lower mantle [*D. Anderson and Jordan*, 1970; *Press*, 1970; *Kumazawa et al.*, 1974] but essentially the same conclusions were reached by *D. Anderson et al.*, [1971] and *Gaffney and Anderson* [1973]. Recent investigation of the mean atomic weight in the mantle [*Watt et al.*, 1975] shows that the mean atomic weight is almost constant throughout the mantle. However, *Watt et al.*'s[1975] result indicated slightly lower mean atomic weight, by 0.1 - 0.9 unit, in the lower mantle than in the upper mantle. If the contents of MgO and SiO_2 are discontinuous, the discontinuity of the other components, such as FeO having larger mean atomic weight, might also be possible.

7. Seismological discontinuity at 650-km depth is the most significant one in the mantle because of the large increases in velocity and density. The observed magnitude of jump in *Vp* from the C1 or C2 layer to the D layer is quite large (~0.7 - 1.2 km/sec) and is hard to explain by a chemically uniform mantle, because the larger estimation of the velocity jump expected from laboratory data is at most 0.5 km/sec both for modified spinel → perovskite + ferropericlase and modified spinel + stishovite → perovskite. Further, the sharpness of the 650-km discontinuity required by the seismological evidence [*Engdahl and Flinn*, 1969; *Fukao*, 1977] is never explained by the phase transformation in the polycomponent system if the chemical composition is uniform around the discontinuity. These facts support the idea that the composition of the major constituent components MgO, SiO_2, and FeO should be discontinuous at the 650-km seismological discontinuity. This leads to an interpretation that there was some process which has resulted in the chemical discontinuity, not only of the major components but also of the minor components. Therefore, the 650-km discontinuity is also the most important horizon from the geochemical view point.

8. There is the possibility of two intermediate phases, mixed oxides and ilmenite, just above the layer characterized by perovskite. These phases are expected to occur around the 650-km discontinuity. However, we do not have sufficient seismological resolution of the 650-km discontinuity to interprete the presence of these possible phases at the present time.

D. Seismic *Vp* Profile and Chemical Composition of the Mantle

Based on the comparison of the physical constants obtained in the laboratory and by the seismological method (Figure 6), the seismic *Vp* and the estimated chemical composition in the mantle are shown in Figure 7. The present result includes

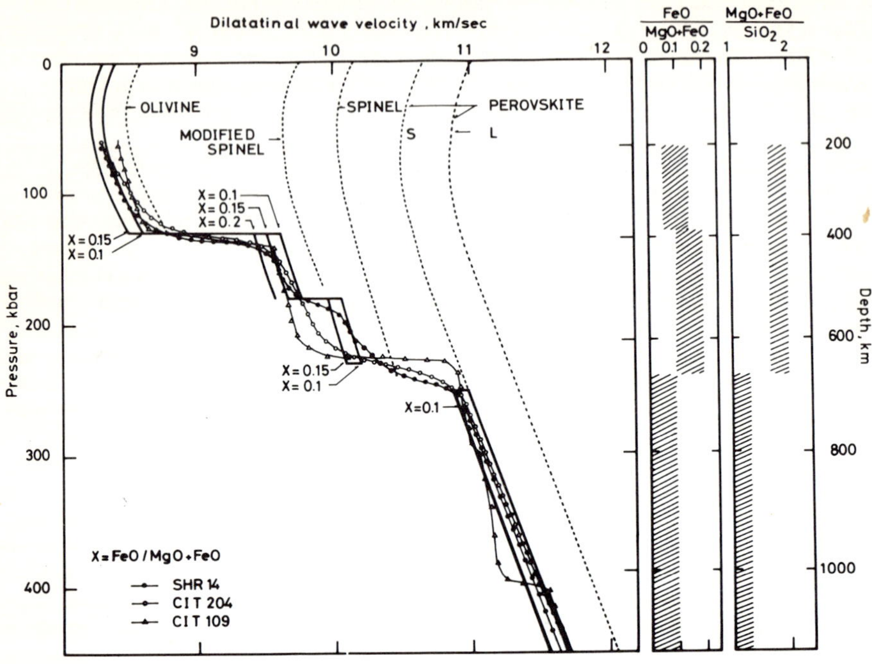

Fig. 7. Profiles of Vp and of inferred chemical composition in the upper mantle and in the uppermost part of the lower mantle. Three representative distributions of seismic Vp are compared with that calculated from the laboratory data. The chemical compositions shown at the right side are those obtained by an interpretation of Figure 6. See text for detail

considerable ambiguity; however, it is noted that the difference of two perovskites (*Liu's* [1976a] and the present study's) does not largely affect the feature described in Figure 7.

In the usual, large differentiated layered igneous bodies of intermediate composition, *i.e.*, Skaergaard, Stillwater, MgO-rich minerals underlie the more acidic (SiO_2-rich) mineral assemblages. Notably the present result is quite opposite from the commonly believed view on the nature of layered crust. However, it should be noted that none of the geophysical, geological, and geochemical evidences is in contradiction to the proposed distribution of chemical composition in the earth's mantle. Rather the present distribution of major elements in the mantle is consistent with other cosmochemical and geochemical constraints.

The major problem to be considered in the future is how such a distribution of elements has been established in the mantle. It is, however, not difficult to consider that some global processes of fractionation and of material transfer have taken place at some stage during the 4.6-billion-year history of the earth.

V. SUMMARY

The $(Mg,Fe)SiO_3$ perovskite has been synthesized and the possible crystal-chemical implication of this mineral is discussed. Based on the properties of the perovskite and the equation of state of the mantle minerals, the constitution of the upper and the lower mantles is proposed. The observed jump in *Vp* at the 650-km depth is not explained if the same composition is assumed throughout the mantle. An overall fit of the experimental data to the seismological data gives rise to the following conclusions: although the mineralogy of the upper mantle is represented mostly by olivine and its high pressure polymorphs with $Fe/(Fe + Mg) = 0.05 - 0.15$, the chemical composition of the lower mantle should be close to pyroxene stoichiometry with $Fe/(Fe + Mg) = 0.00 - 0.10$. In order to account for the significant difference in chemical composition of the upper and lower mantles, some global processes of chemical differentiation, by which magnesium and iron are expelled out of the lower mantle, must have operated during the earth's geological history.

Acknowledgments. The author is grateful to M. Kumazawa for his continuing encouragement and guidance, to T. J. Ahrens for his invaluable comments to improve the manuscript, and to Y. Fukao for his helpful discussions. He wishes to thank S. Akimoto and Y. Sumino for their permission to use the unpublished data on the lattice parameters of β-$(Mg,Fe)_2SiO_4$ and elastic properties of FeO. The present work was financially supported by the grant-in-aid of the Ministry of Education and by the Japanese Geodynamics Project.

REFERENCES

Ahrens, T. J., Petrologic properties of the upper 670 km of the earth's mantle, *Phys. Earth Planet. Interiors, 7*, 167, 1973.

Al'tshuler, L. V., R. F. Trunin, and G. V. Simakou, Shock-wave compression of periclase and quartz and the composition of the earth's lower mantle, *Bull. Acad. Sci., U. S. S. R., Earth Phys. ser., 10*, 657, 1965.

Anderson, D. L., A seismic equation of state, *Geophys. J. Roy. Astron. Soc., 13*, 9, 1967.

Anderson, D. L., Petrology of the mantle, *Mineral. Soc. Amer., Spec. Pap., 3*, 85, 1970.
Anderson, D. L., and T. Jordan, The composition of the lower mantle, *Phys. Earth Planet. Interiors, 3*, 23, 1970.
Anderson, D. L., C. Sammis, and T. Jordan, Composition and evolution of the mantle and core, *Science, 171*, 1103, 1971.
Anderson, O. L., and R. C. Liebermann, Elastic constants of oxide compounds used to estimate the properties of earth's interior, in *The Application of Modern Physics to the Earth and Planetary Interiors*, edited by S. K. Runcorn, John Wiley and Sons, New York, 1969.
Anderson, O. L., E. Schreiber, R. C. Liebermann, and N. Soga, Some elastic constant data on minerals relevant to geophysics, *Rev. Geophys., 6*, 491, 1968.
Anderson, O. L., and N. Soga, A restriction to the law of corresponding states, *J. Geophys. Res., 72*, 5754 1967.
Birch, F., The velocity of compressional waves in rocks to 10 kilobars, Part 2, *J. Geophys. Res., 66*, 2199, 1961.
Burdick, L., and D. L. Anderson, Interpretation of velocity profiles of the mantle, *J. Geophys. Res., 80*, 1070, 1975.
Chung, D. H., Elasticity and equations of state of olivines in the Mg_2SiO_4-Fe_2SiO_4 system, *Geophys. J. Roy. Astron. Soc., 25*, 511, 1971.
Chung, D. H., Birch's law; why is it so good?, *Science, 177*, 261, 1972.
Clark, S. P., and A. E. Ringwood, Density distribution and constitution of the mantle, *Rev. Geophys., 2*, 35, 1964.
Davies, G. F., The estimation of elastic properties from analogue compounds, *Geophys. J. Roy. Astron. Soc., 44*, 625, 1976.
Dziewonski, A. M. Upper mantle models from "pure-path" dispersion data, *J. Geophys. Res., 76*, 2587, 1971.
Engdahl, E. R., and E. A. Flinn, Seismic waves reflected from discontinuities within upper mantle, *Science, 163*, 177, 1969.
Fukao, Y., Upper mantle P-structure and the 650 km discontinuity, in *High Pressure Research: Applications to Geophysics*, edited by M. H. Manghnani and S. Akimoto, Academic Press, New York, 151-161, 1977.
Gaffney, E. S., and D. L. Anderson, Effect of low-spin Fe^{2+} on the composition of the lower mantle, *J. Geophys. Res., 78*, 7005, 1973.
Helmberger, D. V., and R. Wiggins, Upper mantle structure of the midwestern United States, *J. Geophys. Res., 76*, 3229, 1971,
Ito, E., T. Matsumoto, K. Suito, and N. Kawai, High pressure break-down of enstatite, *Proc. Japan Acad., 48*, 412, 1972.
Jeffery, R. N., J. O. Barnet, J. B. Vanfleet, and H. T. Hall, Pressure calibration to 100 kbar based on the compression of NaCl, *J. Appl. Phys., 37*, 3172, 1966.

Jordan, T. H., and D. L. Anderson, Earth structure from free oscillation and travel times, *Geophys. J. Roy. Astron. Soc., 36,* 411, 1974.

Kawai, N., M. Tachimori, and E. Ito, A high pressure hexagonal form of $MgSiO_3$, *Proc. Japan Acad., 50,* 378, 1974.

Kestigian, M., and R. Ward, The lanthanum-titanium-oxygen systems, *J. Amer. Chem. Soc., 77,* 6199, 1955.

Kumazawa, M., Theory of generation of very high static pressure by an external force, *High Temperatures-High Pressures, 5,* 599, 1973.

Kumazawa, M., and O. L. Anderson, Elastic moduli, pressure derivatives, and temperature derivatives of single-crystal olivine and single-crystal forsterite, *J. Geophys. Res., 74,* 5961, 1969.

Kumazawa, M., K. Masaki, H. Sawamoto, and M. Kato, Guide blocks and compressible pads for the practical operation of multiple-anvil sliding system for the production of high pressure, *High Tempratures-High Pressures, 4,* 293, 1972.

Kumazawa, M., H. Sawamoto, E. Ohtani, and K. Masaki, Postspinel phase of forsterite and evolution of the earth's mantle, *Nature, 247,* 356, 1974.

Liebermann, R. C., Velocity-density systematics for the olivine and spinel phase of Mg_2SiO_4-Fe_2SiO_4, *J. Geophys. Res., 75,* 4029, 1970.

Liebermann, R. C., and A. E. Ringwood, Birch's law and polymorphic phase transitions, *J. Geophys. Res., 78,* 6926, 1973.

Liu, L., Post-oxide phase of forsterite and enstatite, *Geophys. Res. Lett., 2,* 417, 1975.

Liu, L., Orthorhombic perovskite phase observed in olivine, pyroxene and garnet at high pressure and temperature, *Phys. Earth Planet. Interiors, 11,* 289, 1976a.

Liu, L., The high pressure phases of $MgSiO_3$, *Earth Planet. Sci. Lett., 31,* (in press) 1976b.

Longo, J. M., P. M. Raccah, and J. B. Goodenough, $Pb_2M_2O_7$ (M=Ru, Ir, Re)-Preparation and properties of oxygen deficient pyrochlores, *Mat. Res. Bull., 4,* 191, 1969.

Masaki, K., H. Sawamoto, E. Ohtani, and M. Kumazawa, High-pressure generation by MASS 3I8-90 type apparatus, *Rev. Sci. Instrum., 46,* 84, 1975.

Ming, L., and W. A. Bassett, High-pressure phase transformations in the system of $MgSiO_3$-$FeSiO_3$, *Earth Planet. Sci. Lett., 27,* 85, 1975a.

Ming, L., and W. A. Bassett, The post spinel phase in the Mg_2SiO_4-Fe_2SiO_4 system, *Science, 187,* 66, 1975b.

Mizutani, H., Y. Hamano, Y. Ida, and S. Akimoto, Compressional wave velocities of Fe_2SiO_4 spinel, and coesite, *J. Geophys. Res., 75,* 2741, 1970.

Piermarini, G. J., and S. Block, Ultrahigh pressure diamond-anvil cell and several semiconductor phase transition pressures in relation to the fixed point pressure scale, *Rev. Sci. Instrum., 46*, 973, 1975.

Press, F., Earth models consistent with geophysical data, *Phys. Earth Planet. Interiors, 3*, 3, 1970.

Rase, D. E., and R. Roy, System $BaO-TiO_2$, in *Phase Diagram for Ceramists*, edited by E. M. Levin, C. R. Robbins, and H. F. McMurdie, The Amer. Ceramic Soc. Inc., Ohio, 1964.

Reid, A. F., and A. E. Ringwood, High pressure scandium oxides and its place in the molar volume relationships of dense structure of M_2X_3 and ABX_3 type, *J. Geophys. Res., 74*, 3238, 1969.

Reid, A. F., and A. E. Ringwood, High pressure modification of $ScAlO_3$ and some geophysical implications, *J. Geophys. Res., 23*, 3363, 1975.

Ridgley, D., and R. Ward, The preparation of a strontium-niobium bronze with the perovskite structure, *J. Amer. Chem. Soc., 77*, 6132, 1955.

Ringwood, A. E., The pyroxene garnet transformation in the earth's mantle, *Earth Planet. Sci. Letters, 2*, 255, 1967.

Ringwood, A. E., Phase transition and the constitution of the mantle, *Phys. Earth Planet, Interiors, 3*, 109, 1970.

Ringwood, A. E., Mineralogy of the deep mantle; current status and future developments, in *The Nature of the Solid Earth*, edited by E. C. Robertson, McGraw-Hill, New York, 1972.

Ringwood, A. E., and A. Major, High pressure transformations in pyroxenes II, *Earth Planet Sci. Letters, 5*, 76, 1968.

Shankland, T. J., Velocity-density systematics: derivation from Debye theory and effect of ionic size, *J. Geophys. Res., 77*, 3750, 1972.

Shimada, M., S. Kume, and M. Koizumi, Possbile existence of dense ferrimagnetic perovskite allotropic form of yttrium iron garnet, *J. Amer. Ceram. Soc., 51*, 713, 1968.

Soga, N., Sound velocity of some germanate compounds and its relation to the law of corresponding state, *J. Geophys. Res., 76*, 3983, 1971.

Swift, C.M., A magnetotelluric Investigation of an electrical conductivity anomaly in the southwestern United States, Ph. D. thesis, Mass. Inst. of Technol., 1969.

Wang, H., and G. Simmons, Elasticity of some mantle crystal structures I. Pleonaste and hercynite spinel, *J. Gephys. Res., 77*, 4379, 1972.

Watanabe, H., Magnetic properties of perovskite containing strontium, *J. Phys. Soc. Japan, 12*, 515, 1957.

Watt, P., T. J. Shankland, and N. Mao, Uniformity of mantle composition, *Geology, 3*, 91, 1975.

THE POST-SPINEL PHASES OF TWELVE SILICATES AND GERMANATES

L. LIU
Research School of Earth Sciences
Australian National University
Canberra, A.C.T. Australia

Abstract

With the exception only of Mn_2SiO_4, the orthosilicates and orthogermanates of Mg, Mn, Fe, Co, Ni, and Zn all possess the spinel structure (or a distorted spinel structure) at some P-T conditions. The post-spinel phases of these compounds have been investigated to loading pressures of about 250-320 kbar at 900-1800°C in a diamond-anvil cell heated by a laser. With the exceptions of Mn_2SiO_4 and Zn-compounds, two distinct groups of the post-spinel phases have been found: (1) Mg_2SiO_4, Mg_2GeO_4, and Mn_2GeO_4 transform ultimately into an assemblage of orthorhombic perovskite plus rocksalt phases without the intervention of the assemblage of mixed oxides; and (2) Fe, Co, and Ni compounds, on the other hand, transform into their component oxides with the rocksalt and rutile structures without transforming further into the assemblage perovskite plus rocksalt phases (the olivine form of Mn_2SiO_4 has also been found to react to the mixed oxides). The β-Zn_2SiO_4 transforms to $ZnSiO_3$ (ilmenite) plus ZnO (rocksalt) and ultimately to its component oxides with the rocksalt and rutile structures. The distorted spinel form of Zn_2GeO_4 transforms to the assemblage of an as yet unidentified orthorhombic phase of $ZnGeO_3$ plus ZnO (rocksalt), and disproportionates into its component oxides at still higher pressure.

I. INTRODUCTION

The earth's upper mantle is predominantly made up of ferromagnesian silicate olivine [$(Mg,Fe)_2SiO_4$]. Olivines containing more than 80% (mole) magnesium have been found to transform into the β-phase (a distorted spinel structure) by *Ringwood and Major* [1970], and the pure magnesium olivine (forsterite) has been found to transform into the spinel phase at a higher pressure by *Ito et al.* [1974*a*]. The 400-km seismic discontinuity of the earth's mantle is believed to be associated with the olivine-spinel (or β-phase) transition. Hence, studies of further phase transformations in the spinel structure have direct implications for the earth's transition zone and lower mantle.

The post-spinel phase of Fe_2SiO_4 as its component oxide mixture was first reported by *Bassett and Takahashi* [1970], whereas the post-spinel phase of Mg_2SiO_4 has been found to be a mixture of the orthorhombic perovskite phase of $MgSiO_3$ plus periclase (MgO) by *Liu* [1976*a*]. Studies of post-spinel phases of other silicates and germanates reveal a general trend of phase transformations in the spinel structure, which is particularly useful to crystal chemistry, and provide insight into the difference of phase changes exhibited by Mg_2SiO_4 and Fe_2SiO_4.

II. EXPERIMENTAL METHODS

Very fine powders of orthosilicates and orthogermanates of Mg, Mn, Fe, Co, Ni, and Zn were compressed in a diamond-anvil press with a lever-and-spring-type assembly and were irradiated by a continuous YAG laser while the samples were under compression. The starting samples of Mg and Zn, both silicates and germanates, were mixed with a few percent of graphite, which serves to absorb the laser irradiation and to heat the samples. The loading pressures of the samples at the central portion of the anvil were estimated from the length of the spring, which was calibrated according to the room temperature NaCl pressure scale. Loading pressures thus estimated are probably accurate to ± 10%. Temperatures were estimated on the basis of the intensity of incandescent light emitted from the samples. Rapid and local laser heating causes the real pressure of the sample to be some 50 kbar higher than that estimated from the room temperature calibration. After quenching and release of pressure, the crystallographic phases of each sample were identified by X-ray diffraction, using a modified 57.3 mm Debye-Scherrer camera with filtered cobalt radiation. The details of the experimental procedure have been described elsewhere [*Liu*, 1975*a*].

III. RESULTS AND DISCUSSION

A. Mg_2SiO_4

The olivine form of Mg_2SiO_4 (forsterite) purchased from Tempres Research Inc. was used as starting material. X-ray diffraction study of the sample recovered from a loading pressure of about 150 kbar and 900-1200°C revealed that the sample is mainly the β-phase with some starting material, forsterite. The spinel phase was first observed at a loading pressure of about 170 kbar. These results are in accordance with those reported by *Ringwood and Major* [1970] and by *Ito et al.* [1974*a*]. The spinel form of Mg_2SiO_4 has been found to react to a mixture of the orthorhombic perovskite phase of $MgSiO_3$ plus periclase (MgO) in the vicinity of 200 kbar loading pressure without the intervention of any other phases or phase assemblages [*Liu*, 1976*a*]. The fact that no component oxide mixture (periclase plus stishovite) has been identified between the spinel and the mixture of perovskite plus periclase is consistent with those of our earlier studies of $MgSiO_3$ [*Liu*, 1976*b*]. *Liu* [1976*b*] found that the ilmenite phase of $MgSiO_3$ transforms directly into the perovskite modification at about 250 kbar and 1000-1400°C without the intervention of its component oxide mixture. The mixture of perovskite plus periclase persists to the highest pressure in this study (320 kbar). The zero-pressure volume change associated with the spinel-postspinel transition is -3.82 cm^3/mole, or -9.6%.

B. $\underline{Mg_2GeO_4}$

The olivine form of Mg_2GeO_4, which is identical to that used by *Ringwood and Reid* [1969], was employed in the present work. The sample quenched from 180 kbar and 900-1200°C comprises mainly the spinel phase of Mg_2GeO_4, the ilmenite phase of $MgGeO_3$, and periclase (MgO), with a trace amount of the starting material. The sample recovered from 260 kbar and 900-1200°C can be indexed mainly as a mixture of the orthorhombic perovskite phase of $MgGeO_3$ and periclase, with minor amounts of the ilmenite phase of $MgGeO_3$. The orthorhombic perovskite phase of $MgGeO_3$ has been confirmed by *Liu* [1976*c*] using the pyroxene form of $MgGeO_3$ as the starting material. The lattice parameters at room temperature and 1-bar pressure are a_o = 4.946 ± 0.002, b_o = 5.100 ± 0.002, and c_o = 7.233 ± 0.003 Å for the perovskite phase. The zero-pressure volume for the mixture of the ilmenite phase plus periclase is 4.3% smaller than that for the spinel phase of Mg_2GeO_4. The zero-pressure volume for the post-postspinel phase, the mixture of the perovskite phase plus periclase, is calculated to be 8.1% smaller than that for the spinel phase.

C. $\underline{Mn_2GeO_4}$

The olivine form of Mn_2GeO_4 has been found to transform to the β phase at about 40 kbar and to the δ phase (the Sr_2PbO_4 structure) at about 60 kbar and 800°C by *Akimoto* [1970]. The latter phase was first reported by *Wadsley et al.* [1968]. A mixture of the orthorhombic perovskite phase of $MnGeO_3$ plus the rocksalt phase of MnO with a trace amount of the δ-Mn_2GeO_4 has been observed by *Liu* [1976*d*] based on the X-ray diffraction data for Mn_2GeO_4 (olivine) recovered from about 250 kbar and

1400-1800°C. The zero-pressure volume change associated with the β-δ transition is calculated to be -9.8% [*Akimoto*, 1970], and that with the β to the mixture of perovskite plus MnO (rocksalt) transition to be -11.6%.

D. Zn_2GeO_4

The phenakite form of Zn_2GeO_4, which is identical to that used by *Ringwood and Major* [1967], was used in this study. This compound was found to transform to a slightly distorted spinel structure (a tetragonal modification) at 20-35 kbar and 1100°C by *Ringwood and Major* [1967], who also reported that the distorted spinel form is stable at least to P = 110 kbar and 900°C. X-ray diffraction data for the sample recovered from about 250 kbar and 1000-1400°C in this study show a mixture of the rocksalt phase of ZnO and an as yet unidentified orthorhombic phase with a minor amount of the distorted spinel phase of Zn_2GeO_4; the sample recovered from about 270 kbar comprises the rocksalt phase of ZnO, the orthorhombic phase, and the rutile phase of GeO_2 in order of decreasing abundance.

The orthorhombic phase has been confirmed to have a chemical formula of $ZnGeO_3$ from the study of an ilmenite form of $ZnGeO_3$ by *Liu* [1976*c*]. Unit cell dimensions for the orthorhombic phase at room temperature and 1-bar pressure are a_o = 5.022 ± 0.002, b_o = 5.194 ± 0.002, and c_o = 7.538 ± 0.003 Å. If 4 molecules per unit cell (the same number as that for the orthorhombic perovskite structure) are assumed for the orthorhombic phase of $ZnGeO_3$, the zero-pressure volume change from the ilmenite to the orthorhombic phase is only -0.2% [*Liu*, 1976*c*].

ZnO crystallizes in the wurtzite structure at standard P-T conditions. The rocksalt phase of ZnO was first reported by *Bates et al.* [1962] at pressures greater than 100 kbar and 200°C; however, complete conversion of the rocksalt phase of ZnO has never been reported in the literature. In this study, no trace of zincite (the wurtzite phase of ZnO) was detected. Presumably, ZnO has all been retained in the rocksalt phase. The lattice parameter for the rocksalt phase of ZnO is a_o = 4.275 ± 0.002 Å at room temperature and 1-bar pressure, which is slightly smaller (0.1%) than the value reported by *Bates et al.* [1962]. The X-ray diffraction data for the six diffraction lines of the rocksalt phase of ZnO observed in this work are listed in Table 1 [only three lines were observed in the work of *Bates et al.*, 1962].

The results of the present work indicate that the distorted spinel phase of Zn_2GeO_4 transforms to the mixture of ZnO (rocksalt) plus $ZnGeO_3$ (orthorhombic phase) with -3.8% change in the zero-pressure volume. This assemblage subsequently transforms to a mixture of its component oxides possessing the rocksalt and rutile structures and a zero-pressure volume 6.5% smaller than the distorted spinel phase.

TABLE 1. X-ray Diffraction Data[a]

I/I_{100}[b]	$d_{(obs)}$, Å	hkl	$d_{(cal)}$, Å[c]
50	2.468	111	2.468
100	2.139	200	2.138
50	1.511	220	1.511
10	1.289	311	1.289
20	1.234	222	1.234
10	0.9560	420	0.9559

a. Room temperature and 1-bar pressure for the rocksalt phase of ZnO quenched from the phenakite phase of Zn_2GeO_4 at 250 kbar and 1000–1400°C (CoKα).
b. Estimated visually.
c. Calculated from a_o = 4.275 Å.

E. Zn_2SiO_4

Synthetic willemite was used as the starting material in this study. Willemite displays several high-pressure modifications with increasing pressure [see *Ito and Matsui,* 1974]. We have studied this compound at loading pressures of 200, 250, and 300 kbar. X-ray diffraction data for the sample recovered at 200 kbar are a mixture of zincite (ZnO), stishovite (SiO_2), the rocksalt phase of ZnO, and the ilmenite phase of $ZnSiO_3$ in the order of their relative abundance based on intensity. The phase transformation is inferred to be Zn_2SiO_4 (β-phase) → $ZnSiO_3$ (ilmenite) + ZnO (rocksalt) → 2ZnO (rocksalt) + SiO_2 (stishovite) with increasing pressure. With the exception of the rocksalt phase of ZnO, this result is equivalent to that observed by *Ito and Matsui* [1974] at 330 kbar and 1000°C. Using Zn_2SiO_4 and $ZnSiO_3$ as starting materials, *Ito and Matsui* [1974] observed that the rocksalt phase of ZnO is not quenchable in their apparatus; they found zincite (ZnO) instead. No further phase changes in Zn_2SiO_4 up to 300-kbar loading pressure have been observed in this study.

Ito and Matsui [1974] have also reported an X phase in their run of $ZnSiO_3$ at 350 kbar and 1000°C. In view of the pressure scale calibrated in their apparatus, the 350-kbar pressure reported by them should be less than the 250-kbar loading pressure estimated in the diamond-anvil press. This is evident from the synthesis of the ilmenite phase of $MgSiO_3$ using the same apparatus and the same pressure scale by them. *Kawai et al.* [1974] have claimed to observe the ilmenite phase of $MgSiO_3$ at about 500 kbar and 1000°C, whereas we have found that the ilmenite phase of $MgSiO_3$ transforms to the perovskite phase in the vicinity of 250 kbar loading pressure [*Liu*, 1976*b*]. Hence, the X phase from $ZnSiO_3$ reported by *Ito and Matsui* [1974] is not confirmed in the present study of Zn_2SiO_4 to a pressure which is much higher than that of their observation.

The zero-pressure volume change from β-Zn_2SiO_4 to the mixture of $ZnSiO_3$ (ilmenite) plus ZnO (rocksalt) is -6.8% and from β-Zn_2SiO_4 to the mixture of 2ZnO (rocksalt) plus SiO_2 (stishovite) is -9.6%.

F. Fe, Co, and Ni orthosilicates and orthogermanates

With the exception of only Fe_2SiO_4, of which both the olivine and spinel forms were used, spinels of Fe_2GeO_4, Co_2SiO_4, Co_2GeO_4, Ni_2SiO_4, and Ni_2GeO_4 have been investigated in this work. Fe_2SiO_4, Co_2SiO_4, and Ni_2SiO_4 have been previously found by *Bassett and Takahashi* [1970] and *Liu* [1975*a,b*] to react to their component oxides with the rocksalt and rutile structures in the pressure region between 140- and 200-kbar loading pressure and 1400 to 1800°C. The same reaction in Co_2GeO_4 and Ni_2GeO_4 has also been found at P = 250 kbar and T = 1400-1800°C by *Liu* [1976*d*].

In this study, we report that spinels of Fe_2SiO_4, Fe_2GeO_4, Co_2SiO_4, Co_2GeO_4, Ni_2SiO_4, and Ni_2GeO_4 have all been found to disproportionate into a mixture of their component oxides with the rocksalt structure for FeO, CoO, and NiO and the rutile structure for SiO_2 and GeO_2 at loading pressures up to 250-280 kbar and temperatures of 1400-1800°C. The zero-pressure volume changes associated with the disproportionation reaction are calculated to be -8.7% for Fe_2SiO_4, -9.0% for Fe_2GeO_4, -8.1% for Co_2SiO_4, - 8.0% for Co_2GeO_4, -8.2% for Ni_2SiO_4, and -7.7% for Ni_2GeO_4.

G. Mn_2SiO_4

Among the various silicates and germanates investigated in this study, Mn_2SiO_4 is the only one not known to crystallize in the spinel or its related structures at any P-T conditions. According to *Ito et al.* [1974*b*], the olivine form of Mn_2SiO_4 transforms to a mixture of $MnSiO_3$ (garnet-like) plus MnO (rocksalt) at 130-140 kbar and to its component oxide mixture at 280-300 kbar and 1000°C. The component oxide mixture for Mn_2SiO_4 (olivine) has been confirmed by a run made at a loading pressure of about 260 kbar and temperatures of 1400-1800 C° in this study.

TABLE 2. Detailed Experiments and Results for Orthosilicates and Orthogermanates Investigated in This Work[a]

Starting Material	P, kbar	T, °C	Sp (or β) Phase →	Post-spinel (or β) Phases	$-\Delta V_o/V_{o,Sp}$,%
Mg_2SiO_4(Ol)	150~320	900~1200	β → Sp	$MgSiO_3$(Pero) + MgO	9.6
Mg_2GeO_4(Ol)	180~260	900~1200	Sp	$MgGeO_3$(Il) + MgO	4.3
				→ $MgGeO_3$(Pero) + MgO	8.1
Mn_2GeO_4(Ol)	250	1400~1800	β	δ-Mn_2GeO_4	9.8
				→ $MnGeO_3$(Pero) + MnO	11.6
Zn_2GeO_4(Ph)	250~270	1000~1400	distorted Sp	$ZnGeO_3$(ortho) + ZnO	3.8
				→ 2ZnO + GeO_2	6.5
Zn_2SiO_4(Ph)	200~300	1000~1400	β	$ZnSiO_3$(Il) + ZnO	6.8
				→ 2ZnO + SiO_2	9.6
Fe_2SiO_4(Ol,Sp)	200~280	1400~1800	Sp	2FeO + SiO_2	8.7
Fe_2GeO_4(Sp)	250	1400~1800	Sp	2FeO + GeO_2	9.0
Co_2SiO_4(Sp)	170~260	1400~1800	β → Sp	2CoO + SiO_2	8.1
Co_2GeO_4(Sp)	250	1400~1800	Sp	2CoO + GeO_2	8.0
Ni_2SiO_4(Sp)	140~260	1400~1800	Sp	2NiO + SiO_2	8.2
Ni_2GeO_4(Sp)	250	1400~1800	Sp	2NiO + GeO_2	7.7
Mn_2SiO_4(Ol)	260	1400~1800	-	2MnO + SiO_2	-

a. For abbreviations: Ol = olivine, Ph = phenakite, Sp = spinel, β = β-phase, Pero = perovskite, δ = Sr_2PbO_4 structure, Il = ilmenite, and ortho = orthorhombic phase. All the divalent metal oxides are in the rocksalt structure, and all the tetravalent metal oxides are in the rutile structure.

IV. SUMMARY AND CONCLUSIONS

The starting materials, the experimetnal pressure and temperature conditions, the spinel (or β) phase, the post-spinel phases, and the zero-pressure volume changes associated with spinel (or β) to post-spinel transitions of all the orthosilicates and orthogermanates investigated in this work are summarized in Table 2.

From Table 2 it is seen that, with the exceptions of Mn_2SiO_4 and Zn-compounds, the post-spinel (or β) phses for the orthosilicates and orthogermanates studied in this work can be divided into two distinct groups: (1) Mg_2SiO_4, Mg_2GeO_4, and Mn_2GeO_4 transform ultimately into an assemblage of perovskite plus rocksalt phases without the intervention of the assemblage of mixed oxides, whereas (2) Fe, Co, and Ni compounds transform to their component oxides with the rocksalt and rutile structures without transforming further into the assemblage perovskite plus rocksalt phases. The different behavior of the post-spinel phases for these two groups cannot be attributed to the difference in the experimental P-T conditions, nor to that of their cation sizes. In fact, the perovskite structure prefers to accommodate larger ions. The ionic sizes for Fe, Co, and Ni are between those of Mg and Mn. It might be the crystal field effects of Fe, Co, and Ni that destabilize the orthorhombic perovskite structure.

Zn atom exhibits a strong tendency toward a lower coordination number of oxygen atoms in Zn-compounds. This tendency is demonstrated by the phenakite structure of Zn_2SiO_4 and Zn_2GeO_4, and the wurtzite structure of ZnO at standard P-T conditions (Zn atoms are surrounded by four oxygen atoms in both the phenakite and the wurtzite structures). These are in marked contrast to the olivine structure (the divalent metals are surrounded by six oxygens) for all the other orthosilicates and orthogermanates (the spinel structure for Fe_2GeO_4) studied in this work, and to the rocksalt structure (metal atom is surrounded by six oxygen atoms) for the divalent metal oxides of these compounds at standard pressure and temperature conditions. In addition, $ZnSiO_3$ and $ZnGeO_3$ are not known to possess the pyroxene-type structure, which requires Zn atom to occupy the octahedral site, at standard P-T conditions. Hence, the Zn-compounds do not tend to crystallize in the perovskite structure (unless an extremely high pressure is applied) because the perovskite structure requires the Zn atom to occupy a site with a coordination number of oxygen greater than six.

Acknowledgement. The author is grateful to A. E. Ringwood and R. S. Coe for their commenting on the manuscript.

REFERENCES

Akimoto, S., High-pressure synthesis of a "modified" spinel and some geophysical implications, *Phys. Earth Planet. Interiors, 3,* 189-195, 1970.

Bates, C. H., W. B. White, and R. Roy, New high-pressure polymorph of zinc oxide, *Science, 137,* 993, 1962.

Bassett, W. A., and T. Takahashi, Disproportionation of Fe_2SiO_4 to 2FeO + SiO_2 at high pressure and high temperature, *Trans. Am. Geophys. Union (EOS), 51,* 828, 1970.

Ito, E., and Y. Matsui, High-pressure synthesis of $ZnSiO_3$ ilmenite, *Phys. Earth Planet. Interiors, 9,* 344-352, 1974.

Ito, E., Y. Matsui, K. Suito, and N. Kawai, Synthesis of γ-Mg_2SiO_4, *Phys. Earth Planet. Interiors, 8,* 342-344, 1974*a*.

Ito, E., T. Matsumoto, and N. Kawai, High-pressure decompositions in manganese silicates and their geophysical implications, *Phys. Earth Planet. Interiors., 8,* 241-245, 1974*b*.

Kawai, N., M. Tachimori, and E. Ito, A high pressure hexagonal form of $MgSiO_3$, *Proc. Japan Acad., 50,* 378-380, 1974.

Liu, L., Disproportionation of Ni_2SiO_4 to stishovite plus bunsenite at high pressures and temperatures, *Earth Planet. Sci. Lett., 24,* 357-362, 1975*a*.

Liu, L., High-pressure disproportionation of Co_2SiO_4 spinel and implications for Mg_2SiO_4 spinel, *Earth Planet. Sci. Lett., 25,* 286-290, 1975*b*.

Liu, L., The post-spinel phase of forsterite, *Nature, 262,* 770-772, 1976*a*.

Liu, L., The high pressure phases of $MgSiO_3$, *Earth Planet. Sci. Lett., 31,* 200-208, 1976*b*.

Liu, L., Post-ilmenite phases, *Trans. Am. Geophys. Union (EOS), 57,* in press, 1976*c*.

Liu, L., High pressure phases of Co_2GeO_4, Ni_2GeO_4, Mn_2GeO_4 and $MnGeO_3$: implications for the germanate-silicate modeling scheme and the earth's mantle, *Earth Planet. Sci. Lett., 31,* 393-396, 1976*d*.

Ringwood, A. E., and A. Major, High pressure transformations in zinc germanates and silicates, *Nature, 215,* 1367-1368, 1967.

Ringwood, A. E., and A. F. Reid, High pressure transformations of spinels (I), *Earth Planet. Sci. Lett., 5,* 245-250, 1969.

Ringwood, A. E., and A. Major, The system Mg_2SiO_4-Fe_2SiO_4 at high pressures and temperatures, *Phys. Earth Planet. Interiors, 3,* 89-108, 1970.

Wadsley, A. D., A. F. Reid, and A. E. Ringwood, The high pressure form of Mn_2GeO_4, a member of the olivine group, *Acta Crystallo., B24,* 740-742, 1968.

PHASE RELATIONS OF PURE Mg_2SiO_4 UP TO 200 KILOBARS

K. Suito
Department of Material Physics
Faculty of Engineering Science, Osaka University
Toyonaka, Osaka-fu, 560, Japan

Abstract

The phase relations of pure Mg_2SiO_4 were investigated to pressures of up to about 200 kbar and at temperatures ranging from 600°C to 1200°C, using a split-sphere type apparatus.

The semi-sintered MgO was employed as a pressure-transmitting medium in the present study. At first, the coesite-stishovite boundary was determined to calibrate pressure at high temperatures. Then the α-β and β-γ phase boundaries were studied. The α-β transition boundary is given by the equation P (kbar) = 108 + 0.035 T (°C). The β-γ phase boundary was then determined. The β-γ phase boundary has positive slope, and the triple point of α, β, and γ phases appears at $T < 500$°C.

I. INTRODUCTION

The olivine-spinel transformation in the system Mg_2SiO_4-Fe_2SiO_4 has been investigated by many authors in connection with the constitution of the earth's mantle. In pure Mg_2SiO_4, *Ringwood and Major* [1966] reported the synthesis of the β phase. *Suito* [1972] reported the synthesis of the γ phase and the α-β phase boundary. *Ito et al.* [1974] studied the detailed crystallographic data of the γ phase. Recently, *Kumazawa et al.* [1974] and *Ming and Bassett* [1975] propounded that the post-spinel phase of pure Mg_2SiO_4 was stishovite and periclase.

However, the detail phase relations among these phases are not clear. In this paper, experimental results for the phase relations among α, β, and γ phases up to about 200 kbar are reported. The new pressure-transmitting medium, namely, semi-sintered MgO, renders the high-pressure and high-temperature experiments successful.

II. EXPERIMENTAL METHODS

A split-sphere type apparatus designed by *Kawai et al.* [1973] was employed in the present study. The cross section of this apparatus is shown in Figure 1.

This apparatus has inner and outer anvil assemblages. The inner anvils are composed of eight WC cubes with a side length of 20.8 mm. A corner of each cubic is truncated to form a triangle with a 3-mm edge length. An octahedral sample made of semi-sintered MgO is put into the specimen chamber surrounded by eight triangles. The outer anvils are composed of a spherical shell of hardened steel segmented in six pieces, and the center of these anvils is so truncated that a hollow cubic space is formed at the center. Three of the six pieces are placed in the lower holder and three in the upper holder. The two opposed holders are compressed in a 2000-ton uniaxial press.

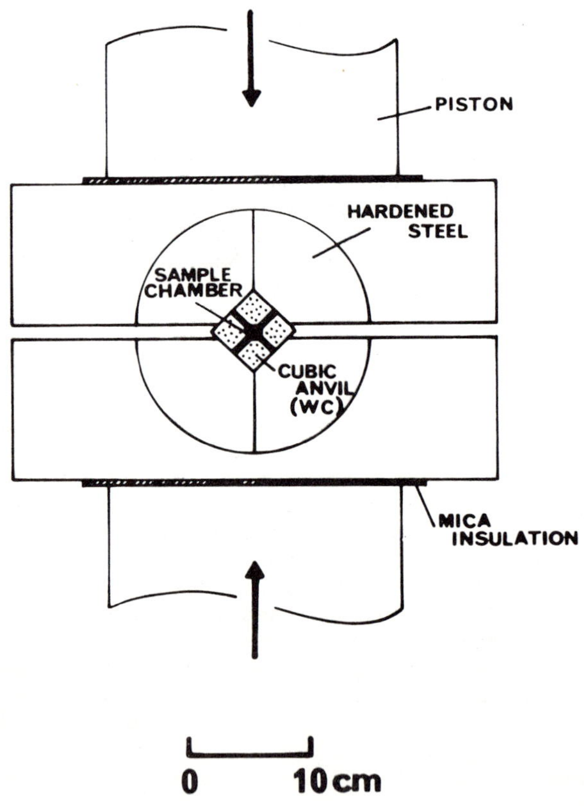

Fig. 1. Cross-section of the split-sphere type apparatus.

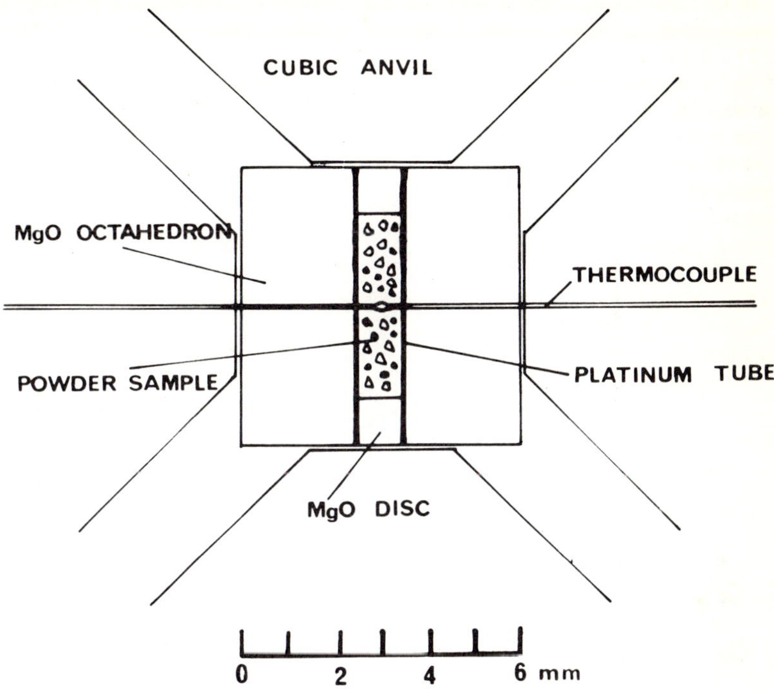

Fig. 2. Configuration of the furnace and the sample assembly.

The sample assembly (Figure 2) consists of octahedron, 6-mm in edge-length, of semi-sintered MgO and a thin platinum tube of 0.02-mm thickness as the heater. While, a Pt-Pt13%Rh thermocouple with a 0.1-mm diameter was employed for measuring temperatures at the center of the sample during heating, the lead wires of the thermocouple were taken out from gaps between anvils. No correction was made for the effect of pressure upon the emf of thermocouples. Synthetic Mg_2SiO_4 (olivine) and the β-phase were used as the starting material. The Mg_2SiO_4 (olivine) was synthesized from the stoichiometric mixture of SiO_2 and MgO at 1600°C for 48 hours. Complete conversion into a single phase of olivine was confirmed by X-ray diffraction. No reactive sample of Mg_2SiO_4 (reactive forsterite) was used. A quenching method was employed to determine the phase relations. Only the central part of the quenched sample close to the junction of the thermocouple was examined by X-ray diffraction to determine the phase present.

The pressure calibration curve at room temperature is shown in Figure 3. In the present study, the following fixed points were assigned: Bi I-II (25.5 kbar), Bi III-V (77 kbar), Pb (130 kbar), ZnS (150 kbar), GaAs (180 kbar), and GaP (220 kbar).

The metallic transition in GaP was not detected by compressions up to 400 tons. In Figure 3, the pressure calibration curve obtained, using pyrophyllite as the pressure-transmitting medium, is also shown. It is evident that the efficiency for producing very high pressure with semi-sintered MgO is better than that with pyrophyllite.

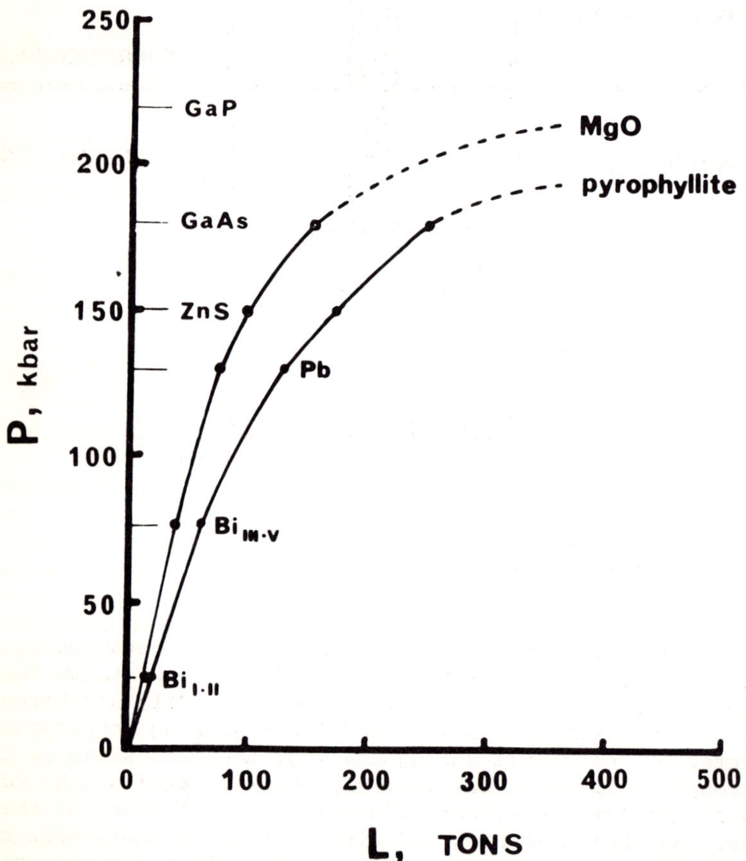

Fig. 3. Pressure calibration curves. MgO: the curve obtained by using the pressure-transmitting medium of semi-sintered MgO; pyrophyllite: the curve obtained by using the pressure-transmitting medium of pyrophyllite.

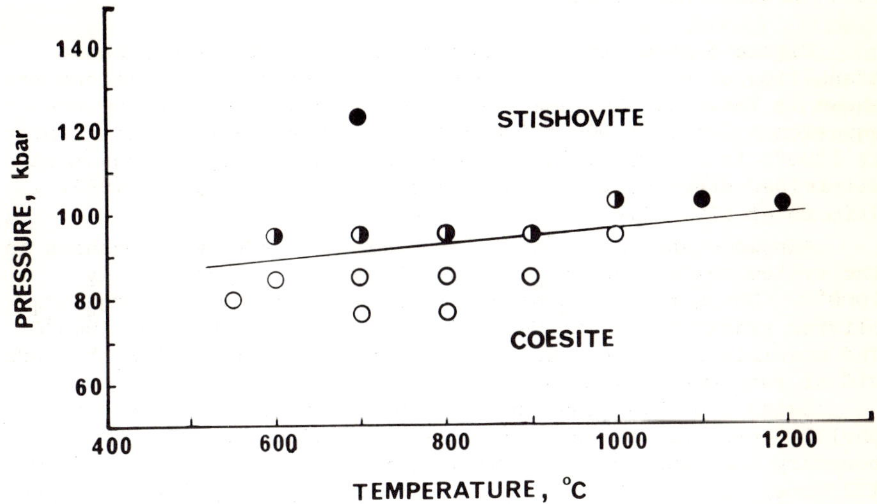

Fig. 4. Coesite-stishovite transition curve.

To calibrate pressure at high temperatures the coesite-stishovite transition boundary was determined, using semi-sintered MgO as the pressure medium. Results are shown in Figure 4. The starting material was α quartz. From the transition boundary, the transition pressure at 1000°C is determined to be about 94 kbar. This value seems to be consistent with the other experimental data (e.g., 95 kbar at 1000°C by *Akimoto and Syono* [1969], 91 kbar at 1000°C by *Yagi* [1975]; *Akimoto et al.*, [1977]). Thus, it was not necessary to make pressure correction at high temperature.

For pressure calibration at 1000°C, the transition point from tetragonal SnO_2 to orthorhombic SnO_2 was used. Based on the results obtained with diamond-anvil cell, *Liu* [1974] reported that an orthorhombic SnO_2 appeared at about 155 kbar and 1000-1400°C. Using a double-stage split-sphere apparatus and pyrophyllite as the pressure medium, *Suito et al.* [1975] determined the phase boundary between the tetragonal SnO_2 and the orthorhombic SnO_2 to be 162 kbar at 1000°C. In the present study, an orthorhombic SnO_2 was synthesized at 156 kbar and 1000°C. Hence the boundary between the tetragonal SnO_2 and the orthorhombic SnO_2 was assumed to lie between 150 and 156 kbar. The difference between the present pressure value at 1000°C and the previous one by *Suito et al.* [1975] for the tetragonal-orthorhombic transition point seems to be due to different pressure media, different thermocouples used and the different pressure scales employed.

III. RESULTS AND DISCUSSION

Figure 5 shows the experimental results of the α-β phase transition of pure Mg_2SiO_4. Detailed experimental conditions are shown in Table 1. The α-β transition boundary is given by the equation P (kbar) = 108 + 0.035 T (°C). The α-β transition point at 1000°C is 143 kbar. This pressure value is higher than that determined previously [*Ringwood and Major*, 1970; *Suito*, 1972; *Akimoto et al.*, 1976].

Ringwood and Major [1970] studied the phase relationships in the system Mg_2SiO_4-Fe_2SiO_4 at 50-200 kbar and approximately 1000°C. Using the Bridgman anvil, they found that pure Mg_2SiO_4 olivine transformed into β-Mg_2SiO_4 at about 120 kbar and 1000°C. The possible error in their temperature measurements for the individual runs was estimated to be less than 200°C.

Using a double-stage split-sphere apparatus and the pyrophyllite pressure medium, *Suito* [1972] studied the α-β phase boundary and reported that the α-β transition point at 1000°C was 132 kbar. It has recently been found that pyrophyllite decomposes into kyanite and coesite and then further transforms into corundum and stishovite at higher pressures and elevated temperatures above

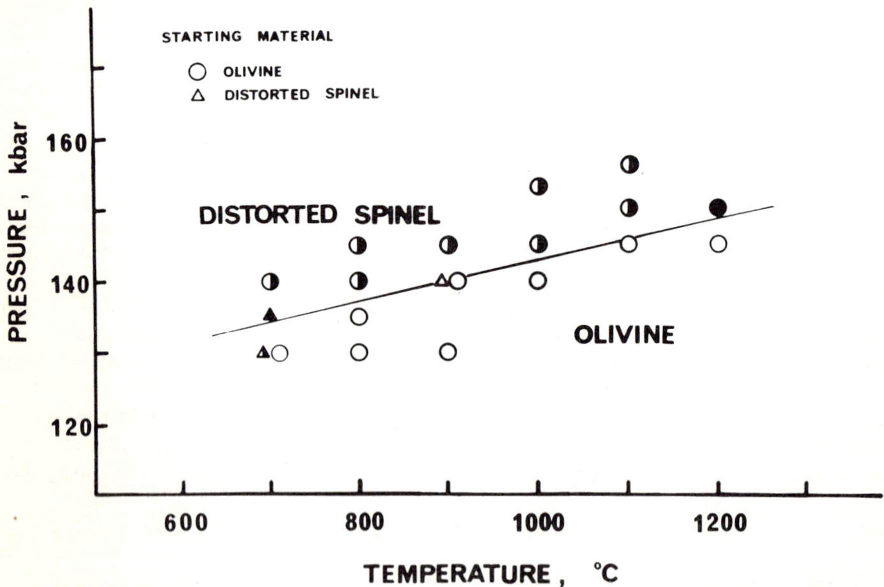

Fig.5. α-β transition curve in pure Mg_2SiO_4

TABLE 1. Experimental Results on α-β Transition

Run Number	Temperature, °C	Pressure, kbar	Time, min	Phase Present
Starting material: olivine (α)				
605	700 ± 10	130	360	α
655	700 ± 20	140	240	α + β
614	800 ± 30	130	10	α
630	800 ± 50	135	130	α
675	800 ± 30	140	120	α + β
611	800 ± 50	145	100	α + β
615	900 ± 50	130	15	α
632	900 ± 30	140	120	α
613	900 ± 30	145	90	α + β
647	1000 ± 50	140	120	α
639	1000 ± 50	145	90	α + β
608	1000 ± 30	153	30	α + β
642	1100 ± 50	145	75	α
659	1100 ± 20	150	5	α + β
638	1100 ± 50	156	15	α + β
644	1200 ± 50	145	5	α
671	1200 ± 50	150	60	β
Starting material: distorted spinel (β)				
666	700 ± 20	130	180	α + β
663	700 ± 30	135	150	β
667	900 ± 50	140	90	α

1000-1200°C, so that these decompositions result in considerably lower pressure values [e.g., *Liu*,1974; *Inoue*, 1975; *Yagi*, 1975]. Thus, pyrophyllite is not a suitable pressure-transmitting medium at high temperatures and at pressures higher than 100 kbar. The experimental results of the α-β phase boundary by *Suito* [1972] should be corrected negatively for pressure at temperatures above 1000°C. The pressure values reported by *Suito* [1972] were overestimated. Using the Bridgman anvil, Akimoto and Kawada [unpublished data, cited in *Ming,* 1974] and *Akimoto et al.* [1976] also reported the α-β transition pressure at 125 kbar and 1000°C.

All the previous investigators used the reactive sample of Mg_2SiO_4 (reactive forsterite) as the starting material. In the present study, pure Mg_2SiO_4 (olivine) which was synthesized from the stoichiometric mixture of SiO_2 and MgO at 1600°C for 48 hours, and the β phase were used as the starting material. No reactive forsterite was used as the starting material. *Ringwood and Major* [1970] reported that pure Mg_2SiO_4 synthesized by solid state reaction was not very reactive and frequently required much higher pressures to transform to the β phase than the reactive sample of Mg_2SiO_4. So the main reason why the present pressure value for the α-β transition at 1000°C are higher than the previous values seems to be due to the different starting materials. It is also certain that the differences among the pressure values for the α-β transition point at 1000°C are due to different pressure media, different apparatus used and the different pressure scale employed. The more detailed experiments will be necessary.

The experimental results for the β-γ transition are shown in Table 2. *Suito* [1972] reported the first synthesis of γ-Mg_2SiO_4 at $P > 200$ kbar and at 1000°C. *Ito et al.* [1974] studied the detailed crystallographic data of γ-Mg_2SiO_4. Figure 6 shows the typical X-ray powder pattern of the γ-Mg_2SiO_4 which was

TABLE 2. Experimental Results on β-γ Transition

Run Number	Temperature, °C	Pressure, kbar	Time, min	Phase Present
648	600 ± 10	153	360	β + γ
606	700 ± 30	153	180	β
625	700 ± 50	163	180	β + γ
621	700 ± 50	184	180	β + γ
18	700[a]	195	70	β + γ
661	800 ± 20	163	150	β + γ
625	800 ± 20	168	35	β + γ
17	800[a]	180	120	γ
13	800 ± 20	210	120	γ
669	900 ± 30	163	90	β
636	900 ± 10	172	20	β + γ
672	1000 ± 20	165	20	β
678	1000 ± 50	170	30	β + γ
26	1000[a]	210	15	β + γ
24	1200[a]	210	15	γ

a. Temperatures were estimated from wattage.

synthesized at about 250 kbar and 1000°C. Detailed X-ray data are summarized in Table 3. Unit-cell dimension (a_o) is determined to be 8.076 ± 0.001 Å.

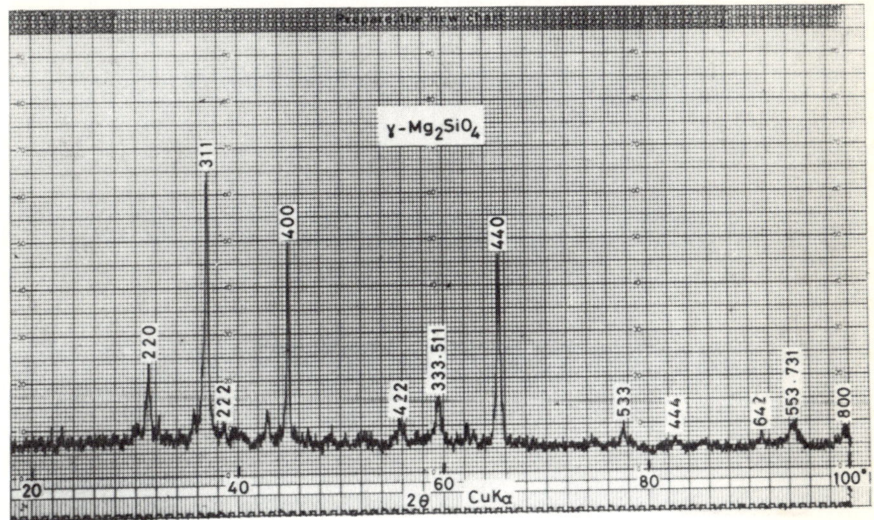

Fig. 6. Powder X-ray diffraction chart of γ-Mg_2SiO_4. [From Ito et al., 1974.]

TABLE 3. X-ray Diffraction Data of γ-Mg_2SiO_4 [From *Ito et al.*, 1974.]

h k l	$d_{obs.}$	$d_{calc.}$	$I_{obs.}$
111	--	4.662	--
220	2.855	2.855	23
311	2.435	2.435	100
222	2.332	2.331	5
400	2.019	2.019	50
331	--	1.8527	--
422	1.6489	1.6484	9
333,511	1.5541	1.5541	25
440	1.4278	1.4276	63
533	1.2314	1.2315	10
444	1.1660	1.1656	6
642	1.0792	1.0791	5
553,731	1.0509	1.0513	12
800	1.0094	1.0094	7

a_o = 8.076 ± 0.001 Å.

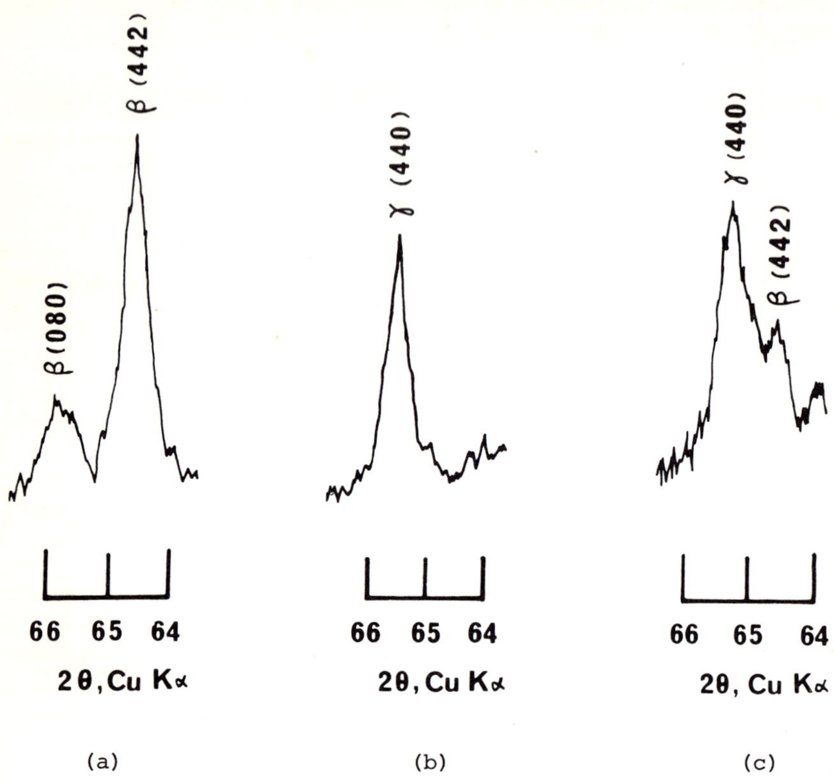

Fig. 7. Parts of the X-ray diffraction patterns for the identification of the γ phase in the presence of the β phase: (a) X-ray diffraction pattern of the β phase alone; (b) X-ray diffraction pattern of the γ phase alone; (c) X-ray diffraction pattern of the γ phase in the presence of the β phase.

Identification of the γ phase in the presence of the β phase was based mainly on comparison of the intensities of the (442) line of the β phase with that of the (440) line of the γ phase. The presence of the γ phase is evident from the fact that the intensity of the line (440) of the γ phase is higher than the intensity of the line (442) in the β phase.

Parts of the X-ray diffraction patterns are shown in Figure 7: (a) shows the β phase alone, (b) shows the γ phase alone, and (c) shows the γ phase in the presence of the β phase. To establish the β-γ phase boundary more precisely, the γ phase should be used as the starting material.

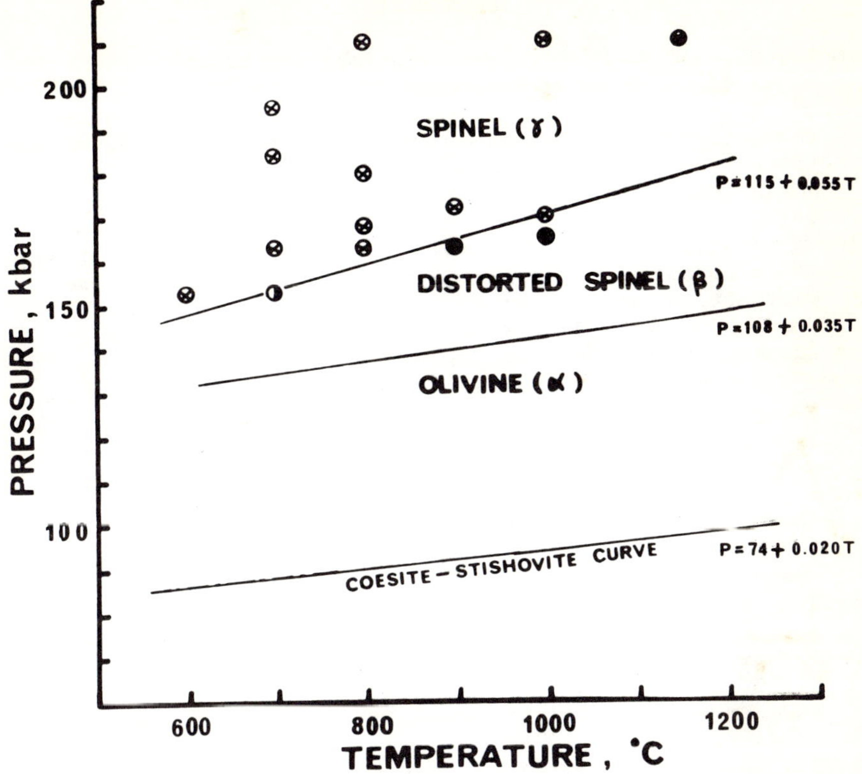

Fig. 8. *β-γ phase transition curve and the α-β phase transition curve in pure Mg_2SiO_4. The coesite-stishovite transition curve is also shown.*

The phase diagram in Figure 8 shows the β-γ phase boundary and the α-β phase boundary. The coesite-stishovite boundary is also shown. From these results, it is evident that the β-γ phase boundary has a positive slope, and the triple point of α, β, and γ phases lies at $T < 500$°C. Tentatively, the equation on β-γ phase boundary is given by P (kbar) = 115 + 0.055 T (°C).

Acknowledgments. The author wishes to express his gratitude to N. Kawai, Faculty of Engineering Science, Osaka University, for his encouragement and guidance and to thank E. Ito, Institute of Thermal Spring Research, Okayama University, for providing synthetic olivine samples. Thanks are due to L. C. Ming, Hawaii Institute of Geophysics, University of Hawaii, for critically reviewing the paper.

REFERENCES

Akimoto, S., and Y. Syono, Coesite-stishovite transition, *J. Geophys. Res., 74,*1653-1659, 1969.

Akimoto, S., Y. Matsui, and Y. Syono, High-pressure crystal-chemistry of orthosilicates and the formation of the mantle transition zone. in *Physics and Chemistry of Minerals and Rocks,* edited by R. G. J. Streus, Wiley Interscience, London, 1976.

Akimoto, S., T.Yagi, and K. Inoue, High temperature-high pressure phase boundaries in silicate systems using *in situ* X-ray diffraction. in *High-Pressure Research: Applications to Geophysics,* edited by M. H. Manghnani and S. Akimoto, Academic Press, New York, 585-602, 1977.

Inoue, K., Developments of high temperature and high pressure X-ray diffraction apparatus with energy dispersive technique and its geophysical applications, Ph. D. thesis, The University of Tokyo, 1975.

Ito, E., Y. Matsui, K. Suito, and N. Kawai, Synthesis of γ-Mg_2SiO_4, *Phys. Earth Planet. Interiors, 8,* 342-344, 1974.

Kawai, N., M. Togaya, and A. Onodera, A new device for pressure vessels, *Proc. Japan Acad., 49,* 623-626, 1973.

Kumazawa, M., H. Sawamoto, E. Ohtani, and K. Masaki, Post-spinel phase of forsterite and evolution of the earth's mantle, *Nature, 247,* 356-358, 1974.

Liu, L., Synthesis of a new high-pressure phase of tin dioxide and some geophysical implications, *Phys. Earth Planet Interiors, 9,* 338-343, 1974.

Liu, L., Disproportionation of kyanite to corundum plus stishovite at high pressure and temperature, *Earth Planet Sci. Letter, 24,* 224-228, 1974.

Ming, L., High pressure phases in the system of FeO-MgO-SiO_2 and their geophysical implications, Ph. D. thesis, The University of Rochester, 1974.

Ming, L., and W. A. Bassett, Post-spinel phases in the system of Mg_2SiO_4-Fe_2SiO_4, *Science, 187,* 66-68, 1975.

Ringwood, A. E., and A. Major, Synthesis of Mg_2SiO_4-Fe_2SiO_4 solid solutions, *Earth Planet. Sci. Letter, 1,* 241-245, 1966.

Ringwood, A. E., and A. Major, The system Mg_2SiO_4-Fe_2SiO_4 at pressures and temperatures, *Phys. Earth Planet. Interiors, 3,* 89-108, 1970.

Suito, K., Phase transitions of pure Mg_2SiO_4 into a spinel structure under high pressures and temperatures, *J. Phys. Earth, 20,* 225-243, 1972.

Suito, K., N. Kawai, and Y. Masuda, High pressure synthesis of orthorhombic SnO_2, *Mat. Res. Bull., 10,* 677-680, 1975.

Yagi, T., Accurate characterization of the high pressure environment by *in situ* X-ray diffraction study and its application to geophysical problems, Ph. D. thesis, The University of Tokyo, 1975.

METALLIC TRANSITION OF OXIDES, H_2O AND HYDROGEN

N. Kawai, M. Togaya, O. Mishima
Department of Physics, Faculty of Engineering Science, Osaka University, Toyonaka, Osaka 560, Japan

Abstract

Very high pressures are generated within a relatively large specimen chamber of a two-stage, split-sphere pressure vessel in which both electrical resistance measurement and internal heating of specimen can be conducted.

Instead of pyrophyllite, semi-sintered MgO is used as the pressure-transmitting medium.

With the vessel and the new medium, H_2O and hydrogen are compressed at room temperature till both become electrical conductors. The data are then compared with the data already obtained such as the insulator-conductor transitions in S, FeO, NiO, Fe_2O_3, Cr_2O_3, TiO_2, SiO_2 and MgO.

I. INTRODUCTION

For a long time, experimentalists have been trying to produce in the laboratory such high static pressure as those existing in the interior of the planet. The techniques usually involve very hard and incompressible materials for the pressure vessel, a large hydraulic press for generating force, and an appropriate pressure-transmitting medium.

Bridgman [1950] used a WC-pipestone combination to generate pressures up to 100 kbar. The General Electric group chose a WC-pyrophyllite combination for synthesizing diamond [*Bundy et al.*, 1955]. This approach has become routine.

Multi-anvil and multi-anvil-sliding systems have been developed to achieve both higher pressure and larger specimen volume. They have overcome the disadvantage of piston-cylinder apparatus which can be used for large volume experiment but limited to lower pressures, and the disadvantage of the Bridgman anvil which can produce very high pressure but only with a small sample.

Recently many studies have been made by using Bridgman's opposed anvils made of single-crystal diamond. It has become possible to carry out many measurements, such as X-ray analysis, laser-heating, and pressure calibration, by use of NaCl lattice parameter or red shift-of ruby R_1 fluorescence, Mössbauer analysis, infrared absorption, elastic moduli determination, and microscopic observation under the applied load. The pressure generated is claimed to be as high as 1 Mbar and temperature as high at 2000°C, although the volume in which the uniform pressure is confined is small and limited to the central part of the opposed anvils.

A two-stage split-sphere pressure vessel has been employed in our laboratory since 1965. The inner assemblage is composed of eight cubic anvils made of hot-pressed sintered Al_2O_3. The outer assemblage is composed of segmented pieces of spherical shell. The vessel can produce very high pressures under which FeO, Fe_2O_3, Cr_2O_3, NiO, TiO_2, SiO_2, and MgO have become metallic [*Kawai and Mochizuki*, 1971a, b; *Kawai and Nishiyama*, 1974a, b]. The transition pressure of FeO is the lowest and that of MgO is the highest among this set of materials. Adding to the above list, the present paper reports recent findings of insulator-conductor transitions in H_2O and in hydrogen.

II. PRESSURE VESSEL AND UNIAXIAL PRESS

For a more than a decade, the split-sphere apparatus has been used to produce ultra-high pressures in the laboratory. However, before 1973 the techniques were rather cumberson. The assemblage had to be covered with a rubber shell and immersed in an oil reservoir. Attachments for measuring the physical properties had to be replaced after each loading. Both handling and arranging the sphere became progressively difficult as its diameter was enlarged. In 1973 *Kawai et al.* [1973] improved and simplified the apparatus to increase its maneuverability. The modified version is described here.

The vessel is composed of cube, sphere, cylinder all split up and assembled in a layered arrangement. A sphere made of sintered WC containing 5% Co binder is equally divided into six wedge-shaped pieces, each with a truncated apex at the center of the sphere, so as to form a six-sided square-faced anvil.

Three out of the six anvils, after being put together at the lateral side surfaces, are placed conformably in a hemi-spherical space in the upper part of the cylinder of hardened steel. One corner of each anvil, exposed above the equator of the hemisphere, is cut off as shown in Figure 1 such that the upper surface of the exposed spheres consists of three inclinig roof planes, a, and one cradle, b, which is surrounded by three mutually perpendicular square surfaces as shown in Figure 2. Similarly the second assemblage can be constructed using the remaining three anvils. The two assemblages are set opposed and then joined so that all roof planes may make respective contact via six faceted blocks of steel, c, shown in the diagram. A cubic cavity whose <111> direction is parallel to the axis of the

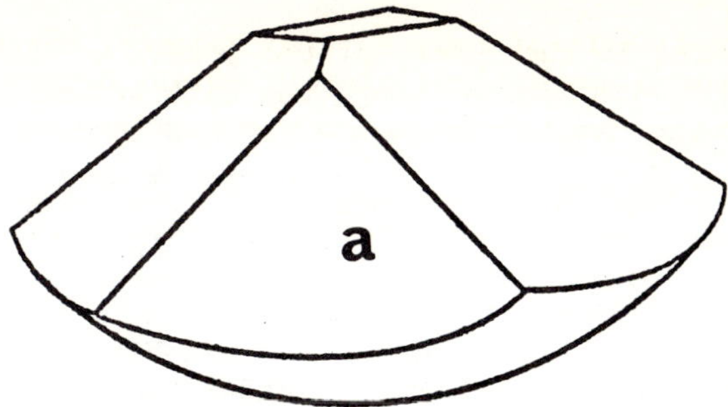

Fig. 1. One of the anvils obtained when a sphere is equally split into six peices. The apex and one corner are truncated.

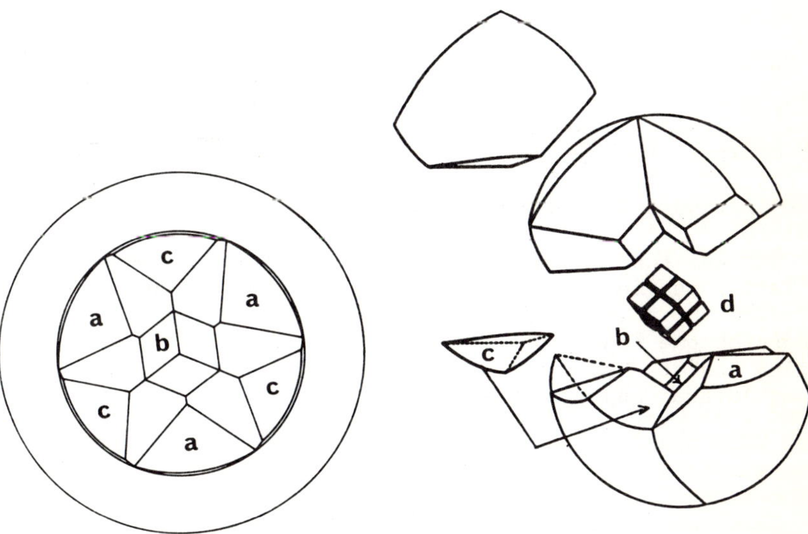

Fig.2. View looking down a set of three anvils in a cylinder. a = inclining roof; b = cradle; c = faceted block; d = innermost assemblage made up of eight cubes.

cylinders remains in the center of the join.

A specimen with similar shape but with volume larger than that of the cubic cavity can be charged and compressed. This cavity can also contain eight cubes piled up to form one large cube. Of the eight cubes, six are made of hot-press-sintered Al_2O_3, and two of WC. The latter are set opposed along the axis of the outer cylinder, and used as the electric leads for the resistance measurement. The sample specimen is placed at the very center of the entire system. A crushable spacer made of cardboard and semi-sintered MgO is sandwiched in the gaps of the eight cubes within the inner cubic assemblage.

The two cylinders with split spheres and cubes are finally compressed in a uniaxial press. As thickness of each spacer decreases with increasing load, the inner anvils move forward uniformly, and high pressure can be generated, as described in Section III.

III. SEMI-SINTERED MgO AS A PRESSURE-TRANSMITTING MEDIUM

Fine particles of MgO are semi-sintered at 1300°C for 10 hours till the density reaches 2.3. The quenched block still contains linked voids distributed uniformly and is machinable with ordinary tools. When a spacer with thickness *a* is made of the new media and placed in front of the cardboard spacer as shown in Figure 3, one obtains a hollow specimen space with volume equal to a^3.

Through the linked voids any gas or liquid is permeable, and the latter can be sent to the space from the exterior. Under pressure, the voids diminish and vanish completely, and the MgO spacers become non-porous. The consolidated spacers act as sealing gasket, keeping fluid sample from leaking. They form a strong lateral support to each cubic anvil as well.

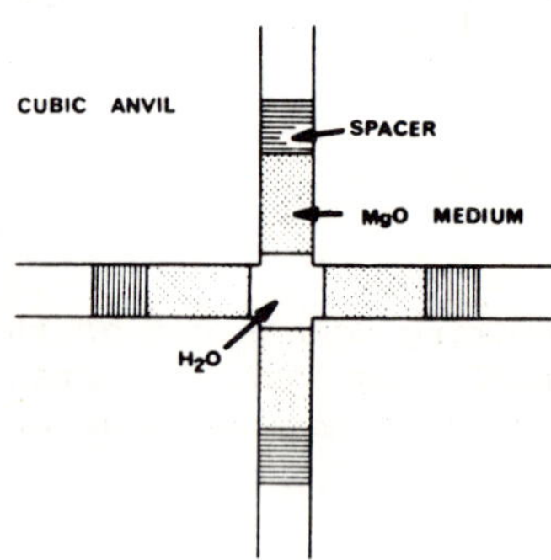

Fig. 3. Cross section of the central anvil assemblage.

With decreasing voids, the thickness *a* reduces, and the sample volume reduces as a^3. With increasing pressure, parts of the MgO spacers intrude into the sample chamber, which further reduces the sample volume. As a result, a rapid reduction of the specimen volume can be expected (Figure 4). In the final stage, the intruded MgO spacers become nonporous and produce large pressure intensification over a very limited area in contact with the specimen. We consider the voidless part that develops toward the center as "growing anvils."

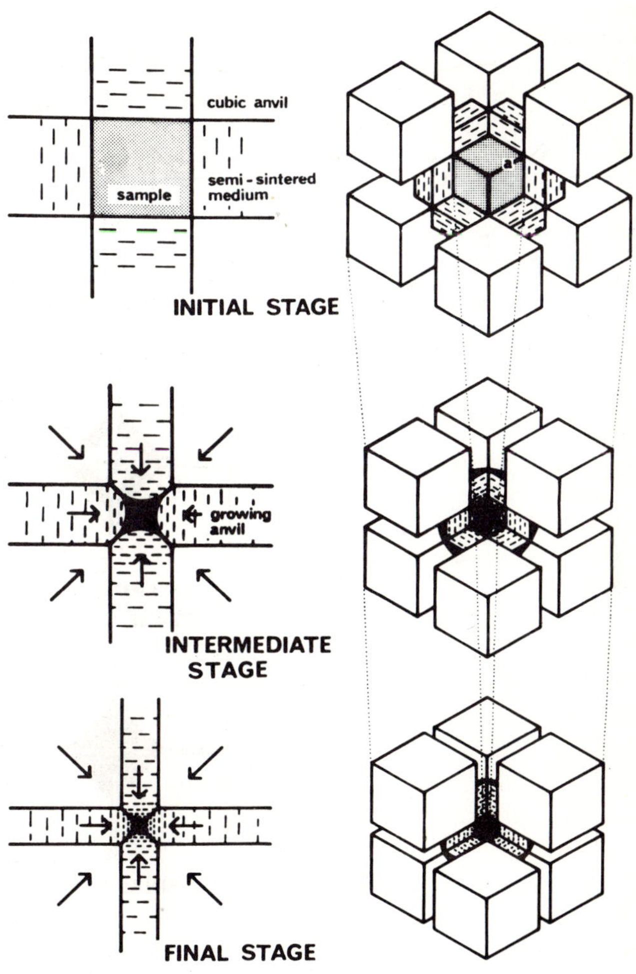

Fig. 4. Rapid volume reduction in specimen chamber and growing anvils whereby the pressure is intensified.

When the octahedral form of the MgO medium is placed in the inner assemblage of eight cubes, each with the corner at the center of the assemblage truncated, and compressed as shown in Figure 5, the reduction in the void space in the octahedron is not uniform. This reduction occurs near the surface of the octahedron and progressively proceeds toward the center as shown in the diagram. The consolidated medium acts as eight growing anvils, trigonal-pyramidal in shape, and the pressure at the center is intensified. The growing MgO anvils protect the cubes from rupture under very high pressures.

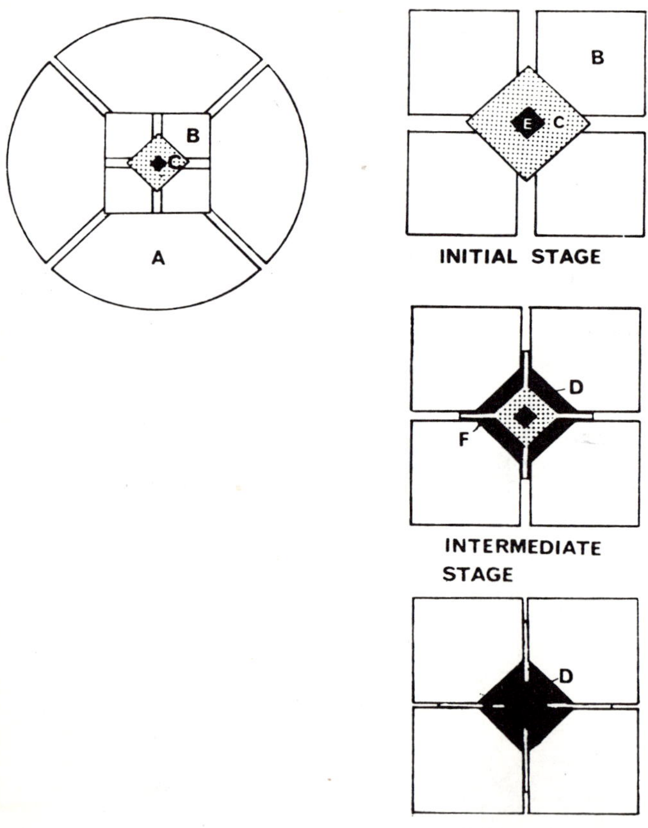

Fig. 5. Cross section of the pressure chamber. A = one part of a sphere. B = inner cubic anvil. C = semi- sintered medium. D = growing anvil. E = cali brant. F = shearing plane.

IV. INSULATOR-CONDUCTOR TRANSITION IN H_2O

Kawai et al. [1975a] compressed pure H_2O at room temperature (23°C) with 1 mm^3 of the ice as the starting material. A change of electric resistance was observed in both the loading and unloading stages, as shown in Figure 6. With increasing pressure, the resistance dropped abruptly to a value less than 10 Ω under loads higher than 950 ton and the specimen became a conductor. In the unloading stage two sudden recoveries of resistance

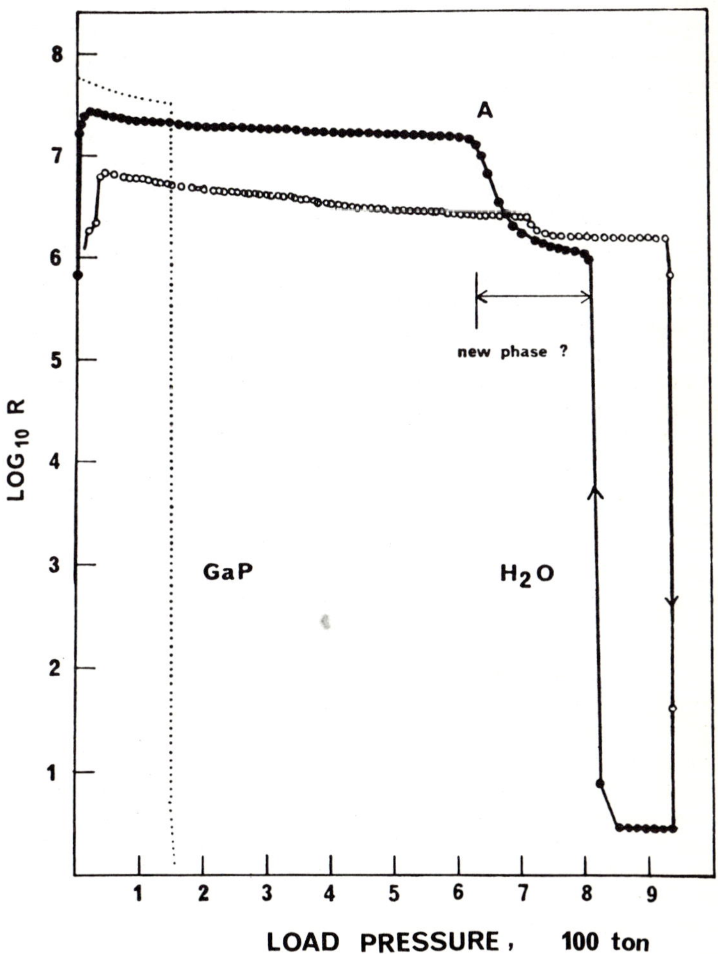

Fig. 6. Change of electrical resistance of H_2O with pressure.

occurred at 810 and 620 ton. Below 620 ton the resistance increased gradually. Just before the load was entirely released, a small but sharp drop took place, and the value of resistance became almost equal to that of pure water. It was possible to pack the water within the medium without leakage. The arrangement of both MgO spacers near the specimen is shown in Figure 3.

Fletcher [1970], using data by *Bridgman* [1937], *Pistorius et al.* [1963], and *Brown and Whalley* [1966], compiled a *P-T* phase diagram of H_2O. Since the present experiment was made at room temperature, we should have first encountered with the ice VI and then with the ice VII, which have been considered by *Kamb* [1968] to possess an interpenetrating diamond structure with an H-O-H angle equal to 109°. Although no phase of H_2O denser than ice VII has been known to exist, the ice that finally appears beyond A in Figure 6 is likely to be a new phase, which is a better electric conductor than ice VII. Alternatively, it is also possible that the liquid water reappeared as the result of melting of ice VII. It is quite plausible that the melting curve has a negative slope close to the insulator-conductor transition. Ionic conduction of Mg^{2+} ion in water due to the expected increase of the solubility of the cation in water may not be responsible for the insulator-conductor transition since the observed conduction is relatively high as compared to the ordinary ionic conduction.

To estimate the pressure at which H_2O metallizes, the transition in GaP was investigated in the same vessel. The insulator-metal transition was observed at a load of only 150 ton, as shown by the dotted curve in Figure 6. Since the pressure required for the Gap transition is 220 kbar, the pressure for the metallic H_2O would be quite high, in the range of 2.5-4 Mbar, although it is difficult to estimate the pressure exactly on the basis of such extrapolation.

According to *Vereshchagin et al.* [1975b], who independently reported the transition of H_2O into the conducting state, the transition pressure is about 1 Mbar at temperatures from -80 to -10°C.

V. INSULATOR-CONDUCTOR TRANSITION IN HYDROGEN

For several years *Kawai et al.*[1975b] have tried to produce metallic hydrogen under very high pressures at room temperature. They have succeeded in making it in a double staged high-pressure vessel which is modified so that any gas can be charged in it and liquefied or solidified therein [*Kawai et al.*, 1975b]. The cross section of the vessel is shown in Figure 7. A rubber ring, a, was placed between the lower and upper cylinder to seal gas at low pressure. A gas channel, b, was built in the upper cylinder so that gas may be sent through from the exterior into a space surrounded by the rubber ring and the two cylinders. Of the eight inner cubes, six were made of hot-press-sintered Al_2O_3

having almost theoretical density, while the two remaining cubes of WC. The cross section of the inner arragement is shown schematically in Figure 8. The semi-sintered MgO spacers which originally contained many linked voids were placed closest to the chamber.

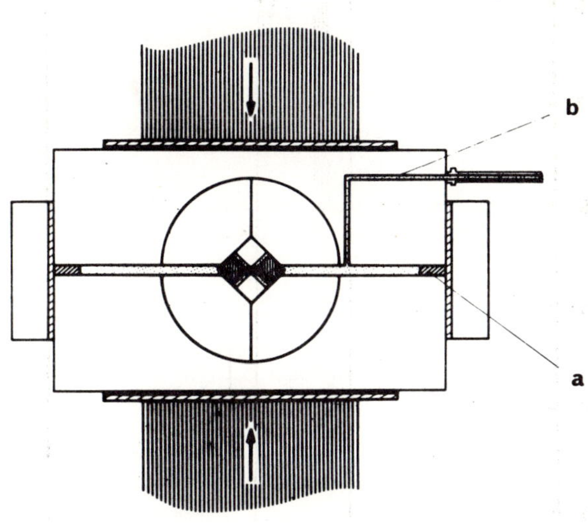

Fig. 7. Cross section of the apparatus.

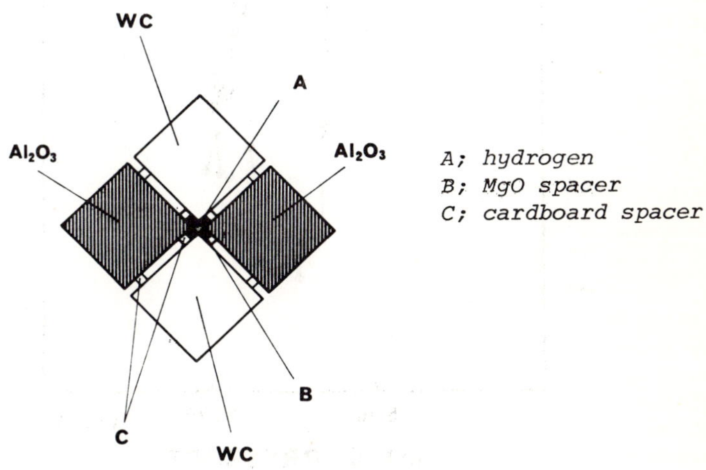

Fig.8. Cross section of the central part of the inner cubes.

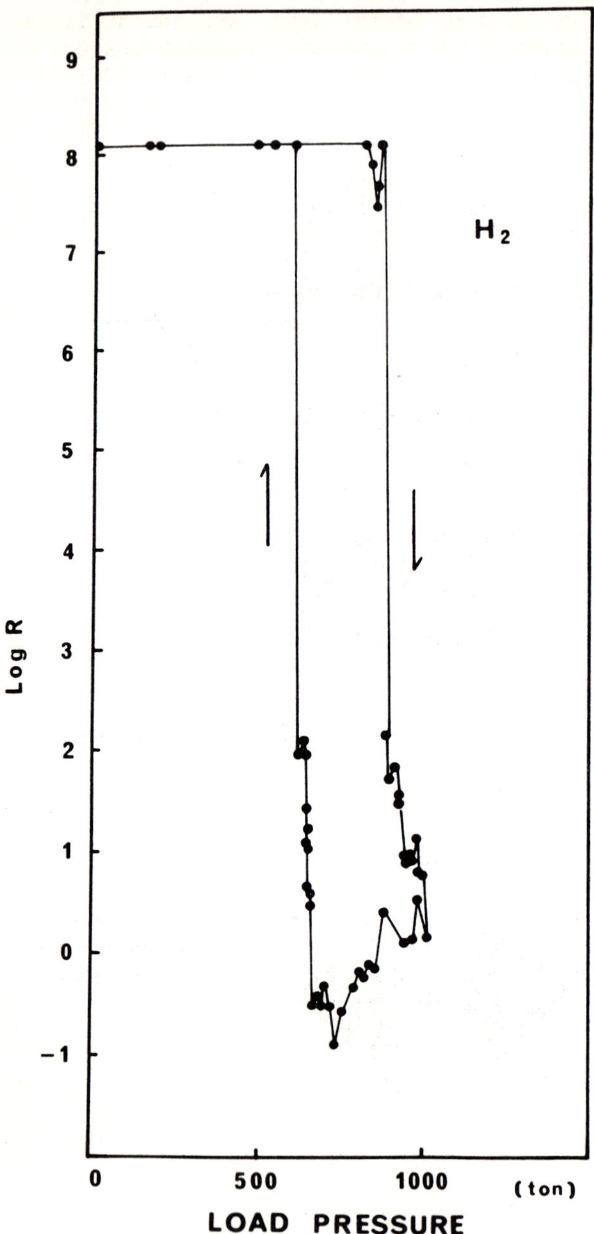

Fig. 9. log R versus applied load relation for hydrogen.

With the permeable MgO spacers, compressed gas could be easily charged into the chamber after the air remaining in the space had been completely evacuated. Before the experiment, N_2 gas was charged and discharged twice to purge the residue air. Then hydrogen was charged and discharged twice also to purge the remaining N_2. About 50 liters of purified hydrogen were finally charged directly from a bomb filled with compressed H_2. The thickness of the MgO spacers was so chosen that the voids would have completely vanished when the original thickness was reduced by one-half. Although the pressure in this stage only increased to 200 bar in the exterior of the specimen chamber, the internal pressure increased at a much higher rate, with the help of the growing anvils. The resistance drop observed in the loading stage and the reversion observed in the unloading stage are shown in Figure 9. The resistance dropped abruptly at the applied load of 855 ton from 126.3 MΩ to 100 Ω, then gradually decreased with fluctuation until the slowly increasing applied load reached 1000 ton. The decrease continued even after releasing the load down to 700 ton. Then the reversion set in and was completed at 600 ton.

There remains a few questions to be solved before we can ascertain that H_2O and hydrogen become conductive under pressure. As we have reported [*Kawai and Nishiyama*, 1974b], MgO itself becomes conductive under very high pressure at room temperature. In repeating the experiment (with the same spacers around the chamber) without H_2O or hydrogen, no drastic change of resistance was observed, even when the applied load was increased above 1100 ton.

Next, N_2 and CH_4 were charged and compressed in the same vessel under the same conditions. As in the case of hydrogen, no such a drop of resistance was observed even when the applied load exceeded 1000 ton. The observed drop of resistance in hydrogen does correspond to the insulator-conductor transition.

The transition pressure for hydrogen still remains difficult to be estimated. *Vereshchagin et al.*[1975a] have succeeded in metallization of hydrogen at approximately 1 Mbar at 4.2 K. *Nakamura et al.* [1976] have calculated the energy of metallic hydrogen, and shown that if the metal assumes to have a deformed simple cubic lattice, with the atoms lined up in a filament structure, the energy decreases significantly from that expected in the bcc structure as pointed out by many authors [*e.g., Winger and Huntington,* 1935]. The transition pressure estimated by them is, however, lower than 1 Mbar.

VI. SUMMARY

Using the same arrangement of both the anvils and spacers ZnS and GaAs were metallized under 35 and 42 ton of external load, respectively.

Load F versus pressure relation was found to lie between the curves

$$P \propto F^{1/0.98} \quad \text{and} \quad P \propto F^{1/1/2}$$

The pressures required for the insulator-metal transitions in Fe_2O_3, Cr_2O_3, TiO_2, NiO, SiO_2, MgO, H_2O and hydrogen extrapolated from $P \propto F^{1/1.1}$ are shown in Figure 10.

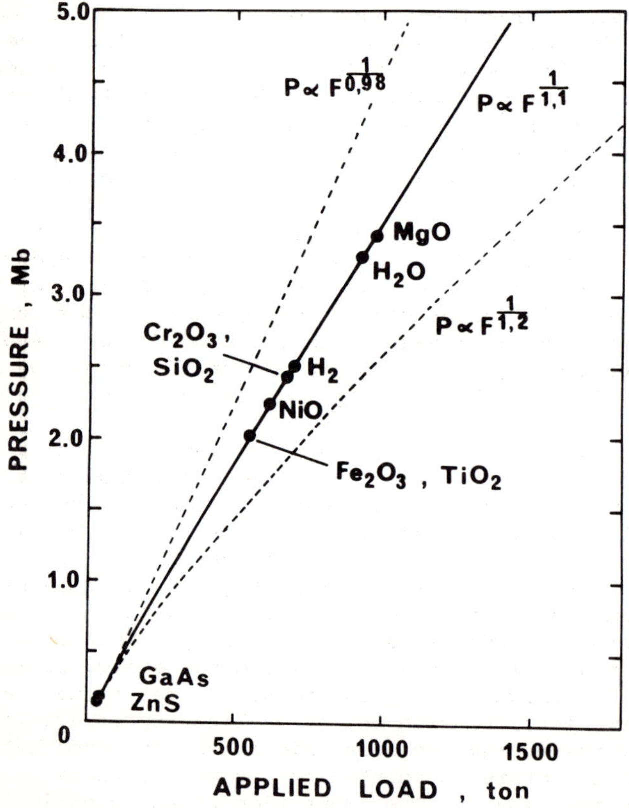

Fig. 10. Extrapolated transition pressures.

Since our estimation is based on the extrapolation of the low-pressure fixed points, the values given in the diagram may change when the data for higher pressures are obtained.

Reference

Bridgman, P. W., The phase diagram of water to 45000 kg/cm^2, *J. Chem. Phys.*, *5*, 964-966, 1937.

Bridgman, P. W., Physics above 2000 kg/cm^2 (Bakerian Lecture of the Royal Society), *Proc. Roy. Astron. Soc.* (London), *[A]203*, L-17, 1950.

Brown, A. J., and E. Whalley, Preliminary investigation of the phase boundaries between ice VI and VII and Ice VI and VIII, *J. Chem. Phy.*, *45*, 4360, 1966.

Bundy, F. P., H. T. Hall, H. M. Strong, and R. H. Wentorf, Jr., Man-made diamonds, *Nature*, *176*, 51-55, 1955.

Fletcher, N. H., in *The Chemical Physics of Ice*, Cambridge University Press, 1970.

Kamb, B., in *Structural Chemistry and Molecular Biology*, edited by A. Rich and N. Davidson, Freeman, 507-542, 1968.

Kawai, N., and S. Mochisuki, Metallic states in the three 3d transition metal oxides, Fe_2O_3, Cr_2O_3 and TiO_2 under static high pressures, *Phys. Lett.*, *36A*, 54-55, 1971a.

Kawai, N., and S. Mochizuki, Insulator-metal transition in NiO, *Solid State Commun.*, *9*, 1393-1395, 1971b.

Kawai, N., M. Togaya, and A. Onodera, A new device for pressure vessels, *Proc. Japan Acad.*, *49*, 623-626. 1973.

Kawai, N., and A. Nishiyama, Conductive SiO_2 under high pressure, *Proc. Japan Acad.*, *50*, 72-75, 1974a.

Kawai, N., and A. Nishiyama, Conductive MgO under high pressure, *Proc, Japan Acad.*, *50*, 634-635, 1974b.

Kawai, N., O. Mishima, M. Togaya, and B. L. Neindre, H_2O rendered metallic under high pressure, *Proc. Japan Acad.*, *51*. 627-629, 1975a.

Kawai, N., M. Togaya, and O. Mishima, A study of metallic hydrogen, *Proc. Japan Acad.*, *51*, 630-633, 1975b.

Nagara, H., H. Miyagi, and T. Nakamura, Stability of metallic hydrogens with cubic structure, Progr. Theoret. Phys., *56*, 396-414, 1976.

Pistorius, C. W. F. T., M. C. Pistorius, J. P. Blakely, and L. J. Admiraal, Melting curve of ice VII to 200 kbar, *J. Chem. Phys.*, *38*, 600-602, 1963.

Vereshchagin, L. F., E. N. Yakovlev, and Yu. A. Timofeev, Possibility of transition of hydrogen into the metallic state, *JETP Lett.*, *21*, 85-86, 1975a.

Vereshchagin, L. F., E. N. Yakovlev, and Yu. A. Timofeev, Transition of H_2O into the conducting state at static pressure $P \approx 1$ Mbar, *JETP Lett.*, *21*, 304-305, 1975b.

Wigner, E., and H. B. Huntingron, On the possibility of a metallic modification of hydrogen, *J. Chem. Phys.*, *3*, 764-770, 1935.

PREDICTION OF HIGH PRESSURE PHASE TRANSITIONS BY ELASTIC CONSTANT DATA

H. H. DEMAREST, JR.
Department of the Geophysical Sciences and
Materials Research Laboratory
University of Chicago
Chicago, Illinois 60637

R. OTA
Institute of Geophysics and Planetary Physics
University of California
Los Angeles, California 90024

and

Department of Industrial Chemistry[1]
Faculty of Engineering
Kyoto University
Yoshida, Kyoto, Japan

O. L. ANDERSON
Institute of Geophysics and Planetary Physics
University of California
Los Angeles, California 90024

Abstract

The Born elastic stability criterion provides a useful method of predicting polymorphic transitions at high pressure. Elastic constants can be extrapolated to high pressure, and a phase transition must occur before any of the shear elastic constants vanish. Many pressure-induced transitions are related to a macroscopic shear of the crystal lattice, in which cases, the dimensionless ratio of the shear elastic constant to the bulk modulus C_t/K is related to the curvature of the free energy along the reaction coordinate leading from the low-pressure to the high-pressure phase, and this ratio is a crude indicator of the Gibbs free energy difference between the two phases.

[1] Present address

We calculate the free energy for a model lattice as it transforms continuously from the NaCl to the CsCl lattice at a number of pressures to demonstrate this connection between the shear elastic constant C_{44} and the pressure-induced phase change. We have measured the single crystal elastic constants of KCl from zero pressure to the phase transition at 20 kbar. These data show that the elastic constants vary linearly with pressure right up to the transition, when the shear constant C_{44} has decreased to the relatively low value of only 20.6% of the bulk modulus. In other alkali halides, we infer similar behavior of the ratio $\alpha = C_{44}/K$ at the transition from the extrapolation of low-pressure elastic constant data, and find similar behavior for other pressure-induced phase transitions with α usually in the range 0.15-0.20. This modification of the Born criterion can be used to predict phase transitions in several compounds when C_{44}/K decreases to a critical value.

I. INTRODUCTION

The experimental study of the polymorphism of solids under pressure is an important tool for learning more about solids and the earth's interior. A good theory to predict phase relations under pressure will be useful in helping to choose experiments that are most likely to yield useful results. Predictions could be applied to minerals or other solids of interest, which are either inaccessible to current experiments or for which results are not yet available.

A comprehensive theory based on a direct calculation of free energy differences is not available. The idea of using the extrapolation of data on the elastic constants at low pressures to predict transitions at high pressure by means of the Born criterion has been around for quite a while. In many cases, the Born criterion was used to predict a transition at the pressure at which a shear elastic constant reached zero; however, experimental evidence now shows that the transition takes place before this pressure.

In Section II of this paper, we show that the transition should take place when the shear constant reaches a critical fraction, α, of the bulk modulus. This new modification of the Born criterion makes possible improved predictions of the transition pressure in cases for which a shear elastic constant decreases under pressure, and also it permits good predictions when a shear constant increases very slowly with pressure.

In Section III, we present new experimental data on the elastic properties of KCl up to the phase transition at 20 kbar. This data is in excellent agreement with our theory.

In Section IV of this article, we gather together an assortment of data on elastic constants and phase transitions to try to test how well the modified Born criterion works, and we make some predictions for phase transitions that have not yet been observed.

II. TRANSITIONS AND ELASTIC CONSTANTS UNDER PRESSURE--THEORY

It follows from fundamental thermodynamic principles that solids will transform to denser, more efficiently packed crystal structures under increasing pressure. Attempts to make detailed predictions of transition pressures based on a calculation of the free energy difference between two phases have only been marginally successful, even when sophisticated models were applied to very simple compounds [e.g., *Cohen and Gordon,* 1975]. It thus does not appear that any successful quantitative theory of phase transformations at high pressure based on thermodynamic equilibrium is available at the present time.

A. Elastic and Thermodynamic Stability

The elastic stability criterion of *Born* [*Born,* 1940; *Misra,* 1940; *Born* and *Huang,* 1954, p. 129] states that for a lattice to be mechanically stable, the Gibbs free energy must be a minimum relative to any states reached by infinitesimal strains. This requires that the eigenvalues of the elastic constant matrix should all be positive. For cubic crystals, these eigenvalues coincide with the bulk modulus and the two shear moduli. We shall refer to "shear moduli" in the following discussion, but it is understood that for applications to crystals of general symmetry, the quantities of interest are the eigenvalues of the elastic constant matrix. *Chang* and *Barsch* [1973] have made an elaborate discussion of this point.

The Born stability criterion is a special case of the general rule that the Gibbs free energy must be a minimum relative to all possible small displacements of the atoms, which requires that the frequencies of all the normal modes be positive. Because there is so much more theoretical and experimental information on the shear constants under pressure than there is for the behavior of other normal modes, we will refer only to the shear constants here, but most of this discussion can also be applied to any mode of vibration.

There always exist reaction paths by which a crystal can be continuously deformed from one crystal structure to another. These reaction paths may be complicated, involving large movements of some atoms, as in the case of reconstructive transitions involving nucleation and growth or disproportionation. Or, these paths may be fairly simple, involving relatively small atomic displacements. In either case, the reaction paths (and there may be several) between crystal structures can be defined in terms of the lattice modes of either structure [*Musgrave,* 1970, p. 274]. The elastic constants are thus related to phase transitions in two ways--through the Born stability criterion, and through the relationship of the elastic constants with the lattice modes which define the reaction coordinate leading from one crystal structure to another.

B. Behavior of Elastic Constants Under Pressure

Anderson [*Anderson*, 1970; *Anderson* and *Liebermann*, 1970; *Anderson* and *Demarest*, 1971] performed simple model calculations for four cubic lattices and showed that in these four cases one of the shear constants vanishes at high pressure. Some important conclusions of this work were that a shear elastic constant of a solid might decrease with pressure, perhaps after a temporary increase, and that the general trend of the less dense phases transforming to the more efficiently packed phases with increasing pressure could be predicted by the Born stability criterion.

Recently, more advanced techniques have been developed to extrapolate elastic constants to high pressure, and these techniques have been used to estimate the values of shear elastic constants at the phase transition. *Thomsen* [1970, 1971, 1972] used finite strain theory to predict that C_{44} for NaCl would vanish at the phase transition to the CsCl structure at 290 kbar. *Demarest* [1972*a*] used an improved lattice model, which included the next nearest neighbor (NNN) anion-anion interaction, to predict the value of C_{44} for seven alkali halides at the pressure of the polymorphic transition. In all cases, the transition took place before the predicted value of C_{44} reached zero. *Demarest* suggested an empirical modification of the Born criterion to predict a phase transition whenever C_{44}/K reached a critical value, α, with $\alpha \approx 0.15$ or 0.20. This modified Born criterion did a fairly good job of predicting all the known transitions occurring in the alkali-halides, including the case of NaCl, in which the new NNN calculation predicted that C_{44} would increase slowly with pressure.

C. Model Calculation of a Phase Change

In order to clarify the relationship between the modified Born criterion and the thermodynamic criterion for phase transformations, we recognize that the energy along the reaction path between the two phases can be expanded in a Taylor series in strain, whose leading coefficients are related to the elastic constants of the two phases [*Devonshire*, 1959]. Figure 1 shows that the NaCl - CsCl transformation can be accomplished by a compression in the (111) direction, which is equivalent to a combination of simple shears associated with the shear constant C_{44}. We used a Born model of the interatomic forces to calculate the energy of the lattice as it is continuously transformed from the NaCl to the CsCl structure at a number of pressures, and the results demonstrate the connection between a relatively weak value of C_{44} and the thermodynamic stability of the CsCl lattice. *Hyde* and *O'Keeffe* [1973] used a similar model to study some aspects of this transition, but they did not discuss C_{44}, and limited their calculation to zero pressure.

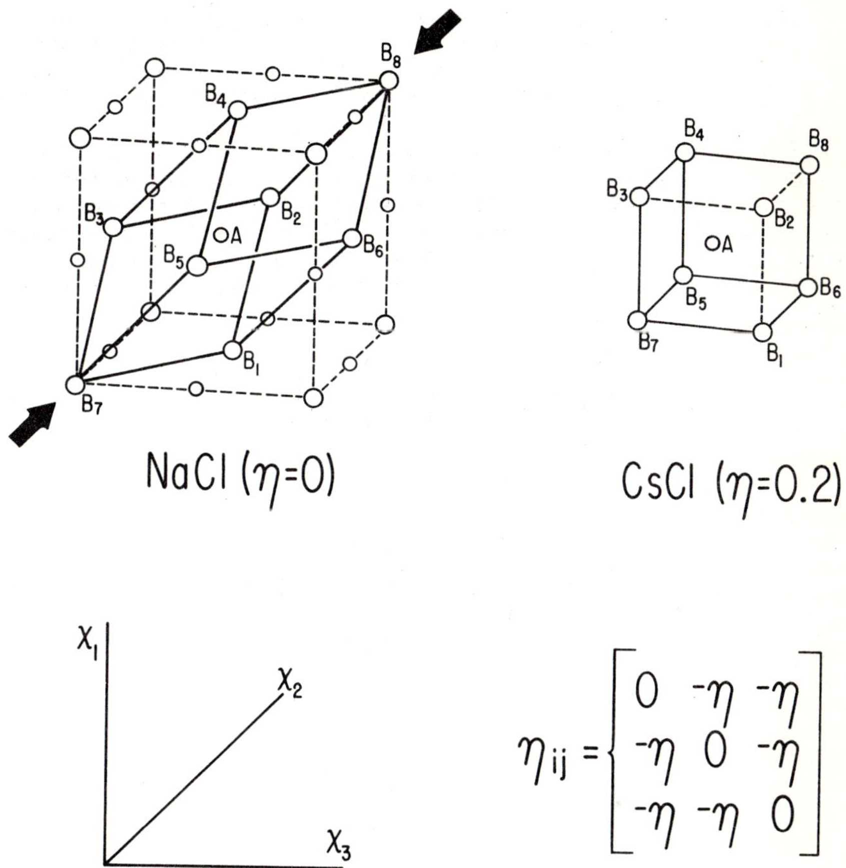

Fig. 1. Transformation from the NaCl to the CsCl lattice by a shear deformation.

In our model calculation, we ignored thermal effects and assumed that the energy of the lattice is given by the expression

$$F = \frac{z^2 e^2 A_r}{r} + \sum_{i=1}^{8} \frac{b}{r_i^n} \tag{1}$$

where r_i is the interatomic distance, r is the nearest neighbor distance, and z^2e^2, b, and n are constants. Summation is over the six nearest-neighbor interactions of the NaCl lattice and the two additional interactions that become nearest neighbors in the CsCl lattice.

The Madelung constant A_r, which gives the sum of all the electrostatic interactions, was calculated by the Ewald method [*Born* and *Huang*, 1954, p. 385] for each value of the reaction coordinate, η. By differentiation of equation (1), expressions for the pressure, P, and the bulk modulus, K, are obtained. It is most useful to write the relevant expressions in dimensionless form

$$\frac{F^*}{K_o V_o} = \frac{9}{n-1}\left[-\frac{r_o}{r^*} + \frac{1}{N}\left(\frac{r_o^*}{r_o}\right)^{n-1}\left(\frac{r_o}{r^*}\right)^n\right]\frac{A_r^*}{A_r} \tag{2}$$

$$\left(\frac{r_o^*}{r_o}\right)^{n-1} = \left(\frac{6 + 2\left(\frac{r^*}{r_2^*}\right)^n}{6 + 2\left(\frac{r}{r_2}\right)^n}\right)\frac{A_r}{A_r^*} \tag{3}$$

$$\frac{V_o^*}{V_o} = \left(\frac{r_o^*}{r_o}\right)^3 \frac{\beta^*}{\beta} \tag{4}$$

$$\frac{V^*}{V_o^*} = \left(\frac{r^*}{r_o}\right)^3 \left(\frac{r_o}{r_o^*}\right)^3 \tag{5}$$

$$\frac{K}{K_o} = \frac{1}{n-1}\left(\frac{V_o}{V}\right)\left[-4\left(\frac{r_o}{r}\right) + (n+3)\left(\frac{r_o}{r}\right)^n\right] \tag{6}$$

where V is the volume and the subscript o refers to zero pressure. Quantities without an asterisk refer to the NaCl lattice, while quantities with an asterisk refer to the deformed lattice. For this deformation, the packing ratio, β is given by $\beta = 2 - 6\eta^2 - 4\eta^3$, and the next nearest neighbor distance, $r_2 = r(3-6\eta)/(3+6\eta^2)$, where η is the reaction coordinate or shear strain defined in Figure 1. We calculated the Gibbs free energy from

$$\frac{G^*}{K_o V_o} = \frac{F^*}{K_o V_o} + (P/K_o) \frac{V^*}{V_o} \tag{7}$$

at several reduced pressures for structures in the series NaCl → CsCl. For the present calculation, we set the adjustable parameter $n = 11$. For each value of η and P/K_o, V^* was adjusted to obtain the minimum value for G^*.

The results are plotted in Figure 2 in the form $\Delta G/KV$. The vertical displacement of each curve is arbitrary. Symmetry requires that the slope of the free energy curve be zero at $\eta = 0$ and $\eta = 0.2$, while the curvature at $\eta = 0$ is related to the elastic constants in the NaCl lattice $\left(\frac{1}{KV}\frac{\partial^2 \Delta G}{\partial \eta^2} = 6\frac{C_{44}}{K}\right)$. This curvature is extrapolated by a dashed line in Figure 1, and the dashed line is a good approximation to the solid curve up to $\eta \approx 0.1$. A large part of the free energy difference between the two phases can be accounted for by this extrapolation. For $P/K_o = 0$, $C_{44}/K = 0.40$, and there is a relatively large free energy difference. (C_{44}/K was calculated using equation (6) and equation A6 of *Anderson* and *Demarest*, 1971.) For $P/K_o = 0.2$, $C_{44}/K = 0.20$, and the free energy difference is quite small. At about $P/K_o = 0.3$ and $C_{44}/K = 0.06$, the free energies of the two phases are equal, and the phase transformation takes place. C_{44} does not vanish until a much higher pressure, $P/K_o \approx 0.48$.

Figure 2 shows that for this model, $C_{44}/K \approx 0.06$ at the transition. We expect that changes in the interatomic forces from this simple model to real crystals will affect both C_{44}/K and ΔG in a similar manner, preserving the result of a finite C_{44}/K at the transition.

There are at least two arguments which suggest that, for all transitions involving a finite displacement, C_t/K will not reach zero at the transition, where C_t is any shear constant.

1. There will always be a free energy barrier between the two stable phases at the transition. This corresponds to a hill on the curve of ΔG versus η (Figure 2). (If there were a valley, this would correspond to a more stable phase.) Since the extrapolation of ΔG based on C_t (the dashed line in Figure 1) will conform closely with the ΔG curve, at least initially, C_t will be

positive. We expect that the size of the energy barrier and hence the value of C_t/K at the transition will be similar for similar compounds undergoing the same transition.

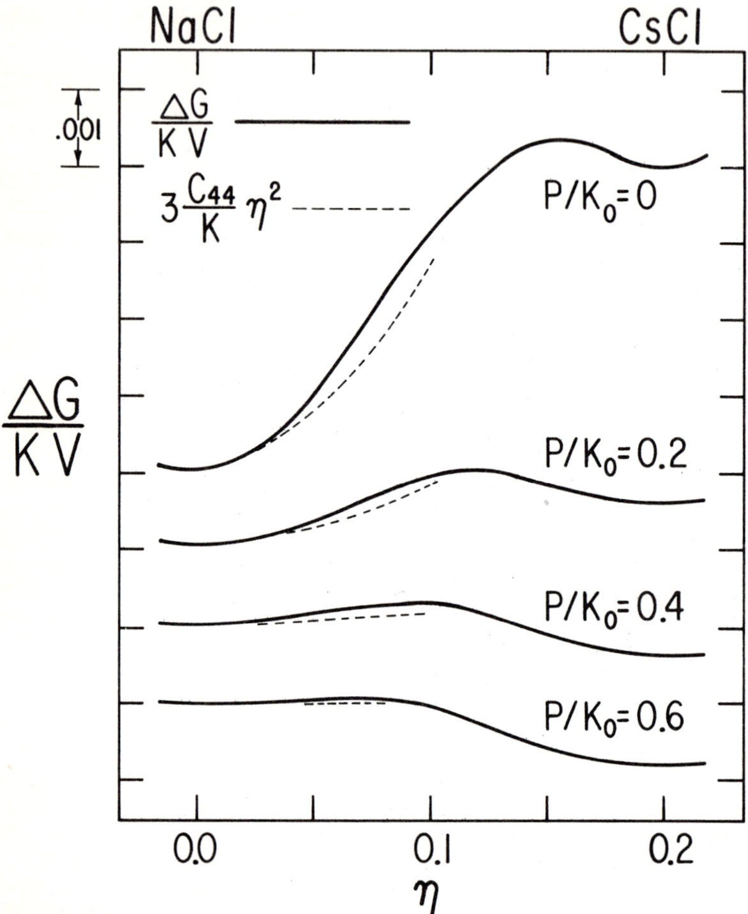

Fig. 2. Free energy change of a model crystal as it transforms from the NaCl to the CsCl structure.

2. If $C_t = 0$ at the transition, there are three stable phases all having the same free energy (the two phases in question, and a shear distortion of the low-pressure phase). This is in violation of the Gibbs phase rule, since we are not including temperature in our analysis.

D. Modified Born Criterion for Phase Transitions

In all phase transitions involving a finite shift of the atoms, the phase transition will take place before any of the shear elastic constants reach zero. The expected critical ratio, $\alpha = C_t/K$, at the transition (where C_t refers to any shear constant with a relatively low value) will depend, to a large extent, on the relationship between the reaction coordinate connecting the low- and high-pressure phase and the shear strain. There are several possibilities.

1. The transformation may be a second order one involving a lowering of symmetry, in which case the shift of atoms at the transition is infinitesimal, and we expect $\alpha = 0$. An example of this is TeO_2 , which transforms continuously from a tetragonal to an orthorhombic structure above 9 kbar. Experiments by *Peercy* and *Fritz* [1974; *Fritz* and *Peercy*, 1975] and the theories of *P. Anderson* and *Blount* [1965] and *Boccara* [1968] confirm that in this case the shear elastic constant associated with this distortion does go to zero at the transition.

2. The transformation involves a macroscopic shear as in the case of the NaCl-CsCl structure transition. In general, we expect α to be larger when the macroscopic shear is larger, and we expect similar values of α for chemically similar compounds. When an empirical value of α is needed to predict a transition, we believe it is best to obtain α from experimental data on a similar compound having the same crystal structure.

3. The transformation primarily involves a macroscopic shear, but an additional distortion is involved (as in the case of the wurtzite → NaCl transition), or the transition can be represented by a transverse acoustic mode at the Brillouin zone boundary (a short-wavelength shear mode), as is the case for calcite [*Merrill*, 1975]. We still expect a relatively low value of C_t/K at the transition, but C_t/K will be a less sensitive measure of ΔG than in (2) above, because it is only approximately equal to the curvature of the free energy along the reaction coordinate. Thus, we expect larger values of α than in (2) above.

4. The transformation may involve optic modes. It may involve large atomic displacements (e.g., disproportionation), or a change of the electronic structure may be involved in addition to atomic displacements (graphite-diamond, or a Mott transition, for example). In these cases, we do not expect any particular value of C_t/K at the transition. Thus, the possibility always exists that a transformation will take place that could not be predicted by the modified Born criterion.

There may be some justification for making α a slowly varying function of pressure. *Anderson* [1975] pointed out that there will be a general tendency for the shear moduli to increase less quickly than the bulk modulus under pressure. This "mode spreading" is a direct result of the Cauchy relation for central forces. For average elastic moduli in the Voigt approximation, and assuming central forces, $C_t/K = 3/5 - 6/5\ P/K$. Since this average decrease of C_t/K will take place for all solids regardless of any phase relationships, it may be appropriate to exclude this effect from consideration. This may be done by making α pressure dependent by the equation

$$\alpha(P) = \alpha_o\ (1-2P/K) \tag{8}$$

If we assume $K' = dK/dP$ is constant, then

$$\alpha(P) = \alpha_o \frac{1+(K'-2)P/K_o}{1+K'\ P/K_o} \tag{9}$$

This change will in most cases be small compared to the uncertainty in α and in the extrapolated elastic constants, and, since we cannot show by experimental data whether it should be included, we shall neglect this effect.

E. Thermal Effects at a Transition

We have so far avoided an explicit treatment of thermal effects. The usual thermal contribution to the free energy and the elastic constants will not alter the qualitative feature of Figure 2. But if α is unusually small there may be an unusual thermal contribution to the free energy and its derivatives as a precursor to the phase transition.

In the quasiharmonic approximation, the thermal contribution to the Helmholtz free energy is given by

$$F_{th} = 1/2\ \sum_i h\nu_i + kT \sum_i \ln\ (1-e^{-h\nu_i/kT}) \tag{10}$$

where ν_i are the phonon (normal mode) frequencies, and

$$P_{th} = 1/V\ \sum \gamma_i h\nu_i\ [1/2 + 1/(e^{h\nu_i/kT} - 1)] \tag{11}$$

where $\gamma_i = \dfrac{-\partial \ln \nu_i}{\partial \ln V}$. When a lattice instability is approached, one of the ν_i's approaches zero, and F_{th} and its derivatives

(P_{th}, K_{th}, etc.) approach negative infinity. While a negative value of the bulk modulus is forbidden by the Born stability criterion, it is possible that there will be a sudden decrease in the bulk modulus and some of the other elastic constants as an immediate precursor to a phase transition. This sort of precursor has been observed in calcite [*Matsushima et al.*, 1975], but in most cases we expect that the phase transition will take place before this sort of thermal precursor can occur.

III. EXPERIMENTAL ELASTIC CONSTANTS OF KCl AT THE PHASE TRANSITION

We wanted to test our theories about the behavior of the elastic constants in the vicinity of a phase transition. KCl is an ideal candidate for this type of test since it has the well-studied NaCl structure and undergoes a phase transition to the CsCl structure at 19.3 kbar [*Vaidya* and *Kennedy*, 1971]. This is within the range of our experiment for the measurement of sound velocities under high hydrostatic pressures.

A. Experimental Technique

A technique similar to that described by *Jayaraman* and *Maines* [1971] was used to generate a hydrostatic environment. This involved a conventional piston cylinder device with a Teflon cell, which contains the sample and the pressure transmitting fluid (a 50/50 mixture of isoamylalcohol and isopentane). The sample was held in a special holder, and the wires from the transducer went out through alumina thermocouple tubing. Further details of the experimental method will be given in a forthcoming paper by *Ota* and *Anderson*.

Since we were unable to fit a manganin coil and our sample holder into the teflon cell at the same time and still allow enough room for the compression of the pressure fluid, we determined the pressure in the cell from the oil pressure driving the piston. The relationship between oil pressure and cell pressure was calibrated by placing a manganin coil and a Bismuth wire in the Teflon cell and by assuming that the resistance change in the manganin coil was linear up to the Bi I→II transition at 25.4 kbar. The effect of friction was about 5% of the total pressure.

A large KCl single crystal was purchased from the Harshaw Chemical Company, and from it were obtained two cubic samples, about 4 mm on a side, with faces oriented to within 0.6 degrees of the (100) and (110) directions. 20 Mhz *P*- or *S*- transducers were mounted to the samples with silver paint, and the travel times for *P* and *S* waves were measured by the method of pulse superposition, with a single transducer being used both to transmit and to receive the reflected pulse.

B. Results

The pulse repetition frequency was measured at zero pressure and as a function of pressure from 0 to 20 kbar for the two longitudinal and two shear waves described in Table 1. This provided enough information for the determination of the three independent elastic constants, with one redundant measurement for a check on our accuracy. Over the entire range, the consistency of our data was excellent.

TABLE 1. Velocities Used for Elastic Constant Determination

Mode	P	S	P	S
Propagation direction	(100)	(100)	(110)	(110)
Polarization direction	(100)	(010)	(110)	$(1\bar{1}0)$
ρv^2	C_{11}	C_{44}	$\frac{1}{2}(C_{11}+C_{12}+2C_{44})$	$\frac{1}{2}(C_{11}-C_{12})$

In Table 2, we compare our zero pressure elastic constants with those of other experimenters. The agreement with *Dobretsov* and *Peresada* [1969] is extremely good, while the agreement with the less recent work [*Lazarus,* 1949; *Haussuhl,* 1960; *Bartels* and *Schuele,* 1965; *Drabble* and *Strathen,* 1967] is only modestly good, with a two percent disagreement in the bulk modulus.

Our data for pulse repetition frequency as a function of pressure were converted to elastic constants as a function of pressure by *Cook's* method [1957]. In this analysis, we assumed $\beta = 110.4 \cdot 10^{-6}$ deg^{-1}, and $C_P = 12.23$ cal mol^{-1} deg^{-1}. The normalized elastic constants $C(P)/C(0)$ are plotted in Figures 3 and 4 from zero to 20 kbar, and are tabulated in Table 3. Our results are remarkably linear. There is no indication of any strange behavior for either C_{44} or K in the neighborhood of the transition. The value of C_{44}/K at the transition is 0.205, a value in good agreement with the modified Born criterion.

There is no evidence of a slight decrease of a K at the transition, which might be expected if one of the normal mode frequencies were approaching zero at or very near to the transition pressure. We can calculate the mode Grüneisen parameter, γ_i, for the mode associated with any elastic modulus C_i by the formula $\gamma_i = -\frac{1}{6} + \frac{1}{2}\frac{K_T}{C_i}\frac{dC_i}{dP}$ [*Schuele* and *Smith,* 1964].

According to our measurements γ_{44} decreases from -0.58 at zero pressure to -1.21 at 20 kbar. Apparently this decrease is not enough to cause a measurable effect on the thermal contribution to P or K.

TABLE 2. Comparison of Elastic Constants of KCl at Zero Pressure, kbar

	C_{11}	C_{12}	C_{44}	C_S	K_S
Present study	405.6	65.5	63.3	170.1	178.9
Dobretsov and *Peresada* [1969]	406.2	65.4	63.9	170.4	179.0
Drabble and *Strathen* [1967]	409.0	70.4	62.7	169.3	183.3
Bartels and *Schuele* [1965]	405.0	69.8	63.0	167.6	181.5
Haussuhl [1960]	407.8	69.0	63.3	169.4	181.9
Lazarus [1949]	409.5	70.6	63.0	169.4	183.6

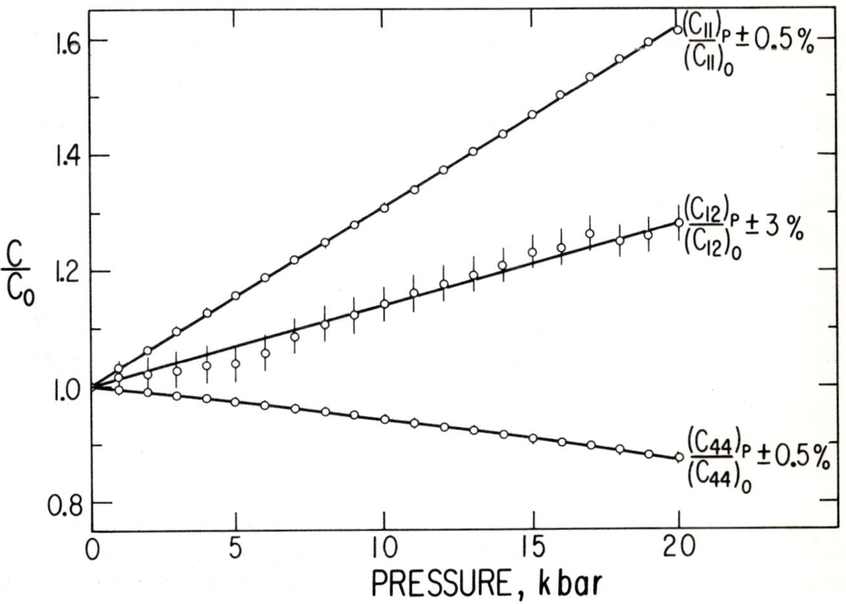

Fig. 3. Variation of elastic constants C/C_o with pressure for KCl single crystal.

TABLE 3. Elastic Constants and Adiabatic Bulk Modulus of KCl Single-Crystal as a Function of Pressure

	Elastic Constant				Bulk Modulus
P, kbar	$C_{11}/C_{11}(0)$	$C_{12}/C_{12}(0)$	$C_{44}/C_{44}(0)$	$C_s/C_s(0)$[a]	$K_S/K_S(0)$
0	1.000	1.000	1.0000	1.000	1.000
1	1.031	1.014	0.9944	1.034	1.027
2	1.061	1.028	0.9889	1.067	1.053
3	1.094	1.035	0.9829	1.105	1.081
4	1.123	1.047	0.9771	1.138	1.106
5	1.151	1.049	0.9714	1.171	1.131
6	1.183	1.068	0.9651	1.205	1.158
7	1.216	1.094	0.9593	1.239	1.187
8	1.247	1.116	0.9530	1.272	1.214
9	1.276	1.129	0.9466	1.304	1.239
10	1.309	1.155	0.9398	1.338	1.267
11	1.337	1.169	0.9328	1.369	1.292
12	1.368	1.185	0.9259	1.403	1.319
13	1.400	1.207	0.9196	1.437	1.346
14	1.433	1.216	0.9127	1.475	1.375
15	1.465	1.234	0.9060	1.509	1.403
16	1.497	1.242	0.8991	1.546	1.430
17	1.527	1.263	0.8922	1.578	1.455
18	1.559	1.237	0.8853	1.621	1.483
19	1.587	1.255	0.8782	1.651	1.508
20	1.614	1.273	0.8712	1.680	1.532

a. $C_s = (C_{11} - C_{12})/2$

In Table 4 we compare the pressure derivatives of the elastic constants which we derive with those of other experimenters. *Dobretsov* and *Peresada* [1969] also measured the elastic constants to 20 kbar. *Wang* [1973] measured only C_{44} to 20 kbar. The other measurements were limited to lower pressures. Our pressure derivatives measure the average change from 0 to 20 kbar, while the others measure the average change at a much lower pressure. We list the data of *Dobretsov* and *Peresada* in two ways: the average change over the whole pressure range, and the initial slope of their quadratic fit. Some of the discrepancies may arise because of the different range of pressures involved. While none of the pressure derivatives that we measure directly differ from the results of *Bartels* and *Schuele* by more than a few percent, these discrepancies combine to yield a 10% disagreement in dK_S/dP; however, our dK_S/dP is in excellent agreement with that of *Dobretsov* and *Peresada*.

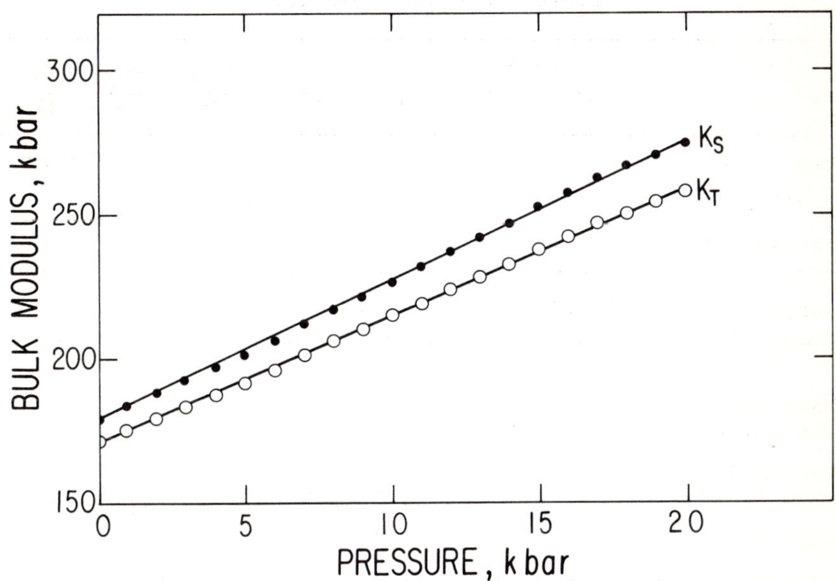

Fig. 4. Variation of adiabatic and isothermal bulk moduli, K_S K_T*, with pressure for KCl single crystal.*

There is only a small amount of data from other compounds relating to the problem of the behavior of shear constants at a phase transition. *Reddy* and *Ruoff* [1965] showed that the elastic constants of RbBr are linear right up to the transition pressure at 4.5 kbar. Recent advances in the measurements of velocities in polycrystalline aggregates to pressure on the order of 100 kbar or more [*Voronov* and *Grigor'ev*, 1969; *Morris, et al.*, 1976; *Frankel, Rich*, and *Homan*, unpublished] are able to provide useful information on the average sound velocities at high pressure. *Voronov* and *Grigor'ev* showed that the average shear velocity in AgCl decreased with pressure prior to the transition, which implies that at least one of the single-crystal shear moduli (C_{44}) decreases toward the transition, but does not provide the much desired measurement of α. *Frankel, et al.*, report sound velocities in polycrystalline NaCl to 27 GPa, just short of the transition, and they show a slight decrease of v_S leading up to the transition. Their data appear to be consistent with a model calculation [*Demarest*, 1972*a*] that predicted C_{44}/K = 0.14 at the transition.

TABLE 4. Comparison of Pressure Derivatives of Elastic Constants of KCl

	Pressure Range, kbar	$\frac{dC_{11}}{dP}$	$\frac{dC_{12}}{dP}$	$\frac{dC_{44}}{dP}$	$\frac{dC_S}{dP}$	$\frac{dK_S}{dP}$
Present study	20	12.51	0.91	-0.41	5.80	4.78
Wang [1973]	20	-	-	-0.40	-	-
Dobretsov and *Peresada* [1969]	20 (average)	12.56	1.14	-0.43	5.14	4.81
Dobretsov and *Peresada* [1969]	linear term of quadratic fit	13.23	0.82	-0.34	5.80	5.12
Drabble and *Strathen* [1967]	0.004	13.00	1.56	-0.56	5.72	5.37
Bartels and *Schuele* [1965]	03.3	12.82	1.60	-0.39	5.61	5.34
Lazarus [1949]	10	12.21	1.09	-0.47	5.56	4.80

IV. EVALUATION, PREDICTIONS, AND CONCLUSIONS

The Born criterion for a phase transition has often been used to predict a transition at the pressure at which a shear constant vanishes. This is only the upper limit for a transition pressure, however, and any reasonably intelligent method of selecting a pressure somewhere between zero and the Born transition pressure would give more accurate predictions. On the basis of our earlier discussion, we believe that the best prediction of a phase transition which we can base on elastic constant data arises from the assumption that a transition will occur when C_t/K reaches a critical value α.

Table 5 displays some selected data relating to the value of α at a transition. We have grouped compounds with the same crystal structure together, and for each compound we either compute the "measured" value of α (usually by extrapolating the elastic constants from low pressure), or we estimate a range of possible values for α and compute the transition pressure, again using extrapolations of the elastic constants. In all cases, the extrapolations were based on the measured pressure derivatives of the elastic constants. For the NaCl-type alkali-halides, we used a NNN lattice model extrapolation formula [*Demarest*, 1972*a*], while in all other cases we assumed that the elastic constants are linear with pressure. We conclude from earlier calculations [*Demarest*, 1972*b*] that this type of extrapolation is adequate for the present purposes up to reduced pressures of $P/K_O \approx 0.5$.

TABLE 5. Evaluations of α and predictions of Phase Transition

Compound	Structure	Soft Shear Mode	Transition Pressure, kbar		α	
			Predicted	Experiment	Assumed	Measured
NaF	NaCl→CsCl	c_{44}		~300		0.14
NaCl				290		0.14
NaBr			>120		<0.20	
NaI			-		<0.20	
KF				17.3		0.26
KCl				19.3		0.20
KBr				17.4		0.19
KI				17.8		0.16
RbF				9.4		0.13
RbCl				5.2		0.22
RbBr				4.5		0.21
RbI				3.4		0.19
MgO	NaCl→CsCl	c_{44}	-		<0.26	
CaO			275-400		0.15-0.20	
SrO			1000-2200		0.15-0.20	
AgCl	NaCl→HgS	c_{44}		76		0.06
AgBr	NaCl→HgS			82		0.05
AgI	NaCl→tetragonal			97		-
Si	Diamond→white Sn	c_S*		110		0.36
Ge				90		0.37
Diamond			-	-		-
CaF_2	Fluorite→$\alpha PbCl_2$	c_S		80		0.54
BaF_2				20		0.36
CdS	Wurtzite→NaCl	c_{44}		25		0.18
ZnS	Wurtzite→Sphalerite	c_{66}		45		(0.29)
	→NaCl	c_{44}		117		0.18
TiO_2	Rutile→Dist. Fluorite	c_S		50		0.16
SnO_2			30-90		0.14-0.18	
GeO_2			150-250		0.14-0.18	
SiO_2	αQuartz→coesite	$\frac{1}{2}\lambda_2$*		20		0.59
	αQuartz→stishovite	$\frac{1}{2}\lambda_2$		85		0.25
$MgAl_2O_4$	Spinel→	c_S	300-700		0.15-0.20	
$MgO \cdot 2.6Al_2O_3$	Spinel→		1000-2000		0.15-0.20	

*$c_S = \frac{1}{2}(c_{11}-c_{12})$ $\qquad \lambda_2 = c_S + c_{44} - \sqrt{(c_S-c_{44})^2 + 4\,c_{14}^2}$

Estimates of α for the NaCl→CsCl transition in the alkali-halides range from 0.13 (RbF) to 0.26 (KF). Most of the data seems to be centered in the range 0.15-0.20. The seven alkali halides for which no transition has been found will probably not transform to the CsCl structure. NaBr may transform above 120 kbar, but this is not clear since C_{44}/K never dips much below 0.20. MgO will probably not have a transformation of this type. SrO may transform to the CsCl structure at extreme pressures, and we think that CaO will probably transform between 270 and 400 kbar.

The silver halides transform to the cinnabar (HgS) structure with $\alpha \cong 0.05$. This transformation is not a pure shear, but it does involve a smaller shift of atom positions than does the NaCl-CsCl transition.

The metallic transitions in Si and Ge are among those which we do not expect to conform well with our theory, because of the change in electronic structure. Here $\alpha = 0.36$. There is no prediction of diamond going through this transition at high pressure. Our theory also fails to predict the high pressure transition in the fluorite structure.

The wurtzite → NaCl transition in CdS and ZnS is in perfect agreement with this theory ($\alpha = 0.18$). We neglect the complication of the sphalerite structure, which for ZnS is stable in an intermediate pressure range, depending upon the temperature and stoichiometry.

The transformation of rutile to distorted fluorite is in excellent agreement ($\alpha = 0.16$). We predict transitions in the range 30-90 kbar for SnO_2 and 150-250 kbar for GeO_2.

Our method fails to predict the α-quartz → Coesite transition. However even in this case, our method shows its usefulness. The Born criterion predicts a lattice instability at 180 kbar, while the modified Born criterion predicts the transition at 95-115 kbar. This is not too far from the transition to Stishovite at 85 kbar.

We have deliberately omitted from Table 5 elements or compounds for which all the shear moduli increase rapidly with pressure. We have selected an assortment of crystal structures for which there was sufficient readily available data to make a test of the modified Born criterion.

Table 5 confirms our theories concerning the interrelationship of pressure-induced phase changes and weak shear constants. Values of α tend to be similar for a particular transformation. In no case is α very small, and the smallest value here is 0.05. The modified Born criterion does not work when the transformation involves a change in the electronic structure of the compound, or when the transformation is in no way related to a shear (these cases were not included in the table). However, in many cases this method will permit good estimates of phase transition pressures with very little effort.

Acknowledgments. We thank G. Kennedy, I. Getting, and R. Chui for their kind cooperation and suggestions on our high-pressure system and pressure calibration. Research performed at the University of Chicago was supported by the Earth Sciences Section, National Science Foundation, NSF Grant EAR 73-00636 A02, and has benefited from the support of the Materials Research Laboratory by the National Science Foundation. Work performed at the University of California was supported by NSF Grant DES 75-04869.

REFERENCES

Anderson, O. L., Elastic constants of the central force model for three cubic structures: pressure derivatives and equation of state, *J. Geophys. Res., 75,* 2719-2740, 1970.

Anderson, O. L., The behavior of shear velocities in solids at extremely high pressures, in *Proc. 4th Internat. Conf. on High Pressure,* edited by Jiro Osugi, pp. 398-403, Physico-Chemical Society of Japan, Kyoto, 1975.

Anderson, O. L., and H. H. Demarest, Jr., Elastic constants of the central force model for cubic structures: polycrystalline aggregates and instabilities, *J. Geophys. Res., 76,* 1349-1369, 1971.

Anderson, O. L., and R. C. Liebermann, Equations for the elastic constants and their pressure derivatives for three cubic lattices and some geophysical applications, *Phys. Earth Planet. Interiors, 3,* 61-85, 1970.

Anderson, P. W., and E. I. Blount, Symmetry considerations on Martensitic transformations: "ferroelectric" metals? *Phys. Rev. Lett., 14,* 217-219, 1965.

Bartels, R. A., and D. E. Schuele, Pressure derivatives of the elastic constants of NaCl and KCl at 295°K and 195°K, *J. Phys. Chem. Solids, 26,* 537-549, 1965.

Boccara, N., Second order phase transitions characterized by a deformation of the unit cell, *Ann. Phys., 47,* 40-64, 1968.

Born, M., On the stability of crystal lattices, l., *Proc. Cambridge Phil. Soc., 36,* 100-165, 1940.

Born, M., and K. Huang, *Dynamical Theory of Crystal Lattices,* Oxford University Press, London, 1954.

Chang, E., and G. R. Barsch, Pressure dependence of single crystal elastic constants and anharmonic properties of wurtzite, *J. Phys. Chem. Solids, 34,* 1543-1563, 1973.

Cohen, A. J., and R. G. Gordon, Theory of the lattice energy, equilibrium structure, elastic constants, and pressure-induced phase transitions in alkali-halide crystals, *Phys. Rev., B 12,* 3228-3241, 1975.

Cook, R. K., Jr., Variation of elastic constants and static strains with hydrostatic pressure: a method of calculation from ultrasonic measurements, *J. Acoust. Soc. Am., 29,* 445-449, 1957.

Demarest, H. H., Jr., Extrapolation of elastic properties to high pressure in the alkali halides, *J. Geophys. Res.*, *77*, 848-856, 1972*a*.

Demarest, H. H., Jr., Lattice model calculation of elastic and thermodynamic properties at high pressure and temperature, *Phys. Earth Planet. Interiors*, *6*, 146-153, 1972*b*.

Demarest, H. H., Jr., Lattice model calculation of Hugoniot curves and Grüneisen parameter at high pressure for the alkali halides, *J. Phys. Chem. Solids*, *35*, 1393-1404, 1974.

Devonshire, A. F., Theory of barium titanate, *Phil. Mag.*, *40*, 1040-1063, 1949.

Dobretsov, A. I., and G. I. Peresada, Dependence of the elastic constants of KCl on pressure, *Soviet Physics--Solid State*, *11*, 1401-1402, 1969.

Drabble, J. R., and R. E. B. Strathen, The third-order elastic constants of potassium chloride, sodium chloride, and lithium fluoride, *Proc. Phys. Soc.*, *92*, 1090-1095, 1967.

Fritz, I. J., and P. S. Peercy, Phenomenological theory of the high pressure phase transition in paratellurite (TeO_2), *Solid State Commun.*, *16*, 1197-1200, 1975.

Haussuhl, S., Thermo-elastische Konstanten der Alkalihalogenide vom NaCl-typ, *Z. Phys.*, *159*, 223-229, 1960.

Hyde, B. G., and M. O'Keeffe, On mechanisms of the B1↔B2 structural transformation, in *Phase Transitions 1973*, edited by Heinz K. Henisch, Pergamon Press, New York, 1973.

Jayaraman, A., and R. G. Maines, Hydrostatic pressures of 50 kbar in a piston-cylinder device: measurement of pressure and characterization of the pressure medium, in *Accurate Characterization of the High Pressure Environment*, edited by E. C. Lloyd, pp. 5-9, Nat. Bur. Std. Spec. Publ. 326, 1971.

Matsushima, S., K. Suito, and S. Kondo, Variation of elastic properties accompanying the calcite 1-2 transition at high pressure and temperature, in *Proc. 4th Internat. Conf. on High Pressure*, edited by Jiro Osugi, pp. 383-388, Physico-Chemical Society of Japan, Kyoto, 1975.

Merrill, L., A soft mode mechanism for the displacive calcite-$CaCO_3$ (II) phase transformation at 1.5 GPa, in *Proc. 4th Internat. Conf. on High Pressure*, edited by Jiro Osugi, pp. 389-392, Physico-Chemical Society of Japan, Kyoto, 1975.

Misra, R. M., On the stability of crystal lattices, 2, *Proc. Cambridge Phil. Soc.*, *36*, 173-182, 1940.

Morris, C. E., J. C. Jamieson, and F. L. Yarger, Ultrasonic measurements at elevated pressures (9 GPa) to determine Poisson's ratio and other elastic moduli of solids, *J. Appl. Phys.*, *47*, 3979-3986, 1976.

Musgrave, M. J. P., *Crystal Acoustics*. Holden-Day, San Francisco, 1970.

Peercy, P. S., and I. J. Fritz, Pressure-induced phase transition in paratellurite (TeO_2), *Phys. Rev. Lett.*, *32*, 466-469, 1974.

Reddy, P. J., and A. L. Ruoff, Pressure derivatives of the elastic constants in some alkali halides, in *Physics of Solids at High Pressures*, edited by C. T. Tomizuka and R. M. Emrick, pp. 510-523, Academic Press, New York, 1965.

Schuele, D. E., and C. S. Smith, Low temperature thermal expansion of RbI, *J. Phys. Chem. Solids, 25,* 801-814, 1964.

Thomsen, L., On the fourth-order anharmonic equation of state of solids, *J. Phys. Chem. Solids, 31,* 2003-2016, 1970.

Thomsen, L., Elastic shear moduli and crystal stability at high P and T, *J. Geophys. Res., 76,* 1342-1348, 1971.

Thomsen, L., The fourth-order anharmonic theory: elasticity and stability, *J. Phys. Chem. Solids, 33,* 363-378, 1972.

Vaidya, S. N., and G. C. Kennedy, Compressibility of 27 halides to 45 kbar, *J. Phys. Chem. Solids, 32,* 951-964, 1971.

Voronov, F. F., and S. B. Grigor'ev, Effect of pressures up to 100 kbar on velocity of sound in silver chloride, *Soviet Physics-Doklady, 13,* 899-901, 1969.

Wang, C. Y., Shear elastic constant C_{44} and phase transformation of KCl, *Nature, 243,* 284-285, 1973.

CHEMICAL COMPOSITION OF THE OUTER CORE

K. Ito
Department of Earth Sciences
Kobe University
Kobe, Japan

Abstract

A hypothesis that the outer cores of terrestrial planets consist of the eutectic melt between Fe and oxides is proposed to explain the following features:

(1) The earth has a liquid outer core and a solid inner core. The density of the outer core corresponds to that of iron with about 20% mixed light elements.

(2) The mantle/core mass ratio of the terrestrial planets increases in the following order: Mercury, Earth, Venus, and Mars.

The presently favored hypothesis that the outer core consists of iron alloyed with light elements such as silicon and sulphur does not satisfactorily explain the second feature. The phase transition theory proposed by Ramsey fits the second feature fairly well but is difficult to explain the large density jump at the mantle-core boundary. The Fe_2O theory proposed by Bullen which also assumes that the mantle-core boundary is a pressure phenomenon avoids the objections to the phase-transition theory, but the stability of Fe_2O is only hypothetical. The present theory is an alternative to Bullen's theory, Fe_2O in the latter being replaced by the eutectic melt between Fe and oxides in the former.

It is well known that insulators and semiconductors tend to transform to metallic phases at high pressures. Therefore, it is reasonable to assume that repulsion forces between Fe and oxides decrease at high pressures. In other words, a Fe-oxide system which is a completely immiscible system at ordinary pressures may become a eutectic system at high pressures reached in the Earth's core. It may further become a completely miscible solid solution system at even higher pressures where the oxide is metallic.

Assume for example, a planet consisting of a core of pure iron and a mantle of pure oxide denoted by MO. The depth and the pressure at the boundary are D_B and P_B, respectively. If the temperature within the planet is anywhere lower than the eutectic temperature of the Fe-MO system, the planet consists of the mantle and a small iron core. Mars is such a case.

If the temperature exceeds the eutectic temperature at D_E and P_E within the primary mantle ($D_E < D_B$, $P_E < P_B$), the mantle between D_E and D_B reacts with a part of the core to form the eutectic melt in the Fe-MO system at the pressure P_E. The mass of the outer core is the mass between D_E and D_B (M_{EB}) divided by the eutectic composition (MO/MO+Fe). For a small planet the mass between D_E and D_B is small, thus the size of the liquid core is small (Venus is such a case). For a relatively larger planet, the size of the outer core is greater (the Earth is such a case).

In a model compatible with observational data and the idea that the overall chemistry of the terrestrial planets (except Mercury) is the same, the primary Fe/Fe+MO ratio is 0.25 when the MO/Fe+MO of the eutectic melt is 0.20. If the former is 0.22, the latter needs to be 0.30. The temperature-depth curve for Venus is estimated to be nearly the same as that for the Earth.

Because the eutectic composition in the Fe-Mo system is expected to shift towards the MO side as pressure increases, the composition of the outer core tends to be more MO-rich, or less dense (at a constant pressure) at greater depths. This gravitational instability may be the cause of the convection in the outer core which is needed to generate the Earth's magnetic field.

SESSION III

EQUATIONS OF STATE AND SHOCK WAVE EXPERIMENTS

EQUATION OF STATE OF MANTLE MINERALS DETERMINED THROUGH HIGH-PRESSURE X-RAY STUDY

Y. SATO
Meteorological College
Asahi-cho, Kashiwa, Chiba, 277, Japan

Abstract

Pressure-volume relationship of the rutile structure (TiO_2, SiO_2, MnO_2) and the spinel structure (γ-Fe_2SiO_4, γ-Co_2SiO_4, γ-Ni_2SiO_4) were determined through X-ray diffraction measurements under hydrostatic compression up to about 120 kbar. The isothermal bulk modulus K_T of stishovite was determined to be 2.81 Mbar for the assumed pressure derivative $K_T' = 5.0$ and 3.01 Mbar for $K_T' = 0$. The values K_T of spinels were determined to be 1.97, 2.06 and 2.23 Mbar for γ-Fe_2SiO_4, γ-Co_2SiO_4 and γ-Ni_2SiO_4, respectively, for an assumed value 4.0 of K_T'.

I. INTRODUCTION

Much effort has been extended toward determination of the equation of state of silicate minerals with a high-pressure form of the crystal structure, such as the rutile or the spinel structure.

From ultrasonic measurements, *Mizutani et al.* [1972] first determined the value of the adiabatic bulk modulus K_S of synthetic polycrystalline stishovite to be 3.46 ± 0.24 Mbar. Static compression done by *Bassett and Barnett* [1970] with the use of a tetrahedral anvil press showed that the isothermal bulk modulus K_T is 3.0 ± 0.3 Mbar for the Birch-Murnaghan equation with assumed value 4.0 of K_T'. *Liu et al.* [1974*a*], in their recent study using a diamond-anvil high-pressure cell, reported that the bulk modulus K_T was calculated as 3.44 ± 0.27 Mbar, assuming K_T' to be in the range 2-7.

There are several reports of static compression experiments on spinels (γ-Fe_2SiO_4, *Mao et al.* [1969]; γ-Co_2SiO_4, *Liu et al.* [1974*b*]; γ-Ni_2SiO_4, *Mao et al.* [1970]) and ultrasonic measurements (γ-Fe_2SiO_4, *Mizutani et al.* [1970] and *Liebermann* [1975]; γ-Ni_2SiO_4, *Liebermann* [1975]).

Recent high-pressure X-ray diffraction studies suggest that the static compression curves obtained with the aid of solid-pressure media should be reexamined under hydrostatic pressures.

Sato et al. [1975] showed that pressure inhomogeneity was generated in a mixture of two materials with different elastic properties, and this results in a marked hysteresis phenomenon in the pressure-volume relationships of various materials. *Piermarini et al.* [1973] revealed properties of a solid- and a liquid-pressure medium in the diamond-anvil pressure cell by line-broadening and line-shift measurements of the sharp R_1 ruby fluorescence line.

In ultrasonic measurements on a hot-pressed polycrystalline sample, ambiguities are always caused by porosity or anisotropic properties of the sample.

In this study, the equations of state of the rutile and the spinel structures were reexamined using X-ray diffraction and a liquid-pressure medium.

II. EXPERIMENTAL METHODS

For the high-pressure X-ray diffraction study, a cubic-anvil high-pressure apparatus was used in conjunction with a high-power, molybdenum, X-ray source operated at 55 kV and 160 mA, and a scintillation counter measuring system.

The sample powder and a pressure standard were mixed and put into a pressure capsule made of polyethylene tubing with 1.5-mm diameter and 3.0-mm length. After putting a sufficient amount of the liquid-pressure medium (methanol:ethanol = 4:1 by volume), the pressure cell was sealed with epoxy resin and inserted into the center hole of the outer cubic pressure medium made of amorphous boron solidified with polyester resin. A cubic pressure medium with 6-mm edge length was used in the experiment with cemented tungsten carbide anvils of $4 \times 4\ (\mathrm{mm})^2$ cube face. A pressure medium of 5-mm edge length was used with $3 \times 3\ (\mathrm{mm})^2$ anvils.

Measurements of the diffracted X-rays were made by a step-scanning method, with each step width 0.05°. The 2Θ-angle of a diffraction line was determined within the accuracy of ±0.02% in a least-squares curve fitting of the measured intensities to a Gaussian profile. In order to reduce the error caused by a possible shift of the center of diffraction, each diffraction line was measured on both sides of the direct beam.

In runs on stishovite and spinels, a NaF internal pressure standard was used in order to measure as many diffraction lines as possible from the sample. The pressure-volume relationship of NaF shown in Figure 1 was determined in this study under hydrostatic compression on the basis of Decker's pressure scale for NaCl [1971]. This agrees well with the result by *Olinger and Jamieson* [1970] or *Spieglan and Jamieson* [1974] if the Decker's pressure scale [1971] is adopted for NaCl.

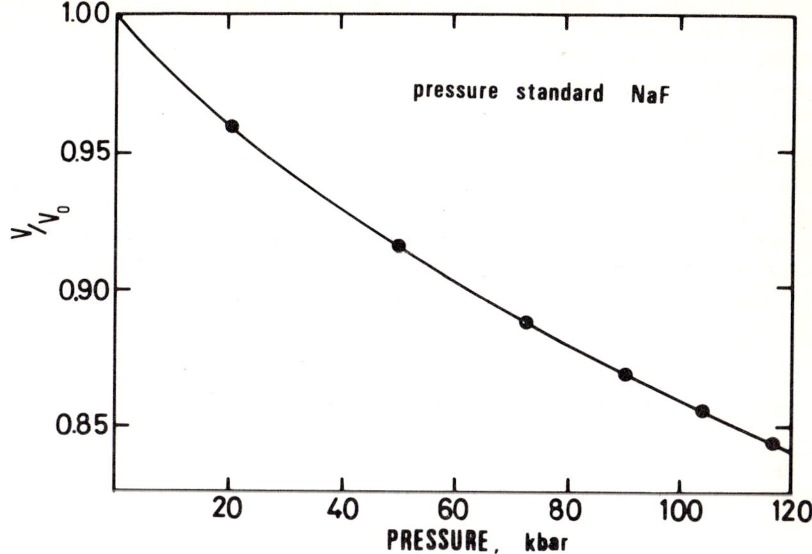

Fig. 1. Pressure-volume relationship of NaF used as an internal pressure standard. The solid line shows equation (1).

For the calculation of the pressure, the second-order polynomial equation

$$P = 429\left(1 - \frac{V}{V_o}\right)\left[1 + 4.65\left(1 - \frac{V}{V_o}\right)\right] \quad (P \text{ in kbar}) \qquad (1)$$

was used. The solid line in Figure 1 shows equation (1).

III. RESULTS AND DISCUSSIONS

A. TiO_2

Changes in cell parameters and cell volumes were measured for TiO_2 using Decker's pressure scale of NaCl [1971]. The experimental results are shown in Table 1 and Figure 2. Five diffraction lines--(110), (101), (111), (211), and (301)-- were measured and used in the calculation of cell parameters of TiO_2, and two diffraction lines--(200) and (220)-- were measured for the NaCl pressure standard. A least-squares calculation using the second-order Birch-Murnaghan equation

$$P = \frac{3}{2} K_T \left[\left(\frac{V_o}{V} \right)^{7/3} - \left(\frac{V_o}{V} \right)^{5/3} \right] \cdot \left\{ 1 - \frac{3}{4} (4 - K_T') \left[\left(\frac{V_o}{V} \right)^{2/3} - 1 \right] \right\} \qquad (2)$$

gives the isothermal bulk modulus K_T = 1.88 ± 0.05 Mbar and its pressure derivative K_T' = 10.6 ± 2.2. We get K_T = 1.97 Mbar if we assume K_T' = 6.84, which was obtained by *Manghnani* [1969] from the ultrasonic measurements on single-crystal TiO_2. There is, however, a slight discrepancy between the present static compression data and the ultrasonic K_T = 2.136 Mbar reported by *Manghnani* [1969].

TABLE 1. Experimental Results for TiO_2

Run	TiO_2			NaCl	
	V/V_o	a/a_o [a]	c/c_o [a]	$-\Delta a/a_o$	P, kbar [b]
TI 16	0.9969	0.9986	0.9997	0.0081	6.1
	0.9944	0.9970	1.0003	0.0142	11.3
	0.9925	0.9963	0.9999	0.0201	16.7
	0.9887	0.9954	0.9979	0.0258	22.5
	0.9858	0.9949	0.9960	0.0320	29.3
	0.9828	0.9937	0.9953	0.0379	36.4
	0.9788	0.9920	0.9946	0.0445	45.0
	0.9755	0.9903	0.9946	0.0507	54.0
	0.9727	0.9888	0.9948	0.0550	60.6
	0.9702	0.9881	0.9937	0.0584	66.2
	0.9682	0.9871	0.9937	0.0620	72.4
TI 19	0.9935	0.9966	1.0001	0.0136	10.8
	0.9916	0.9959	0.9997	0.0190	15.7
	0.9895	0.9953	0.9989	0.0236	20.2
	0.9882	0.9952	0.9977	0.0286	25.5
	0.9849	0.9940	0.9969	0.0332	30.7
	0.9825	0.9938	0.9948	0.0368	35.0
	0.9797	0.9922	0.9952	0.0420	41.7
	0.9772	0.9916	0.9937	0.0482	50.3
	0.9721	0.9889	0.9940	0.0545	59.8

a. Cell parameters are accurate to ±0.05% and ±0.1% for *a* and *c*, respectively.
b. Pressure values are accurate to ±1.5%.

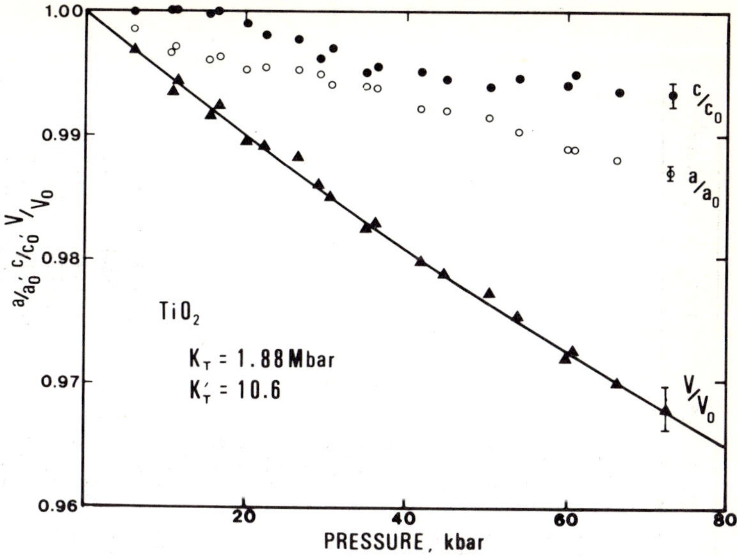

Fig. 2. Changes in cell parameters and volume of TiO_2. Solid line is Birch-Murnaghan curve with K_T = 1.88 Mbar and K_T' = 10.6.

For an assumed value $K_T' = 4.0$, we get $K_T = 2.03$ Mbar. This agrees well with the value $K_S = 2.03$ Mbar with $K_S' = 4.9$ given by *McQueen et al.* [1967] in the shock compression experiment.

B. Stishovite

Five diffraction lines--(110), (111), (210), (211), and (220)-- were measured for stishovite. Cell parameters were obtained within the accuracy of ±0.05% for the *a*-axis and ±0.1% for the *c*-axis. The volumes are accurate to ±0.2%. Only the (220) line of NaF was measured to use as standard. The (200) line of NaF was not used for the calculation of the cell parameter because the (200) line was overlaid by the (101) line of stishovite.

For five runs--ST 14, 15, 28, 35, and 40-- experimental results of cell parameter changes (a/a_o, c/c_o) and volume change (V/V_o) are shown in Table 2, with the volume change of NaF and pressure obtained using equation (1). Figure 3 shows the pressure-cell parameter and the pressure-volume relationships for all five runs.

TABLE 2. Experimental Results for Stishovite

Run	SiO_2			NaF		MgO	
	V/V_o	a/a_o	c/c_o	V/V_o	P, kbar[a]	V/V_o	P, kbar[b]
ST 14	0.9971	0.9989	0.9994	0.9829	7.9		
	0.9937	0.9971	0.9996	0.9629	18.7		
	0.9908	0.9961	0.9986	0.9483	27.5		
	0.9874	0.9951	0.9971	0.9332	37.6		
	0.9847	0.9938	0.9970	0.9201	47.0		
	0.9822	0.9929	0.9962	0.9113	53.8		
	0.9795	0.9922	0.9949	0.9010	62.1		
	0.9778	0.9913	0.9949	0.8942	67.7		
ST 15	0.9981	0.9991	0.9999	0.9838	7.5		
	0.9955	0.9978	1.0000	0.9701	14.6		
	0.9923	0.9964	0.9995	0.9531	24.5		
	0.9876	0.9950	0.9975	0.9331	37.6		
	0.9839	0.9933	0.9971	0.9171	49.3		
	0.9744	0.9899	0.9944	0.8825	78.0		
	0.9718	0.9886	0.9943	0.8729	86.8		
	0.9690	0.9875	0.9937	0.8653	94.0		
	0.9760	0.9902	0.9954	0.8873	73.7		
	0.9850	0.9933	0.9983	0.9205	46.7		
ST 28	0.9976	0.9991	0.9994	0.9845	7.1	0.9963	6.0
	0.9949	0.9979	0.9990	0.9665	16.6	0.9894	17.5
	0.9908	0.9959	0.9990	0.9498	26.6	0.9846	25.7
	0.9885	0.9954	0.9976	0.9350	36.3	0.9777	38.0
	0.9852	0.9943	0.9965	0.9204	46.8	0.9713	49.7
	0.9786	0.9918	0.9949	0.8946	67.4	0.9608	70.0
	0.9759	0.9908	0.9942	0.8850	75.8	0.9563	79.1
	0.9734	0.9894	0.9944	0.8755	84.4	0.9526	86.7

TABLE 2 (continued)

ST 35	0.9890	0.9954	0.9982	0.9420	31.6		
	0.9833	0.9932	0.9968	0.9716	48.9		
	0.9784	0.9914	0.9955	0.8956	66.6		
	0.9735	0.9894	0.9945	0.8785	81.6		
	0.9704	0.9883	0.9937	0.8671	92.3		
	0.9687	0.9875	0.9934	0.8617	97.5		
	0.9667	0.9868	0.9929	0.8559	103.3		
	0.9650	0.9862	0.9923	0.8504	108.9		
ST 40	0.9960	0.9988	0.9987	0.9807	9.0	0.9943	9.3
	0.9897	0.9960	0.9974	0.9495	26.8	0.9823	29.7
	0.9853	0.9945	0.9965	0.9276	41.5	0.9742	44.3
	0.9808	0.9927	0.9955	0.9090	55.6	0.9669	58.1
	0.9761	0.9915	0.9940	0.8953	66.8	0.9611	69.4
	0.9744	0.9905	0.9932	0.8830	77.5	0.9561	79.5
	0.9710	0.9888	0.9929	0.8711	88.5	0.9491	94.0
	0.9677	0.9882	0.9920	0.8616	97.6	0.9439	105.2
	0.9648	0.9873	0.9908	0.8536	105.6	0.9408	112.0
	0.9631	0.9859	0.9906	0.8487	110.6	0.9372	120.1

a. Uncertainty of pressure value is ±2.5%.
b. Uncertainty of pressure value is ± 6 %.

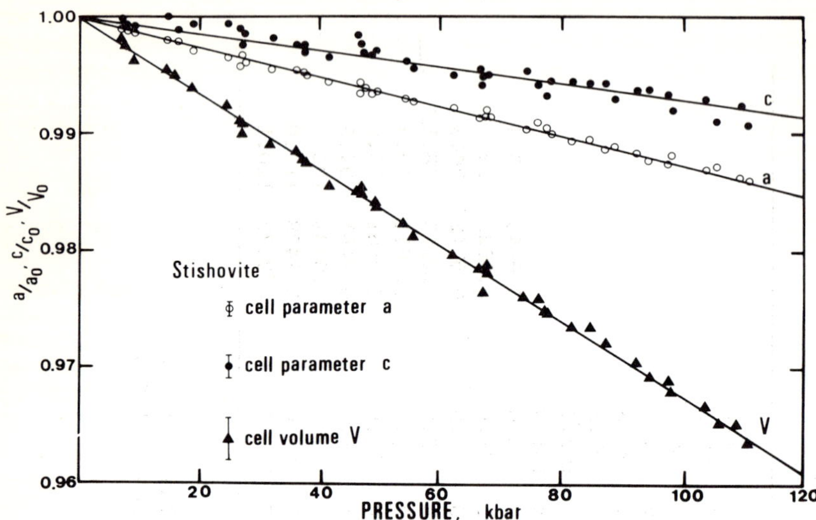

Fig. 3. Changes in cell parameters and volume of stishovite. Solid lines are $1 - a/a_o = 12.8 \times 10^{-5}P$, $1 - c/c_o = 7.3 \times 10^{-5}P$ (P in kbar), and a Birch-Murnaghan curve with $K_T = 3.01$ Mbar and $K_T' = 0$.

The compression of the cell parameters can be represented as

$$1 - \frac{a}{a_o} = 12.8 \times 10^{-5}P \tag{3}$$

and

$$1 - \frac{c}{c_o} = 7.3 \times 10^{-5}P \text{ (P in kbar).} \tag{4}$$

A least-squares calculation with the second-order Birch-Murnaghan equation (2) gives 2.98 ± 0.05 Mbar and 0.7 ± 1.1 as the most probable values of K_T and K_T' at zero pressure, respectively. The Murnaghan equation

$$P = \frac{K_T}{K_T'}\left[\left(\frac{V_o}{V}\right)^{K_T'} - 1\right] \tag{5}$$

gives $K_T = 2.99 \pm 0.05$ Mbar and $K_T' = 0.5 \pm 1.2$. The values K_T and K_T' do not depend on choice of the equation of state.

If we calculate the value K_T for a given parameter K_T' from -10.0 to +10.0 in the least-squares curve fit to the Birch-Murnaghan equation, we obtain the K_T versus K_T' relationship shown in the solid line in Figure 4. The equation of state with the values K_T and K_T' on the solid line can successfully represent the experimentally obtained pressure-volume data with the residual $R = \Sigma(P_{obs} - P_{calc})^2/\Sigma P_{obs}^2$ shown in the dotted line in Figure 4. The residual R has a broad minimum near the most probable value $K_T = 2.98$ Mbar.

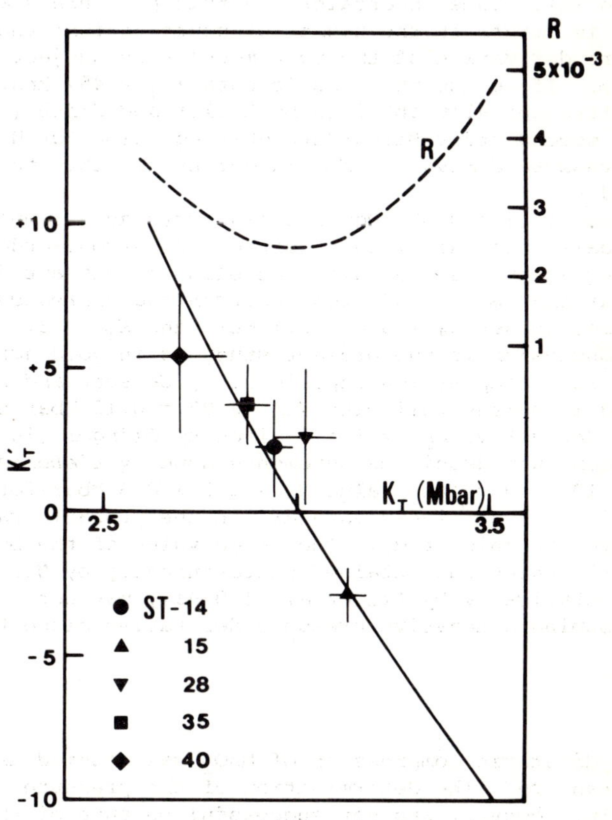

Fig. 4. K_T-K_T' relationship of stishovite.

When we calculate the values K_T and K_T' for each run using the second-order Birch-Murnaghan equation (2), there can be recognized a disagreement among values obtained as is shown in Figure 4. These values lie along the solid line and range from 2.70 Mbar to 3.15 Mbar for K_T and from 5.4 to -2.5 for K_T'.

Since the residual R has a broad minimum and since the values K_T and K_T' for each run scatter widely, it can be suggested that it is difficulty to accurately determine K_T and K_T'. In this study, we can safely conclude that the isothermal bulk modulus K_T takes a value between 2.7 Mbar and 3.15 Mbar.

In this study, a NaF pressure standard was used on the basis of equation (1). Some uncertainty in this pressure standard exists and is larger in the higher pressure region; therefore, the pressure derivative of the bulk modulus is subjected to uncertainty. If we use the elastic data (K_T = 457 kbar, K_T' = 5.183) ultrasonically obtained by *Miller and Smith* [1964] and assume the second-order Birch-Murnaghan equation for NaF, we obtain a pressure derivative K_T' larger by 2.6 than those shown in Figure 4.

In runs ST 28 and 35, MgO was also used as a complementary pressure scale. To calculate pressure, the second-order Birch-Murnaghan equation was used with the elastic constant data K_T = 1.599 Mbar and K_T' = 4.52 from *Anderson and Andreatch* [1966]. In this case, we get K_T = 3.0 ± 0.1 Mbar and K_T' = 3.1 ± 2.4.

The conclusion in the present study is in good agreement with the static compression experiment by *Bassett and Barnett* [1970], and it agrees well with K_T = 2.88 ± 0.13 Mbar based on a pressure derivative K_T' = 6 ± 1 given by *Olinger* [in press, 1976]. Recent ultrasonic measurements done by *Liebermann et al.* [in press, 1976] give the value K_S = 2.5 ± 0.3 Mbar for the assumed K_S' = 6 ± 2. These data are in the probable range of K_T and K_T' given in this study. The large value of the bulk modulus (∿3.45 Mbar), which was obtained ultrasonically by *Mizutani et al.* [1972] and statically by *Liu et al.* [1974*a*], may not be accepted without assuming a negative pressure derivative close to -10.

C. MnO_2

The hydrostatic compression of MnO_2 was studied using a LiF pressure standard. The determination of the pressure-volume relationship, however, was not successful because of the broadening of the diffraction line (211). At the highest pressure 125 kbar, splitting of the diffraction line (211) into two lines was clearly recognized. Figure 5 shows the ratio of the maximum half width $\Delta\Theta$ of the diffraction line (211) to that of (110) and the ratio of the d-spacing of (211) to that of (110). Those changes shown in Figure 5 are reversible with pressure change and may be attributed to a phase transformation that is different from the phase transformation found by *Liu* [1976] at 220-250 kbar and 1000°C.

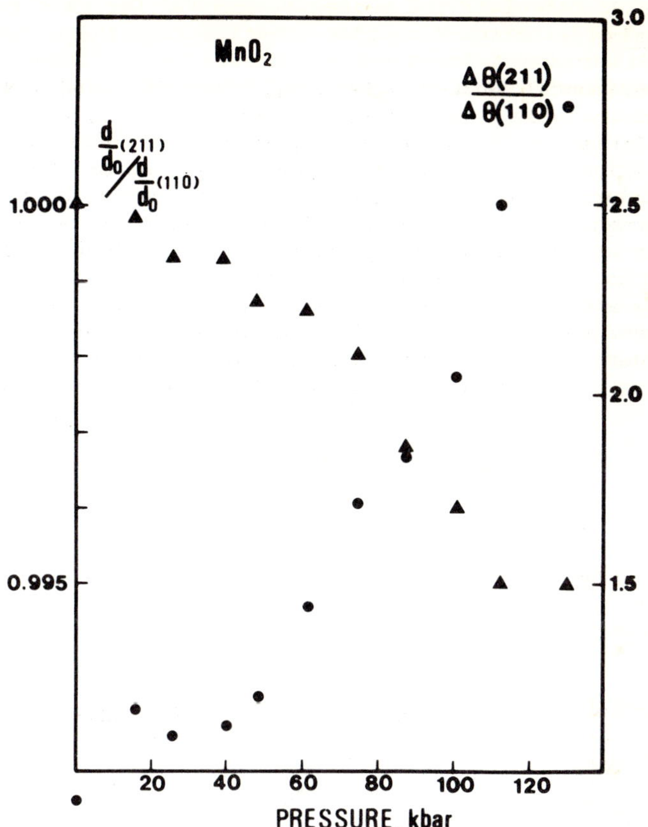

Fig. 5. The ratio of the maximum half width Δθ of the diffraction line (211) to that of (110), and the ratio of the normalized d-spacing of (211) to that of (110) in MnO_2.

The line broadening or splitting was not recognized in other diffraction lines (110), (101), (111), and (220) in the present study using Mo Kα radiation. Thus, there are not enough data to determine the crystal structure of the high-pressure phase found in this study.

In the static compression experiment of MnO_2, *Clendenen and Drickamer* [1966] gave a pressure-volume relationship in which a rapid decrease in volume can be recognized near 80 kbar. This abnormal behavior in volume change of MnO_2 may not be intrinsic behavior but could be due to some unknown experimental error. *Clendenen and Drickamer* [1966] may have overlooked the line broadening or splitting.

D. Spinels

Experimental data on γ-Fe_2SiO_4, γ-Co_2SiO_4, and γ-Ni_2SiO_4 are shown in Table 3 and plotted in Figures 6, 7, and 8. Diffraction lines (311), (400), (333), (511), and (440) were measured for the spinels; (200) and (220) were measured on the NaF pressure standard. The solid line in each figure is the second-order Birch-Murnaghan curve with K_T and K_T' obtained in a least-squares curve fit and shown in each figure. The experimental data were also fit to the Birch-Murnaghan equation with $K_T' = 4.0$. The values K_T and K_T' obtained in this study by using the second-order Birch-Murnaghan equation and the values K_T obtained by using the first-order Birch-Murnaghan equation are shown in Table 4 with the static compression data K_T and ultrasonic data K_S by other workers.

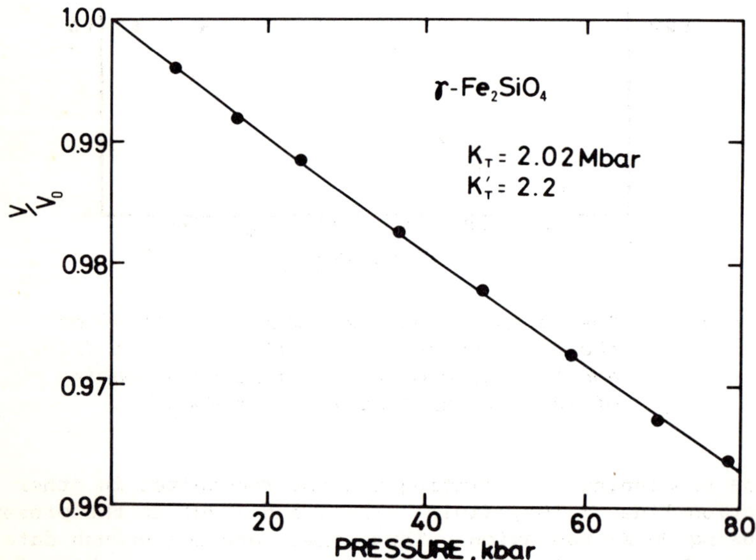

Fig. 6. Pressure-volume relationship of γ-Fe_2SiO_4.

TABLE 3. Experimental Data for Three Spinels Compared to NaF

γ-Fe_2SiO_4 (SP 22)			γ-Co_2SiO_4 (SP 18)			γ-Ni_2SiO_4 (SP 17)		
	NaF			NaF			NaF	
V/V_o [a]	V/V_o	P, kbar [c]	V/V_o [b]	V/V_o	P, kbar [c]	V/V_o [a]	V/V_o	P, kbar [c]
0.9960	0.9821	8.3	0.9963	0.9859	6.4	0.9963	0.9845	7.1
0.9917	0.9674	16.0	0.9947	0.9734	12.8	0.9936	0.9721	13.5
0.9880	0.9531	24.5	0.9910	0.9636	18.3	0.9902	0.9596	20.6
0.9824	0.9345	36.7	0.9840	0.9354	36.1	0.9864	0.9439	30.4
0.9779	0.9199	47.2	0.9803	0.9259	42.8	0.9830	0.9331	37.6
0.9726	0.9055	58.4	0.9755	0.9145	51.3	0.9810	0.9223	45.4
0.9669	0.8927	69.0	0.9745	0.9090	55.6	0.9776	0.9130	52.4
0.9638	0.8822	78.3	0.9716	0.9011	62.0	0.9750	0.9045	59.2
			0.9690	0.8900	71.4	0.9722	0.8953	66.8
						0.9700	0.8869	74.1

a. Volumes are accurate to ±0.05%.
b. Volumes are accurate to ±0.08%.
c. Pressure values are accurate to ±2.5%.

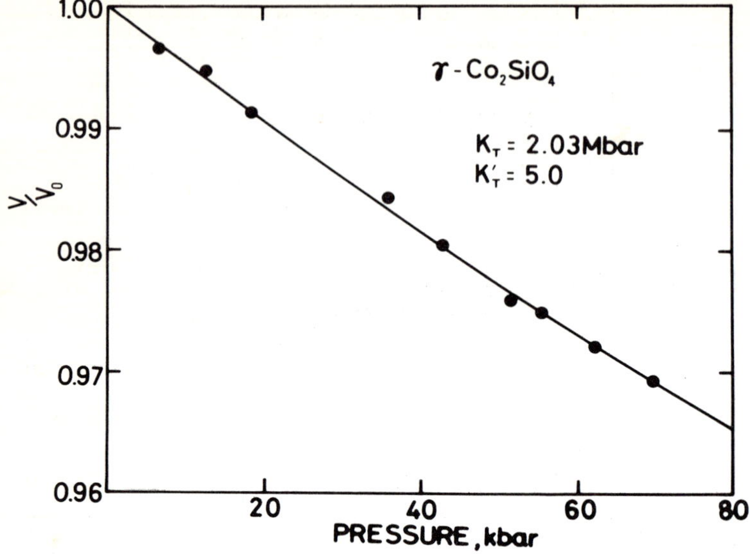

Fig. 7. Pressure-volume relationship of γ-Co_2SiO_4.

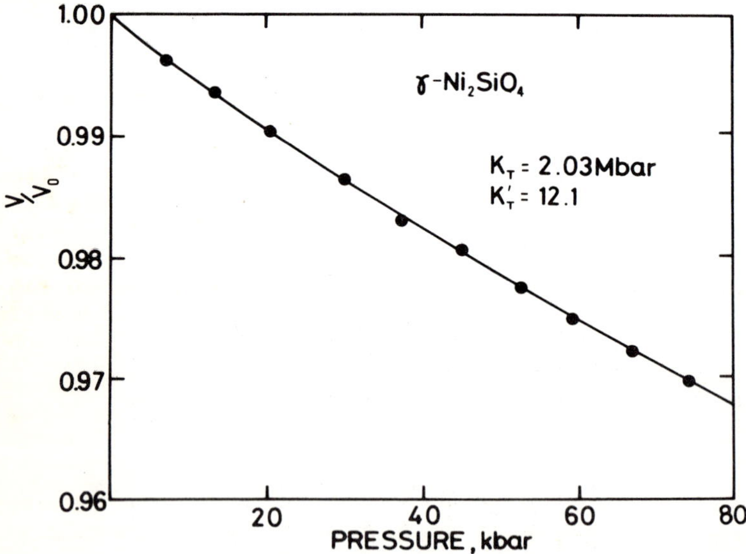

Fig. 8. Pressure-volume relationship of γ-Ni_2SiO_4.

TABLE 4. Bulk Modulus Values for Three Spinels

	This work			References
	K_T,[a] Mbar	K_T'[a]	K_T, Mbar[b] (K_T'= 4.0)	
γ-Fe_2SiO_4	2.02 (±0.06)	2.2 (±1.7)	1.97 (±0.02)	K_T = 2.12 Mbar (*Mao et al.*, 1969) K_S = 2.05 Mbar (*Mizutani et al.*, 1970) K_S = 1.93 Mbar (*Liebermann*, 1975)
γ-Co_2SiO_4	2.03 (±0.13)	5.0 (±4.6)	2.06 (±0.02)	K_T = 2.10 Mbar (*Liu et al.*, 1974*b*)
γ-Ni_2SiO_4	2.03 (±0.06)	12.1 (±2.1)	2.23 (±0.02)	K_T = 2.14 Mbar (*Mao et al.*, 1970) K_S = 2.16 Mbar (*Liebermann*, 1975)

a. The second-order Birch-Murnaghan equation was used for the calculation of K_T and K_T'.
b. The first-order Birch-Murnaghan equation was used for the calculation of K_T.

Acknowledgments. The author wishes to express her gratitude to S. Akimoto for his kind acceptance of the author's proposal to continue her study in the Institute for Solid State Physics, University of Tokyo. The author is grateful to Y. Ida for his constant encouragement and assistance and to E. Ito for his kindness in providing a sample of stishovite and to J. C. Jamieson for his critical reviewing the paper.

REFERENCES

Anderson, O. L. and P. Andreatch, Pressure derivatives of elastic constants of single-crystal MgO at 23° and -195.8°C, *J. Amer. Ceramic Soc.*, *49*, 404-409, 1966.

Bassett, W. A., and J. D. Barnett, Isothermal compression of stishovite and coesite up to 85 kilobars at room temperature by X-ray diffraction, *Phys. Earth Planet. Interiors, 3,* 54-60, 1970.

Clendenen, R. L., and H. G. Drickamer, Lattice parameters of nine oxides and sulfides as a function of pressure, *J. Chem. Phys., 44,* 4223-4228, 1966.

Decker, D. L., High-pressure equation of state for NaCl, KCl, and CsCl, *J. Appl. Phys., 42,* 3239-3244, 1971.

Liebermann, R. C., Elasticity of olivine (α), beta (β), and spinel (γ) polymorphs of germanates and silicates, *Geophys. J. Roy. Astron. Soc., 42,* 899-929, 1975.

Liebermann, R. C., A. E. Ringwood, and A. Major, Elasticity of polycrystalline stishovite, *Earth Planet. Sci. Lett.,* in press, 1976.

Liu, L., Synthesis of a new high-pressure phase of manganese dioxide, *Earth Planet. Sci. Lett., 29,* 104-106, 1976.

Liu, L., W. A. Bassett, and T. Takahashi, Effect of pressure on the lattice parameters of stishovite, *J. Geophys. Res., 79,* 1160-1164, 1974*a*.

Liu, L., W. A. Bassett, and T. Takahashi, Isothermal compression of a spinel phase of Co_2SiO_4 and magnesian ilmenite, *J. Geophys. Res., 79,* 1171-1174, 1974*b*.

Manghnani, M. H., Elastic constants of single-crystal rutile under pressures to 7.5 kilobars, *J. Geophys. Res., 74,* 4317-4328, 1969.

Mao, H. K., T. Takahashi, and W. A. Bassett, Isothermal compression of the spinel phase of Ni_2SiO_4 up to 300 kilobars at room temperature, *Phys. Earth Planet. Interiors, 3,* 51-53, 1970.

Mao, H. K., T. Takahashi, W. A. Bassett, J. S. Weaver, and S. Akimoto, Effect of pressure and temperature of the molar volumes of wüstite and three $(Fe,Mg)_2SiO_4$ spinel solid solutions, *J. Geophys. Res., 74,* 1061-1069, 1969.

McQueen, R. C., J. C. Jamieson, and S. P. Marsh, Shock-wave compression and X-ray studies of titanium dioxide, *Science, 155,* 1401-1404, 1967.

Miller, R. A., and C. S. Smith, Pressure derivatives of the elastic constants of LiF and NaF, *J. Phys. Chem. Solid, 25,* 1279-1292, 1964.

Mizutani, H., Y. Hamano, and S. Akimoto, Elastic-wave velocities of polycrystalline stishovite, *J. Geophys. Res., 77,* 3744-3749, 1972.

Mizutani, H., Y. Hamano, Y. Ida, and S. Akimoto, Compressional wave velocities of fayalite, Fe_2SiO_4 spinel and coesite, *J. Geophys. Res., 75,* 2741-2747, 1970.

Olinger, B., and J. C. Jamieson, Relative compression of NaF and NaCl to 130 kilobars, *High Temp.-High Press., 2,* 513-520, 1970.

Olinger, B., The compression of stishovite, *J. Geophys. Res.,* in press, 1976.

Piermarini, G. J., S. Block and J. D. Barnett, Hydrostatic limits in liquids and solids to 100 kbar, *J. Appl. Phys.*, *44*, 5377-5382, 1973.

Sato, Y., T. Yagi, Y. Ida, and S. Akimoto, Hysteresis in the pressure-volume relation and stress inhomogeneity in composite material, *High Temp.-High Press.*, *7*, 315-323, 1975.

Spieglan M., and J. C. Jamieson, Relative compression of NaF and NaCl to 110 kilobars; a redetermination, *High Temp.-High Press.*, *6*, 479-481, 1974.

COMPRESSION STUDIES OF FORSTERITE (Mg_2SiO_4) AND ENSTATITE ($MgSiO_3$)

B. Olinger
Los Alamos Scientific Laboratory
Los Alamos, New Mexico 87545

Abstract

From hydrostatic, high-pressure, X-ray diffraction studies of forsterite, Mg_2SiO_4, and enstatite, $MgSiO_3$, the following zero pressure isothermal bulk moduli and their pressure derivatives were determined.

	ρ_o	K_{to}	K'_{to}
forsterite	3.234 g/cm^3	1.20 Mbar	5.6
enstatite	3.227 g/cm^3	1.25 Mbar	5 (assumed)

The cell edge (linear) and volume compressions are compared with those calculated from elastic constants [*Graham and Barsch*, 1969; *Frisillo and Barsh*, 1972] and with shock compression data [*Ahrens et al.*, 1971; *Ahrens and Gaffney*, 1971].

I. INTRODUCTION

Forsterite and enstatite are endmembers of two important mineral series thought to make up a large portion of the earth's upper mantle. Evidence for this hypothesis can be found in comparisons of seismic data and mass distribution models of the earth with the densities and sound velocities of these minerals under pressure. Recently, there have been several determinations of both densities and sound speeds of these minerals under both static and dynamic stress. *Kumazawa and Anderson* [1969] and *Graham and Barsch* [1969] measured the elastic constants of forsterite, Mg_2SiO_4, and *Ahrens et al.* [1971] determined the dynamic compression of forsterite. *Kumazawa* [1969] and *Frisillo and Barsch* [1972] measured the elastic constants of orthopyroxenes, $(Mg_{.85}, Fe_{.15}) SiO_3$ and $(Mg_{.8}, Fe_{.2}) SiO_3$ respectively, and *Ahrens and Gaffney* [1971] determined the dynamic compression of orthopyroxene, $(Mg_{.86}, Fe_{.14}) SiO_3$. Here are reported some recent measurements of the linear and volume compression of both enstatite and forsterite using a hydrostatic, high-pressure, X-ray diffraction technique. The linear compressions are compared with the values calculated for forsterite [*Graham and Barsch*, 1969] and for bronzite [*Frisillo and Barsch*, 1972] from elastic constant data. The bulk moduli are also compared with those derived from ultrasonic data. The isothermal compression fits are then converted to shock-particle velocities along the Hugoniot and compared with the results of *Ahrens et al.* [1971] and *Ahrens and Gaffney* [1971].

II. EXPERIMENTAL METHOD

The technique used here has been described earlier [*Halleck and Olinger*, 1974]. Briefly, an annulus of beryllium 2.5-mm in diameter, 0.2-mm thick with an 0.2-mm diameter hole at its center is pressed between two tungsten carbide Bridgman anvils. The hole is filled with a mixture of powdered sample and powdered aluminum or NaF (the pressure indicators), and a mixture of 4:1 methanol-ethanol that remains liquid to 100 kbar [Piermarini et al., 1973]. A Cu Kα X-ray beam is directed through the annulus, and the diffraction is recorded on a 114.6-mm diameter cylindrical film surrounding the annulus.

The pressures were determined from the diffraction patterns of the NaF or Al mixed with the sample. The two lines read from the NaF were the (200) and (220). The three lines read from the Al were the (111), (220), and (311). The standard patterns for the NaF and Al are given by *Swanson and Tatge* [1953] (a_o = 4.6342 Å and 4.0494 Å, respectively). The relative volumes, V/V_o, of these two materials are correlated with pressures to 13 GPa (130 kbar) elsewhere [*Olinger and Halleck*, 1976].

III. EXPERIMENTAL RESULTS

The forsterite, Mg_2SiO_4, studied here was a synthesized, hot-pressed compound. Microprobe analysis indicated good homogeneity. There was minor interstitial contamination from carbon, but this did not affect the diffraction patterns of the material. The diffraction pattern of the forsterite at ambient conditions yielded the following orthorhombic cell edge parameters: $a = 4.7470 \pm 0.0018$ Å; $b = 10.1800 \pm 0.0035$ Å; and $c = 5.9771 \pm 0.0012$ Å. The calculated crystal density is 3.234 g/cm^3. The enstatite, $MgSiO_3$, came from the Mt. Egerton meteorite, Australia (an enstatite achondrite). The chemical analysis of the enstatite is given by *Reid and Cohen* [1967]. The major impurities are Ca (0.33 wt%), excess SiO_2 (0.33%), Al (0.02 wt%) and Mn (0.02 wt%). The diffraction pattern of the enstatite at ambient condition yielded the following orthorhombic cell edge parameters; a = 8.8189 ± 0.0038 Å, b = 18.2205 ± 0.0095 Å, and c = 5.1803 ± 0.0028 Å. The calculated crystal density is 3.227 g/cm^3.

The quality of the forsterite X-ray diffraction patterns was usually good. The d-spacings were determined at high pressures for the following planes: (021), (101), (130), (131), (112), (004), and (062). The quality of the enstatite diffraction patterns at high pressures was not good, and the d-spacings measured had 2-theta diffraction angles of less than 40° (up to 65° for forsterite). The lines measured at high pressures were for the following planes: (211), (240), (160), (311), and (022). The problems for enstatite were compounded by no agreement among the crystallographic literature for the indexing of higher angle lines and the necessary use of an overground sample that could not be readily replaced. (Too finely ground powder causes poor coherent diffraction.)

Tables 1 and 2 list the data in the form of the relative volumes of the pressure indicators, the correlated pressures, the relative cell edges, the relative volumes and the shock-particle velocity analogues, U_{st} and U_{pt} [*Olinger and Halleck*, 1976]. Figures 1 and 3 show the relative volumes and cell edges calculated from elastic constants of forsterite [*Graham and Barsch*, 1969], and of bronzite [*Frisillo and Barsch*, 1972]. In Figures 2 and 4 the U_{st}, U_{pt} values are plotted along with the loci of the shock-particle velocities, U_s-U_p, calculated from the same elastic constants. Also plotted in Figures 2 and 4 are the U_s-U_p data of *Ahrens et al.* [1971] and *Ahrens and Gaffney* [1971].

IV. DISCUSSION

Figure 1 shows that the relative volumes, $V(P)/V_o$, for forsterite calculated from the elastic constants [*Graham and Barsch*, 1969] are in excellent agreement with the present X-ray data. However, the X-ray data show the b and c axes or cell edges to be less compressible than that calculated from the elastic constants, and the a axis to be more compressible. From *Graham and Barsch* [1969] for forsterite,

$$a/a_o = 1 - 1.863 \times 10^{-4}\,P + 4.42 \times 10^{-7}\,P^2,$$
$$b/b_o = 1 - 3.526 \times 10^{-4}\,P + 7.36 \times 10^{-7}\,P^2,$$
$$c/c_o = 1 - 2.730 \times 10^{-4}\,P + 5.20 \times 10^{-7}\,P^2, \text{ P in kbars.}$$

From the present study for forsterite,

$$a/a_o = 1 - 1.976 \times 10^{-4}\,P,$$
$$b/b_o = 1 - 2.761 \times 10^{-4}\,P,$$
$$c/c_o = 1 - 2.480 \times 10^{-4}\,P, \text{ P in kbars.}$$

TABLE 1. The Compression of Forsterite

V/V_o [a] N: NaF A: Al		P [b], kbar	a/a_o [a]	b/b_o [a]	c/c_o [a]	V/V_o [a]	U_{st} [b], km/s	U_{pt} [b], km/s
0.9663 (02)	N	17.5 (0.1)	0.9946 (08)	0.9957 (03)	0.9956 (02)	0.9860 (09)	6.21(0.20)	0.087(0.003)
0.9609 (04)	A	33.0 (0.4)	0.9939 (06)	0.9905 (03)	0.9911 (03)	0.9758 (08)	6.49(0.11)	0.157(0.003)
0.9370 (02)	N	35.9 (0.2)	0.9918 (14)	0.9894 (05)	0.9904 (04)	0.9719 (15)	6.28(0.17)	0.177(0.005)
0.9211 (17)	N	47.3 (1.4)	0.9901 (12)	0.9853 (04)	0.9883 (03)	0.9641 (13)	6.38(0.15)	0.229(0.005)
0.9280 (19)	N	47.5 (1.5)	0.9905 (17)	0.9863 (06)	0.9877 (05)	0.9650 (18)	6.43(0.19)	0.227(0.007)
0.9394 (01)	A	54.6 (0.1)	0.9908 (05)	0.9849 (03)	0.9861 (04)	0.9623 (07)	6.69(0.07)	0.252(0.002)
0.9072 (19)	N	58.2 (1.7)	0.9901 (08)	0.9819 (03)	0.9853 (02)	0.9579 (08)	6.54(0.12)	0.275(0.005)
0.9004 (02)	N	63.7 (0.1)	0.9884 (11)	0.9816 (04)	0.9836 (03)	0.9542 (11)	6.56(0.08)	0.300(0.004)
0.8998 (02)	N	64.2 (0.1)	0.9897 (17)	0.9808 (06)	0.9837 (04)	0.9549 (18)	6.63(0.14)	0.299(0.006)
0.8825 (15)	N	79.8 (1.4)	0.9844 (14)	0.9777 (05)	0.9819 (04)	0.9450 (15)	6.70(0.11)	0.368(0.006)
0.9164 (12)	A	80.9 (1.3)	0.9826 (11)	0.9766 (04)	0.9808 (03)	0.9412 (12)	6.52(0.08)	0.383(0.005)
0.9108 (05)	A	87.8 (0.7)	0.9848 (13)	0.9762 (07)	0.9800 (06)	0.9421 (16)	6.85(0.10)	0.396(0.006)
0.9021 (01)	A	99.1 (0.2)	0.9829 (07)	0.9726 (05)	0.9784 (04)	0.9354 (09)	6.88(0.05)	0.445(0.003)

a. Standard deviation x 10^4 in parentheses.
b. Standard deviation in parentheses.

TABLE 2. The Compression of Enstatite

V/V_o [a] N: NaF A: Al		P [b], kbar	a/a_o [a]	b/b_o [a]	c/c_o [a]	V/V_o [a]	U_{st} [b], km/s	U_{pt} [b], km/s
0.9749 (08)	A	20.3 (0.6)	0.9949 (03)	0.9940 (03)	0.9959 (02)	0.9848 (04)	6.44 (0.13)	0.098 (0.002)
0.9658 (05)	A	28.5 (0.4)	0.9924 (15)	0.9931 (16)	0.9936 (14)	0.9792 (26)	6.52 (0.41)	0.136 (0.008)
0.9586 (07)	A	35.2 (0.6)	0.9915 (09)	0.9911 (09)	0.9916 (08)	0.9744 (15)	6.53 (0.21)	0.167 (0.005)
0.9362 (15)	N	36.5 (1.1)	0.9895 (13)	0.9914 (07)	0.9916 (30)	0.9727 (33)	6.44 (0.39)	0.176 (0.011)
0.9495 (04)	A	44.1 (0.4)	0.9893 (04)	0.9902 (04)	0.9896 (04)	0.9694 (07)	6.68 (0.08)	0.205 (0.002)
0.9240 (05)	N	45.1 (0.3)	0.9827 (19)	0.9920 (20)	0.9897 (18)	0.9647 (33)	6.30 (0.30)	0.222 (0.010)
0.9444 (12)	A	49.4 (1.4)	0.9866 (05)	0.9886 (06)	0.9882 (05)	0.9639 (09)	6.51 (0.12)	0.235 (0.004)
0.9376 (11)	A	56.5 (1.2)	0.9837 (07)	0.9882 (07)	0.9905 (06)	0.9628 (11)	6.86 (0.13)	0.255 (0.005)
0.9027 (10)	N	61.8 (0.8)	0.9834 (18)	0.9855 (19)	0.9833 (16)	0.9529 (29)	6.38 (0.20)	0.300 (0.010)

a. Standard deviation x 10^4 in parentheses.
b. Standard deviation in parentheses.

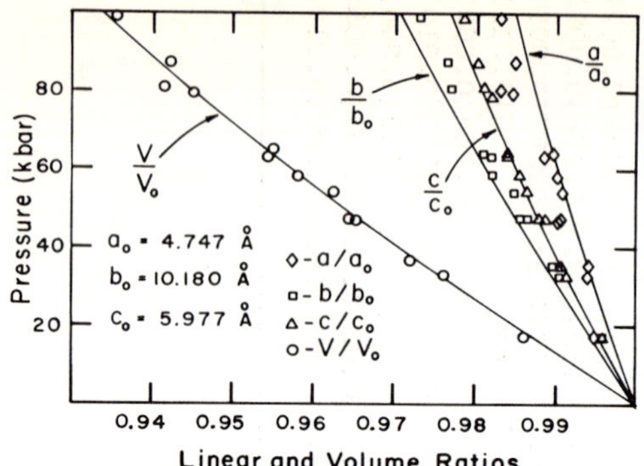

Fig. 1. The relative cell edges and volumes of forsterite at pressures to 10 GPa from the present work are shown as symbols. The ambient cell edge values are listed in the figure. The same relative values calculated by Graham and Barsch [1969] from ultrasonic data are shown as solid curves.

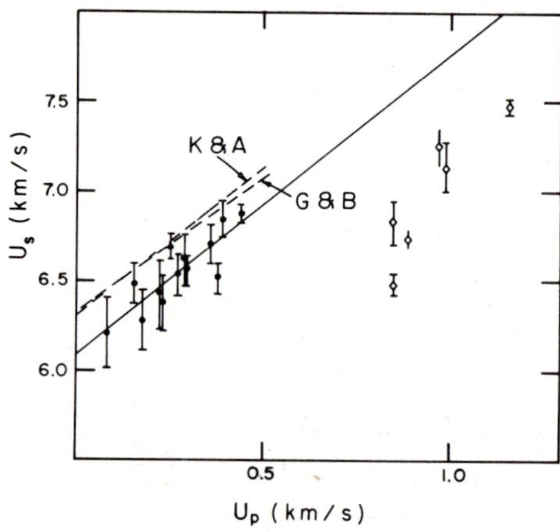

Fig. 2. The volume compression data for forsterite in the shock-particle velocity plane. The X-ray volume compression data are shown as the dots with error bars; the solid line is a linear least squares fit to the data. The data of Ahrens et al. [1971] are the diamonds with error bars. The Hugoniots calculated from the ultrasonic work of Graham and Barsch [1969] and Kumazawa and Anderson [1969] are labeled "G & B" and "K & A."

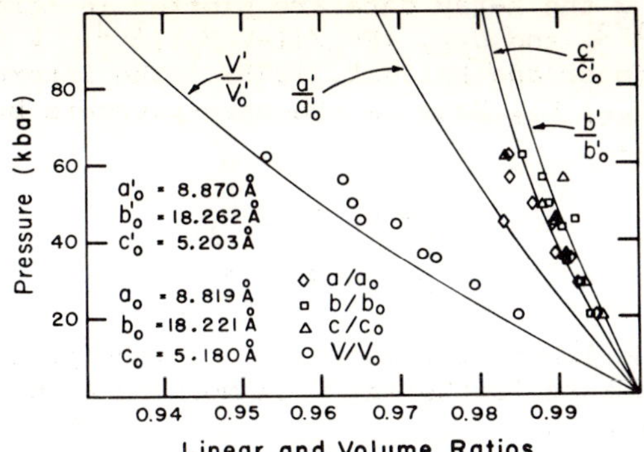

Fig. 3. The relative cell edge and volumes of enstatite at pressures to 10 GPa from the present work are shown as symbols. The ambient cell edge values are listed in the bottom left portion of the figure. The same relative values calculated by Frisillo and Barsch [1972] from ultrasonic data for bronzite ($Mg_{.8}$,$Fe_{.2}$) SiO_3 are shown as solid curves. The ambient cell edge values for the bronzite are listed in the middle left portion of the figure.

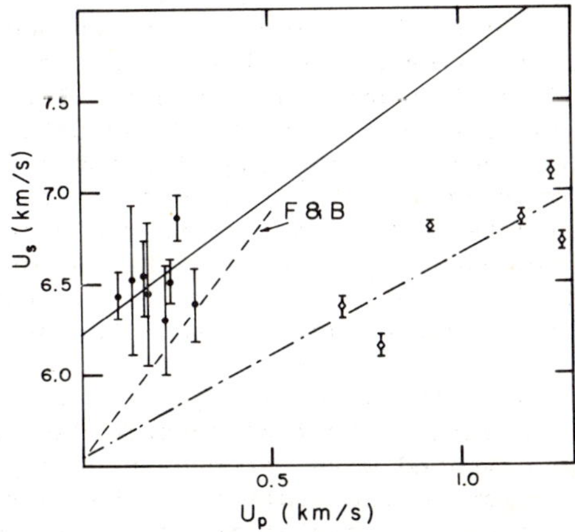

Fig. 4. The volume compression data for enstatite in the shock-particle velocity plane. The X-ray volume compression data are shown as the points with error bars; the solid line is an average linear fit with a forced slope of 1.5. The data of Ahrens and Gaffney [1971] are the diamonds with error bars. The Hugoniot calculated from the ultrasonic work of Frisillo and Barsch [1972] for bronzite is labeled "F & B." The Hugoniot having the bulk sound speed as an intercept and an average slope to the shock data of Ahrens and Gaffney [1971] is shown as a broken line.

In Figure 2 the X-ray data are plotted in terms of U_{st}, $(PV_o/(1-V/V_o))^{1/2}$, and U_{pt}, $(PV_o/(1-V/V_o))^{1/2}$. V is the specific volume [see *Olinger and Halleck*, 1975]. The intercept of a linear fit to U_{st}, U_{pt} data is the zero-pressure bulk sound speed of the material.

$$U_{st} = c_t + s_t U_{pt}, \tag{1}$$

$$K_{to} = \rho_o c_t^2. \tag{2}$$

The slope of the linear fit is related to the pressure derivative of the zero pressure isothermal bulk modulus as given by the following expression:

$$K'_{to} = 4 s_t - 1. \tag{3}$$

The linear fit to the present forsterite data is

$$U_{st} = 6.09 + 1.65 U_{pt}. \tag{4}$$

From this fit, the isothermal zero-pressure bulk modulus is calculated to be 1.20 Mbars. This value compares well with the values calculated from the ultrasonic data, 1.280 Mbar [*Graham and Barsch*, 1969] and 1.275 Mbar [*Kumazawa and Anderson*, 1969]. The pressure derivative of the modulus determined from the slope is 5.6, and this also compares well with the values determined from ultrasonic data, 5.0 and 5.4, respectively.

Along with the linear fit to the X-ray data in Figure 2, are shown the Hugoniots derived from the ultrasonic data and the low pressure shock compression data of *Ahrens et al.* [1971]. The shock compression data do not agree with the ultrasonic-derived or X-ray diffraction-derived results. The reason for the disagreement seems to be caused by the porosity of the samples of *Ahrens et al.* [1971], which was not less than 4% for the data in Figure 2.

In Figure 3 the relative cell edges and volumes of bronzite, $(Mg_{.8}Fe_{.2})\,SiO_3$, calculated from ultrasonic data [*Frisillo and Barsch*, 1972] are compared with the present X-ray data for enstatite, $MgSiO_3$. Despite the scatter in the X-ray data, it is apparent that the *a* axis in enstatite is not as compressible as that determined for bronzite. It is of interest to note that with increasing FeO content, the *a* axis expands more than the other cell edges. The *a* axis is possibly a direction of weak bonding and repulsion. Because of the less compressible *a* axis, the volume of enstatite is found to be less compressible than that of bronzite. *Frisillo and Barsch* [1972] calculated the following relative cell edge fits for bronzite as a function of pressure.

$$a/a_o = 1 - 4.375 \times 10^{-4} P + 1.801 \times 10^{-6} P^2,$$
$$b/b_o = 1 - 2.344 \times 10^{-4} P + 0.929 \times 10^{-6} P^2,$$
$$c/c_o = 1 - 3.303 \times 10^{-4} P + 2.205 \times 10^{-6} P^2, \text{ P in kbars.}$$

From the present data on enstatite, the following fits are calculated.

$$a/a_o = 1 - 2.780 P,$$
$$b/b_o = 1 - 2.451 P,$$
$$c/c_o = 1 - 2.263 P, \text{ P in kbars.}$$

In Figure 4 the X-ray data are plotted in terms of U_{st} and U_{pt}. The scatter is great enough to prohibit any reasonable linear fit directly to the data. Instead, a value for the slope is assumed, 1.5 ($K'_{to} = 5$), and an intercept is calculated for each datum. The average of these is 6.22 ± 0.19 km/s. Thus the linear fit estimated for the X-ray data is

$$U_{st} = 6.22 + 1.5\, U_{pt}. \tag{5}$$

This implies a bulk modulus of 1.25 ± 0.08 Mbar assuming K'_{to} equals 5. For bronzite, $(Mg_{.8}Fe_{.2})\ SiO_3$, *Frisillo and Barsch* [1972] measured a bulk modulus of 988 kbar and a derivative of 9.5 (see Figure 4). (From our data, K_o = 1.15 ± 0.08 Mbar if K_o' = 9.5.) The large pressure derivative of the bulk modulus is not substantiated by the Hugoniot data of *Ahrens and Gaffney* [1971]. They studied nearly the same mineral as *Frisillo and Barsch* [1972], and for that study the samples had no porosity. The low-pressure data (below 300 kbar) are plotted in Figure 4. An equilibrium Hugoniot can be calculated by using the adiabatic bulk sound speed of *Frisillo and Barsch* [1972] as the intercept and averaging the slope through the Hugoniot data. The Hugoniot is

$$U_s = 5.55 + 1.10\, U_p. \tag{6}$$

The pressure derivative for the zero pressure bulk modulus is 3.4 for this fit. This result is also supported by the shock compression of a bronzitite rock (Stillwater complex, Montana) [*McQueen et al.* 1967] where the composition of the bronzitite was 94% bronzite, $(Mg_{0.9}Fe_{0.1})\ SiO_3$. The linear fit to that data before the onset of a 350 kbar phase transition is

$$U_s = 6.10 + 1.01\, U_p. \tag{7}$$

The pressure derivative of the bulk modulus in this case is 3.0. Thus, the estimated slope in equation (5) from the X-ray data seems reasonable, and the bulk modulus pressure derivative of *Frisillo and Barsch* [1972] appears to be too large.

Acknowledgments. This work was supported by the U.S. Energy Research and Development Administration.

REFERENCES

Ahrens, T. J. and E. S. Gaffney, Dynamic compression of enstatite, *J. Geophys. Res., 76,* 5504-5513, 1971.

Ahrens, T. J., J. H. Lower, and P. L. Lagus, Equation of state of forsterite, *J. Geophys. Res., 76,*518-528, 1971.

Graham, E. K., Jr., and G. R. Barsch, Elastic constants of single-crystal forsterite as a function of temperature and pressure, *J. Geophys. Res., 74,* 5949-5960, 1969.

Frisillo, A. L. and G. R. Barsch, Measurement of single-crystal elastic constants of bronzite as a function of pressure and temperature, *J. Geophys. Res., 77,* 6360-6384, 1972.

Halleck, P. M. and B. Olinger, A method for the accurate measurement of lattice compressions of low-Z materials at pressures up to 12 GPa by x-ray diffraction, *Rev. Sci. Instrum., 45,* 1408-1410, 1974.

Kumazawa, M., The elastic constants of single-crystal orthopyroxene, *J. Geophys. Res.,* 5973-5980, 1969.

Kumazawa, M. and O. L. Anderson, Elastic moduli, pressure derivatives and temperature derivatives of single-crystal olivine and single-crystal forsterite, *J. Geophys. Res., 74,* 5961-5972, 1969.

McQueen, R. G., S. P. Marsh, and J. N. Fritz, Hugoniot equation of state of twelve rocks, *J. Geophys. Res., 72,* 4999-5036, 1967.

Olinger, B. and P. M. Halleck, Compression and bonding of ice VII and an empirical linear expression for the isothermal compression of solids, *J. Chem. Phys., 62,* 94-99, 1975.

Olinger, B. and P. M. Halleck, The compression of α-quartz, *J. Geophys. Res., 81,* in press, 1976.

Piermarini, G. J., S. Block, and J. D. Barnett, Hydrostatic limits in liquids and solids to 100 kbar, *J. Appl. Phys., 44,* 5377-5382, 1973.

Reid, A. M. and A. J. Cohen, Some characteristics of enstatite from enstatite achondrites, *Geochim. Cosimochim. Acta, 31,* 661-672, 1967.

Swanson, H. E. and E. Tatge, *Standard X-ray Diffraction Powder Patterns, I,* p. 11 and p. 63, (NBS Cir. 539, U.S. Govern. Print. Off., 1953).

A COMPARISON OF α-QUARTZ SHOCK COMPRESSION DATA WITH RECENT DETERMINATIONS OF THE BULK MODULUS OF STISHOVITE

B. OLINGER
Los Alamos Scientific Laboratory
Los Alamos, New Mexico 87545

Abstract

Three trial bulk moduli (3.5, 3.0, and 2.5 Mbar) and two pressure derivatives of the moduli (6 and 3) for stishovite are transposed to the shock-velocity, particle-velocity plane. The phase transformation from α-quartz to stishovite is taken into account. Best agreement is found between shock compression measurements on α-quartz and calculated Hugoniots based on the trial modulus and derivatives K_o = 3 Mbar and K_o' = 3 to 6 for stishovite. This trial modulus agrees well with the moduli determined by *Bassett and Barnett* [1970], *Chung* [1974], *Olinger* [1976], and *Sato* [1976].

I. INTRODUCTION

There is interest in the equation of state of stishovite, the stable solid phase of SiO_2 at some pressure above 80 kbars, because substantial evidence shows that silicates decompose to oxide states at conditions inside the earth; a major constituent to these oxides is SiO_2. Data has been gathered for stishovite at deep earth-interior conditions using shock compression techniques, [*Wackerle*, 1972; *Altshuler et al.*, 1965; *Trunin et al.*, 1970; *Trunin et al.*, 1971], but the results are complicated by having to start with α-quartz or glass from which the stishovite phase is obtained only after undergoing a phase transformation involving large energy and volume changes.

More recently there has been substantial work on the thermodynamic properties of pure stishovite. The heat capacities and enthalpies of transformation were determined by *Holm et al.* [1967] for α-quartz and stishovite. Also, there have been two determinations of the volume thermal expansion [*Weaver et al.* 1973; *Ito et al.*, 1974]. And finally, measurements of the zero pressure bulk modulus, $V_o(\partial P/\partial V)_{P=0}$, have been made using ultrasonic techniques [*Mizutani et al.*, 1972; *Chung*, 1974; *Liebermann et al.*, 1977] and using high pressure X-ray diffraction techniques [*Bassett and Barnett*, 1970; *Liu et al.*, 1974; *Olinger*, 1976;

Sato, 1976]. Here various trial bulk moduli and their pressure derivatives for crystal density stishovite are compared with the shock compression data using the heat capacity, transformation enthalpy, and thermal expansion data to transform the moduli to the shock compression Hugoniots of α-quartz. The results of this comparison are discussed in light of the various moduli determinations.

II. DISCUSSION

In Table 1 are listed the six sets of K_o (zero pressure bulk modulus) and K_o' (pressure derivative of the zero pressure bulk modulus) values that are to be compared with the shock data. They range from K_o = 3.50 Mbar, K_o' = 6 to K_o = 2.50 Mbar K_o' = 3, and span the experimentally determined values. Here the difference between adiabatic and isothermal moduli is ignored since there is only about 1/2% difference between them. The shock compression Hugoniots of crystal density stishovite are easily calculated from these moduli. The relation between the shock velocity (U_s) and the particle velocity (U_p) behind the shock wave has been found to be nearly linear [*McQueen et al.*, 1967]

$$U_s = c_o + sU_p. \tag{1}$$

the coefficients can be related to the K_o and K_o'

$$c_o = (K_o/\rho_o)^{1/2} \tag{2}$$

$$s = (K_o' + 1)/4 \tag{3}$$

The coefficients calculated from the trial moduli are listed with them in Table 1. The P, V values along the Hugoniot defined by the K_o and K_o' values are calculated from

$$P_h = \rho_o U_s U_p , \tag{4}$$

$$V_h = (U_s - U_p)/(\rho_o U_s) . \tag{5}$$

TABLE 1. Trial Moduli for Stishovite

Curve in Fig. 1	K_o, Mbar	K_o'	c_o, km/s	s	γ_o
A	3.5	6	9.04	1.75	1.52
B	3.5	3	9.04	1.00	1.52
C	3.0	6	8.37	1.75	1.30
D	3.0	3	8.37	1.00	1.30
E	2.5	6	7.64	1.75	1.09
F	2.5	3	7.64	1.00	1.09

Once the P, V conditions have been defined along the Hugoniot starting with crystal density stishovite, then new pressure values can be calculated for stishovite at volume, V_h, to account for the energy increase resulting from the transformation from α-quartz under shock compression. The transformation equation was described earlier [*McQueen et al.*, 1963; *McQueen et al.*, 1967]

$$P_h'\{V_h\} = \frac{P_h\{V_h\}[1-(\gamma/V)\ (V_{os}-V_h)/2] + (\gamma/V)\ [\Delta E_o]}{1-(\gamma/V)\ (V_{oq}-V_h)/2} \ . \qquad (6)$$

The brackets mean "evaluated at." V_{oq} is the ambient specific volume of α-quartz from ρ_o = 2.647 g/cm^3 [*Olinger and Halleck*, 1976]. V_{os} is the ambient specific volume of stishovite from ρ_o = 4.280 g/cm^3 [*Olinger*, 1976]. γ is the Grüneisen constant and is calculated from

$$\gamma_o = \alpha_v c_o^2/C_p \ . \qquad (7)$$

α_v is the thermal expansion, 13.1 x 10^{-6}/K [*Ito et al.*, 1974]. The thermal expansion of *Ito et al.* [1974] was chosen instead of the value from *Weaver et al.* [1973] because more information was available about the experimental technique used by the former, and the experiment appeared to be carefully done. The heat capacity, C_p, is 0.704 J/gK [*Holm et al.*, 1967]. The Grüneisen constants used with each trial moduli are listed with them in Table 1. The ratio, γ/V, is assumed here to remain constant along the Hugoniot [*McQueen et al.*, 1967]. The transformation enthalpy between α-quartz and stishovite, ΔE, is 822.1 J/g [*Holm et al.*, 1967]. Having calculated $P_h'\{V_h\}$, the U_s, U_p values can then be calculated from the following relations

$$U_s = (P_h'\ V_{oq}\ (1-V_h/V_{oq}))^{1/2} \qquad (8)$$

$$U_p = (P_h'\ V_{oq}\ (1-V_h/V_{oq}))^{1/2} \ . \qquad (9)$$

The Hugoniots calculated from the trial moduli for crystal density stishovite but based on crystal density α-quartz are illustrated in Figure 1.

The U_s, U_p shock compression data for stishovite transformed from near crystal density α-quartz are listed in Table 2 with their sources. As can be seen in the table, the data credited to *Al'tshuler et al.* [1965] are repeated by *Trunin et al.* [1970]. These data are also plotted in each of the illustrations in Figure 1 to be compared with the calculated Hugoniots.

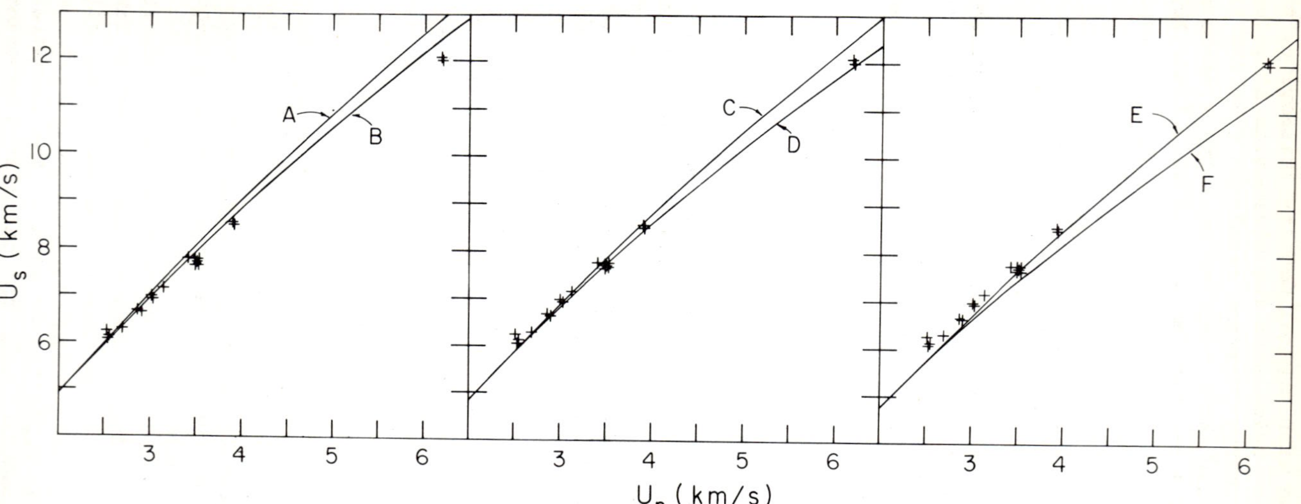

Fig. 1. The crosses are the shock compression data of α-quartz listed in Table 2. The curves labeled with letters are the Hugoniots of stishovite having trial moduli listed in Table 1 transformed from α-quartz starting at ambient conditions.

TABLE 2. α-Quartz U_s,U_p Hugoniot Data Between U_p = 2.5 to 6.5 km/s Centered on Near Crystal Density α-Quartz at Ambient Conditions

U_p, km/s	U_s, km/s	U_p, km/s	U_s, km/s
Wackerle [1962]		*Al'tshuler et al.* [1965]	
2.55	6.12	3.13	7.18
2.70	6.29	3.92	8.54
2.89	6.66	6.20	12.01
2.89	6.66		
3.03	6.95	*Trunin et al.* [1970]	
3.03	6.95	2.52	6.27
3.42	7.76	2.54	6.10
3.49	7.76	3.13	7.18[a]
3.50	7.63	3.91	8.56
3.50	7.72	3.92	8.54[a]
3.52	7.70	6.18	12.12
3.52	7.75	6.20	12.01[a]

a. These data may be the same data published in *Al'tshuler et al.* [1965]. Although *Trunin et al.* [1970] do not so indicate, they are considered here to be from two different experiments.

An examination of Figure 1, where the Hugoniots for the trial moduli of stishovite are compared with the Hugoniot data of α-quartz, reveals the following. The moduli, K_o = 3.50 Mbar, K_o' = 3 to 6, match only the lower pressure (2.5 km/s < U_p < 3 km/s, 0.39 to 0.55 Mbar) Hugoniot data. The moduli, K_o = 2.50 Mbar, K_o' = 3 to 6 include only several of the higher pressure Hugoniot within their range. However, the moduli K_o = 3.00 Mbar, K_o' = 3 to 6 include most of the Hugoniot data within their range and give the best fit over the range of the Hugoniot data.

In Table 3 are listed the results of recent ultrasonic and high pressure X-ray measurements of the bulk modulus of stishovite. The high pressure X-ray measurement of *Bassett and Barnett* [1970], *Olinger* [1976], and *Sato* [1976], and the ultrasonic measurement of *Chung* [1974] all have determined a bulk modulus close to the 3.00 Mbar trial value. Obviously, these measurements best agree with the Hugoniot data. The ultrasonic measurement of *Mizutani et al.* [1972] and the high pressure X-ray measurement of *Liu et al.* [1974] both determined the modulus to be approximately 3.50 Mbar. *Liebermann et al.* [1976], who used a technique very similar to that of *Mizutani et al.* [1976], determined the modulus to be close to 2.50 Mbar trial value. These latter measurements thus agree only in part with the shock compression data.

TABLE 3. Experimentally Determined Bulk Moduli for Stishovite

Source	K_o,Mbar	K_o', (assumed or estimated)
Bassett and Barnett [1970]	2.92	0
(high pressure X-ray diffraction)	2.62	8
Mizutani et al. [1972] (ultrasonic velocities)	3.46 ± 0.24	
Chung [1974] (ultrasonic velocities in mixtures)	2.93	
Liu et al.[1974]	3.63 ± 0.07	2
(high pressure X-ray diffraction)	3.19 ± 0.05	8
Olinger [1976] (high pressure X-ray diffraction)	2.88 ± 0.13	6 ± 1
Sato [1976]	3.01	0
(high pressure X-ray diffraction)	2.81	5
Liebermann et al. [1977] (ultrasonic velocities)	2.50 ± 0.30	

Acknowledgments. I wish to thank, once again, Joseph Fritz who aided me in understanding the concepts and assumptions incorporated in this paper. I also want to thank Bob Liebermann, Ted Ringwood, Alan Major, and Yoshiko Sato for sending me their results prior to their being published. This work was supported by the U. S. Energy Research and Development Administration.

REFERENCES

Al'tshuler, L. V., and R. F. Trunin, and G. V. Simakov, Shock compression of periclase and quartz and the composition of the lower mantle, *Izv. Acad. Sci. USSR, Phys. Solid Earth, 10, Engl. Transl.*, 657-660, 1965.

Bassett, W. A. and J. D. Barnett, Isothermal compression of stishovite and coesite up to 85 kilobars at room temperature by X-ray diffraction, *Phys. Earth Planet. Interiors, 3*, 54-60, 1970.

Chung, D. H., General relationship among sound speeds, *Phys. Earth Planet. Interiors, 8*, 113-120, 1974.

Holm, J. L., O. J. Kleppa, and E. F. Westrum, Jr., Thermodynamics of polymorphic transformations in silica. Thermal properties from 5 to 1070 K and pressure-temperature stability fields for coesite and stishovite, *Geochim. Cosmochim. Acta 31*, 2289-2307, 1967.

Ito, H., K. Kawada, and S. I. Akimoto, Thermal expansion of stishovite, *Phys. Earth Planet. Interiors, 8*, 277-281, 1974.

Liu, L., W. A. Bassett, and T. Takahashi, Effect of pressure on the lattice parameters of stishovite, *J. Geophys. Res., 74*, 4317-4328, 1969.

Liebermann, R. C., A. E. Ringwood, and A. Major, Elesticity of polycrystalline stishovite, *Earth Planet. Sci. Lett.*, 1977.

McQueen, R. G., J. N. Fritz, and S. P. Marsh, On the equation of state of stishovite, *J. Geophys. Res., 68*, 2319-2322, 1963.

McQueen, R. G., S. P. Marsh, and J. N. Fritz, Hugoniot equation of state of twelve rocks, *J. Geophys. Res., 72*, 4999-5036, 1967.

Mizutani, H., Y. Hamano, and S. Akimoto, Elastic-wave velocities of polycrystalline stishovite, *J. Geophys. Res., 77*, 3744-3749, 1972.

Olinger, B., The compression of stishovite, *J. Geophys. Res., 81*, in press, 1976.

Olinger, B. and P. M. Halleck, The compression of α-quartz, *J. Geophys. Res., 81*, in press, 1976.

Sato, Y., Pressure-volume relationship of stishovite under hydrostatic compression, a preprint, 1976.

Trunin, R. F., M. A. Podurets, and G. V. Simakov, Compression of porous quartz by strong shock waves, *Izv. Acad. Sci. USSR, Phys. Solid Earth, Engl. Transl. 1*, 8-12, 1970.

Trunin, R. G., G. V. Simakov, M. A. Podurets, B. N. Moeseyev, and L. V. Popov, Dynamic compressibility of quartz and quartzite at high pressure, *Izv. Acad. Sci. USSR, Phys. Solid Earth, Engl. Transl. 2*, 102-106, 1971.

Wackerle, J., Shock-wave compression of quartz, *J. Applied Phys., 33*, 922-937, 1962.

Weaver, J. S., T. Takahashi, W. A. Bassett, Thermal expansion of stishovite, *EOS (Trans. Am. Geophys. Union), 54*, 475, 1973.

SOME COMMENTS ON THE ELASTICITY OF STISHOVITE AS DETERMINED BY ULTRASONIC AND HIGH PRESSURE X-RAY DIFFRACTION TECHNIQUES

R. C. LIEBERMANN and A. E. RINGWOOD
Research School of Earth Sciences
Australian National University
Canberra, A.C.T. 2600, Australia

Abstract

Recent experiments in our laboratory using ultrasonic techniques and in Olinger's and Sato's laboratories using high pressure X-ray diffraction techniques indicate that significant revision to previous determinations of the bulk modulus of stishovite is necessary. The combined evidence from the data o these two experimental methods suggests that the most probable values of the bulk modulus and its pressure derivative evaluated at zero pressure for stishovite are K_o = 2.7-2.8 Mbar and $K_o' \geq 5$. These new stishovite data would make it impossible to satisfy the elasticity-density data of the lower mantle by usin an oxide mixture with either olivine or pyroxene stoichiometry.

I. DISCUSSION

Polycrystalline specimens of stishovite have been hot-pressed in our laboratory by subjecting samples of natural quart powder to pressures of 120 kbar at T = 900°C for about 30 min. These cylindrical specimens are 2 mm in diameter and 0.9-1.4 mm long, and they have a grain size less than 10 μm and bulk densities of about 98% of the X-ray density. Specimens were polished for velocity measurements in both the axial (Z) and transverse (X) directions; this polishing revealed the existence of thin laminar cracks oriented perpendicular to the cylinder axis, but no other crack orientations were observed.

We have measured the compressional (v_p) and shear (v_s) velocities in these polycrystalline specimens as a function of pressure to 10 kbar at room temperature using ultrasonic techniques described elsewhere [*Liebermann et al.*, 1975, 1976]. The velocity-pressure data for specimen S27 are plotted in Figure 1. Five acoustic modes were studied, including all possible propagation ($\tilde{k}$) and polarization ($\tilde{s}$) directions in the cylindrical specimens.

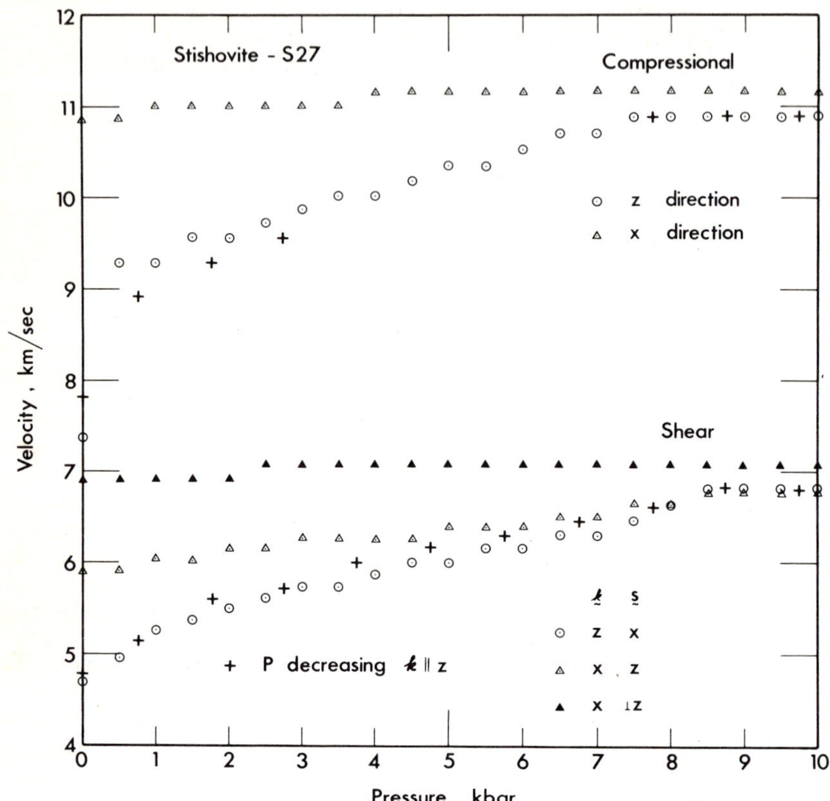

Fig. 1. Compressional and shear velocities in polycrystalline stishovite specimen S27 to P = 10 kbar. Z-direction is the cylindrical axis of uniaxial hot-pressing; X-direction is perpendicular to Z. For the shear modes, the propagation (k) and (s) directions are indicated.

Both v_p and v_s for various modes converge to respective common values at $P > 8$ kbar, demonstrating the intrinsic isotropy of the polycrystalline specimen and indicating that the anisotropy at low pressures is entirely due to the laminar cracks that close on application of pressure. The velocity data at $P = 10$ kbar (corrected for length change upon compression) and the X-ray density ($\rho = 4.287$ g/cm^3) give the following values of the isotropic elastic properties of stishovite:

Compressional velocity	v_p = 11.0 ± 0.2 km/sec
Shear velocity	v_s = 6.9 ± 0.3 km/sec
Bulk sound velocity	v_ϕ = 7.6 ± 0.5 km/sec
Poisson's ratio	σ_S = 0.18 ± 0.05
Bulk modulus	K_S = 2.5 ± 0.3 Mbar
Shear modulus	μ = 2.0 ± 0.2 Mbar

A more complete account of these new experimental results has been given elsewhere [*Liebermann et al.*, 1976].

Our velocity data are most directly comparable with those from the ultrasonic experiment of *Mizutani et al.* [1972]: the compressional velocities in the two independent studies are identical (11.0 km/sec), but our shear velocity (6.9 km/sec) is very much higher than theirs (5.5 km/sec). We have discussed the possible causes for the shear velocity discrepancy with H. Mizutani and S. Akimoto and have concluded that the most likely cause is the possible presence of crack orientations in their specimen which affect v_s but not v_p.

The only other previous ultrasonic data on the elasticity of stishovite came from a study by *Chung* [1974] who measured the elastic wave velocities of a hot-pressed mixture (80/20 volume ratio) of stishovite with AgCl. On the basis of certain assumptions about the behavior of composite materials, *Chung* [1974] extrapolated the mixture data to estimate the elastic moduli of stishovite as K_S = 2.93 Mbar and μ = 1.69 Mbar. There is, however, no experimental or theoretical justification for the assumptions made [*Liebermann et al.*, 1976; *Watt et al.*, 1976] and alternative models indicate that the uncertainties in such estimates may be greater than ±1 Mbar. These calculations demonstrate the inherent weakness of the mixture technique for two materials whose shear moduli are very different (such as stishovite and AgCl), and we conclude that Chung's data do not provide a meaningful estimate of the elastic properties of stishovite.

The previous static compression data for stishovite of *Bassett and Barnett* [1970] and *Liu et al.* [1974] were obtained in solid-media, high-pressure devices and implied zero-pressure bulk moduli for stishovite of K_o = 3.0 ± 0.3 Mbar and K_o = 3.44 ± 0.27 Mbar. It is now well known that substantial nonhydrostatic stresses are present in the sample region of such apparati that make it impossible to determine the correct pressure-volume relationship even when NaCl is used as the solid-pressure medium [*e.g.*, *Olinger*, 1976; *Sato*, 1976]. After our ultrasonic experiments were completed, the results of the two new compression studies under hydrostatic conditions became available, both of which indicate that stishovite is considerably more compressible in the pressure range to 110 kbar than the earlier work had suggested. Both *Sato's* [1976] and *Olinger's* [1976] static data imply that K_o = 2.8 Mbar if K_o' = 5-6 (a value which is in the range of experimental determinations for single-crystal specimens of the isostructural rutile oxides, GeO_2, TiO_2 and SnO_2).

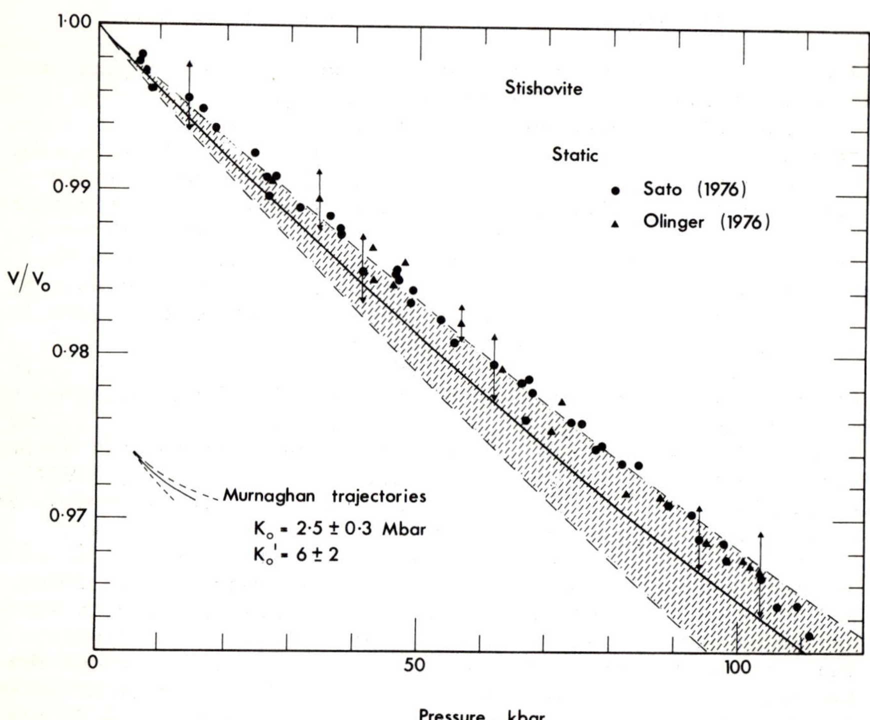

Fig. 2. Volume compression data for stishovite from static isothermal studies to P = 110 kbar in hydrostatic media [Sato, 1976; Olinger, 1976]. Vertical bars on some of the data points are the experimental uncertainties in volume determination given by the original authors. Both sets of data have been plotted using Decker's [1971] pressure scale for NaCl. Also plotted are the band of Murnaghan P-V trajectories derived from the ultrasonic bulk modulus K_o *= 2.5 ± 0.3 Mbar and assuming* K_o' *= 6 ± 2.*

A Murnaghan *P-V* trajectory based on our ultrasonic data (K_o = 2.5 ± 0.3 Mbar, K_o' = 6 ± 2) and the hydrostatic compression data of Sato and Olinger are plotted together in Figure 2. The static data and the extrapolated curve based on our ultrasonic data agree within the mutual uncertainties over the entire pressure range to 110 kbar. Neither our ultrasonic data nor the compression data alone permit a unique determination of K_o and K_o'. The combined evidence from the data of the two experimental techniques, however, suggests that the most probable values of the bulk modulus and its pressure derivative evaluated at zero pressure for stishovite are K_o = 2.7-2.8 Mbar and $K_o' \geq 5$.

TABLE 1. Elasticity of Various Model Lower-Mantle Mineral Assemblages at Zero Pressure[a]

Stoichiometry	Mineralogy[a]	ρ, g/cm^3	K_S, Mbar	ϕ, (km/sec)^2
Olivine (Mg_2SiO_4)	Spinel	3.56	2.1	59
	Periclase+Ilmenite	3.75	1.9	51
	Periclase+Stishovite	3.85	2.0	52
	Periclase+Perovskite	3.93	2.4	61
Pyroxene ($MgSiO_3$)	Ilmenite	3.82	2.1	55
	Periclase+Stishovite	3.97	2.2	55
	Perovskite	4.08	2.8	69
Earth models[b]	?	4.15	2.7	66

[a] Calculations performed using the Voigt-Reuss-Hill procedure and elastic moduli measured or predicted for the various minerals [*see Liebermann et al.*, 1976, for references], including K_S = 2.8 Mbar as upper limit for stishovite. Uncertainties in the moduli of the mixed-phase assemblages due to the Voigt-Ruess bounds are less than 0.1 Mbar.

[b] As extrapolated to P = 1 bar, T = 25°C for the lower-mantle (depth 670-2885 km) regions of the B1 [*Jordan and Anderson,* 1974] and PEM [*Dziewonski et al.*, 1975] models by *Davies and Dziewonski* [1975].

Many previous authors have attempted to draw inferences about the chemical composition of the earth's lower mantle by comparing the elastic properties of various model mineral assemblages with the properties of the mantle from seismic data extrapolated to zero pressure and room temperature. We may illustrate the impact of the new stishovite data on such discussions by referring to the diagrams of seismic parameter (ϕ) versus density (ρ) popularized by *Anderson and Jordan* [1970]. For example, *Davies and Dziewonski* [1975] have shown that the elastic properties of the lower mantle (depth interval = 670-2885 km) for both the B1 model of *Jordan and Anderson* [1974] and the PEM model of *Dziewonski et al.* [1975] extrapolate to $\phi = 66\ (\text{km/sec})^2$ and $\rho = 4.15\ \text{g/cm}^3$ at $P = 1$ bar, $T = 25°\text{C}$. From Table 1 we can see that the periclase + stishovite mixtures with either olivine or pyroxene stoichiometry have ϕ and ρ values that are much too low to satisfy the earth models (addition of Fe would improve the density agreement but enhance the ϕ contrast). From Table 1, it appears that only the mineral assemblages that contain $MgSiO_3$-perovskite have sufficiently high densities and seismic parameters to be compatible with the earth models.

Acknowledgments. We have profited from discussions with S. Akimoto, G. F. Davies, H. Mizutani, B. Olinger, and Y. Sato, and we also thank J. P. Watt, G. F. Davies, R. J. O'Connell, B. Olinger, and Y. Sato for the opportunity to examine and discuss their papers in advance of publication.

REFERENCES

Anderson, D. L., and T. H. Jordan, The composition of the lower mantle, *Phys. Earth Planet. Interiors, 3*, 23-35, 1970.

Bassett, W. A., and J. D. Barnett, Isothermal compression of stishovite and coesite up to 85 kilobars at room temperature by X-ray diffraction, *Phys. Earth Planet. Interiors, 3*, 54-60, 1970.

Chung, D. H., General relationships among sound speeds. 1. New experimental information, *Phys. Earth Planet. Interiors, 8*, 113-120, 1974.

Davies, G. F., and A. M. Dziewonski, Homogeneity and constitution of the Earth's lower mantle and core, *Phys. Earth Planet. Interiors, 10*, 336-343, 1975.

Decker, D. L., High-pressure equation of state for NaCl, KCl, and CsCl, *J. Appl. Phys., 42*, 3239-3244, 1971.

Dziewonski, A. M., A. L. Hales, and E. R. Lapwood, Parametrically simple Earth models consistent with geophysical data, *Phys. Earth Planet. Interiors, 10*, 12-48, 1975.

Jordan, T. H., and D. L. Anderson, Earth structure from free oscillations and travel times, *Geophys. J. Roy. astr. Soc, 36*, 411-459, 1974.

Liebermann, R. C., A. E. Ringwood, D. J. Mayson, and A. Major, Hot-pressing of polycrystalline aggregates at very high pressure for ultrasonic measurements, in *Proc. 4th International Conf. High Pressure*, edited by J. Osugi, pp. 495-502, Physico-Chemical Society of Japan, Kyoto, 1975.

Liebermann, R. C., A. E. Ringwood, and A. Major, Elasticity of polycrystalline stishovite, *Earth Planet. Sci. Letters, 32*, in press, 1976.

Liu, L., W. A. Bassett, and T. Takahashi, Effect of pressure on the lattice parameters of stishovite, *J. Geophys. Res., 79*, 1160-1164, 1974.

Mizutani, H., Y. Hamano, and S. I. Akimoto, Elastic-wave velocities of polycrystalline stishovite, *J. Geophys. Res., 77*, 3744-3749, 1972.

Olinger, B., The compression of stishovite, *J. Geophys. Res., 81*, in press, 1976.

Sato, Y., Pressure-volume relationship of stishovite under hydrostatic compression, *Earth Planet. Sci. Letters, 32*, in press, 1976.

Watt, J. P., G. F. Davies, and R. J. O'Connell, The elastic properties of composite materials, *Rev. Geophys. Space Phys., 14*, in press, 1976.

bcc TRANSITION METALS UNDER PRESSURE: RESULTS FROM ULTRASONIC INTERFEROMETRY AND DIAMOND-CELL EXPERIMENTS

K. W. KATAHARA, M. H. MANGHNANI, L. C. MING
Hawaii Institute of Geophysics
University of Hawaii
Honolulu, Hawaii 96822

E. S. FISHER
Argonne National Laboratory
Argonne, Illinois 60439

Abstract

Hydrostatic pressure derivatives of the single-crystal elastic moduli, dC_{ij}/dP, have been measured ultrasonically for bcc niobium-molybdenum and tantalum-tungsten solid solutions. The composition dependence of various electronic properties of these alloys is known to be reasonably well approximated by a rigid-electron-band-filling model where *e/a*, the electron per atom ratio, is the primary parameter. The results indicate that the elastic moduli and their pressure derivatives may also be calculated in such a model. In particular, the dC_{ij}/dP show relatively sharp increases at *e/a* compositions of 5.4 for Nb-Mo and 5.7 for Ta-W. Both compositions correspond to changes in Fermi surface topology, as deduced from existing band calculations using the rigid band assumption. The results are discussed in the light of related electronic properties and possible geophysical implications.

A comparison is also made between ultrasonic results and X-ray diffraction data for Nb. Using the diamond-anvil pressure cell, the compression of Nb was determined by X-ray diffraction up to 55 kbar in a liquid medium under purely hydrostatic conditions, and up to 175 kbar in a solid medium under nonhydrostatic conditions. The data obtained under hydrostatic conditions agree well with the ultrasonic equation-of-state and shock-wave data. The nonhydrostatic results are offset toward higher volumes, probably because of uniaxial stress in the sample.

I. INTRODUCTION

Band structure calculations have recently been used to study the elastic properties of geophysically important materials [*Thomsen*, 1975; *Bukowinski and Knopoff*, 1976, 1977]. *Bukowinski and Knopoff* [1976, 1977] have evaluated the density and bulk modulus of iron at core pressures. It would be of interest to compute the shear elastic moduli of highly compressed iron in order to compare them with seismic values for the inner core as obtained, for instance, by *Gilbert and Dziewonski* [1975] and *Anderson and Hart* [1976]. Unfortunately, efficient theoretical methods have not yet been developed for calculating shear moduli in transition metals, and it is only recently that attempts have been made in this direction [*e.g.*, *Peter et al.*, 1974; *Oli and Animalu*, 1976]. The research reported here may be regarded as an experimental approach to this problem. The elastic moduli, C_{ij}, and their pressure derivatives, dC_{ij}/dP, have been measured ultrasonically for single crystal Nb-Mo and Ta-W alloys. The results for these materials are in some respects ideal for testing and refining any theoretical methods that might be applied to calculating the shear moduli of iron at core pressures.

The electronic band structures of the transition metals are characterized by narrow *d*-electron bands crossing and hybridizing with a broader, nearly free electron band, and the properties of these metals are largely dominated by the behavior of the electrons in the *d*-bands. The bcc transition metals of groups VB (V, Nb, Ta) and VIB (Cr, Mo, W) have particularly strong band structure contributions to the cohesive energy [*Pettifor*, 1970], the elastic moduli [*Fisher*, 1975], and the pressure derivative of the bulk modulus [*Fisher et al.*, 1975]. Thus, any theoretical models that can successfully explain the present results might be expected to be applicable to other transition metals where *d*-band effects are not as pronounced.

Since the bcc transition metals form randomly disordered solid solutions over a wide range of compositions, the population of electrons in the bands can be varied at constant crystal structure and slowly varying atomic volume. In this report, the alloy compositions are characterized by the *e/a* ratio, defined as the average number of conduction electrons per atom, which is 5 and 6 for the group VB and VIB elements, respectively. At atmospheric pressure and room temperature, these elements form bcc alloys with each other and with elements of groups IVB and VIIB (and VIII to some extent) over the *e/a* range shown in Table 1 [*Pearson*, 1958, 1966]. The stability of the bcc structure over a wide *e/a* range becomes useful when it is recognized that the rigid band model works reasonably well for these metals, as shown by comparisons between experimental and theoretical densities of states [*McMillan*, 1968; *Pickett and Allen*, 1974]. The rigid band model is here intended not to include the assumption, which is known to fail [*Posternak et al.*, 1975], that the

TABLE 1. Structure of Alloys Based on Transition Elements of Groups IV, V, and VI

e/a Range	Structure
4.0 < e/a < 4.1	hcp
4.1 < e/a < 4.2	hexagonal (omega phase)
4.2 < e/a < 6.3	bcc
6.3 < e/a	hcp

bands remain rigid when the lattice is deformed. Rather it is only assumed that within the Brillouin zone of the lattice, which may be either deformed or undeformed, the density of states and the shapes of the bands do not change very much as an element is alloyed with neighboring elements. Thus, by varying the alloy composition, and hence changing e/a and the Fermi energy, E_F, the properties of different energy regions of the electronic band structure can be experimentally examined. A theoretical calculation of the strain derivatives of the electronic energy levels would then need to be done for only one bcc element of group VB or VIB, and the results for its solid solutions could be obtained by adding or subtracting electrons. Any valid calculational scheme should thus be capable of at least qualitatively reproducing the present results.

Some of the more interesting aspects of these results and their connection with the electronic band structure are discussed in the next section. A more complete discussion, together with a description of the samples and the experimental details, will be published elsewhere [*Katahara et al.*, in preparation].

The last section of this paper is concerned with the accuracy of the present ultrasonic measurements and of diamond-anvil compression studies. Equations of state obtained by different methods for the pure elements are compared and the sources of disagreement are discussed.

II. RESULTS AND DISCUSSION FOR THE SOLID SOLUTIONS

The adiabatic single crystal C_{ij} and dC_{ij}/dP were measured ultrasonically for 6 Nb-Mo alloys, 5 Ta-W alloys, and the pure elements Mo and W. The experimental methods have been described in a previous paper [*Katahara et al.*, 1976] in which data on the other two endmembers, Nb and Ta, were reported.

Figures 1 and 2 show the present C_{ij} values for the alloys, the previous results of *Hubbell and Brotzen* [1972] for the Nb-Mo

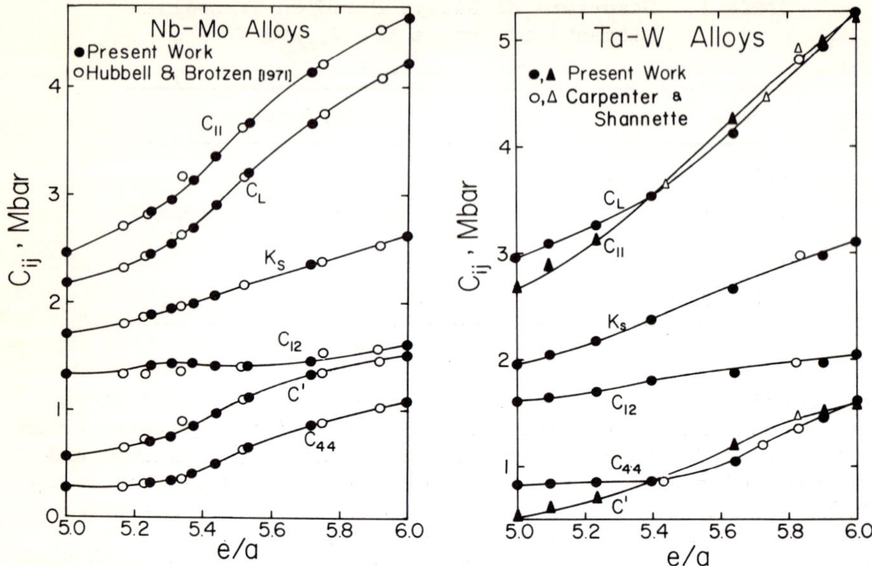

Fig. 1. Adiabatic single-crystal elastic moduli of Nb-Mo alloys versus e/a from present work and from Hubbell and Brotzen 1972 . K_S is the adiabatic bulk modulus, $C_L = (C_{11} + C_{12})/2 + C_{44}$ and $C' = (C_{11} - C_{12})/2$.

Fig. 2. Adiabatic single-crystal elastic moduli of Ta-W alloys versus e/a from present work and from unpublished research of M. L. Carpenter and G. W. Shannette.

system, and the unpublished results of *M. L. Carpenter and G. W. Shannette* [personal communication, 1975] on some rather inhomogeneous Ta-W solid solutions. $C_L = (C_{11} + C_{12})/2 + C_{44}$ is the effective elastic coefficient for longitudinal waves propagating along [110], K_S is the adiabatic bulk modulus, and $C' = (C_{11} - C_{12})/2$ is the effective elastic coefficient for shear waves propagating along [110] and polarized along [1$\bar{1}$0]. Note that the bulk modulus increases nearly linearly with *e/a* (as does the density [see *Pearson*, 1966, and references given there]). The shear moduli, C_{44} and C', and the longitudinal moduli, C_{11} and C_L, show more structure in their *e/a* dependence, and this is more so for C_{44} and C_L than for C' and C_{11}. Since longitudinal strains include both shear and volume components, it is clear that it is the response of the electrons to shear that is causing the marked variation of these moduli with *e/a*; thus, the main emphasis in this report is on the shear moduli. These quantities increase slowly at first for low *e/a*, then they begin to increase more rapidly with the curves reaching inflection points at $e/a \sim 5.4$ for the Nb-Mo alloys and at $e/a \sim 5.7$ for the Ta-W alloys.

As discussed extensively by *Fisher* [1975], simple, nearly free electron metals crystallizing in the bcc structure generally have C_{44} much greater than C'. In Figure 1 we see that the opposite occurs for the Nb-Mo alloys, indicating strong band structure effects on the shear moduli. On the other hand, C_{44} is either greater or slightly less than C' in Figure 2. This is interpreted to mean that the Ta-W alloys are in a sense more free-electron-like than the Nb-Mo alloys. In fact, the 5*d*-bands in Ta are broader than the 4*d*-bands in Nb [*Mattheiss*, 1970], indicating less localization of the conduction electrons in the Ta-W alloys.

Figures 3 and 4 show the dC_{ij}/dP for the Nb-Mo and Ta-W alloys, respectively. The dK_S/dP values, like the bulk moduli themselves, vary relatively smoothly with e/a, when compared with the pressure derivatives of the shear moduli. In both alloy systems, dC_{44}/dP and dC'/dP at first increase slowly near $e/a \sim 5$, then increase more rapidly, after which the curves level out rather abruptly. These trends are enhanced in plots of $\pi_{ij} = d \ln C_{ij}/d \ln V$ versus e/a in Figures 5 and 6. The pronounced

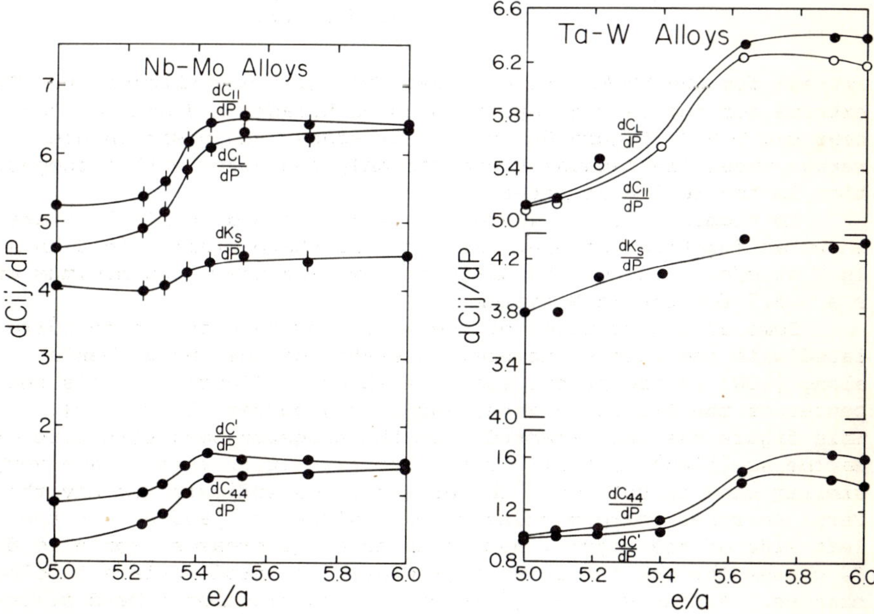

Fig. 3. Pressure derivatives of the elastic moduli of Nb-Mo alloys versus e/a.

Fig. 4. Pressure derivatives of the elastic moduli of Ta-W alloys versus e/a.

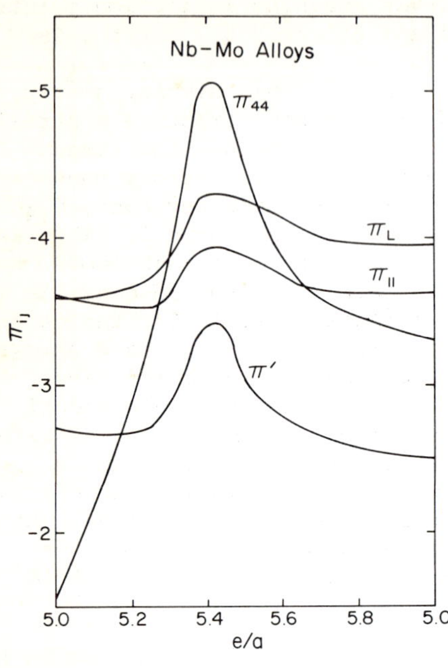

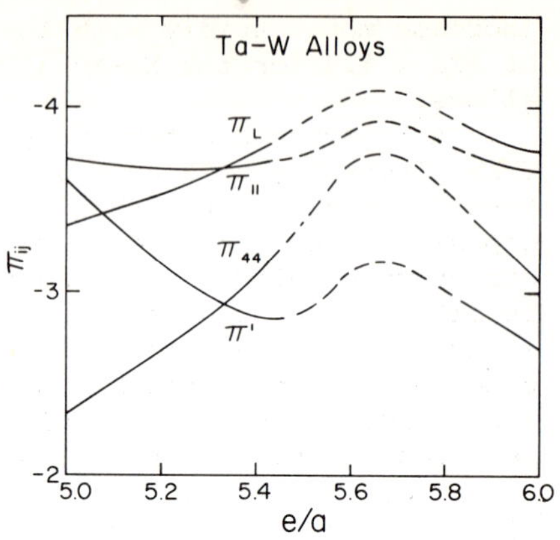

Fig. 6. π_{ij} *versus e/a for Ta-W alloys.*

Fig. 5. π_{ij} *versus e/a for Nb-Mo alloys.* ($\pi_{ij} = d \ln C_{ij}/d \ln V$).

extrema for the Nb-Mo alloys appear at $e/a \sim 5.4$ (Figure 5). The extrema for the Ta-W alloys seem to be broader and are located near $e/a \sim 5.7$ (Figure 6), but the broadness and position are rather uncertain because there was only one alloy with a composition in the region of interest.

In Figure 7, ln A (where $A = C_{44}/C'$) and d ln A/d ln V are shown as functions of e/a. We note the sharp minimum in d ln A/d ln V at $e/a = 5.4$ for the Nb-Mo system, and the broad minimum at $e/a \sim 5.7$ for the Ta-W system.

Some of the features of the preceding results can be correlated with the band structure. A sketch of the energy bands along [100] in reciprocal space is shown in Figure 8. Γ is the center of the Brillouin zone, and H is a corner along [100]. This figure has been adapted from the tungsten band structure of *Mattheiss* [1965], but all of the bcc transition metals have very similar band structures. The broken lines indicate roughly the Fermi energies corresponding to e/a values of 5 and 6. On the left side of the figure, note that as e/a increases from 5 to 6, E_F passes through the level Γ'_{25} where the Fermi surface topology changes. *Mattheiss* [1970] has calculated electronic band structures for Nb and Ta. We assume a rigid band model, as discussed earlier, in which these band structures and the associated densities of states also hold for alloys based on Nb and Ta. The variation of E_F with e/a then follows directly from the integrated densities of states given in Figure 3 of *Mattheiss* [1970].

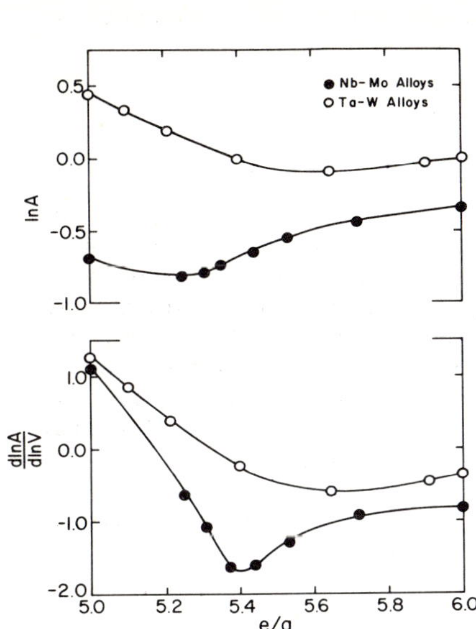

Fig. 7. ln A and d ln A/d ln V versus e/a for Nb-Mo and Ta-W alloys. $A \equiv C_{44}/c'$.

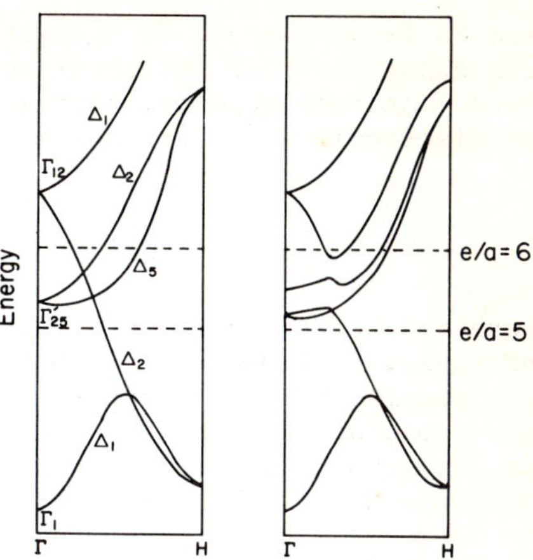

Fig. 8. Electronic energy bands along [100] in the Brillouin zone for bcc transition metals, after Mattheiss [1965]. Γ is the zone center and H is a corner. The dashed lines indicate the approximate positions of the Fermi energy at e/a = 5 and 6. The bands are shown without spin-orbit splitting on the left and with spin-orbit splitting on the right.

The compositions at which $E_F = \Gamma'_{25}$ are thus found to be *e/a* ∿ 5.4 for the Nb-Mo alloys and *e/a* ∿ 5.7 for the Ta-W alloys. These are the *e/a* values at which the inflection points in the C_{ij} and the extrema in the π_{ij} are observed. It thus appears that the energy levels near Γ'_{25} contribute strongly to the shear moduli and their pressure derivatives. The relative broadness of the extrema for the Ta-W π_{ij} and d ln *A*/d ln *V* may be partly due to spin-orbit splitting, which is much stronger in the 5*d*-elements and is illustrated on the right-hand side of Figure 8. The point Γ'_{25}, which is triply degenerate, is split, and the result may be to smear out the effects on the shear moduli.

Although it is possible to qualitatively correlate the results of this study with the band structure, a fundamental explanation of how shear strains affect the electronic energy levels requires the framework of a theoretical model. It is hoped that the present results will both stimulate and provide a good test for further theoretical work in this direction.

The present study also illustrates a point possibly applicable to the earth's core: relatively small changes in *e/a* can lead to large changes in transition metal properties. This has been shown here for the shear moduli and their pressure derivatives. Another quantity that may be geophysically important is the electronic Grüneisen parameter,

$$\gamma_e = \frac{d \ln N(E_F)}{d \ln V}$$

where $N(E_F)$ is the density of states at the Fermi energy. *Smith and Finlayson* [1976] have demonstrated that for the bcc transition metals, γ_e varies appreciably with *e/a*. One might therefore speculate that these properties will also vary rapidly with *e/a* for iron solid solutions in the core. If, for instance, a small amount of sulfur were dissolved in iron at core pressures, the iron conduction bands would presumably be distorted slightly, but the major effect might well be due to the creation of sulfur 3*p*-bands below the conduction bands. The *e/a* ratio would decrease, and the shear moduli, γ_e, and other properties would be substantially different from those calculated for pure iron.

III. EQUATIONS OF STATE

A problem that frequently arises in equation of state studies is that *P-V* measurements by different methods, or by different workers using the same method, do not agree. Figures 9 and 10 illustrate this point for the metals of groups VB (Nb and Ta) and VIB (Mo and W), respectively. The curves shown are (1) the shock Hugoniots of *McQueen et al.* [1970] centered at 20°C and zero pressure, (2) 20°C isotherms obtained by *McQueen et al.* [1970] from the Hugoniots, and (3) 25°C isotherms extrapolated from the ultrasonic measurements of the present study and of *Katahara et al.* [1976] according to the Birch-Murnaghan equation [*Birch*, 1938, 1947]:

$$P = \frac{3}{2} K_T \left[\left(\frac{V_o}{V}\right)^{7/3} - \left(\frac{V_o}{V}\right)^{5/3} \right] \left\{ 1 - \frac{3}{4} (4 - K_T') \cdot \left[\left(\frac{V_o}{V}\right)^{2/3} - 1 \right] + \ldots \right\} \quad (1)$$

Here K_T, the isothermal bulk modulus, and $K_T' = (\partial K_T/\partial P)_T$, are evaluated at zero pressure and have been derived from the corresponding adiabatic values, measured ultrasonically, according to equation 10.15 of *Thurston* [1965, p. 1331]. Terms in (1) involving second and higher order derivatives of the bulk modulus

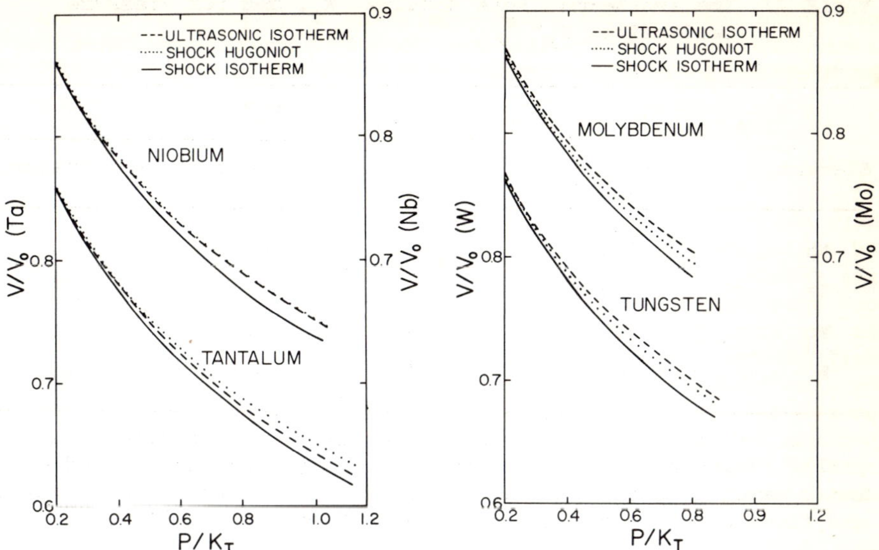

Fig. 9. V/V_o *versus* P/K_T *for Nb and Ta. The ultrasonic curves are from Katahara et al. [1976]. The shock Hugoniots and isotherms are from McQueen et al. [1970]. The ultrasonic* K_T *values in Table 2 were used to normalize the pressures.*

Fig. 10. V/V_o *versus* P/K_T *for Mo and W. The ultrasonic curves are from the present results. The shock Hugoniots are from McQueen et al. [1970]. As in Figure 9, the pressures have been divided by the ultrasonic* K_T *values given in Table 2.*

have been neglected. Table 2 shows the ultrasonic results for K_T and K_T' and also gives K_T' values obtained by fitting (1) to the isotherms of *McQueen et al.* [1970] and assuming the ultrasonic K_T values. The fourth column shows the percentage difference between the two K_T' values.

Note that the ultrasonic isotherms give uniformly higher volumes than the shock compression isotherms. This is not totally unexpected from an experimental point of view since errors in the present ultrasonic measurements due to densification of the pressure medium [*Katahara et al.*, 1976] and to transducer and bond effects [*McSkimin and Andreatch*, 1962; *Davies and O'Connell*, 1977] tend to give systematically high values for K_T'. The K_T' differences in Table 2 are somewhat larger than rough estimates for the maximum errors (∿ 5%) in the ultrasonic measurements. These errors may well have been underestimated, but it is also possible that some of the difference is due to faulty assumptions in deriving the shock isotherms from the Hugoniots,

TABLE 2. The Isothermal Bulk Modulus, K_T, and its Pressure Derivative, K_T' [a]

Element	K_T, Mbar	K_T' Ultrasonic	K_T' Shock Wave	% Difference
Nb	1.690	4.1	3.7	9
Ta	1.942	3.8	3.6	5
Mo	2.610	4.7	3.9	15
W	3.084	4.5	3.9	13

a. K_T and ultrasonic K_T' values are from *Katahara et al.* [1976] for Nb and Ta, and from the present study for Mo and W. Shock wave K_T' were obtained by assuming the ultrasonic K_T values and fitting (1) to the isotherms of *McQueen et al.* [1970].

or to neglect of higher order terms in (1). There is additional uncertainty arising from the choice of (1) for extrapolation of the ultrasonic data since other equations of state give different results. In the high-pressure region, the Birch-Murnaghan equation usually predicts higher volumes than are obtained from shock-wave experiments [*Ahrens and Thomsen*, 1972]. On the other hand, an equation of state based on a Lagrangian finite strain formalism predicts smaller volumes than the Birch-Murnaghan equation [*e.g.*, see *Thomsen*, 1970; *Davies*, 1973], but it does not necessarily give better agreement with the shock-wave data [*Ahrens and Thomsen*, 1972].

Regardless of the cause of disagreement at high pressures, the shock compression and ultrasonic isotherms agree quite well at lower pressures and indeed can provide a good check on other types of experiments. With this in mind, the compression of Nb in a diamond-anvil pressure cell has been studied by the X-ray diffraction technique under both hydrostatic and nonhydrostatic conditions in order to determine the effects of nonhydrostatic stresses.

In the present study, the Bassett-type diamond cell [*Bassett et al.*, 1967] was used. The hydrostatic *P-V* measurements were carried out up to 55 kbar by gasketing the sample immersed in a 4:1 methanol:ethanol mixture between the anvils and by using the ruby fluorescence pressure calibration method [*Piermarini et al.*, 1973, 1975]. In addition, some hydrostatic measurements used NaCl as the pressure indicator instead of ruby. In the nonhydrostatic experiments, which extended up to 175 kbar, a fine mixture of Nb and NaCl (1:4 in volume) was gasketed between the diamond anvils without any liquid pressure medium. When NaCl was used, the isothermal compression curve of *Weaver et al.* [1970] determined the pressure.

The results are shown in Figure 11, together with an extrapolation of the ultrasonic results [*Katahara et al.*, 1976]. In this pressure region, the shock isotherm of *McQueen et al.* [1970] falls very slightly below the ultrasonic curve. Equation (1) has been fitted both to the hydrostatic and to the nonhydrostatic data by a least-squares procedure in which K_T' has been assumed to be equal to the ultrasonic value. The K_T values obtained are given in Table 3, and the corresponding curves are also shown in Figure 11.

Given the scatter and the limited pressure range for the hydrostatic data, about all that can be said is that they are consistent with both the ultrasonic and the shock-wave measurements. On the other hand, the nonhydrostatic points tend to fall above the ultrasonic curve (and hence the shock compression curve also). In Table 3 it can be seen that the nonhydrostatic K_T is higher than the ultrasonic K_T by three standard errors.

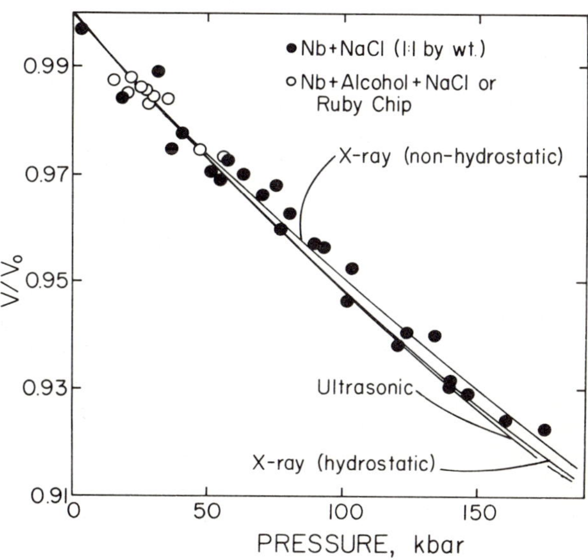

Fig. 11. V/V_0 versus pressure for Nb. Open and solid circles are diamond-anvil X-ray data for hydrostatic and nonhydrostatic conditions, respectively. Solid lines are equation (1) with K_T and K_T' as given in Table 3. For the X-ray data, $K_T' = 4.08$ has been assumed from the ultrasonic results, and K_T has been obtained from a least-squares fit.

TABLE 3. Comparison of K_T and K_T' of Nb from Different Methods

K_T, Mbar	K_T'	Equation of State Used	Experimental Method	Reference
1.690	4.1		Ultrasonic	*Katahara et al.* [1976]
1.71 ± 0.07	4.1[a]	Birch-Murnaghan	X-ray (Hydrostatic)	Present study
1.79 ± 0.03	4.1[a]	Birch-Murnaghan	X-ray (Non-hydrostatic)	Present study

a. Assumed.

Pressure intensification effects as suggested by *Jamieson and Olinger* [1971] and *Sato et al.* [1973] would cause lower rather than higher observed volumes for the nonhydrostatic case. The discrepancy is more likely caused by the existence of a uniaxial stress in the sample parallel to the load axis of the cell. In a study of the strain configuration of different materials in an ungasketed diamond-anvil cell, *Kinsland* [1974] observed that strains in NaCl were virtually isotropic up to 200 kbar. On the other hand, he found that materials with greater yield strengths, such as MgO and Mg_2SiO_4 (forsterite), showed significantly higher elastic strains parallel to the load axis than perpendicular to it. This is shown schematically in Figure 12. Lattice planes parallel to the load axis have larger lattice spacings than planes perpendicular to the axis (*i.e.*, $a_{||} > a_{\perp}$). In cases where the incident X-ray beam is parallel to the load axis, only those lattice planes with large spacings participate in the observed Bragg diffraction. The volume inferred from the assumption that $V = a_{||}^3$ is thus larger than the true volume, $V = a_{||}^2 \cdot a_{\perp}$, at the pressure indicated by the NaCl standard. On the basis of hydrostatic measurements on Fe_2SiO_4 (spinel), *Wilburn* [1975] concluded that this kind of effect was present in the earlier ungasketed experiments of *Mao et al.* [1969] on the same substance. It is not unreasonable to suppose that the yield strength of Nb is large enough to cause the present nonhydrostatic data to be similarly offset toward higher volumes. This study, therefore, provides further evidence that nonhydrostatic equation of state experiments can be significantly in error, especially for materials with yield strengths and low compressibilities.

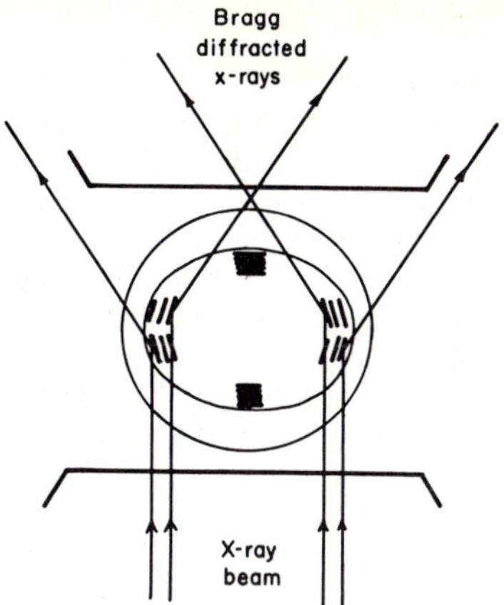

Fig. 12. Schematic diagram of anisotropic strain in the diamond cell after Kinsland [1974]. A circle in the unstressed sample is compressed into an ellipse. Lattice planes which are nearly parallel to the X-ray beam have larger lattice spacings than the planes which are more nearly perpendicular to the beam. Because of a limited aperture for observing diffracted X-rays, the larger lattice spacings are the ones seen experimentally and the calculated volumes are thus too large. Reproduced with author's permission.

Acknowledgments. We are indebted to Gary W. Shannette for the use of his Ta-W alloy crystals and for showing us his data before publication. Thanks are due M. S. T. Bukowinski, L. Thomsen and J. C. Jamieson for critically reviewing the paper. This research was carried out under the auspices of the U. S. Energy Research and Development Administration contract E(04-3)-235. This paper constitutes Hawaii Institute of Geophysics contribution No. 780.

REFERENCES

Ahrens, T. J., and L. Thomsen, Application of the fourth order anharmonic theory to the prediction of equations of state at high compressions and temperatures, *Phys. Earth Planet. Interiors, 5*, 282-294, 1972.

Anderson, D. L., and R. S. Hart, An earth model based on free oscillations and body waves, *J. Geophys. Res., 81,* 1461-1475, 1976.

Bassett, W. A., T. Takahashi, and P. W. Stook, X-ray diffraction and optical observations on crystalline solids up to 300 kbar, *Rev. Sci. Instrum., 38,* 37-42, 1967.

Birch, F., the effect of pressure upon the elastic properties of isotropic solids, according to Murnaghan's theory of finite strain, *J. Appl. Phys., 9,* 279-288, 1938.

Birch, F., Finite elastic strain of cubic crystals, *Phys. Rev., 71,* 809-824, 1947.

Bukowinski, M. S. T., and L. Knopoff, Electronic structure of iron and models of the earth's core, *Geophys. Res. Letters, 3,* 45-48, 1976.

Bukowinski, M. S. T., and L. Knopoff, Physics and chemistry of iron and potassium at lower mantle and core pressures, in *High-Pressure Research: Applications to Geophysics,* edited by M. H. Manghnani and S. Akimoto, Academic Press, New York, 367-387, 1977.

Davies, G. F., Quasi-harmonic finite strain equations of state of solids, *J. Phys. Chem. Solids, 34,* 1417-1429, 1973.

Davies, G. F., and R. J. O'Connell, Transducer and bond phase shifts in ultrasonics, and their effect on measured pressure derivatives, in *High Pressure Research: Applications to Geophysics,* edited by M. H. Manghnani and S. Akimoto, Academic Press, New York, 533-562, 1977.

Fisher, E. S., A review of solute effects on the elastic moduli of bcc transition metals, in *The Physics of Solid Solution Strengthening,* edited by E. W. Collins and H. G. Gegel, Plenum, New York, 1975.

Fisher, E. S., M. H. Manghnani, and K. Katahara, Application of hydrostatic pressure and shock wave data to theory of cohesion in metals, *Rev. Phys. Chem. Japan,* 393-397, 1975.

Gilbert, F., and A. M. Dziewonski, An application of normal mode theory to the retrieval of structural parameters and source mechanisms from seismic spectra, *Phil. Trans. Roy. Soc. London Ser. A., 278,* 187, 1975.

Hubbell, W. C., and F. R. Brotzen, Elastic constants of niobium-molybdenum alloys in the temperature range -190 to +100°C *J. Appl. Phys., 43,* 3306-3312, 1972.

Jamieson, J. C., and B. Olinger, Pressure inhomogeneity: a possible source of error in using internal standards for pressure gages, in *Accurate Characterization of the High-Pressure Environment,* edited by E. C. Lloyd, N.B.S. Special Publication No. 236, 321-323, U.S. GPO, Washington, D.C., 1971.

Katahara, K. W., M. H. Manghnani, and E. Fisher, Pressure derivatives of the elastic moduli of niobium and tantalum, *J. Appl. Phys., 47,* 434-439, 1976.

Kinsland, G. L., Yield strength under confining pressures to 300 kbar in the diamond anvil cell, Ph.D. thesis, University of Rochester, Rochester, N.Y., 1974.

Mao, H. K., T. Takahashi, W. A. Bassett, J. S. Weaver, and S. Akimoto, Effect of pressure and temperature on the molar volume of wustite and of three $(Fe, Mg)_2SiO_4$ spinel solid solutions, *J. Geophys. Res., 74,* 1061-1069, 1969.

Mattheiss, L. F., Fermi surface in tungsten, *Phys. Rev., 139,* A1893-A1904, 1965.

Mattheiss, L. F., Electronic structure of niobium and tantalum, *Phys. Rev. B, 1,* 373-380, 1970.

McMillan, W. L., Transition temperature of strong-coupled superconductors, *Phys. Rev., 167,* 331-344, 1968.

McQueen, R. G., S. P. Marsh, J. W. Taylor, J. N. Fritz, and W. J. Carter, The equation of state of solids from shock wave studies, in *High Velocity Impact Phenomena,* edited by R. Kinslow, pp. 293-417, Academic Press, New York, 1970.

McSkimin, H. J., and P. Andreatch, Jr., Analysis of the pulse superposition method for measuring ultrasonic wave velocities as a function of temperature and pressure, *J. Acoust. Soc. Am., 34,* 690-614, 1962.

Oli, B. A., and A. O. E. Animalu, Lattice dynamics of transition metals in the resonance model, *Phys. Rev. B, 13,* 2390-2410, 1976.

Pearson, W. B., *Handbook of Lattice Spacings and Structures of Metals and Alloys,* vol. 1, Pergamon, New York, 1958.

Pearson, W. B., *Handbook of Lattice Spacings and Structures of Metals and Alloys,* vol. 2, Pergamon, New York, 1966.

Peter, M., W. Klose, G. Adam, P. Entel, and E. Kudla, Tight-binding model for transition metal electrons-1, *Helv. Phys. Acta, 47,* 807-832, 1974.

Pettifor, D. G., Theory of the crystal structures of transition models, *J. Phys. C., 3,* 367-377, 1970.

Pickett, W. E., and P. B. Allen, Density of states of Nb and Mo using Slater-Koster interpolation, *Phys. Letters, 48A,* 91-92, 1974.

Piermarini, G. J., S. Block, and J. D. Barnett, Hydrostatic limits in liquids and solids to 100 kbar, *J. Appl. Phys., 44,* 5377-5382, 1973.

Piermarini, G. J., S. Block, J. D. Barnett, and R. A. Forman, Calibration of the R_1 ruby fluorescence line to 195 kbar, *J. Appl. Phys., 46,* 2774-2780, 1975.

Posternak, M., W. B. Waeber, R. Griessen, W. Joss, W. van der Mark, and W. Wejgaard, The stress dependence of the Fermi surface of molybdenum: I. The electron lenses, *J. Low Temp. Phys., 21,* 47-74, 1975.

Sato, Y., S. Akimoto, and K. Inoue, Pressure intensification in the composite material, *High Temp.-High Press., 5,* 289-297, 1973.

Smith, T. F., and T. R. Finalyson, Thermal expansion of bcc solid solution alloys in the system Zr-Nb-Mo-Re, *J. Phys. F: Metal Phys.*, *6*, 709-724, 1976.

Thomsen , L., On the fourth-order anharmonic equation of state of solids, *J. Phys. Chem. Solids*, *31*, 2003-2016, 1970.

Thomsen, L. Towards an equation of state for the lower mantle (abstract), *EOS Trans. A.G.U.*, *56*, 1063, 1975.

Thurston, R. N., Ultrasonic data and the thermodynamics of solids, *Proc. I.E.E.E.*, *53*, 1320-1336, 1965.

Weaver, J. S., T. Takahashi, and W. A. Bassett, Calculation of the P-V relation for sodium chloride, up to 300 kilobars at 25°C, in *Accurate Characterization of the High-Pressure Environment*, edited by E. C. Lloyd, N.B.S. Special Publication No. 326, pp. 189-200, U.S. GPO, Washington, D.C., 1971.

Wilburn, D. R., Isothermal compression of spinel (Fe_2SiO_4) up to 100 kbar with the gasketed diamond cell, M.S. thesis, University of Rochester, Rochester, N.Y., 1975.

PHYSICS AND CHEMISTRY OF IRON AND POTASSIUM AT LOWER-MANTLE AND CORE PRESSURES

M.S.T. Bukowinski and L. Knopoff
Institute of Geophysics and Planetary Physics
University of California, Los Angeles
Los Angeles, California 90024

Abstract

The effect of pressure on the electronic band structure and density of face-centered cubic iron has been calculated via the Augmented Plane Wave method. The results indicate that, at core pressures, the electronic structure of iron is very similar to that at surface conditions. The postulated electronic transition does not occur until fourfold compression. The calculated equation of state is consistent with the latest seismic inner core models and predicts a Grüneisen ratio of 1.8 ± 0.4 and a temperature of 5500 ± 500°K with an adiabatic gradient.

A similar calculation for body-centered cubic potassium shows that the latter undergoes a series of electronic transitions which transfers it into a typical transition metal at pressures as low as 300 kbar (at 0°k). In addition, the collapses in volume which accompany these transitions further decrease the ionic size of potassium which, due to a large compressibility, decreases very rapidly with pressure. These results show that low-pressure experiments on the miscibility of potassium in iron melts are not relevant to the problem of the apparent depletion of K in the earth. In fact, at pressures as low as 500 kbar, potassium should be compatible with iron to the extent that it might substitute for the latter in certain alloys.

I. INTRODUCTION

To assist in the identification of the chemical composition of inaccessible parts of the earth, seismic data may be inverted to provide a pressure-density relationship for the interior of the earth, and to interpret the results properly, the equations of state of the candidate earth-forming materials must be determined. Static laboratory data, limited to pressures of about 100 kbar, are useful only in studies of the uppermost layers of the earth: rather simple geophysical arguments point to the inescapable conclusion that the pressure at the center of the earth is approximately 3.6 Mbar. Thus, it is necessary to resort to some method of estimating the appropriate equations of state.

These estimates are generally arrived at through three different approaches: (1) theoretical calculations based on the statistical Thomas-Fermi method and its variants, (2) extrapolation of experimental data taken in the laboratory to pressures and temperatures of the inaccessible regions, and (3) reduction of data obtained from shockwave experiments.

High-pressure equations of state may be computed by the Thomas-Fermi [*Slater and Krutter*, 1935; *Feynman et al.*, 1949; *Latter*, 1955] or the Thomas-Dirac model [*Feynman et al.*, 1949; *Metropolis et al.*, 1953; *Cowan and Ashkin*, 1957]. Both theories predict pressures at low density that are much too high principally because the Thomas-Fermi model neglects the atomic shell structure. As a matter of fact, both models are constructed on the assumption that the electronic density is dense and smooth enough to be treated statistically. Obviously, the approximation becomes better the higher the pressure and the higher the atomic number; but, even for large atoms, it is not expected to hold for pressures lower than 10 Mbar. Attempts to improve the theory by including quantum and exchange corrections [*Kirzhnits*, 1957, 1959] failed to make the theory applicable to low densities. In particular, the predicted pressures at low densities are still too high [*Kalitkin*, 1960].

Most (published) attempts to provide equations of state for pressures of several Mbar have been semi-empirical extrapolations from relatively low-pressure laboratory measurements. Although they appear on the surface to be radically different, a careful analysis reveals that most of these methods amount to a particular choice of an interatomic potential. Models of the interatomic potential have often been made with considerable physical insight. However, the material properties of interest, e.g., pressures, densities, and elastic moduli -- correspond to high-order derivatives of the potential function. Thus, small errors in the potential function in the range of experimental observations may lead to large errors in the derivatives of these functions at pressures well outside the laboratory range. Furthermore, an empirical potential usually masks many fundamental properties of materials. A good example, among others, is given by the potential function

$$\frac{3A}{b\rho_o} \ell^{b(1-X^{1/3})} - \frac{3A}{\rho_o} X^{1/3}$$

from [*Zharkov and Kalinin*, 1971]. In the expression $x = \rho_o/\rho$, ρ_o is the zero-pressure density and b and A are adjusted to the bulk modulus and its pressure derivative at zero pressure. *Zharkov and Kalinin*, find that this potential gives a good description of ionic crystals, metals and silicates [*Zharkov and Kalinin*, 1968]. A single potential function, which describes equally well such diverse materials as metals and ionic crystals, cannot shed much light on the fundamental properties of matter.

The theory of finite strain is a macroscopic theory that takes the compressibility of a material as an empirical parameter and then develops an equation of state from an assumed relationship between strain and the strain energy. It was originally proposed by *Murnaghan* [1973, 1941, 1951] and later applied, chiefly by *Birch* [1938, 1947, 1952] to geophysical problems. Although the theory has a purely geometrical basis, it, too, may be cast in terms of power-law interatomic potentials [*Knopoff*, 1963].

In recent years, the finite-strain theory has been extended to include the effect of temperature and fourth-order terms in the potential [*Thomsen*, 1970, 1972; *Davies*, 1972]. The accuracy of the extended theories depends on which definition of strain is used: as has been shown [*Knopoff*, 1963], the definition of strain in low-order theories is ambiguous. Although agreement with high-pressure experimental data appears to be improved, this is achieved at the cost of introducing additional empirical parameters, which require additional precise measurements for their evaluation, so that the extrapolation can be made.

By far the greatest shortcoming common to all the methods discussed above is their inability to take into account the pressure-induced changes of the electronic structure of the constituent atoms. It was shown by *Lifshitz* [1960] that compression may drastically alter the constant energy surfaces of the electrons in a solid. Thus, a semi-conductor may become a conductor when the originally empty conduction band is forced to overlap the full valence band. Similarly, discontinuous changes in the electronic density at the Fermi level of a metal may occur where an empty band drops below the Fermi surface. Such changes would be reflected in all properties sensitive to the electronic density of states. When large extrapolations are made in pressure, the discrete character of the electronic levels must be accounted for if we are to obtain a quantitative description of the effects of comrpession.

Electronic transitions have been proposed for a number of phase transformations that have been observed in the laboratory. An electronic transition was first proposed [*Lawson and Tang*, 1949; *Herman and Swenson*, 1958; *Likhter et al.*, 1958; *Lawson*, 1966] to explain a discontinuous drop in resistance in Ce at 5 kbar, accompanied by a discontinuous change in volume, but without change in crystal structure. It was postulated that under pressure, a 4*f*-electron is promoted to the 5*d*-band.

A large increase in the resistance of Cs at 42 kbar is also accompanied by an isostructural change in volume. *Sternheimer*, [1950] proposed that the transition is due to the promotion of a 6*s*-electron to the empty 5*d*-band. Although *Sternheimer's* analysis is at best qualitative, his main conclusion has since been shown to be essentially correct [*Yamashita and Asano*, 1970; *Averill*, 1971].

Since the pioneering work on Ce and Cs, numerous other electronic transitions have been observed. A review of the state of the experimental evidence on electronic transition is given by *Drickamer* [1965]. A careful study of the observations, plus some simple theoretical arguments, indicate that all elements with unfilled inner electron bands must eventually undergo pressure-induced electronic transitions under sufficiently high pressures. In the limit of infinite pressures, any atomic system must break down into a Fermi sea of electrons and nuclei.

The variable character of the atom certainly influences the observed properties of materials under shock-wave conditions. Since shock-waves are our only experimental link with lower-mantle and core pressures, their proper interpretation and reduction to isothermal equations of state is of great importance. Geophysical materials are sufficiently complex to display a wide spectrum of pressure-induced phase transformations and other, more gradual changes that cannot be understood on the basis of shock-wave data alone. Although this is a rather obvious point, it is one well worth emphasizing when we consider the fact that very little theoretical effort is aimed at understanding high-pressure phenomena.

In the remainder of this article we summarize the results of two investigations of the sort advocated above: (1) a calculation of the electronic band structure and equation of state of face-centered cubic Fe, and (2) a similar calculation for body-centered cubic K. In section II we sketch the methods used. Details of the calculations will not be given; the interested reader is referred to the original papers, which are referenced. Our purpose here is to summarize the results obtained thus far, with the hope of demonstrating that first-principles calculations are practical and provide a powerful tool for interpreting both seismic and laboratory high pressure data.

Fe and K were chosen partly because of their relative simplicity when compared to most minerals, and partly because we believed that the results would help to clarify the structure of the core of the earth. The results of the Fe calculation and some of the implications to the structure of the core are presented in section III. Section IV summarizes the results for K, which is seen to have some very interesting properties that almost certainly invalidate the conclusions about its miscibility in Fe and FeS melts based on low-pressure experiments.

II. OUTLINE OF THE METHOD OF CALCULATIONS

The electronic band structures were calculated self-consistently via the Augmented Plane Wave (APW) method [*Slater*, 1937], which is closely related to the self-consistent Hartree-Fock technique for free atoms. Other than Green's function method, it is the only one capable of dealing with transition-metal crystals in a relatively simple, systematic manner.

Numerous calculations have shown that the APW method is capable of producing results that are in good agreement with experiments [*Loucks*, 1967; *Mattheiss et al.*, 1968].

The basic problem involved in obtaining the electronic structure of a crystal as a function of pressure is the solution of Schrodinger's equation in a periodic potential. The APW method accomplished this by assuming that the potential is of the "muffin-tin" form (Figure 1). In this approximation, the crystal is subdivided into two types of regions. Around each atom a sphere is drawn, within which the potential is assumed to be spherically symmetric; between the spheres the potential is constant, usually chosen to be the spatial average of the intersphere potential. The spheres must not overlap, but are otherwise arbitrary. Non-"muffin-tin" corrections to the potential are small for close-packed metals but may be significant in ionic solids and particularly in covalent solids.

Having assumed a muffin-tin potential, the APW scheme continues by using different expansions of the electronic wave functions in each of the crystal regions. Inside the spheres, the base states are expressed as superpositions of atom-like solutions of Schrodinger's equations. In the constant potential region, the base states are given by a single plane wave. The requirements that the base functions be continuous across the sphere boundary, and that the total energy of the system be stationary with respect to arbitrary variations of the coefficients of the base states, leads to a secular equation that must be solved numerically.

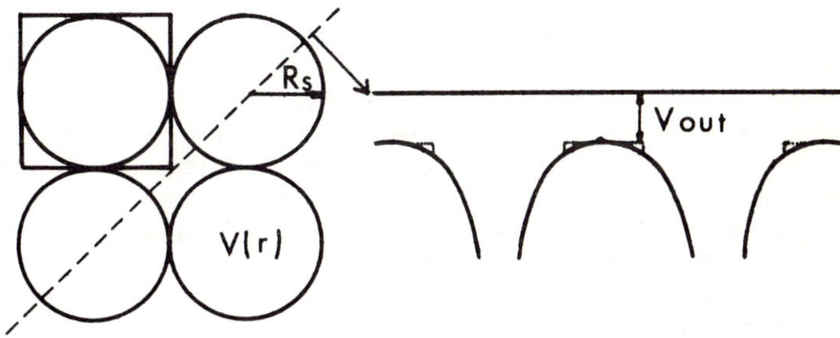

Fig. 1. A "muffin-tin" potential. V(r) is spherically symmetric and Vout is a constant obtained by averaging the potential over the intersphere region. The square on the left side is the unit cell for this hypothetical square lattice.

The solution yields the eigenvalues and eigenfunctions of the electrons, which may then be used to calculate the total energy and pressure of the lattice.

The numerical procedure used to obtain a set of energy bands and an equation of state is illustrated schematically in Figure 2. A starting potential is constructed from atomic potentials and charge densities. The Coulomb interaction V_C is simply a superposition of the atomic Coulomb potentials, while the exchange interaction V_{ex} is computed from the superposed atomic densities via the Kohn-Sham-Slater approximation [*Slater*, 1951; *Kohn and Sham*, 1965] . Details of the procedure may be found in *Loucks* [1967] and *Mattheiss et al* [1968].

Given the potential, the secular matrix is constructed, and the electronic eigenstates and eigenvalues are obtained. From the latter, the total electronic density is constructed, which, in turn, yields a new value of the potential. At this point, the total energy and pressure of the crystal are computed. If one is interested in the equation of state, then the convergence of the pressure is used as the self-consistency criterion. It is interesting to note that when the pressure is self-consistent within a kbar, the energy eigenvalues are self-consistent to 10^{-6} Rydbergs.

III. IRON AND THE CORE

Elsasser and Isenberg [1949] suggested that, at core pressures, ordinary iron might undergo a transition to the $3d^8$ state. At that time, the phenomenon of pressure-induced electronic transitions was largely unknown, and the suggestion was quickly forgotten.

In recent years, theoretical studies of the earth's core have been greatly intensified by the availability of high-resolution data from seismic arrays [*Julian et al.*, 1972] and from studies of the inverse problem for the earth's free oscillations and seismic travel times [e.g., *Gilbert and Dziewonski*, 1975] . The structure of the core began to display properties that were difficult to explain in terms of simple solid Fe-Ni inner-core and liquid Fe-Ni outer-core models. These strange properties included an apparent sub-adiabatic temperature gradient in the outer core [*Higgins and Kennedy*, 1971] and an extremely high Poisson's ratio for the inner core.

About the same time, the observations of several experimentally verified pressure-induced electronic transitions [*Drickamer*, 1965] led to a resurrection of *Elsasser and Isenberg's* original idea. *McLachlan and Ehlers* [1971] pointed out that the occurrence of an electronic transition in Fe in the core would invalidate existing extrapolations of the melting curve of Fe. *Stacey* [1972] rejected this idea on the grounds that it would make the resistivity of the outer core too large to maintain a magnetic field. However, it is possible to construct a model of the core in which the electronic transition is responsible for

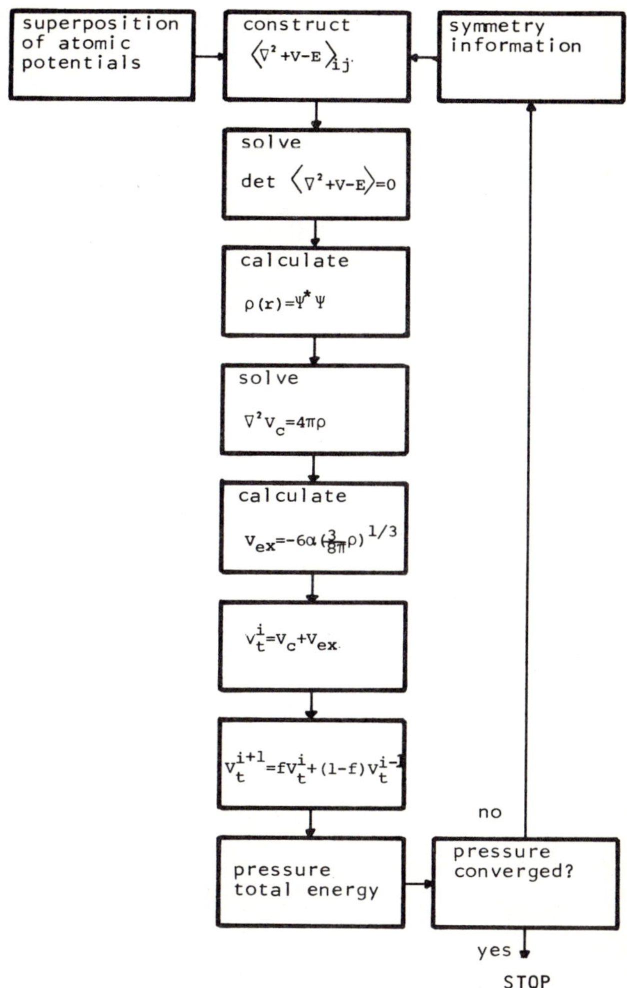

Fig. 2. Outline of the numerical calculation $\rho(r)$ is the electronic density, Ψ is the wavefunction, V_C is the Coulomb potential, V_{ex} is the exchange potential, V_T is the total potential obtained in the ith iteration, and f is a weighting number with value between 0 and 1.

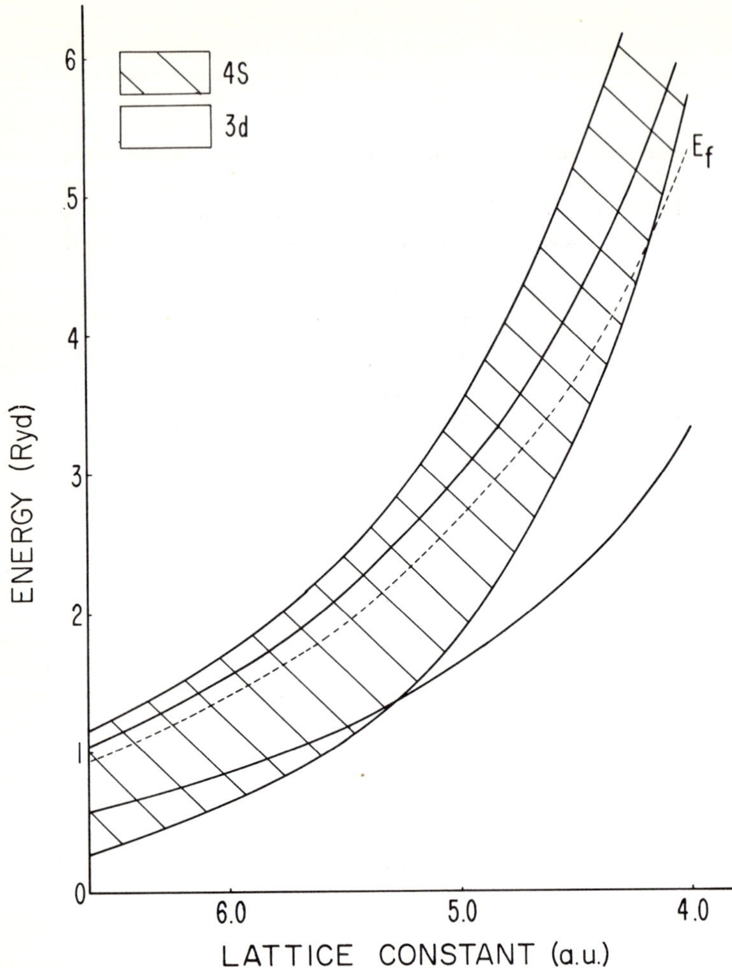

Fig. 3. Energy bands of fcc Fe that arise from 3d and 4s atomic states. Note that the bottom of the 4s band is above the Fermi energy after fourfold compression.

the inner core-outer core discontinuity in an entirely liquid Fe core [*Bukowinski and Knopoff*, 1976a] . Notwithstanding the validity of this model, the major issue was whether an electronic transition in Fe takes place in the pressure and temperature regime of the core.

To investigate the possible existence of an electronic collapse of Fe at core pressures, a fairly complete band structure calculation was carried out [*Bukowinski* 1976 ; *Bukowinski and Knopoff*, 1976b] . It was found that the electronic structure of Fe is stable to at least twofold compression. Figure 3 shows the Fermi energy E_F and electronic bands arising from the $3d$ and $4s$ atomic levels. The electronic transition occurs at the point where the bottom of the $4s$-band moves above the Fermi surface, namely, at fourfold compression. While this transition is continuous in terms of the fractional occupation of the $4s$-band, it causes a discontinuous change in the topology of the Fermi surface. Figure 4 shows part of the energy bands just before and just after the transition. Before the transition, the band joining Γ_1 to X_1 is completely filled and is a $3d$-$4s$ hybrid. Just after the transition, the band loses most of its s-character (Γ_{12} and X_1 are mainly d-like), and a hole develops around the center of the Brillouin zone corresponding to this band. Since this process is discontinuous, it should give rise to discontinuities in those properties of the lattice that depend upon the electronic density of states [*Lifshitz,* 1960].

The theoretical result indicates that, at 0°K, the electronic structure of Fe is very stable well past core densities. A consideration of temperature effects shows that the high temperatures of the core render the possible occurrence of a first-order phase transition within the core even less likely [*Bukowinski,* 1976]. Since the predicted transition density is approximately 2½ times as large as the inner-core density, it is difficult to escape the conclusion that there are no electronic transitions within the earth's core. This lends strong support to core models where, on the assumption that Fe is the main component in the core, the inner-outer core boundary is on the phase boundary between solid Fe-Ni and a liquid phase containing Fe and 10-20% lighter elements.

An equation of state for Fe may be obtained from the results of the band structure calculation. If $kT/E_F \ll 1$, the pressure is given by [*Bukowinski,* in preparation],

$$P = P_o(V) + (3k\gamma_G + \frac{1}{2}\gamma_e \beta T)\frac{T}{V} \tag{1}$$

where γ_G is the lattice Grüneisen parameter, β is the electronic specific heat coefficient, given by

$$\beta = \beta_o \, \ell \int_{V_o}^{V} \gamma_e \frac{dV}{V} \tag{2}$$

where $\gamma = \partial \ell n\beta/\partial \ell nV$ and β may be computed from the electronic structure of Fe. When $kT/E_F \ll 1$, we have

$$\beta = \frac{1}{3}\pi^2 k^2 D(E_F) \tag{3}$$

where k is Boltzman's constant, and $D(E_F)$ is the electronic density of states at the Fermi surface. The latter may be estimated from the computed band structure of Fe. We find

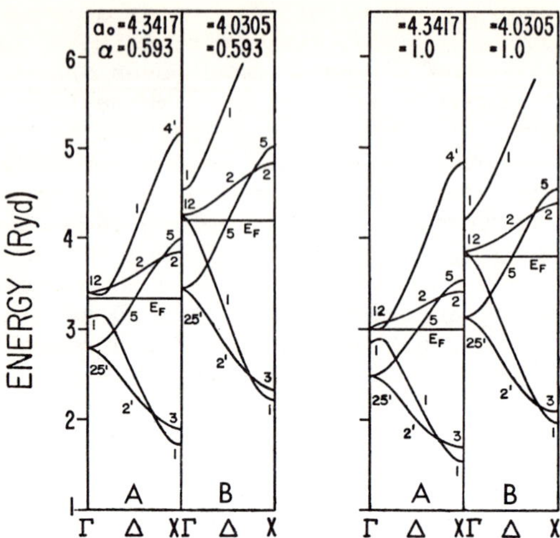

Fig. 4. Energy bands of Fe just before (A), and after (B), the 4s band goes above Fermi energy. Notation is that of Bouckaert, L. C., R. Smoluchowski, and E. Wigner, Phys. Rev. 50, 58, 1936.

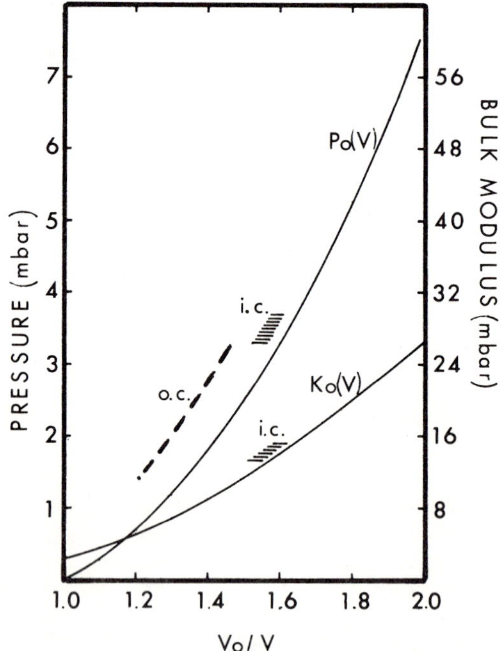

Fig. 5. Equation of state and bulk modulus of $0^{o}K$ fee Fe.

[*Bukowinski*, in preparation] .

$$\beta = (801 \pm 26) \left(\frac{V}{V_o}\right)^{1.5 \pm .1} \quad \text{ergs/deg}^2\text{-g} \tag{4}$$

Note that this is just equation (2) with β_o = 801 ± 26 ergs/deg^2-g and γ_e = 1.5 ± .1.

$P_o^e(V)$ was computed from the electronic density and with help of the virial theorem for contained systems [e.g., *Slater*, 1974]. Figure 5 shows $P_o(V)$ and the corresponding bulk modulus, compared with the values predicted by seismic models of the core. We may now ask whether there is a reasonable temperature distribution and γ_G that will bring the theoretical equation of state of fcc Fe into agreement with the ones deduced from seismic data.

The pressure is given in terms of two unknown variables: γ_G and T. These may be determined by fitting equation (1) to available seismic models of the core. Five models were used: model B1 [*Jordan and Anderson*, 1974] , model C2 [*Anderson and Hart*, 1976] , models 1066A and 1066B [*Gilbert and Dziewonski*, 1975] , and the PEM model [*Dziewonski et al.*, 1975].

Dziewonski et al. [1975] claim that core data do not require any deviations from homogeneity and adiabaticity within the inner core. If we make the assumption that the core is indeed adiabatic, then we have a strong constraint on the possible temperature distribution and γ_G in the inner core. Equation (1) plus the condition that the inner core be adiabatic leads to the following expression [*Bukowinski*, in preparation] :

$$\gamma_G = \frac{\partial \ln T}{\partial \ln \rho}\bigg|_S \left(1 + \frac{C_{VE}}{C_{VL}}\right) - \gamma_e \frac{C_{VE}}{C_{VL}} \tag{5}$$

where C_{VE} and C_{VL} are the electronic and lattice specific heats at constant volume, respectively.

Equation (5) may be used together with equation (1) to obtain T and γ_G from seismic models. Because of the small density change in the inner core, γ_G is treated as an adjustable constant. The temperature is also expected to be a slowly varying function within the inner core. Since at most three parameters could be resolved unambiguously by the seismic models, T was expressed as a simple linear relationship,

$$T = T_o + T_1 \left(\frac{\rho - \rho_T}{\rho_C - \rho_T}\right) \tag{6}$$

where ρ_T and ρ_C are the densities at the top and center of the inner core, respectively. Then T_o is the temperature at the inner core-outer core boundary, and T_1 is the temperature difference across the inner core.

The values of T_o, T_1, and γ_G were obtained by first fixing γ_G and finding the corresponding T_o and T_1 that gave a least-squares fit to the seismic models. The procedure was then iterated, with values of γ_G obtained from equation (5), until γ_G changed by less than (0.01) between consecutive iterations.

Models 1066A and 1066B could not be fitted with an adiabatic temperature gradient. In fact, all attempts to fit the theoretical equation of state to the P-ρ relations given by these models resulted in negative average-temperature gradients. Since the uncertainties in the theoretical results are almost certainly much smaller than the uncertainties in the seismic models, we conclude that models 1066A and 1066B give unsatisfactory representations of the inner core. Their apparent deviations from adiabaticity are probably due to improper treatment of biased seismic data [*Dziewonski et al.*, 1975] .

The values of the adjustable parameters for the remaining models are shown in Table 1. Models C2 and B1 are very similar, and this is reflected in the numbers in the table. The temperatures obtained from these models are rather low and would result in a small gradient across the outer core. The values of γ_G are larger than any observed values for materials on the surface of the earth. Since γ_G decreases with pressure, it is certain that the values demanded by these models are unacceptable. In addition, the large RMS of the residuals represents a systematic difference between the slope of the seismic and theoretical pressure curves.

Model PEM gave acceptable values of all parameters. The temperature of approximately 5.5×10^3 °K gives a reasonable gradient in the outer core and is safely below estimates of the melting temperature of pure Fe. A Grüneisen parameter of 1.87 is not inconsistent with fcc Fe at core densities since typical fcc metals have γ_G's in the range of 2-2.5 at surface conditions. Furthermore, the fit between the PEM model and theory is excellent, as indicated by the extremely small residuals.

The errors shown in Table 1 are due to the uncertainties in the theoretical equation of state and do not reflect uncertainties in the seismic models. A 1% uncertainty in the seismic density gives rise to the following error bounds on the adjustable parameters: ±500°K for T_o; ±20°K for T_1 and ±0.3 for γ_G. By comparison, the theoretical errors are small.

The five seismic models of the inner core were evaluated on the assumption that the equation of state of pure Fe is relevant to the inner core. Although it is possible that the inner core contains other elements in addition to Fe, the excellent agreement between PEM and the equation of state of fcc Fe provides us with a very simple model of the inner core. On this basis, we make the tentative conclusion that the inner core consists of essentially pure Fe at a temperature of 5500 ± 500°K. The temperature gradient is estimated to be approximately 0.9°K/km, and the Grüneisen parameter is 1.8 ± 0.4. Larger data scatter will increase the uncertainties in the parameters but should not alter the conclusion that the PEM model is the most suitable.

TABLE 1. Temperature and Grüneisen Parameter obtained for Three Seismic Models

Model		T_0 (°K)	T_1 (°K)	RMS (Mb)
PEM	1.87 ± 0.1	5450 ± 65	234 ± 2	0.37 ± 10^{-3}
B1	3.88 ± 0.1	4400 ± 40	328 ± 2	0.39 ± 10^{-1}
A-H	3.93 ± 0.1	4700 ± 40	350 ± 2	0.47 ± 10^{-1}

IV. POTASSIUM

Arguments in favor of the hypothesis that the earth is depleted in K range from geochemical studies of earth rocks [*Gast*, 1960; *Wasserburg et al.*, 1964; *Hurley* 1968a, 1968b] to laboratory experiments on the miscibility of K in Fe-FeS melts [*Oversby and Ringwood*, 1972, 1973; *Ganguly and Kennedy*, 1966] to geochemical models of the earth [e.g., *Ringwood*, 1966] . However, none of these arguments is conclusive. Geochemical studies establish only that the crust and upper mantle are depleted in K; *Ringwood's a priori* assumption that this depletion extends to the center of the earth is not justifiable. In fact, more recent models of planetary condensation [*Lewis*, 1972a, 1972b, 1973] predict that the earth should have the full solar abundance of K, while some models of core formation indicate that most of the earth's K has segregated into the core [*Murthy and Hall*, 1970, 1972; *Lewis*, 1971; *Hall and Murthy*, 1971] . The controversy is not resolved by the laboratory data. Experiments by *Goettel* [1972, 1975, in preparation] led him to conclude that the solubility of K in FeS liquid is sufficiently large to predict that a significant amount of K would be partitioned into the outer core, in contradiction to the other experiments.

Disputes based on results of laboratory experiments carried out at low pressures are quite pointless, since the data themselves are of little relevance to the actual physics and chemistry in the lower mantle and core. The negative experiments attract many believers because of an underlying assumption that the large ionic size of K renders it immiscible in Fe, but this argument ignores the well-known fact that K is extremely compressible and that the chemical properties of elements may be drastically affected by large pressures.

There is some experimental evidence that the 42-kbar electronic transition in Cs significantly increases its solubility in Fe [*F. O. Stacey*, private communication] . It is likely that the effect is due to both the decreased ionic size and a change of the electronic structure of Cs to a transition metal-like configuration. The electronic configuration of K suggests that it too should undergo pressure-induced electronic collapses. This possibility was investigated by the APW method.

The band structure, total energy, pressure, and enthalpy

of bcc K at $0^{o}K$ were computed for 15 values of the lattice parameter, covering a range of one- to fivefold compression.

Figure 6 shows the occupied and some unoccupied energy bands for three different compressions in the H(1,0,0), N(½,½,0), and P(½,½,½) directions. Note that at 2.5-fold compression, the region around point N is occupied, whereas it was empty in the uncompressed state. Similarly, the region around H is about to be occupied at a compression slightly larger than 2.5-fold. Two bands are coming under the Fermi level at point H; this corresponds to the creation of electron pockets around H in the first and second Brillouin zones. Similarly, the complete occupation in the N direction corresponds to the extension of necks perpendicular to the N faces, somewhat akin to the well-known necks in Cu.

As in the case of Fe, the topological changes in the Fermi surface of bcc K are expected to cause a discontinuous behavior of its macroscopic properties. That this is indeed the case is shown by the plot of enthalpy as a function of pressure (Figure 7). The two points marked by circles indicate the presence of first-order transitions at pressures of approximately 245 kbar and 290 kbar. It is possible that a discontinuity observed in the resistance of K at 280 kbar and $77^{o}K$ [*Drickamer,* 1965] is related to these transitions.

Figure 8 shows the computed pressure-density equation of state at $0^{o}K$. This result may be compared with the equation of state at $0^{o}K$ derived from shock-wave experiments [*Bakanova* et al., 1965] and static compression experiments [*Monfort and Swenson,* 1965]. Note that the electronic transitions cause density collapses of approximately 7 and 11%. The accuracy of the theoretical isotherm is reflected by the close agreement of the computed uncompressed bulk modulus of 36.41 kbar and the experimental value of 35.9 kbar [*Monfort and Swenson,* 1965], a difference of only 1.5%.

A major part of the discrepancy between the theoretical and static curves and the equation of state derived from shock-wave experiments is probably due to an inadequate reduction of the Hugoniot. *Bakanova et al.* [1965] assumed that Grüneisen's parameter is given by the Dugdale-MacDonald formula; this is an unjustified assumption. Furthermore, *Bakanova et al.* ignore the fact that the Hugoniot for K intersects the melting curve at approximately 15 kbar [*Rice,* 1965]. Melting and the extremely high temperatures in the shock (about $3 \times 10^{4}\,{}^{o}K$ at $\rho/\rho_o = 3.5$) would certainly smear out the electronic transitions through the mechanism of interband scattering. Some average pressure curve, which is shifted from the low-pressure $0^{o}K$ isotherm in the direction of the post-transition curve, might then be expected. This in turn, affects the deduced uncompressed bulk modulus, which is

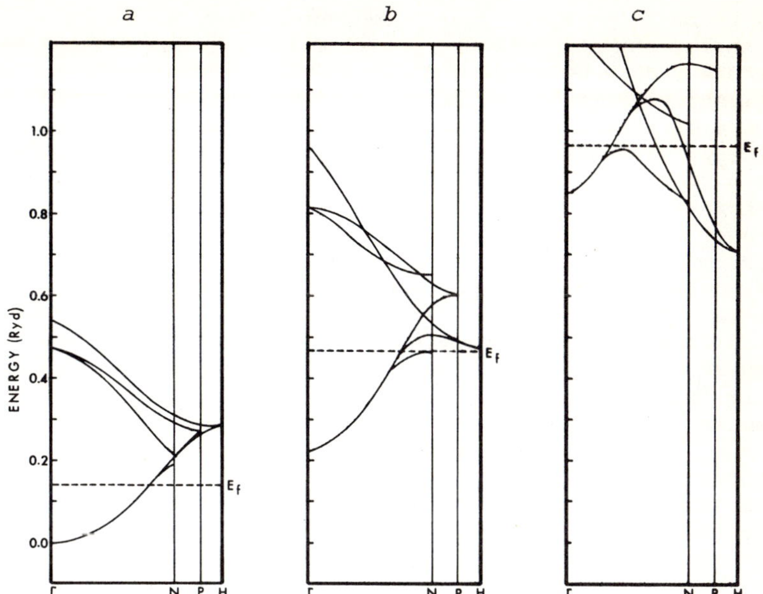

Fig. 6. Occupied and selected unoccupied bands of bcc K, at three compressions: a) $\rho/\rho_o = 1.0$; b) $\rho/\rho_o = 2.5$; c) $\rho/\rho_o = 5.0$.

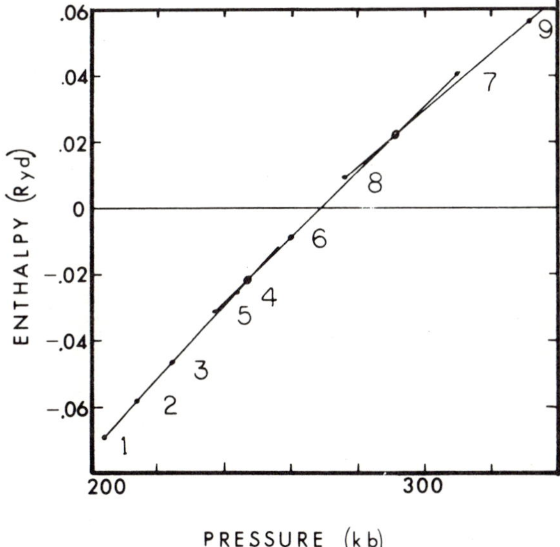

Fig. 7. Theoretical enthalpy of K at 0^oK. The two small circles mark the first order transitions. (From Bukowinski, 1976a).

found to be 54 kbar, in poor agreement with static data.

An investigation of the pressure-induced changes in the band structure and Fermi surface indicates that at higher pressures additional electronic transitions will occur; there is evidence that one of them occurs between four- and fivefold compression. No attempt was made to locate the precise location of these transitions, partly because of the cost involved, and partly because K is expected to go into an fcc structure at high pressures. In any case, the results presented here are sufficient to show that predictions concerning the high-pressure behavior of K, based solely on low-pressure experimental data, cannot possibly be accurate.

The crystal field experienced by a K atom in the lower mantle or outer core is certainly not that of a bcc configuration, but that of the major local chemical components. However, the properties of bcc K reflect the nature of the free K atom, and they are not a result of the particular geometrical symmetry of the lattice. The large compressibility of the crystal is due to the very diffuse 4*s*-atomic state, while the complicated electronic structure arises from the relatively low-lying 3*d*-states.

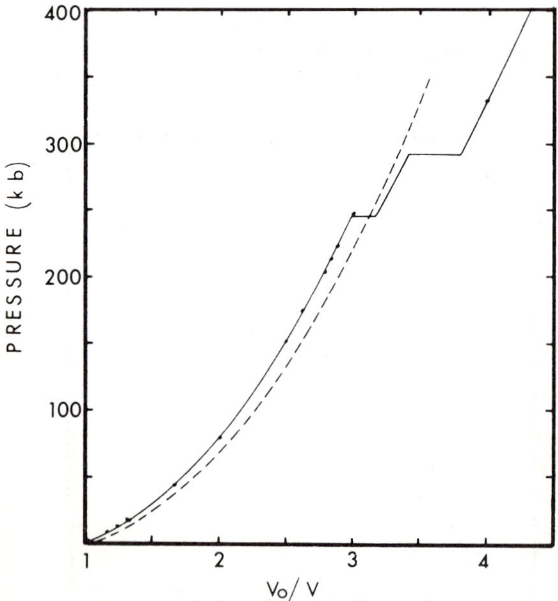

Fig. 8. Equation of state of K at 0°K. The dashed line was obtained from Bakanova et al., (1965). The static data of Monfort et al., is denoted by the small crosses near the origin (From Bukowinski, 1976a).

Calculations by [*Yamashita et al.* 1969] show that bcc and fcc K have qualitatively similar electronic bands at high pressures, with a structure typical of a transition metal. These considerations lead us to the conclusion that, as long as K finds itself in a fairly close-packed environment, its electronic structure is determined by the volume available to its outer electronic shells. Thus, in the lower mantle and outer core, K almost certainly has a transition metal character. This, in turn, should make K siderophile. The possible implications for the geophysics of the lower mantle and outer core are obvious: the presence of a K^{40} component in the outer core might provide the means for regeneration of the magnetic field by serving as a source of power for the magnetohydrodynamics of the core.

V. CONCLUSIONS

The availability of modern high-speed computers has made it possible to carry out detailed first-principles theoretical investigations of the high-pressure properties of materials. We believe that the usefulness of such calculations has been amply demonstrated by the results obtained for Fe and K. Further improvements in the theoretical methods should lead to similar calculations for more complicated materials of geophysical interest. The results may provide useful information for the inversion of seismic data and the reduction of shock-wave equations of state.

REFERENCES

Anderson, D. L., and R. S. Hart, An earth model based on free oscillations and body waves, *J. Geophys. Res., 81,* 1461, 1976.

Averill, F. W., Calculation of the total energy and pressure of compressed cesium, *Phys. Rev., 178,* 3315, 1971.

Bakanova, A. A., I. P. Dudoladov, and R. F. Trunin, Compression of alkali metals by strong shock waves, *Soviet Phys. Solid State, 7,* 1307, 1965.

Birch, F., The effect of pressure upon the elastic parameters of isotropic solids, according to Murnaghan's theory of finite strain, *J. Appl. Phys., 9,* 279, 1938.

Birch, F., Finite elastic strain of cubic crystals, *Phys. Rev., 71,* 809, 1947.

Birch, F., Elasticity and constitution of the earth's interior, *J. Geophys. Res., 57,* 227, 1952.

Bukowinski, M. S. T., On the electronic structure of iron at core pressures, *Phys. Earth. Planet. Interiors* (in press) 1976.

Bukowinski, M. S. T., The effect of pressure on the physics and chemistry of potassium, *Geophys. Res. Letts., 3,* 491, 1967a.

Bukowinski, M. S. T., and L. Knopoff, Electronic transitions in iron and the properties of the core, in *Physics and Chemistry of Minerals and Rocks*, (Proc. 1974 NATO Conf. on Petrophysics), edited by R. G. J. Strens, Wiley, New York, 1976*a*.

Bukowinski, M. S. T., and L. Knopoff, Electronic structure of iron and models of the earth's core. *Geophys. Res. Lett.*, *3*, 45, 1976*b*.

Cowan, R. D., and J. Ashkin, Extension of the Thomas-Fermi-Dirac statistical theory of the atom to finite temperatures, *Phys. Rev.*, *105*, 144, 1957.

Davies, G. F., Quasi-harmonic finite strain equations of state of solids, *J. Phys. Chem. Solids*, *34*, 1417, 1972.

Drickamer, H. G., The effects of high pressure on the electronic structure of solids, *Solid State Phys.*, *17*, 1, 1965.

Dziewonski, A., A. L. Hales and E. R. Lapwood, Parametrically simple earth models consistent with geophysical data, *Phys. Earth Planet. Interiors*, *10*, 12, 1975.

Elsasser, W. M., and I. Isenberg, Electronic phase transition in iron at extreme pressures, *Phys. Rev.*, *76*, 469, 1949.

Feynman, R. P., N. Metropolis, and E. Teller, Equations of state of elements based on the generalized Fermi-Thomas theory, *Phys. Rev.*, *75*, 1561, 1949.

Ganguly, J., and G. C. Kennedy, Solubility of K in Fe-S liquid, silicate-K-$(Fe\text{-}S)^{liq}$ equilibria, and their planetary implications, *J. Geophy. Res.* (in press).

Gast, P. W., Limitation of the composition of the upper mantle, *J. Geophys. Res.*, *65*, 1287, 1960.

Gilbert, F., and A. M. Dziewonski, An application of normal mode theory to the retrieval of structural parameters and source mechanisms from seismic spectra, *Phil. Trans. Roy. Soc. London*, *278*, 187, 1975

Goettel, K. A., Partitioning of potassium between silicates and sulphide melts: Experiments relevant to the earth's core, *Phys. Earth Planet. Interiors*, *6*, 161, 1972.

Goettel, K. A., Partitioning of potassium between solid silicates and Fe-FeS melts, *EOS, Trans. Am. Geophys. Un.*, *56*, 387, 1975.

Hall, H. T., and V. R. Murthy, The early chemical history of the earth: some critical elemental fractionations, *Earth Planet. Sci. Lett.*, *11*, 239, 1971.

Herman, R., and C. A. Swenson, Temperature dependence of the phase transition in cerium, *J. Chem. Phys.*, *29*, 398, 1958.

Higgins, G., and G. C. Kennedy, The adiabatic gradient and the melting point gradient in the core of the earth, *J. Geophys. Res.*, *76*, 1870, 1971.

Hurley, P. M., Absolute abundance and distribution of Rb, K, and Sr in the earth, *Geochim. Cosmochim. Acta*, *32*, 273, 1968*a*.

Hurley, P.M., Correction to: Absolute abundance and distribution of Rb, K, and Sr in the earth, *Geochim. Cosmochim. Acta, 32,* 1025, 1968*b*.

Jordan, T. H., and D. L. Anderson, Earth structure from free oscillations and travel times, *Geophys. J. Roy. Astron. Soc., 36,* 411, 1974.

Julian, B. R., D. Davies, and R. M. Sheppard, PKJKP, *Nature, 235,* 317, 1972.

Kalitkin, N. N., The Thomas-Fermi model of the atom with quantum and exchange corrections, *Soviet Phys., JETP 11,* 1106, 1960.

Kirzhnits, D. A., Quantum corrections to the Thomas-Fermi equation, *Soviet Phys., JETP 5,* 64, 1957.

Kirzhnits, D. A., The limits of applicability of the quasi-classical equation of state of matter, *Soviet Phys., JETP 8,* 1081, 1959.

Knopoff, L., The theory of finite strain and compressibility of solids, *J. Geophys. Res., 68,* 2929, 1963.

Kohn, W., and L. J. Sham, Self-consistent equations including exchange and correlation effects, *Phys. Rev., 140,* A1133, 1965.

Latter, R., Thomas-Fermi model of compressed atoms, *J. Chem. Phys., 24,* 280, 1955.

Lawson, A. W., The effect of pressure on electrical resistivity of metals, in *Progress in Metal Physics,* Vol. 6 edited by B. Chalmers and R. King, p. 1062, Pergamon Press, New York, 1966.

Lawson, A. W., and T. Y. Tang, Concerning the high pressure allotropic modification of cerium, *Phys. Rev., 76,* 301, 1949.

Lewis, J. S., Consequences of the presence of sulfur in the core of the earth, *Earth Planet. Sci. Lett., 11,* 130, 1971.

Lewis, J. S., Metal/silicate fractionation in the solar system, *Earth Planet. Sci. Lett., 15,* 286, 1972*a*.

Lewis, J. S., Low temperature condensation from the solar nebula, *Icarus, 16,* 241, 1972*b*.

Lewis, J. S., Chemistry of the planets, *Ann. Rev. Phys. Chem., 24,* 339, 1973.

Lifshitz, I. M., Anomalies of electron characteristics of a metal in the high pressure region, *Soviet Phys. JETP 11,* 1130, 1960.

Likhter, A. W., N. Riabinin, and L. F. Vereschaguin, The phase diagram for cerium, *Soviet Phys. JETP 6,* 469, 1958.

Loucks, T., *Augumented Plane Wave Method,* W. A. Benjamin, Inc., New York and Amsterdam, 1967.

Mattheiss, L. F., J. H. Wood, and A. C. Switendick, A procedure for calculating electronic energy bands using symmetrized augmented plane waves, in *Methods in Computational Physics* Vol. 8, edited by B. Alder et al., 63, 1968.

McLachlan, D. W., and E. Ehlers, Effect of pressure on the melting temperature of metals, *J. Geophys. Res., 76,* 2780, 1971.

Metropolis, N., A. Rosenbluth, M. N. Rosenbluth, A. H. Teller, and E. Teller, Equation of state calculations by fast computing machines, *J. Chem. Phys.*, *21*, 1087, 1953.

Monfort, C. E., and C. A. Swenson. An experimental equation of state for potassium metal, *J. Phys. Chem. Solids*, *26*, 291, 1965.

Murnaghan, F. D., Finite deformations of an elastic solid, *Amer. J. Math.*, *59*, 235, 1937

Murnaghan, F. D., *Theodor von Karman Anniversary Volume*, p. 121, California Institute of Technology, Pasadena, 1941.

Murnaghan, F. D., *Finite Deformation of an Elastic Solid*, Wiley, New York, 1951.

Murthy, V. R., and H. T. Hall, The chemical composition of the earth's core: Possibility of sulfur in the core. *Phys. Earth Planet. Interiors*, *2*, 276, 1970.

Murthy, V. R., and H. T. Hall, The origin and chemical composition of the earth's core, *Phys. Earth Planet. Interiors*, *6*, 23, 1972.

Oversby, V. M., and A. E. Ringwood, Potassium distribution between metal and silicate and its bearing on the occurrence of potassium in the earth's core, *Earth Planet. Sci. Lett.*, *14*, 345, 1972.

Oversby, V. M., and A. E. Ringwood, Reply to comments by K. A. Goettel and J. S. Lewis, *Earth Planet. Sci. Lett.*, *18*, 151, 1973.

Rice, M. H., Pressure-volume relations for the alkali metals from shock-wave experiments, *J. Phys. Chem. Solids*, *26*, 483, 1965.

Ringwood, A. E., Chemical solution of the terrestrial planets, *Geochim. Cosmochim. Acta*, *30*, 41, 1966.

Slater, J. C., Wave functions in a periodic potential, *Phys. Rev.*, *51* 151, 1937.

Slater, J. C., A simplification of the Hartree-Fock method, *Phys. Rev.*, *81*, 385, 1951.

Slater, J. C., *The Self-Consistent Field for Molecules and Solids*, Vol. 4,, *Quantum Theory of Molecules and Solids*, McGraw-Hill, New York, 1974.

Slater, J. C., and H. M. Krutter, The Thomas-Fermi method for metals, *Phys. Rev.*, *47*, 559, 1935.

Stacey, F. O., Physical properties of the earth's core, *Geophys. Survey*, *1*, 99, 1972.

Sternheimer, R., On the compressibility of metallic cesium, *Phys. Rev.*, *78*, 238, 1950.

Thomsen, L., On the fourth-order anharmonic equation of state of solids, *J. Phys. Chem. Solids*, *31*, 2003, 1970.

Thomsen, L., On the fourth-order anharmonic theory: Elasticity and stability, *J. Phys. Chem. Solids*, *31*, 363, 1972.

Wasserburg, G. J., G. J. F. MacDonald, F. Hoyle, and W. A. Fowler, Relative contributions of uranium, thorium, and potassium to heat production in the earth, *Science*, *143*, 465, 1964.

Yamashita, J., and S. Asano, Band structure of metals under high pressure. II. Fermi surface of Cs, *J. Phys. Soc., Japan, 29,* 264, 1970.

Yamashita, J., S. Wakoh, and S. Asano, Band structure of metals under high pressure. I. Fermi surface of Na and K, *J. Phys. Soc., Japan, 27,* 1153, 1969.

Zharkov, V. N., and V. A. Kalinin, The equations of state of solids at high pressure, *Izvestiya Physics of the Solid Earth,* (English Edition), 1968.

Zharkov, V. N., and V. A. Kalinin, *Equations of State for Solids at High Pressures and Temperatures,* Academic Press, New York and London, 1971.

PROCESSES OCCURRING IN SHOCK WAVE COMPRESSION OF ROCKS AND MINERALS

D. E. GRADY
Sandia Laboratories
Albuquerque, New Mexico 87115

Abstract

Recent advances in time-resolved shock-wave instrumentation and improved methods for analyzing continuous stress-wave data are providing a clearer picture of the processes of yielding and phase transformation occurring during shock compression of rocks and minerals. The present report reviews the advances in time-resolved shock-wave instrumentation which are currently being used and describes the analysis techniques which have been developed to relate shock-wave profiles to the thermomechanical processes occurring during shock compression. Evidence for complicating features resulting from yielding in silicates and oxides is reviewed and a physical model for phase transitions based on heterogeneous deformation and thermal activation is discussed. New supporting data on the shock compression and relief properties of periclase and calcite are presented and considered in terms of a mechanism of heterogeneous deformation.

I. INTRODUCTION

Shock-wave studies on rocks and minerals have made a substantial contribution to our current understanding of the earth's mantle. With increasing awareness of mantle dynamics and its coupling within the total earth system, it is clear that improved shock-wave techniques could continue to provide useful data on high pressure physical properties. There is, for instance, a need for measurements of temperature, transport properties, and higher order thermodynamic properties of minerals at pressures existing in the earth's mantle.

A clearer understanding of the physical state achieved by the process of shock compression is required, however. Recent results have rendered the simplifying assumption of an equilibrium thermodynamic state at Hugoniot pressures questionable.

How, for instance, does a mineral compressed to a given Hugoniot pressure and temperature compare with a mineral at the same state achieved by static compression techniques? Is pressure and temperature equilibrium achieved in the brief microsecond or less during which a high pressure Hugoniot state can be maintained? Is the material single crystalline or polycrystalline? Are petrographic features peculiar to the shock compression process introduced? Shock compression is a uniaxial phenomenon. Can the compression process introduce anisotropy into an initially isotropic material? Observations by *Tyunyaev et al.* [1972] of electrical conductivities parallel and perpendicular to the shock front in NaCl differing by several orders of magnitude are particularly unsettling in this respect. Can careful preparation of the initial material assist in achieving equilibrium in a shock compression experiment? For instance, would samples composed of sub-micron mineral powder be better than monocrystals?

Such questions concerning the physical state of the shock compressed material are extremely difficult to answer due to its very transient existence. In practice, the nature of the Hugoniot state must be inferred, indirectly, from experimental observations made during and after the shock compression process. Experiments which have been conducted include magnetic, electric, and optical observations during shock compression; flash x-ray studies at Hugoniot pressures; studies of the details of shock and release wave propagation; examination of material recovered from shock wave studies; and others. Such diverse studies place rigid constraints on models describing the thermomechanical processes occurring during shock compression and will eventually lead to a clearer understanding of the states achieved at Hugoniot pressures.

This report is focused primarily on the use of shock compression and release wave studies in assessing the material response of rocks and minerals. In recent years, large strides have been made in the development of instrumentation capable of accurately measuring wave profiles during shock compression and release. Evolution of the measured wave profiles relate directly to the thermomechanical models governing material response during shock compression, and considerably more complexity has been observed in high pressure wave propagation than was originally envisioned. Complex elastic-plastic response is observed, stress relaxation and attenuation indicative of strain rate effects occur, and details in both compression and release waves due to phase transition kinetics appear.

In section II, the experimental methods for measuring time-resolved, large-amplitude wave profiles which have been, or are becoming, fairly widely accepted, are reviewed and contrasted. In section III, methods that have been developed to analyze more complex wave propagation experiments are discussed. In section

IV, evidence for complicating features resulting from dynamic yielding in silicates and oxides are reviewed and a physical model for processes of phase transformation and deformation occurring during shock compression is discussed. New supporting data on the shock compression and relief properties of periclase and calcite are also presented.

II. CONTINUOUS SHOCK-WAVE MEASUREMENTS

A large portion of Hugoniot data available on rocks and minerals was obtained with discrete experimental techniques such as electrical discharge pins [*Minshall*, 1955] and argon flash gaps [*Walsh and Christian*, 1955]. These methods, although quite reliable, provided only the final shock velocity-particle velocity state and, in some cases, similar information on the elastic precursor wave. These methods depended on the fundamental assumption of a stable Hugoniot state to interpret the high pressure thermodynamic properties.

Increased interest in total large-amplitude loading and relief wave behavior both for improved understanding of the deformation process and for additional information on high pressure properties has served as impetus for development of continuous recording techniques. Of the various methods that have been explored, some have received fairly wide acceptance and are currently used in a number of shock wave laboratories.

Several of these methods (which will not be discussed here) are concerned exclusively with the measurement of free surface response, include the inclined mirror [*Duvall and Fowles*, 1963] or inclined prism [*Eden and Wright*, 1966], the inclined wire [*Barker et al.*, 1964], and the capacitor [*Rice*, 1961; *Ivanov and Novikov*, 1963] techniques. Other methods have been developed to measure in-material or quasi-in-material wave profile structures, including the piezoelectric, piezoresistant, electromagnetic, and laser interferometry techniques. Since these techniques are currently being used to investigate the shock compression phenomenon in rocks and minerals, it is important to understand the relative strengths and weaknesses of each method including frequency response limitations, stress limitations, calibration accuracies, and impedance matching difficulties. In this section we review these techniques and contrast their various features. Further discussion of these techniques are given by *Keeler and Royce* [1971], *Fowles* [1972], and *Graham and Asay* [1976].

A. Piezoelectric Techniques

A number of materials have been investigated for shock wave transducers because of their piezoelectric property, including quartz [*Graham et al.*, 1965] and lithium niobate [*Graham and Jacobson*, 1973]. The quartz gauge has become widely used and will be focused on here. The transducer capabilities of quartz have been extensively studied over the past decade [*Graham et al.*, 1965; *Ingram and Graham*, 1970] and it is presently one of the most accurately calibrated shock-wave instruments available.

The quartz gauge consists of a disc of X-cut single crystal quartz with a large diameter-to-thickness ratio to insure one-dimensional strain in the active gauge region during transit of the stress wave. The gauge is placed on the back surface of the sample under test so that the stress wave in the sample enters the gauge and propagates along the X-axis as illustrated in Figure 1. The transducer has vapor-deposited electrodes and

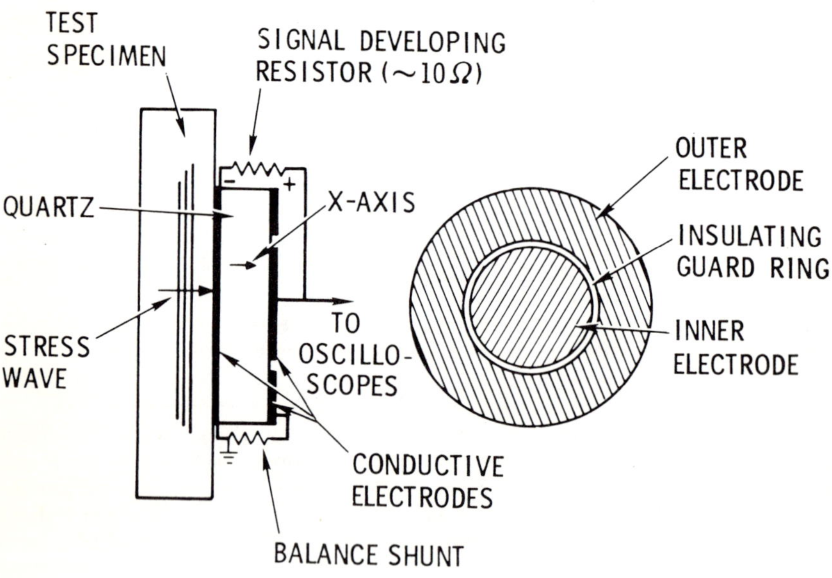

Fig. 1. Quartz gauge typical electrode configuration.

accurate use requires a guard ring electrode [*Graham et al.*, 1965] to assure that measurements are made in a region where one-dimensional strain and electric fields persist.

Current is monitored by measuring the voltage drop across a resistive shunt. In a linear approximation, the current produced during shock transit by the piezoelectric effect in the quartz is given by

$$i(t) \simeq kA \frac{c_l}{d} \sigma(t) \qquad (1)$$

where A is the area of the inner electrode, c_l the wave speed in quartz, d the gauge thickness, σ the stress at the sample-gauge interface, and k is the stress-coefficient for quartz. Equation 1 is valid for times less than one transit across the gauge thickness.

The quartz gauge has an upper stress limitation of 40 kbar [*Graham*, 1974]. X-cut quartz has a Hugoniot elastic limit of 60 kbar [*Graham*, 1974], although shock-induced conductivity limits the upper stress for accurate gauge response. The stress level which can be accurately detected in the sample can be higher than 40 kbar if the mechanical impedance of the sample is higher than that of the quartz. Al_2O_3 (corundum) is an example of a mineral with a significantly higher shock impedance.

Time resolution of the quartz gauge system in reproducing the stress profile is limited only by the rise time of the recording instrumentation and by tilt (nonplanarity) of the stress wave sweeping over the finite active area (typically 4 to 12 mm diameter) of the transducer. Impedance differences between the quartz and sample material introduce difficulties in interpretation since the measured wave profile differs from the original wave propagating in the sample material, although it usually retains the essential features. Approximate techniques have been developed for determining the undisturbed profile from the measured profile.

In Figure 2, a stress wave profile, measured with a quartz gauge, in [100] oriented MgO (periclase) is shown [*Grady*, unpublished, 1970] which illustrates the measurement capability of this technique. It is of interest to note the character of the initial loading precursor and subsequent stress relaxation which is significantly more complex than simple elastic-plastic response would predict. The measured elastic precursor amplitude is about 19 kbar, which corresponds to about 31 kbar in the periclase due to the significant impedance difference between the two materials. The finite slope in the late time part of the deformation wave is due to finite strain in the transducer quartz and must be accounted for in the data analysis.

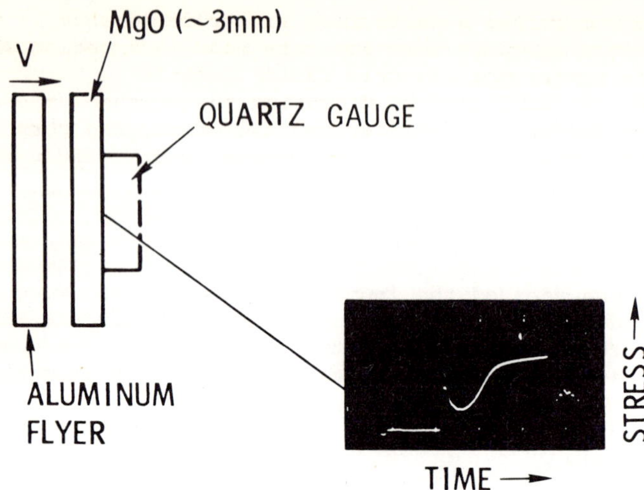

Fig. 2 Stress-wave profile obtained with quartz gauge on MgO. A step stress-wave input was provided by impact of an aluminum flyer (0.41 km/sec) on the specimen. The stress wave after 3-mm propagation distance was measured at the MgO-quartz interface. 0.2 μsec time marks are shown.

B. Piezoresistivity Techniques

A number of materials are known to exhibit a stress-induced resistance change, however, only two, ytterbium and manganin, have received serious attention as material for stress wave transducers. The former undergoes a phase transition at about 40 kbar restricting its use to stress levels below this value, however, the latter seems ideally suited for high pressure wave profile measurements in rocks and minerals. The manganin gauge has been used routinely at stress levels to 500 kbar [*Grady et al.*, 1974*a*; *Murri et al.*, 1975] and more recently stress wave measurements in granite in excess of 1000 kbar have been obtained [*DeCarli*, 1975].

Early studies on manganin as a shock wave transducer material were conducted by *Fuller and Price* [1964] and by *Bernstein and Keough* [1964]. Manganin, an alloy with a nominal composition of 84% Cu, 12% Mn, and 4% Ni, exhibits a positive pressure coefficient of resistance and an extremely small temperature coefficient of resistance. *Keough et al.* [1964] have shown that the pressure coefficient of resistance is independent of temperature between 25 and 500°C for pressures to at least 150 kbar. These properties are particularly favorable for stress wave transducer applications.

Although manganin wire has been used, present methods favor gauge grid patterns photoetched from 25 to 50 μm manganin foil. The grids are usually mounted between slabs of the test sample with the grid-face oriented normal to the direction of wave propagation as illustrated in Figure 3. A four lead gauge configuration, two for current and two for measuring the voltage signal induced when the stress wave passes through the sensitive gauge element, is also currently used. Since the current is maintained constant, the transducer provides a voltage-time history which is directly related to the stress-time history through the transducer stress-resistance calibration curve.

The calibration of manganin has not received the exhaustive study subjected to the quartz transducer and is complicated by sensitivity to the medium surrounding the element. *Bridgman* [1950] reported a linear resistance with static pressure up to 30 kbar. Shock loading stress-resistance data to 400 kbar [*Lyle et al.*, 1969; *Keough and Wong*, 1970] appears to be best fit by a cubic stress-resistance expression. Satisfactory agreement was observed on a similar manganin by *Dremin and Kanel'* [1972]. *Keough* [1968] observed that the coefficient during unloading differs from that on loading and a residual net resistance change upon complete unloading can occur. Recent studies on ytterbium [*Ginsberg et al.*, 1973] and silver

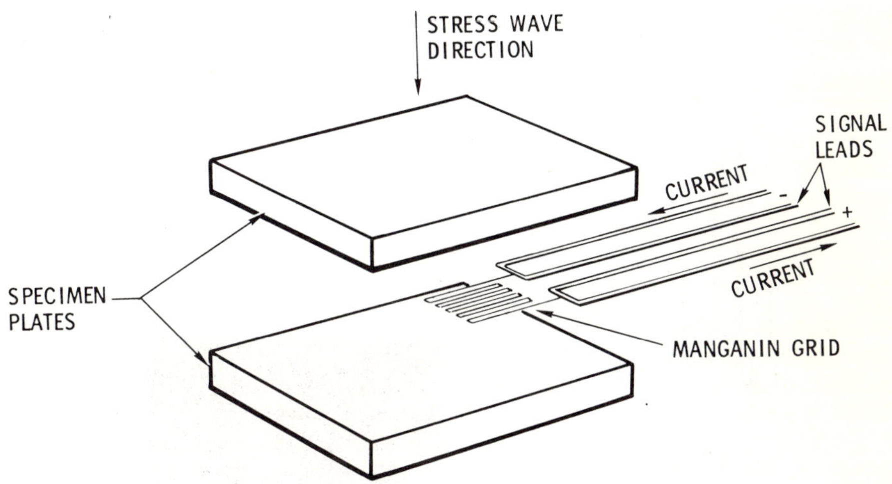

Fig. 3. Manganin transducer design showing a typical assembly geometry. Bonding is accomplished with epoxy, and the gauge plane is normally 25-50 μm.

[*Dick and Styris*, 1975] have shown that the loading stress-resistance coefficient can be separated into a part due to an intrinsic piezoresistivity effect and a part due to damage-induced defects introduced during plastic yielding at the transducer material. It was further observed that the damage-induced portion saturates, becoming small at high stress levels. Work by *Murri et al.* [1975] with gauges in rock indicated saturation for manganin and that above about 200 kbar, the loading and unloading coefficients are essentially the same.

Installation of gauges capable of surviving the 500 to 1000 kbar stress environment for the required recording time of several μs requires considerable care. In granular rocks, differential motion of grains can cause dimpling of the gauge material and provide erroneous stress profile measurements. Also, the piezoelectric property of some mineral types can cause significant noise problems. It has been found that isolating the gauge from the sample with thin sheets of mica can significantly mitigate both of these effects [*Murri et al.*, 1975].

Gauge planes between test samples thinner than 25 to 50 μm are difficult to achieve, and, since this is a region of mechanical impedance different from the sample, it provides for the most serious signal degrading factor in the manganin gauge technique. This problem has been analyzed by *Murri et al.* [1975] who find that for a 50 μm gauge plane in a typical rock material, a characteristic rise time of 0.035 μs occurs in a

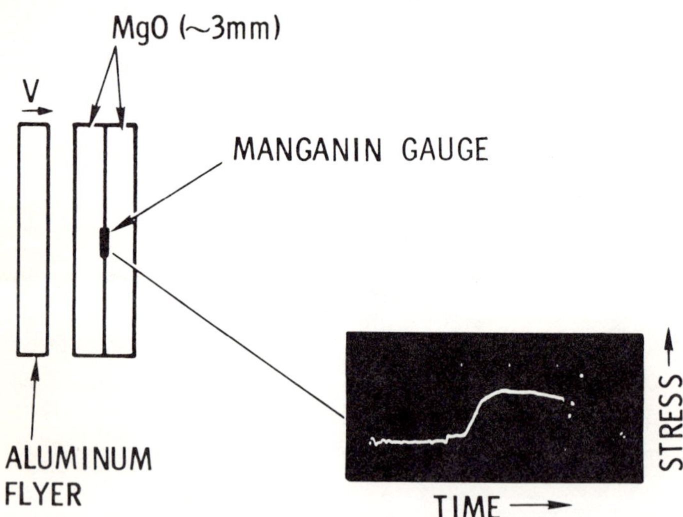

Fig. 4. Stress-wave profile obtained with manganin gauge in MgO. A step-stress input wave was provided by impact of an aluminum flyer (0.72 km/sec). The stress-wave profile was measured at an in-material point approximately 3 mm from the impact interface. 0.2 μsec time marks are shown.

sharp shock loading.

In Figure 4, a stress wave profile obtained from a 1 Ω manganin gauge grid mounted in MgO is shown [*Grady*, unpublished, 1970]. Maximum stress achieved in the experiment is approximately 65 kbar. Comparison of the initial wave shape with the quartz-gauge stress profile shown in Figure 2 illustrates the profile degradation that can occur because of the finite thickness of the gauge plane. This comparison somewhat unfairly represents the manganin gauge, however, in that thin sheets of aluminum foil plus mylar were required on each side of the manganin gauge element to shunt a very strong stress-induced polarization signal and resulted in an approximately 100 μm gauge plane thickness. The manganin gauge profiles shown in Figure 5 better illustrate the capability of the technique [*Grady et al.*, 1974*a*]. All six profiles were obtained in one experiment to a peak stress of about 250 kbar in a quartz rock. Three gauge planes (< 50 μm thickness) at increasing distance from the impact interface illustrate dispersion of the wave

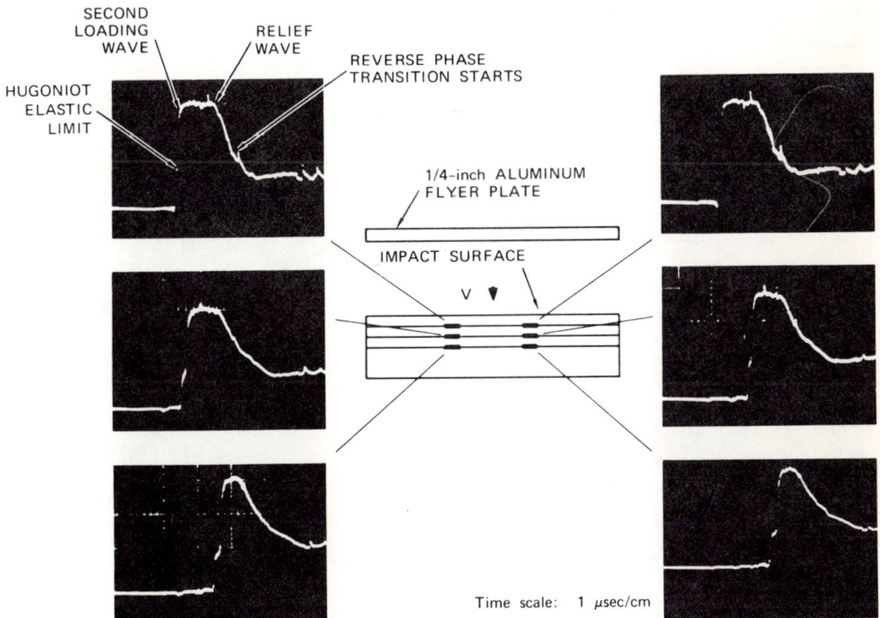

Fig. 5. Manganin stress-wave profiles obtained in quartz rock (Arkansas novaculite). The maximum stress achieved was 250 kbar. (After Grady et al. [1974].)

with propagation distance and also indicate reproducibility of the manganin-gauge technique. The break in slope of the relief wave has been attributed to the transformation of SiO_2 in octahedral coordination to SiO_2 in tetrehedral coordination upon unloading [*Grady et al.*, 1974*a*].

C. Electromagnetic Techniques

An electromagnetic transducer based on the principle of a moving conductor in a stationary magnetic field has been used extensively to measure particle velocity in shock-loaded solids [*Dremin and Adadurov*, 1964; *Dremin et al.*, 1965; *Petersen et al.*, 1970]. The method has been extended to the measurement of total wave profiles in rock to stress levels of 400 kbar [*Grady et al.*, 1974*a*; *Murri et al.*, 1975]. Some features of the electromagnetic gauge are similar to the manganin stress gauge: both gauges are normally installed in-material (between two slabs of the test sample); both can be used over the same stress range; and both have the same resolution difficulties due to wave degradation from the finite thickness of the gauge plane.

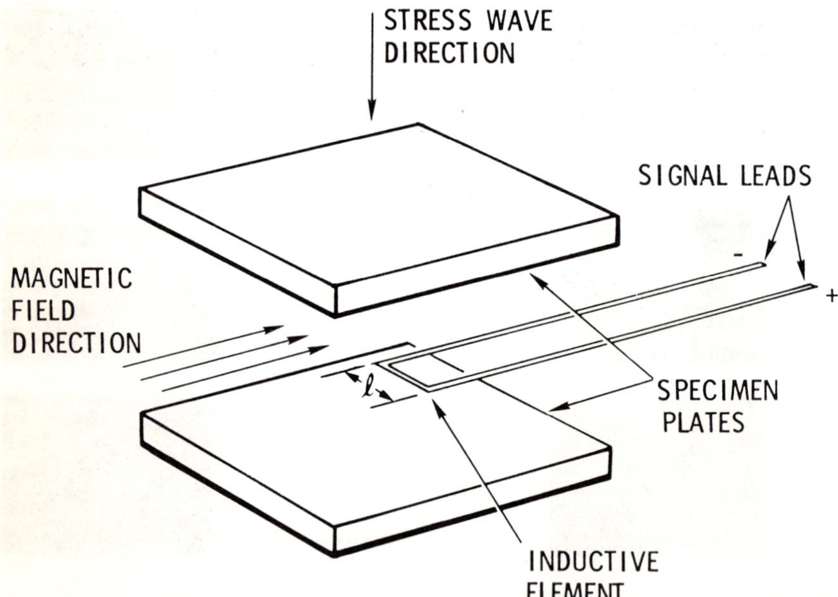

Fig. 6. Electromagnetic transducer design showing typical assembly geometry. l is the length of the active gauge element. The stress-wave direction, magnetic field, and active gauge element are mutually orthogonal.

The transducer element is typically photoetched from 25-50 μm copper foil in a U shape as is illustrated in Figure 6. The gauge is sandwiched between slabs of the test material and oriented so that the signal leads are parallel to the applied magnetic field and the active gauge element of length l is perpendicular to the applied field. Wave propagation is directed perpendicular to both the applied field and the active gauge element. Other workers have used a similar gauge geometry but with the signal lead exiting through the back of the test sample [*Altshuler et al.*, 1965]. In either case, the particle velocity history as the stress wave sweeps across the gauge plane is linearly related to the induced voltage $\varepsilon(t)$ through the relation

$$u(t) = \frac{\varepsilon(t)}{Hl} \tag{2}$$

where l is the center-to-center active element length and H is the magnetic field intensity. The magnetic field (typically 500 to 1000 gauss) can be generated with pulsed Helmholtz coils or with permanent magnets.

There are several attractive features about the electromagnetic gauge. No calibration is required; the gauge factor is obtained analytically through Equation 2 and measurement accuracy reduces to careful orientation of the gauge leads, applied field, shock direction, and accurate measurement of the magnetic intensity and active gauge length. The electromagnetic effect is not as sensitive to dimpling caused by differential grain motion and, since the gauge is a low impedance source, piezoelectric noise effects are small.

In Figure 7, particle velocity profiles obtained by the electromagnetic technique to peak stresses of approximately 124 kbar in Salem limestone are shown [*Murri et al.*, 1975]. Note that stress relief which begins when the incident shock emerges at the material free surface appears in the particle velocity record as a further acceleration of the material. The breaks in the relief wave are due to several phase transitions occurring in the calcite during unloading.

D. Laser Interferometry Techniques

The most accurate technique which has been developed for measuring particle velocity resulting from impulsive loading of solids utilizes laser interferometry. With this technique, either the particle velocity history at the specimen free surface or at an interface between specimen and an optical window material can be measured. Several improvements have followed the inception of the technique. Early instruments measured displacement history and velocity was obtained by differentiation of the records [*Barker and Hollenbach*, 1965]. Later, a

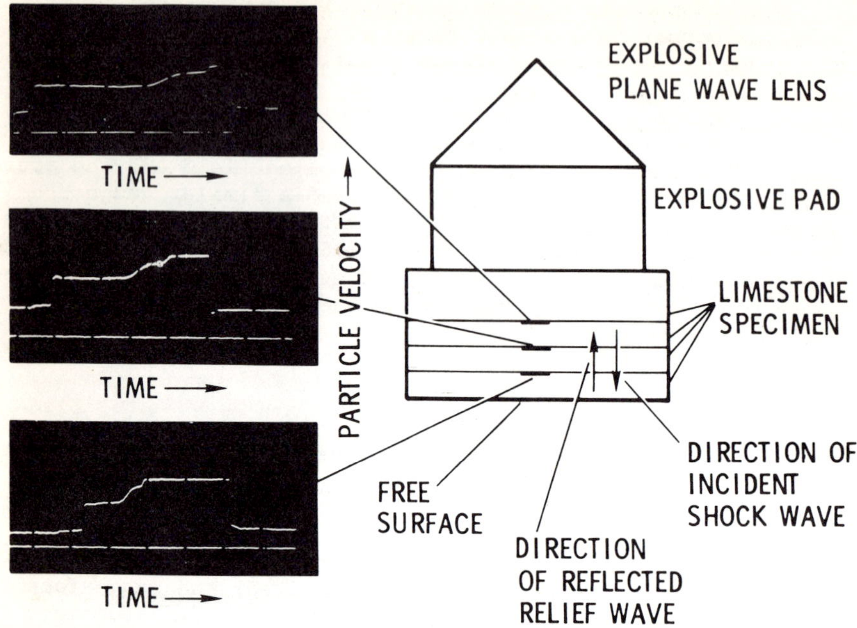

Fig. 7. Particle-velocity profiles obtained in Oakhall limestone. Peak stress equals 124 kbar. First and second waves correspond to arrival of the initial shock-wave and free-surface relief wave, respectively. (After Murri et al. [1975].) 1.0 μsec time marks are shown.

velocity interferometer technique which measured the particle velocity directly was developed [*Barker*, 1968]. Recently, the need for a specular reflecting surface or interface, required in early systems, has been eliminated with the development of a diffuse surface velocity interferometer system [*Barker and Hollenbach*, 1972]. The ability to monitor the motion of diffuse surfaces is a particularly important feature since most geological materials cannot be polished to a spectral finish, and, even if a mirror can be introduced by artificial means, loss of reflectivity usually results from passage of a strong shock wave.

The laser velocity interferometer (specular or diffusely reflecting) is based on the principle of temporally coherent light interference resulting from the moving reflecting surface. The technique uses a beam of laser light, which is reflected from a surface on the test specimen. The beam is then split and one portion is delayed a short time τ with respect to the other (on the order of a few ns) and then recombined.

Recombination of the beams effectively differences the displacement at time t and that at time t - τ and provides an accurate measure of the velocity averaged over the time τ. *Barker and Hollenbach* [1970] have derived the following expression for the velocity:

$$v(t - \frac{1}{2}\tau) = \frac{\lambda}{2\tau} F(t) \tag{3}$$

where λ is the wave length of the laser light and $F(t)$ is the fringe count at time t.

The diffuse surface interferometer operates on the same principle as the specular velocity interferometer with the exception that spatial coherence of the reflected light beam is not required. Interferometry with spatially incoherent light can be accomplished provided that the two legs of the interferometer are nearly the same length. To satisfy this requirement and also achieve a relative delay time τ required for velocity interferometry, fused silica etalons are used in one leg of the light beam. The different index of refraction of the etalons makes the apparent path length of the two legs identical, but produces a net delay time. A schematic of the currently used diffuse surface velocity interferometer system is shown in Figure 8.

In the study of large amplitude wave propagation in solids, it is less ambiguous to measure the in-material wave properties rather than the free surface motion which is complicated by the reflected wave. This can be accomplished with laser interferometry by using transparent window materials and measuring the particle motion at a diffusely reflecting interface between the specimen and window material. Although reflected wave effects are not totally eliminated, they are significantly reduced by careful selection of the impedance of the window material. Stress-induced index of refraction changes occur when the stress wave enters the window material and this must be corrected for in the velocity interferometer equation, Equation 3. The corrected equation which has been derived by *Barker and Hollenbach* [1970] is

$$v(t - \frac{1}{2}\tau) = \frac{\lambda F(t)}{2\tau(1 + \Delta\nu/\nu_o)} \tag{4}$$

where $\Delta\nu/\nu_o$ is the index of refraction correction. A correction due to doppler shift of the reflected beam and wave length dependence of the index of refraction of the etalon material must also be made [*Barker and Schuler*, 1974]. Three window materials have been calibrated, PMMA (polymethyl methacrylate), fused quartz, and Z-cut single crystal sapphire, which provide a good selection of different mechanical impedances [*Barker and Hollenbach*, 1970]. Fused quartz remains transparent to at

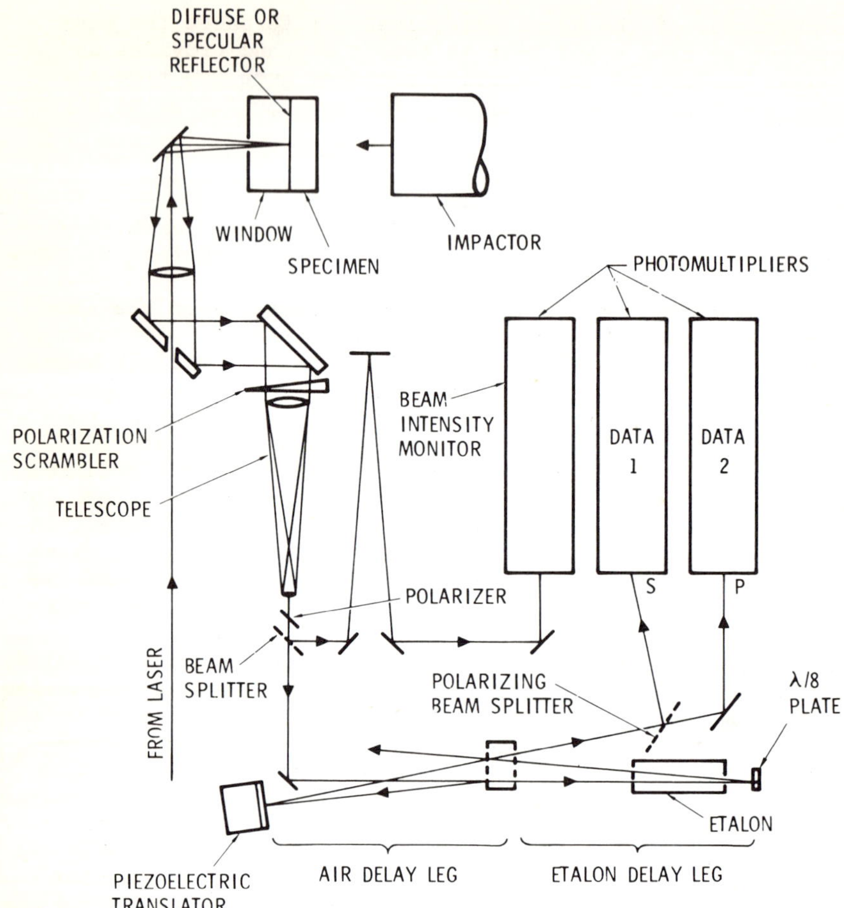

Fig. 8. System schematic of the diffuse-surface velocity interferometer, illustrating the optical path to and from the test specimen, system optics, and data-collecting photomultiplier tubes. Surface acceleration can be distinguished from deceleration by separation of return beam into S and P polarized light displaced 90° in phase. (After Barker and Hollenbach [1972].)

least 65 kbar and sapphire appears to lose transparency at about 130 kbar [*Barker and Hollenbach*, 1970] although *Gaffney and Ahrens* [1973] have observed transparency in sapphire at approximately 530 kbar. The upper stress limit for PMMA has not yet been ascertained although it has been used successfully by *Asay and Hayes* [1975] to 110 kbar.

Laser interferometry is the most accurate stress wave instrumentation currently available and has the capability of resolving features in the wave structure which are usually not observed with other methods. For measurement of free surface velocity there is, in principle, no upper stress limitation, although with strong shocks local surface jetting can complicate results. With window materials, the high pressure behavior has not been studied thoroughly although efforts are currently underway in this area.

In Figure 9, a particle velocity profile obtained with a diffuse surface velocity interferometer on single crystal periclase is shown. The maximum stress achieved in the

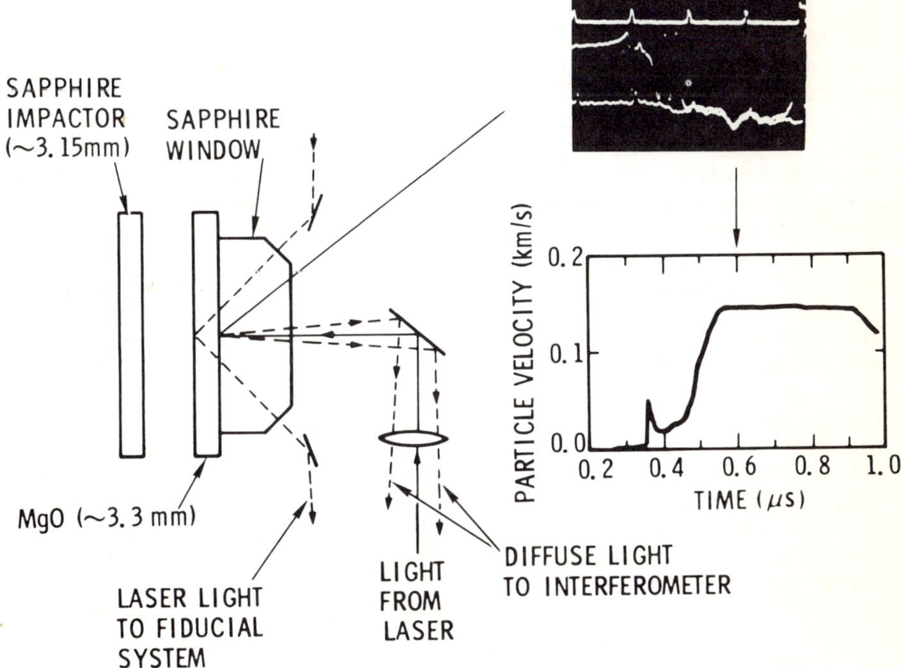

Fig. 9. Interferometer fringe pattern and the resulting particle-velocity profile obtained in MgO with diffuse-surface velocity interferometry. Maximum stress achieved in MgO was approximately 48 kbar. Second laser beam provides impact fiducial.

periclase was approximately 48 kbar. This result can be compared with the profile obtained with a quartz gauge

(Figure 2) and with a manganin gauge (Figure 4). Also shown in the figure is a second laser beam which undergoes total internal reflection at the center of the MgO front surface. At the time of projectile impact at the target center, the laser beam is diverted from its preimpact optical path and provides a very accurate fiducial for determining the transit time across the test specimen [*Nunziato et al.*, 1974].

III. ANALYSIS METHODS

In early shock-wave studies on solids, shock velocity and particle velocity were the most readily measured quantities. They were of little intrinsic interest, however, and the Rankine-Hugoniot relations [*Duvall and Fowles*, 1963] were used to calculate thermodynamic quantities of the high pressure state; namely, the pressure P, the specific volume V, and the specific internal energy E. Application of these relations assumed that transition from the initial state occurred by a single discontinuous shock wave.

With improved instrumentation, it has become recognized that, in many cases, the shock transition process is continuous and complex, and that application of the Rankine-Hugoniot relations is not reliable. Other methods are needed to analyze continuous wave propagation data and to obtain the stress, volume, and internal energy histories. These quantities are not easily obtained from continuous experimental stress-time or particle velocity-time profiles. However, the problem has prompted a number of studies [*Fuller and Price*, 1964; *Fowles and Williams*, 1970; *Cowperthwaite and Williams*, 1971; *Grady*, 1973; *Seaman*, 1974] which have resulted in methods for calculating constitutive relations among stress, particle velocity, specific volume, and specific energy directly from the experimental profiles and the conservation equations of mass, momentum, and energy. These results relate to the thermomechanical processes occurring during shock compression and release and assist in understanding and in establishing models to describe these processes.

A. Conservation Equations

The conservation laws of mass, momentum, and energy for continuous one-dimensional flow are, respectively,

$$\left(\frac{\partial V}{\partial t}\right)_h - \frac{1}{\rho_o}\left(\frac{\partial u}{\partial h}\right)_t = 0 \ , \tag{5}$$

$$\left(\frac{\partial u}{\partial t}\right)_h + \frac{1}{\rho_o}\left(\frac{\partial \sigma}{\partial h}\right)_t = 0 \ , \tag{6}$$

$$\left(\frac{\partial E}{\partial t}\right)_h + \frac{\sigma}{\rho_o}\left(\frac{\partial u}{\partial h}\right)_t = 0 \ , \tag{7}$$

where V is the specific volume, u the particle velocity, σ the stress in the direction of wave propagation, E the specific internal energy, and ρ_o is the initial density. The independent variables are the time t and the material or Lagrangian position h. These equations are generally valid and no recourse to equilibrium states is implied. The only restriction is in Equation 7 where heat flow and radiation have been neglected.

In terms of the experimental conditions, which are to measure continuous stress-time or particle velocity-time profiles at discrete material positions, it is convenient to express the conservation equations in the following integrated form,

$$V(h,t) = V(h,t_o) + \frac{1}{\rho_o}\int_{t_o}^{t}\left(\frac{\partial\sigma}{\partial h}\right)_t dt \ , \tag{8}$$

$$u(h,t) = u(h,t\ ) - \frac{1}{\rho_o}\int_{t_o}^{t}\left(\frac{\partial\sigma}{\partial h}\right)_t dt \ , \tag{9}$$

$$E(h,t) = E(h,t_o) - \frac{1}{\rho_o}\int_{t_o}^{t}\sigma\left(\frac{\partial u}{\partial h}\right)_t dt \ , \tag{10}$$

where the subscripted variables refer to a reference state usually selected in a region of constant state downstream from the profile data to be evaluated. In each of these integral expressions, the term within the integral sign is evaluated from experimental profile data. For instance, if stress histories at several gauge locations are available, the data are used to estimate $(\partial\sigma/\partial h)_t$ and equation 9 can be evaluated. Then, in turn, Equations 8 and 10 may be evaluated to provide the total stress, volume, energy, and particle velocity history through the compression or relief wave. If particle velocity profiles are measured, Equation 9 must be formulated differently.

Frequently, various simplifications in the wave propagation occur which simplify evaluation of the conservation equations. Such simplifications may result from a priori information about the material constitutive response or experimental boundary conditions, or they may be required due to experimental limitations such as the number or accuracy of the gauges used.

B. Shock Waves

Even with the increased resolution of continuous in-material gauges, discontinuities in the wave profiles are often observed. Elastic shocks can occur and deformation shocks can also form due to positive curvature of the stress-strain response. Rarefaction shocks due to phase transitions are also observed during unloading.

If a discontinuity in h-t space occurs in the first order flow variables, V, u, σ, and E, say along a path $h = \zeta(t)$, then the Rankine-Hugoniot relations apply and can be derived from the continuous conservation integral expressions in Equations 8, 9 and 10. If the velocity of the discontinuity (the Lagrangian shock velocity) is given by $U = d\zeta/dt$ and if the integrands in Equations 8, 9 and 10 are transformed by the differential expression $\partial/\partial h = (d/dt - \partial/\partial t)/U$, then application of the mean-value theorem to the resulting integrals and proceeding to the limit results in

$$V_1 = V_2 - \frac{1}{\rho_o U}(u_1 - u_2) \;, \tag{11}$$

$$u_1 = u_2 + \frac{1}{\rho_o U}(\sigma_1 - \sigma_2) \;, \tag{12}$$

$$E_1 = E_2 + \frac{\sigma_1 + \sigma_2}{2\,\rho_o U}(u_1 - u_2) \;, \tag{13}$$

where the subscripts 1 and 2 refer to values in front of and behind the discontinuity, respectively. When discontinuous flow is observed, then measurement of any two of the flow variables, say shock velocity and the change in particle velocity, are sufficient to determine the change in the remaining flow variables across the discontinuity.

C. Steady Waves

Steady waves occur frequently in the study of large amplitude compression waves. Such a wave propagates unchanged in form and at a fixed velocity into a region of constant state. They can result from a balancing of viscous stresses, which tend to diffuse the wave profile, and positive curvature of the stress-strain relation which tends to steepen the profile.

In steady waves, the flow variables are functions only of the single independent variable $\zeta = t - h/U$ where U is the constant velocity of the wave. With this functional constraint, the integral Equations 8, 9 and 10 can be evaluated directly and result again in the Rankine-Hugoniot form,

$$V(t) = V_o - \frac{1}{\rho_o U} (u(t) - u_o) , \tag{14}$$

$$u(t) = u_o + \frac{1}{\rho_o U} (\sigma(t) - \sigma_o) , \tag{15}$$

$$E(t) = E_o + \frac{1}{2 \rho_o U} (\sigma(t) + \sigma_o)(u(t) - u_o) , \tag{16}$$

which apply from the reference state to any point on the wave profile. In σ-V-u space the material path from the reference state to the final state is a chord (Rayleigh line). Gauge records of either stress or particle velocity obtained at two positions are required to experimentally verify steady wave propagation.

D. Simple Waves

Another common simplification results when the material response can be regarded as rate-independent and propagation is into a region of constant state. For such waves, called simple waves, the theory of characteristics shows that the space and time derivatives are related by the differential expression $dx = c\, dt$ where c is the Lagrangian sound speed and depends on the current state of stress at each point in the wave. The conservation Equations 8, 9, and 10 then reduce to the Riemann integral form,

$$V(t) = V_o - \frac{1}{\rho_o} \int_{u_o}^{u(t)} \frac{du}{c} , \tag{17}$$

$$u(t) = u_o + \frac{1}{\rho_o} \int_{\sigma_o}^{\sigma(t)} \frac{d\sigma}{c} , \tag{18}$$

$$E(t) = E_o + \frac{1}{\rho_o} \int_{u_o}^{u(t)} \frac{\sigma\, du}{c} \tag{19}$$

Consequently, two continuous stress gauges, or particle velocity gauges, can be used to determine $c(\sigma)$, or $c(u)$, and the integrals evaluated to determine $V(t)$, $u(t)$, or $E(t)$. Although two gauges are sufficient to estimate c, three gauges are necessary to verify that the paths of constant stress or particle velocity are straight and that simple-wave theory is valid.

E. Centered Waves

A special case of simple wave propagation occurs when the wave is centered at the impact interface of the target specimen. The relief wave, as well as the loading wave, can be centered since some impactor materials, such as fused silica, can produce unloading shock waves which will introduce discontinuous unloading at the impact interface when a thin flyer plate is used. In h-t space, the characteristics are straight lines focused at the point of origin of the discontinuous imput wave, and there is a one-to-one correspondence between the continuous stress $\sigma(t)$ or particle velocity, $u(t)$ and the Lagrangian sound speed, $c = h/(t - t_o)$. The integral Equations 8, 9 and 10 can be evaluated with one gauge record although two are required to verify that the wave propagation is centered.

F. Unsteady Waves

When none of the foregoing simplifications are justified, experimenters have used the integral Equations 8, 9 and 10 directly to interpret wave propagation data. Specifically, consider three stress gauge records equally spaced a distance Δh apart. Since the data are continuous in time and discrete in space, this suggests the following centered finite difference approximation [*Herrmann*, 1976].

$$\left(\frac{\partial \sigma}{\partial h}\right)_{12} = \frac{\sigma_2(t) - \sigma_1(t)}{\Delta h} - \frac{1}{24}\left(\frac{\partial^3 \sigma}{\partial h^3}\right)_{12} \Delta h^2 + \ldots \tag{20}$$

where the subscript 12 refers to the midway point between gauges 1 and 2. Ignoring second and higher order terms, Equation 9 can be evaluated to obtain the particle velocity history at the 12 midway point. An expression similar to Equation 20 can be used to determine the particle velocity history at the 23 midway point and the approximation,

$$\left(\frac{\partial u}{\partial h}\right)_2 = \frac{u_{23}(t) - u_{12}(t)}{\Delta h} - \frac{1}{24}\left(\frac{\partial^3 u}{\partial x^3}\right)_2 \Delta h^2 + \ldots , \tag{21}$$

can be truncated to the first term and used to evaluate the specific volume and specific internal energy histories through Equations 8 and 10. When particle velocity profiles are used, a different analytic approach must be taken [*Seaman*, 1974].

When stress relaxation is observed in the data or when severe wave attenuation occurs due to rapid overtaking of the loading wave by the relief wave, it has been observed [*Grady*,

1972; *Grady*, 1973; *Seaman*, 1974] that the integrands in Equations 8, 9 and 10 can be expressed to make better use of the experimental data. Namely, the space derivative can be written

$$\frac{\partial}{\partial h} = \frac{1}{c_\zeta}\left(\frac{d}{dt} - \frac{\partial}{\partial t}\right) \tag{22}$$

where d/dt is a directional derivative along a path $h = \zeta(t)$ chosen to make optimum use of the wave profile data and $c_\zeta = \dot{\zeta}(t)$. The integral Equations 8, 9 and 10 then transform to

$$V(h,t) = V(h,t_o) - \frac{1}{\rho_o}\int_{u_o}^{u} \frac{du}{c_\zeta} + \frac{1}{\rho_o}\int_{t_o}^{t} \left(\frac{du/dt}{c_\zeta}\right)dt\ , \tag{23}$$

$$u(h,t) = u(h,t_o) + \frac{1}{\rho_o}\int_{\sigma_o}^{\sigma} \frac{d\sigma}{c_\zeta} - \frac{1}{\rho_o}\int_{t_o}^{t} \left(\frac{d\sigma/dt}{c_\zeta}\right)dt, \tag{24}$$

$$E(h,t) = E(h,t_o) + \frac{1}{\rho_o}\int_{u_o}^{u} \frac{\sigma\ du}{c_\zeta} - \frac{1}{\rho_o}\int_{t_o}^{t} \left(\frac{\sigma\ du/dt}{c_\zeta}\right)dt\ . \tag{25}$$

In practice, the paths should be chosen to connect similar features in the gauge records such as the peaks of elastic precursors, wave maxima, or the initiation of rarefaction shocks. When paths have been selected the experimental profiles are smoothed and incremented in time, the directional derivatives are evaluated similar to Equations 20 and 21, and Equations 23, 24 and 25 are integrated numerically.

When attenuation is not observed in the wave features, paths of constant stress or constant particle velocity are used, and $c_\zeta = c_\sigma$, or c_u, which are the wave velocities at constant stress, or constant particle velocity, first discussed by *Fowles and Williams* [1970]. Under this specialization, the last term in Equations 23, 24 and 25 vanishes and they take the form

$$V(t) = V_o - \frac{1}{\rho_o}\int_{u_o}^{u} \frac{du}{c_u} \tag{26}$$

$$u(t) = u_o + \frac{1}{\rho_o}\int_{\sigma_o}^{\sigma} \frac{d\sigma}{c_\sigma} \tag{27}$$

$$E(t) = E_o + \frac{1}{\rho_o} \int_{u_o}^{u} \frac{\sigma \, du}{c_u}, \tag{28}$$

which represent a generalization of the Riemann integral expressions, Equations 17, 18 and 19. *Fowles and Williams* [1970] and *Cowperthwaite and Williams* [1971] have derived a number of relations concerning c_σ and c_u. In particular, they have shown that these velocities are equal for simple and steady wave propagation.

G. Impedance Mismatch Corrections

In both the quartz-gauge and the velocity-interferometer techniques, the material downstream from the Lagrangian point at which the wave profile measurement is made differs from the test material. Although care is usually exercised to select a backing material with mechanical impedance close to that of the test specimen, the match is never exact and wave reflections occur. The measured profile is then a distorted version of the profile originally propagated in the undisturbed test medium. To analyze these profiles, it is necessary to first account for the impedance mismatch and determine the wave profiles in the undisturbed medium.

A scheme frequently used is based on an incremental impedance matching technique and results in relations similar to linear acoustic theory [*Grady*, 1976]. The method is readily appreciated through consideration of the Riemann invariants,

$$J_+ = u + \int^{\sigma} \frac{d\sigma}{\rho_o c} \tag{29}$$

$$J_- = u - \int^{\sigma} \frac{d\sigma}{\rho_o c} \tag{30}$$

defined in the test specimen upstream from the interface at which the wave profile is measured. The Riemann invariants J_+ and J_- are constant on right and left-facing characteristic paths, respectively.

The backing material, either crystal quartz or a laser window material such as fused silica, has known material properties. That is, if a velocity profile $U(t)$ is measured then the stress profile $P(t)$ is readily determined from the governing constitutive equation. Since both stress and particle velocity must be continuous, it follows that, at the measurement interface,

$$J_{+} = U(t) + \int^{P(t)} \frac{d\sigma}{\rho_{o}c} , \tag{31}$$

$$J_{-} = U(t) - \int^{P(t)} \frac{d\sigma}{\rho_{o}c} . \tag{32}$$

Due to invariance, the value of J_{+} on a given characteristic would be the same if no interface were present. However, since reflections would not occur in this case, the value of J_{-} would be zero. Coupling this with Equations 29 and 30 results in the following expression for the corrected (undisturbed) particle velocity profile,

$$u(t) = \frac{1}{2}\left(U(t) + \int^{P(t)} \frac{d\sigma}{\rho_{o}c}\right). \tag{33}$$

Application is usually in differential form in which case corrections to the differential particle velocity and stress are, respectively,

$$du = \frac{1}{2}\left(d\,U + \frac{d\,P}{\rho_{o}c}\right), \tag{34}$$

$$d\sigma = \frac{1}{2}\,(d\,P + \rho_{o}c\;d\,U) . \tag{35}$$

Neglected in this analysis is the slight refraction of J_{+} characteristics when they enter the interaction region due to wave reflections, which causes slight arrival time errors upon reaching the measurement interface. These time errors are small, however, if propagation time in the region of interaction is maintained small compared to propagation time in the noninteraction region and if the impedance mismatch is kept small so that bending of characteristics is minimized.

In application of Equations 34 or 35, the value of c is approximated from the measured Lagrangian wave velocity corresponding to a given stress or particle velocity level in the wave profile. Since the value of c should correspond to the corrected stress or particle velocity profile, one or two iterations in application of Equation 34 or 35 can improve the results.

This method of analysis is strictly applicable to rate independent material response. Although frequently used on wave profiles occurring in rate dependent material, it is not clear how representative the corrected profiles are.

H. Factors Affecting Accuracy

There are a number of error sources in this type of wave form analysis which make the total accuracy difficult to assess. The analysis itself is subject to approximation errors due to truncated terms in the estimation of spatial derivatives. Also smoothing and digitizing the profile records and the process of numerical interpolation, differentiation, and integration are possible error sources to which some attention must be given. In addition, there are experimental errors in the gauge records themselves. Discrepancies in gauge calibration and in time correlation between gauges occur. The time resolution problems with imbedded gauges are significant sources of error and if the gauge is backed by a different impedance material, uncertainties arise due to wave reflections.

Analysis on idealized profiles obtained from known solutions [*Cowperthwaite and Williams*, 1971; *Grady*, 1973; *Seaman*, 1974] have shown that, provided derivatives are estimated along properly chosen paths, truncation errors do not result in significant loss of accuracy. The major problem in this regard seems to occur when severe attenuation between gauge profiles is encountered. This can be minimized by proper placement of the gauges.

The dominant source of error resides in the gauge records themselves. Time correlation between records is extremely important since the wave velocity appears to the second power in the stress-volume response. Provided similar gauges are used (for instance, multiple manganin gauges), errors due to calibration or frequency response limitations will appear to the same order in the evaluated stress-strain or stress-particle velocity response. Although similar analysis can be completed using combinations of gauges, say stress and particle velocity gauges, this should probably be avoided due to the different accuracy problems of the gauge types which can lead to excessive error amplification.

However, if proper attention is given to the numerical details, and if the accuracy limitations of the gauge type used are understood, such analysis can provide a useful tool in evaluating material response from large amplitude wave propagation data. Details of the stress, strain, strain rate and energy history are more transparent than the measured stress-time or particle velocity-time profiles and provide further insight into the thermomechanical process active during dynamic deformation.

IV. YIELDING AND PHASE TRANSITIONS IN MINERALS

Certain features in the Hugoniots obtained from shock compression studies on silicate and oxide minerals have avoided a consistent explanation for some time. Notable complicating features are the metastable portion of the silicate Hugoniots in the mixed phase region and the significant reduction in strength observed above the Hugoniot elastic limit in both silicates and oxides. Recent application of newer shock wave techniques, combined with careful interpretation of earlier shock wave and recovery experiments, has resulted in a plausible description of the physical processes occurring in the shock compression of minerals. The basic concepts were reached independently by *Ananin et al* [1974], *Graham* [1974], and *Grady et al.* [1974*b*, 1975]. It would be premature to state that the shock compression process in minerals is now understood. It is not clear whether the description spans all minerals of interest, although certain similarities in experimental results suggest that some aspects of the proposed description occur in most minerals.

The process of dynamic yielding when shock stresses in excess of the Hugoniot elastic limit are achieved seems to be the crucial factor in determining the subsequent compressibility. It is not yet clear whether yielding is a brittle process (complete shearing of atomic planes) or a plastic process due to shear activation and multiplication of dislocations in the lattice. The choice is not critical to the present argument. The important hypothesis is that yielding is a heterogeneous process and that large regions remain virtually undamaged during shock loading. This tenet is supported by a considerable body of experimental evidence which will be discussed in section IV A. Since yielding is localized to shear zones, energy dissipation due to nonconservative forces in the shock process will also be localized to these zones and lead to large temperature gradients in the shocked material. Energy considerations on the Hugoniot show that temperatures in these shear zones can be sufficient to cause local melting even though average temperature estimates are still well below melting. Due to low thermal conductivity in these minerals, the nonuniform temperature distribution can persist for some time after passage of the shock front.

A process of adiabatic shear and localized melting can account for most of the curious shock compression effects observed in rocks and minerals including metastable Hugoniots, reduction in strength, release wave behavior, and petrographic features identified in samples subjected to strong shock waves.

A. Experimental Observations in Shock Compression Studies

Reduction in Strength Above the Hugoniot Elastic Limit

The behavior of metals during shock compression is readily explained within the framework of elastic-plastic theory. The Hugoniot above the elastic limit is usually observed within experimental error to maintain a constant offset from the hydrostat equal to 4/3 τ_Y where τ_Y is the resolved shear stress at yield. Exceptions are normally explained by work hardening or strain rate dependence. The situation for the silicates and oxides is less clear. Studies on polycrystalline corundum [*Ahrens et al.*, 1968; *Gust and Royce*, 1970] indicate that considerable strength is retained above the Hugoniot elastic limit. On the other hand, single crystal periclase [*Ahrens*, 1966], corundum [*Graham and Brooks*, 1971], and quartz [*Wackerle*, 1962; *Fowles*, 1967; *Graham*, 1974] show a substantial loss of strength.

It seems reasonable that, at stress levels where significant adiabatic shearing causes softening or melting in localized planes, the bulk viscosity of the material will be sharply reduced accounting for the observed reduction in strength. There are several reasons why the same behavior might not be expected in fine grain polycrystalline material. First, polycrystals normally exhibit lower Hugoniot elastic limits than the equivalent single crystal, dependent, usually, on the amount of porosity in the material. The lower elastic limit would provide for less elastic shear energy available during the yield process. Second, the separation of shear zones, on the order of 5-20 μm suggested by recovery work in single crystal quartz [*Ananin et al.*, 1974], could be significantly reduced due to grain boundary constraints. *Borg* [1974] noted one or more sets of planar features, presumably post shock signatures of shear zones, localized to quartz grains in material shocked to 150 kbar, suggesting that the deformation energy in polycrystalline material is more uniformly distributed resulting in less energy per shear zone. Hence, at the same stress level, softening, or melting in the polycrystal should be less than in the single crystal, resulting in greater residual strength.

Metastable Hugoniot

Perhaps the most notable experimental feature observed in the minerals which undergo a primary coordination phase transition under shock compression is the metastable Hugoniot in the mixed phase region [*McQueen et al.*, 1967; *Ahrens et al.*, 1969]. The α-quartz to stishovite transition is representative. According to equilibrium thermodynamics, this transition should traverse the mixed phase region on the Hugoniot within a few tens of kbar. Instead the transition, which initiates in the neighborhood of 100 kbar, does not reach completion until shock pressures on the order of 400 to 500 kbar are achieved.

A number of the primary coordination phase transitions which occur during shock compression have also been observed under static conditions [*Stishov and Popova*, 1961; *Ringwood et al.*, 1967]. They are reconstructive, hence thermally-activated, and are observed to proceed very slowly. When shock compression of sufficient intensity to cause local melting and loss of shear strength occurs, the material state in undamaged regions between the planes of melt should differ little from the static experiment with the exception of a slight temperature rise due to near isentropic volume compression. It is unreasonable to expect that the phase transition rate in these regions should differ significantly from the static case. However, in the shear zones, temperatures are easily sufficient to transform material to the high pressure phase or a liquid form of the high pressure phase by a thermally activated nucleation and growth process, even on the sub-microsecond scale of a shock-wave experiment. It seems, then, that at a given stress level, the transformation will proceed within the shock front only in those regions which are sufficiently hot. Continued transformation in cooler regions will proceed on a considerably longer time scale and give the illusion, within the duration of a shock-wave measurement that an equilibrium Hugoniot state has been attained.

Reduced Sound Velocities on the Hugoniots

The velocity of the initial break of the relief wave following a strong shock wave, or the rate at which radial relief waves propagate into a shock compressed specimen, provide values of the longitudinal sound velocity at high pressure. These methods have been used with some success in metals [*Al'tshuler et al.*, 1960]. Recent studies on rock forming minerals differ from the results on metals. *Grady et al.* [1975] have observed initial relief wave velocities very close to the expected bulk sound velocity in quartz and feldspar, and *Bless and Ahrens* [1976] have observed relief velocities intermediate between the bulk and longitudinal velocities in corundum. These studies provide further evidence for loss of strength during shock compression and further suggest that the loss of strength can persist for some duration after passage of the shock wave.

Ferrimagnetic Effects Under Shock Compression

Shaner and Royce [1968] have observed that yttrium iron garnet remains ferrimagnetic to at least 400 kbar, substantially above the Hugoniot elastic limit of approximately 50 kbar in this material. Continued ferrimagnetism, a physical phenomenon requiring long range crystalline order, further attests to crystalline integrity of a significant portion of the material at large Hugoniot pressures.

The reduction in magnetization observed by *Shaner and Royce* [1968] was originally explained strictly by magneto-elastic effects. However, if a condition of thermal heterogeneity persisted at the Hugoniot state, then hot zones would be easily in excess of the Neél temperature and provide for reduction of the bulk saturation magnetization. This alternative explanation is consistent with the observed magnetization versus bias field intensity data which indicates an increasingly reduced stress-induced magnetic anisotropy effect with increased stresses above the Hugoniot elastic limit. This would be expected if reduction in material strength is occurring above the Hugoniot elastic limit.

Petrographic Features in Pre-Shocked Material

Extensive studies have been conducted on minerals which have been shock-loaded to stress levels in excess of the Hugoniot elastic limit. A large portion of this work has been on specimens recovered from the vicinity of meteorite impact or nuclear explosions [*Bunch*, 1968; *Chao*, 1968; *Borg*, 1974] and neither the stress level nor the duration of the stress pulse is well determined, although the times of stress application are appreciably longer than can be achieved in a laboratory experiment. Some work on samples which have been recovered from laboratory shock wave experiments has been documented where both the stress level and stress duration are fairly accurately known [*DeCarli*, 1968].

Planar features (shock lamellae) have been identified in recovered samples of shocked minerals [*Stöffler*, 1972]. They have been studied most extensively in quartz. The lamellae are observed in materials shocked between about 100 to 350 kbar and are identified optically as features separating otherwise undamaged minerals and having spacings of between 2 to 20 μm. They occur preferentially on lattice planes of densest packing. Planar features have been categorized according to differing optical characteristics, however, it can be assumed that they correspond to shear zones which originate due to yielding during shock compression and differences occur due to differing stress levels and thermal and strain-rate histories.

Ananin et al. [1974] observed that samples of single crystal quartz shock loaded to just above the Hugoniot elastic limit broke into small rectangular blocks of dimensions 100-200 μm perpendicular to the direction of shock propagation and 5-20 μm parallel to the shock direction. The stresses achieved were substantially lower than the stress region in which post shock planar features are observed, however, the results are suggestive of the same nonuniform yield process.

Klein [1965] conducted recovery experiments on single crystal periclase at shock stresses near 80 kbar. Planar features observed in the recovered samples were noted to increase in separation with increasing distance from the explosive-specimen interface suggesting that, as the shock front widened with increased propagation distance, a coarser shear band structure was favored. He also noted that at large distances, an initial [110] direction preference for the features shifted to a [100] preference.

Shock-recovered specimens of quartz and feldspar are also observed to contain regions of amorphous short-range-order phases. Diaplectic glass [*Engelhardt and Stöffler*, 1968] is observed to predominate in the lower shock pressure region (200 - 400 kbar) and fused glass more characteristic of those quenched from a liquid state predominate at higher pressures. Diaplectic glass differs from ordinary quartz glass in density, index of refraction, and its character of preserving the morphological signature of the parent crystalline state. Diaplectic glass is thought to be the reversion product of a high density, short-range-order solid state phase during pressure release [*DeCarli and Milton,* 1965]. Alternatively it could be the reversion product of the melted high density phase where freezing occurred prior to total pressure release. In such short times, liquid diffusion processes would be insufficient to totally erase the morphology of the previous crystal state. *Boganov et al.* [1971] have demonstrated the possibility of such instantaneous freezing of quartz. In quartz, traces of the high density polymorph, stishovite have also been identified in recovered specimens of natural and laboratory shocked material [*DeCarli and Milton,* 1965].

Phase Transition on Release

Ahrens et al. [1968], using liquid reflection techniques, and *Grady et al.* [1974*a*], using manganin gauges, have identified the transition from high density to low density state in quartz on pressure released from Hugoniot states in the mixed phase region. The transition, occurring between 80 to 100 kbar, is well-defined but does not exhibit a rarefaction shock, suggesting some rate dependence in the transition process. Similar studies on feldspar do not show a well-defined transition break; rather, relief paths suggest a continuous transition toward the low density phase [*Ahrens et al.,* 1969; *Grady and Murri,* 1976]. This is probably a consequence of the more complex mineral content of the feldspar specimens studied.

Triangular stress wave input pulses have been used to study the behavior of quartz rock in the mixed phase region [*Grady et al.,* 1974*a*]. By analyzing the wave profiles, it was determined that, after the initial rapid transformation to a Hugoniot state in the mixed phase region, a continuing transition from low-to-high density proceeded at a considerably slower rate. A possible explanation for this second transformation rate is thermal diffusion from initial hot spots into regions of untransformed material.

B. Dynamic Yielding

The silicate and oxide minerals are a class of brittle compounds which, under shock loading conditions, exhibited exceptionally high elastic limits. Theoretical studies on the ultimate strength of solids place probable lower limits of 0.03 G (G = shear modulus) as the shear stress at which failure occurs. Although most solids yield at considerably lower stress levels, the silicates and oxides are observed to yield under shock loading at stresses closely approaching the theoretical limit. Notable examples are single crystal corundum and quartz which exhibit maximum shear stress values of 0.056 C_{44} and 0.11 C_{44}, respectively [*Graham and Brooks*, 1971].

It has been argued that brittle failure (shear fracture) in these materials is unlikely due to the very large isotropic stress component which occurs in a shock-wave experiment, and that an energetically more favorable mechanism of dislocation motion and multiplication must occur. The very high observed elastic limits provide no support for this argument, however, and at present there is little experimental base for choosing a ductile mechanism over a brittle mechanism. Perhaps the most suggestive experimental evidence for a ductile yield process in minerals is the strong similarity between stress relaxation occurring in periclase (Figures 2 and 9), and probably in quartz as suggested by elastic precurser decay [*Ahrens and Duvall*, 1966], with that observed in LiF. The latter material has been studied extensively under shock-loading conditions, and strong correlations between precursor characteristics and dislocation dynamics have been observed [*Asay et al.*, 1972].

The reduction in shear stress observed in single crystal silicates and oxides above the Hugoniot elastic limit is certainly linked to the yield process. This reduction in shear stress is frequently equated with loss of strength at the Hugoniot state and carries the implication that subsequent material response would be fluid like. Initial release wave measurements on quartz and feldspar near the expected bulk sound speed seem to support this [*Grady et al.*, 1975]. However, measurements of release wave velocities on corundum [*Bless and Ahrens*, 1976] suggest reduced but not total loss of strength.

Material softening or melting during shock compression is the most reasonable explanation for the initial and persistent loss of strength, except that estimates of bulk temperature rise during shock compression are far below those required to produce the observed effects. This objection is mitigated somewhat if localization of softened or melted regions occur.

It is reasonable to expect a process of heterogeneous yielding leading to localization of shear. The processes of slip occurring during yielding are nonconservative and the energy dissipated must lead to local temperature increase. Plastic processes such as dislocation motion and multiplication are enhanced by increased temperatures and can lead to accelerated yielding in local hot regions. Analysis of constitutive relations with temperature dependence, coupled with the thermomechanical equations of motion, show that homogeneous yielding can be subject to Taylor instabilities [*Bellman and Pennington*, 1955]. As an example, *Holtzman and Cowan* [1961] have used a continuum description of dynamic yielding in solids of the form

$$\tau = \tau_Y + \eta \frac{\partial u}{\partial x} \tag{36}$$

where τ is the shear stress, τ_Y, the shear stress at yield, η, the dynamic viscosity, and u, the particle velocity normal to the direction x. A reasonable form for temperature dependence of the viscosity is

$$\eta = \eta_o e^{-a(T-T_o)} \ . \tag{37}$$

Under conditions of flow normal to the x direction due to the shear stress τ, the equation of motion is

$$\frac{\partial u}{\partial t} - \frac{1}{\rho_o} \frac{\partial \tau}{\partial x} = 0 \tag{38}$$

and the energy equation, assuming Fourier heat conduction, is

$$\frac{\partial T}{\partial t} - \chi \frac{\partial^2 T}{\partial x^2} = \frac{1}{\rho_o c} \tau \frac{\partial u}{\partial x} \ . \tag{39}$$

T, χ, c, and ρ_o are the temperature, thermal diffusivity, specific heat, and density, respectively. A first variation solution to Equations 36, 37, 38 and 39 [*Grady*, unpublished, 1976] show that the solution is unstable to homogeneous deformation. Analysis indicates that, if τ_Y is assumed small, the wave length of perturbations at maximum growth rate is

$$\lambda_m = \frac{2\pi}{\dot{\gamma}} \left[\frac{\rho_o \chi c^2}{a^2 \eta_o^2} \right]^{1/4} , \tag{40}$$

the maximum growth rate is

$$\alpha_m = \frac{a \ \eta_o}{\rho_o \ c} \dot{\gamma}^2 , \tag{41}$$

and the cutoff wave length below which perturbations will not grow is

$$\lambda_c = \frac{2\pi}{\dot{\gamma}} \left[\frac{\rho_o \chi c}{a \eta_o} \right]^{1/2} , \tag{42}$$

where $\dot{\gamma}$ is the plastic strain rate.

Expressions for finite τ_Y are similar but more complicated. Comparison of the optimum wave length with the periodicity of shear banding observed in shock-loaded aluminum are in reasonable agreement. Material property data are presently insufficient to make similar comparisons in minerals, however, the variations in planar feature separation with propagation distance observed by *Klein* [1965] appears consistent with this explanation. Taylor instabilities, therefore, seem to provide a plausible explanation for the heterogeneous yield process apparently occurring in these materials.

Some previously unreported experimental results by the author on single crystal periclase, although preliminary, provide some interesting observations on the process of yielding in oxide minerals. In this work, samples of periclase 3.3 mm in thickness were shock-loaded and unloaded in the [100] direction by planar impact of [0001] oriented sapphire plates (3.15 mm thickness) to stress levels between 48 and 112 kbar. [0001] sapphire response is elastic in this region. Diffuse surface velocity interferometry was used to measure the back interface particle velocity profile between the periclase sample and a sapphire window material. A second laser beam was reflected internally off of the center of the sample impact face. Upon impact the beam is diverted and provides a very accurate time of arrival fiducial as was described in section II. In the present experiments, wave transit times within 1% could be measured. The experimental set-up was previously illustrated in Figure 9.

Profiles obtained in the three experiments conducted are shown in Figure 10. The initial elastic loading wave consists of a sharp rise to a stress level of about 25 kbar followed by significant stress relaxation. The measured precursor velocity of 9.34 ± 0.05 km/s is consistent with ultrasonic velocities measured by *Spetzler* [1970]. Subsequent loading to the Hugoniot state proceeds in the deformation wave. Note that the width of the deformation wave decreases with increasing peak stress. Two features in the relief wave structure are identified in Figure 10. The first will be called the initial elastic release, the second, the initial plastic release. The structure of the release wave is similar to elastic-plastic response observed in release waves for certain metals [*Asay and Hayes*, 1975] and strongly suggests that material strength persists, or has recovered, in periclase at the Hugoniot state.

In Figure 11, elastic precursor stress-strain points and the final Hugoniot states are identified in the stress-volume plane. Also shown is the periclase hydrostatic compression curve estimated with a Murnaghan equation extrapolation of ultrasonic single crystal data to 8 kbar [*Spetzler*, 1970]. Within experimental uncertainty, the Hugoniot stress volume data indicate total collapse to the hydrostat. Shock velocities lower than the local bulk sound speed were measured above the

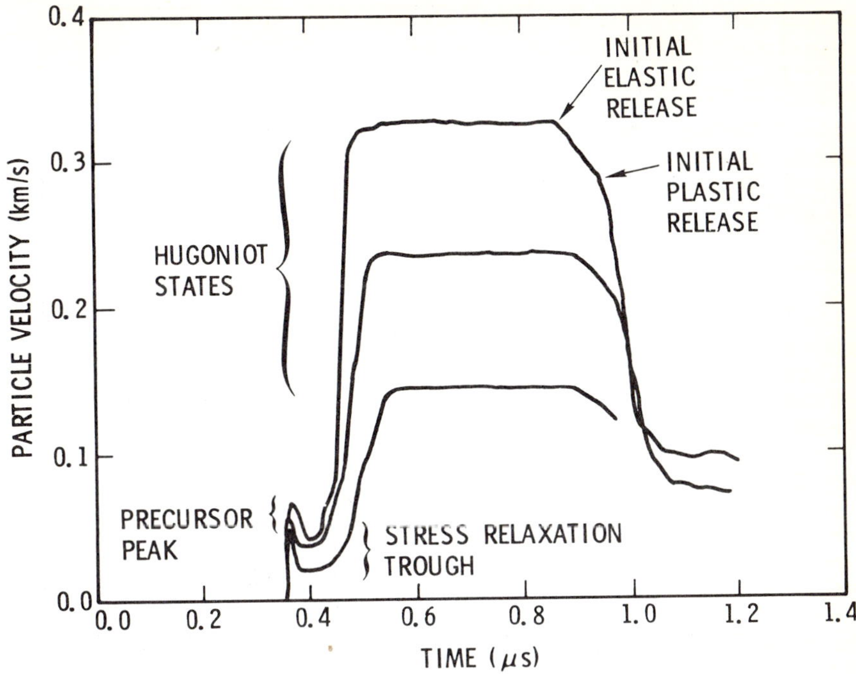

Fig. 10. Particle-velocity profiles obtained in [100]-oriented MgO after approximately 3.3-mm propagation distance. Essential features are the initial elastic precursor, stress relaxation, the subsequent deformation wave, and indications of elastic-plastic response in the release wave. Peak stresses are 48, 82, and 112 kbar.

Hugoniot elastic limit which also suggests collapse toward the hydrostat. Reduced shock velocities have also been noted by *Graham* [1974] in single crystal sapphire and quartz.

Release wave velocities obtained from the shock-wave experiments are compared with extrapolation of ultrasonic data [*Spetzler*, 1970] in Figure 12. The initial elastic release velocities are similar to those reported by *Bless and Ahrens* [1976] on corundum. Velocities are below expected [100] longitudinal velocities but are substantially above expected bulk sound velocities. On the other hand, the plastic release velocities are very close to the bulk sound velocity.

In Figure 11, the initial release paths are shown for the two highest Hugoniot states. The release paths deviate significantly from the hydrostat and imply a shear stress build-up through an axial stress drop of 14.5 ± 1.0 kbar at

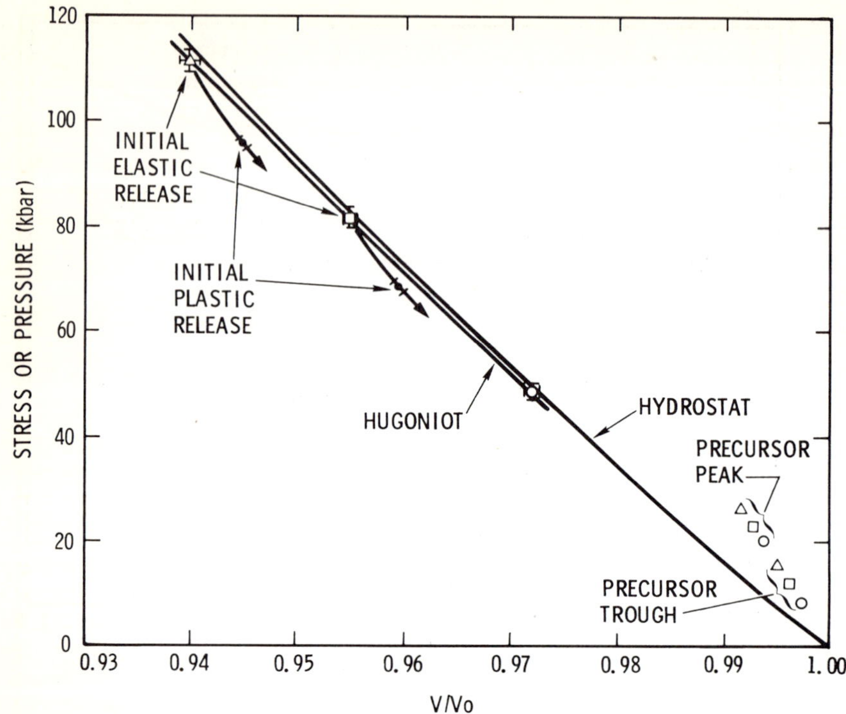

Fig. 11. Comparison of the experimental Hugoniot for MgO and the hydrostat estimated from ultrasonic data [Spetzler, 1970]. Also shown are theoretical release stress-volume paths, which indicate substantial recovery of material strength at the Hugoniot state.

which point subsequent unloading appears to parallel the hydrostat.

It appears then that single-crystal periclase, under shock compression, collapses to the hydrostat and exhibits, perhaps briefly, loss of shear strength. However, subsequent material response during the release wave, which arrives about 0.5 μs after the Hugoniot state is attained, suggests substantial recovery of material strength. The observations in periclase are not inconsistent with a process of local shearing and local heat generation leading to loss of strength during shock compression. Energy dissipated during shock compression in periclase is about a factor of 20 less than occurs in quartz at the same stress level. Calculations show that melted regions could not persist in periclase for the 0.5 μs separating the shock and relief wave due to thermal conduction away from those regions, although residual thermal heterogeneity

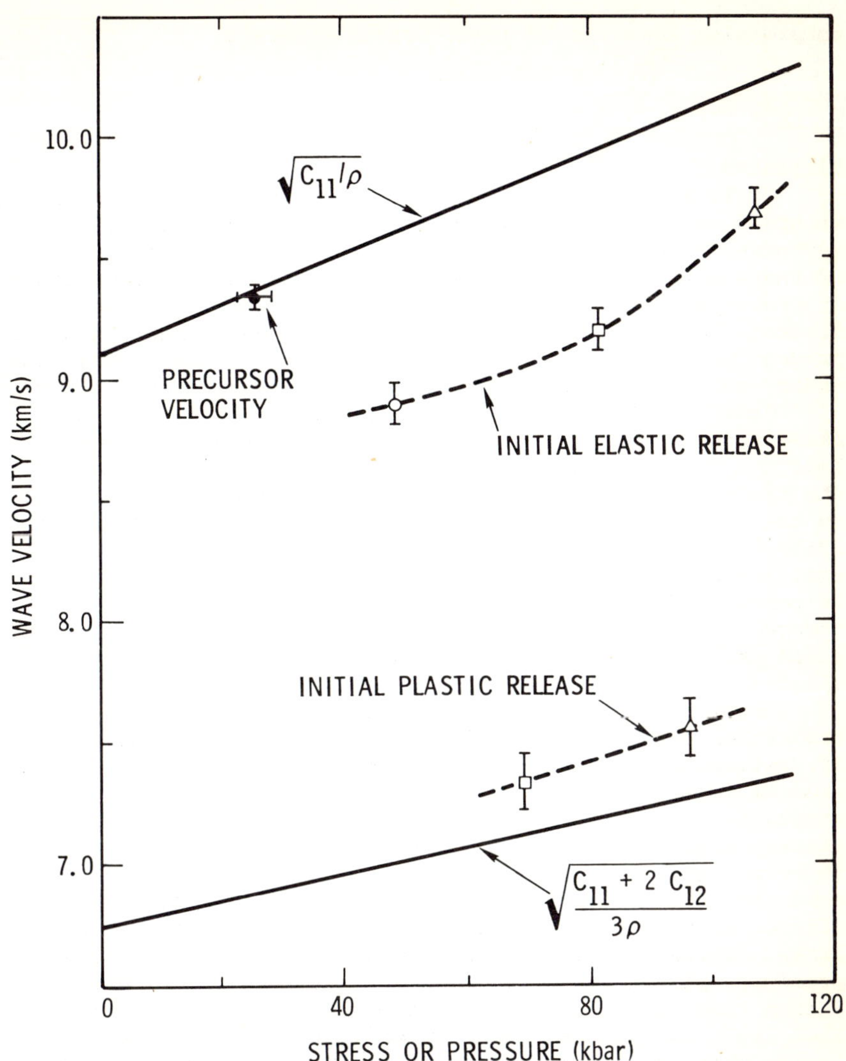

Fig. 12. Elastic and plastic Hugoniot release-wave velocities are compared with single-crystal MgO longitudinal and bulk sound speeds extrapolated from ultrasonic data [Spetzler, 1970].

could result in a reduced shear modulus. Hence, recovery of material strength prior to relief wave arrival would seem reasonable. Observations such as these, in which the shear strength is time-dependent, illustrate the complexity of the compression features of shock-loaded minerals.

C. The Phase Transformation

Silicates

It is fortunate that the phase transitions in the silicate minerals can proceed within the submicrosecond time scale of a shock-wave experiment. If this were not the case, the use of shock waves to investigate physical properties of minerals at pressures existing in the lower mantle would be severely limited. One is justified in asking how these transitions can proceed on such a rapid time scale, when the same transitions, under static conditions, are observed to be extremely sluggish.

Considerable effort has been devoted to the study of phase transitions occurring under conditions of shock-wave loading, and a fairly extensive literature has developed on the theory of shock-induced phase changes and their consequences on the propagation of finite amplitude waves [*Duvall and Horie*, 1965; *Andrews*, 1973; *Hayes*, 1974]. In most of these studies equilibrium thermodynamics is assumed and, when rate dependence is considered, it is usually assumed, at least tacitly, that a homogeneous process occurs. The consequences of a heterogeneous yield process on phase transition kinetics have not been considered.

Dremin and Breusov [1968] have argued that a polymorphic transition cannot proceed yielding during shock compression. If yielding proceeds on localized planes by an adiabatic shear process in silicate minerals and if near or total loss of strength has occurred due to near or total melting on these planes, then the regions between planes will experience close to hydrostatic pressure. It is difficult to see, then, how conditions differ significantly from static compression conditions except for the potentially quite different thermal state. *Teller* [1962] concluded that shear phenomena at high pressures play an important role in chemical processes. Rather than direct shear activation, however, he suggested that the shear process contributes to dissipative energy accumulation which aids thermal activation in the chemical process. This last idea seems worth pursuing with regard to the completion of reconstructive phase transitions in silicate minerals under shock-loading conditions.

We will focus on the α-quartz-stishovite transformation for which considerable shock-wave data is available and make the simplifying assumption that yielding under shock compression is confined to an equidimensional rectangular lattice with lattice dimension d. Assume also that the dissipative energy incurred during shock compression is uniformly distributed on this lattice as thermal energy and that the regions or blocks between planes undergo isentropic compression only. The dissipated energy can be estimated from the measured Hugoniot and isentropic compression properties of the shocked mineral. Assume further that the processes of yielding and phase transition are separate in time with yielding proceeding on the shorter time scale as suggested by *Dremin and Breusov* [1968]. This will allow attention to be focused strictly on the phase transition process.

At this point consider one deformation plane located at the Lagrangian coordinate $h = 0$ with the h axis, $h > 0$, normal to the plane. At time $t = 0$ a certain amount of thermal energy, consistent with energy dissipated in the shock process, is deposited on the plane $h = 0$. A state of uniform hydrostatic pressure P exists throughout. For $t > 0$ heat will flow from the deformation plane into the adjacent material which is initially at a temperature $T_s(P)$ resulting strictly from isentropic compression. Assume that heat transfer is governed by Fourier conduction with constant specific heat, c, and thermal diffusivity, χ. Estimates of radiative heat transfer at the calculated temperatures are found small within the time scale of interest.

The primary coordination phase change is assumed to be thermally-activated with a frequency factor and activation energy independent of the pressure and a simplified from of the Johnson-Mehl equation [*Burke*, 1965] will be used to describe the reaction rate of the transition. This description then leads to the equations

$$\frac{\partial T}{\partial t} = \chi \frac{\partial^2 T}{\partial h^2} + \frac{l}{c} \frac{\partial \lambda}{\partial t} , \tag{43}$$

$$\frac{\partial \lambda}{\partial t} = F e^{-\frac{A}{kT}} (1 - \lambda) , \tag{44}$$

which govern heat flow and the transition reaction rate in the neighborhood of one deformation plane. λ is the mass fraction of the high density phase, F and A are the frequency factor and activation energy, respectively, and l is the latent heat of transition. Equations 43 and 44 are coupled and not readily solvable. However, if the second term on the right hand side of Equation 43 is ignored, the heat flow equation can be solved separately. This is justified on two accounts. First, this term is small in the time scale of interest. Second, the dissipative energy at the Hugoniot state, which was estimated in this work by simply subtracting the isentropic elastic energy of quartz from the total Hugoniot energy, already includes the latent heat of transition.

With this simplification, solution of the thermal conduction equation results in the temperature distribution at time t of

$$T = T_s(P) + \frac{Q}{\sqrt{4\pi \chi t}} e^{-h^2/4\chi t} \tag{45}$$

where $Q = \varepsilon/\rho_o c$ and ε is the thermal energy per unit area initially deposited at $h = 0$. Some temperature profiles at

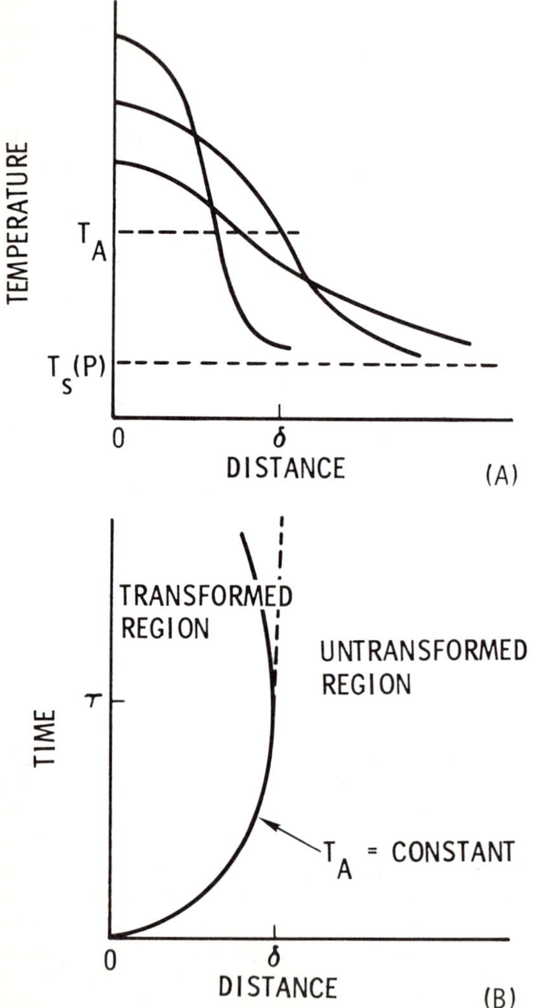

Fig. 13. (A) Temperature profiles at increasing time due to thermal diffusion from the plane $h = 0$. (B) A path of constant temperature, T_A and the region of material transformed due to temperatures in excess of this value.

increasing times are illustrated in Figure 13*a*.

Due to the exponential form of the Boltzmann factor, reactions which proceed very slowly at a given temperature can proceed very rapidly when the temperature is raised a few hundred degrees. If a temperature T_A is considered such that the coordination transformation can proceed by thermal activation within the time scale of a shock compression wave, say $\dot{\lambda} \simeq 10^8/S$, then the solution, Equation 45, acts as a thermal wave and carries the temperature T_A to the untransformed material. A path of constant temperature is governed by the differential equation,

$$\frac{dh}{dt} = \frac{h}{t}\left(1 - \frac{\chi}{h^2/2t}\right), \tag{46}$$

and is shown in Figure 13*b*. Note that a given temperature level does not continue to propagate away from $h = 0$ but reaches a maximum distance,

$$\delta = \sqrt{2\ \tau\chi} \tag{47}$$

in a time

$$\tau = \frac{Q^2}{4\pi\ e\chi(T_A - T_s(P))^2}, \tag{48}$$

and then recedes, determining the domain of material which is transformed rapidly by the thermal wave from that which is not.

If a specific energy E_D, due to energy dissipated in the Hugoniot process is distributed uniformly over the deformation lattice with lattice parameter d, a value

$$Q = \frac{E_D\ d}{3\ c} \tag{49}$$

is obtained. If transformation proceeds to a maximum distance δ from the deformation plane and if the lattice parameter is d, the resulting mass fraction of transformed material is

$$\bar{\lambda} = \frac{6\delta}{d}. \tag{50}$$

Also, τ, Q, and δ can be eliminated from Equations 47-50 resulting in an expression for the temperature,

$$T_A = T_s(P) + \sqrt{\frac{2}{\pi e}}\ \frac{E_D}{c\bar{\lambda}}, \tag{51}$$

capable of transforming the material within the shock front at the expressed reaction rate. E_D can be estimated from the quartz Hugoniot as previously described. The fraction of the high density phase $\bar{\lambda}$ in the mixed phase region can also be estimated from the relation of the mixed phase Hugoniot state to the high and low density phase boundaries. This was accomplished for quartz between P = 150-350 kbar. We found T_A to have a constant or slightly increasing value between 1250-1350°K. Using a value of $\dot{\lambda} = 10^{-4}/S$ for the quartz-stishovite reaction rate at T = 500°K, a reasonable value of 50 kcal/mole is estimated for the activation energy. The calculation also suggests that transformation cannot proceed instantaneously but is controlled by propagation of the thermal wave into the untransformed material. Equation 48 predicts a value of τ = 0.25 μs at a pressure of 250 kbar, although better than 75% of the transformation occurs within less than 0.05 μs. This value is consistent with shock-wave rise times which have been observed [*Wackerle*, 1962; *Grady et al.*, 1974].

Calcite

The mineral calcite is observed to experience two phase transitions below stress levels of 20 kbar under both static and shock-loading conditions. At these lower stress levels, very accurate shock-wave measurements can be made, providing increased understanding of the processes of shock-induced phase changes in rocks and minerals. The calcite I-II and calcite II-III transitions occur statically at stress levels of about 14.5 and 17.5 kbar, respectively [*Bridgman*, 1939; *Singh and Kennedy*, 1974]. Recent shock-wave studies on Solenhofen limestone [*Schuler and Grady*, 1976] using interferometry techniques have revealed the effect of both transitions in the structure of compression and relief wave propagation. Shock-wave profiles obtained on Vermont marble have been particularly distinct in this respect, however, and one such profile is shown in Figure 14.

In the experiments on Vermont marble, samples approximately 8 mm in thickness were impacted with fused silica plates approximately 4.8 mm in thickness. The resulting impact provided a square wave input stress pulse of approximately 1.6 μs duration and a stress amplitude dependent on the impact velocity. Maximum impact stress achieved in the experiment shown in Figure 14 was 35 kbar. Samples were backed with fused silica laser window material and the particle velocity profile which evolved over 8 mm of propagation distance was measured at the marble-fused silica interface. The marble and fused silica have quite similar mechanical impedance characteristics.

In the particle-velocity profile shown in Figure 14, features resulting from different physical mechanisms are quite distinct. The dynamic stress-volume path, also shown in Figure 14, was determined assuming self similarity of both the

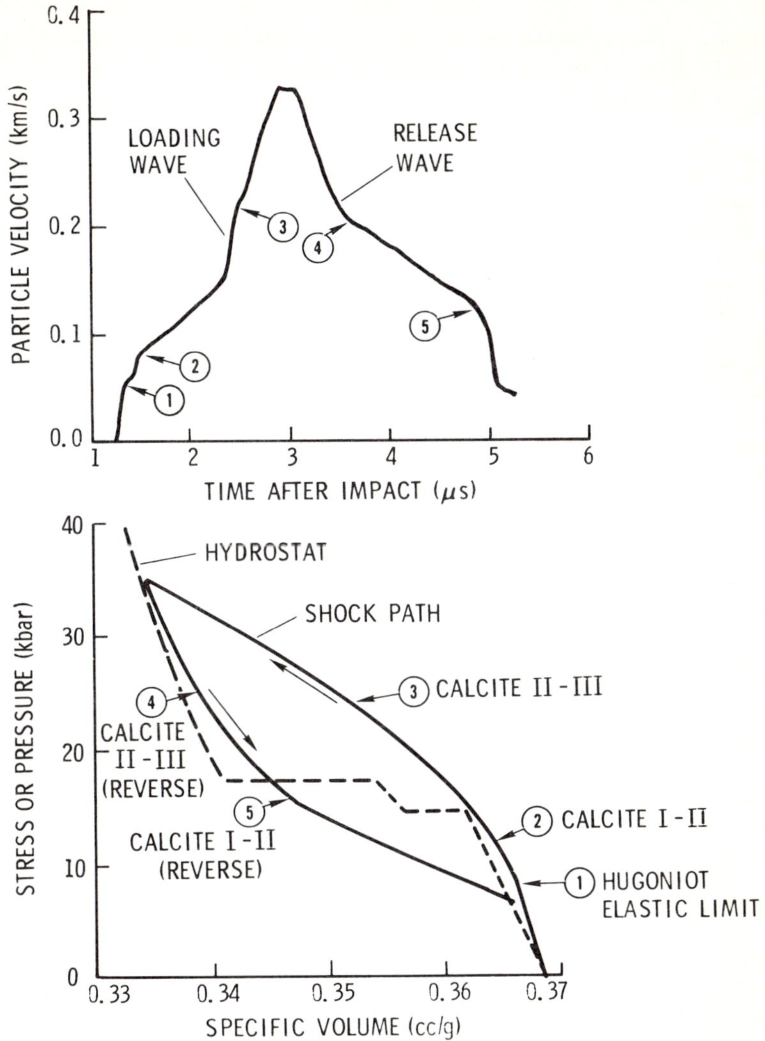

Fig. 14. Particle-velocity profile and the resulting stress-volume path obtained on polycrystalline calcite (Vermont marble). The single-crystal calcite hydrostat is also shown [Singh and Kennedy, 1974]. The numbered points coordinate various features in the wave profile with corresponding points in the stress-volume curve.

compression and release wave. Although self similarity has not yet been strictly verified for marble, wave propagation results obtained in limestone suggest that it is a reasonable approximation. The dynamic stress-volume path is compared with the calcite hydrostat [*Singh and Kennedy*, 1974]. Similar features occurring in the profile and in the stress-volume path are numbered in Figure 14.

The Hugoniot elastic limit (point 1 in Figure 14) was achieved in marble prior to onset of the calcite I-II phase transition (point 2 in Figure 14). This differs from observations in limestone where no separation of the plastic and transition waves are observed [*Schuler and Grady*, 1976]. The calcite I-II transition initiates at an axial stress of about 12 kbar, somewhat less than the 14.5 kbar transition pressure observed statically in single crystal calcite [*Singh and Kennedy*, 1974]. The shock-wave transition stress is consistent, however, with ultrasonic studies on marble where local stress concentrations are believed responsible for the early initiation of the I-II transition. In the loading wave profile, the calcite II-III transition (point 3 in Figure 14) is believed to initiate at a stress of about 24 kbar, somewhat higher than the observed static value. Note also that the dynamic loading path (representative of the Hugoniot in this material) is significantly higher than the hydrostat, similar to the metastable Hugoniots observed in the silicates.

Features in the release wave also mirror characteristics of the calcite phase transitions. Point 4 in Figure 14 is thought to be initiation of the reverse calcite II-III transition and point 5 the reverse calcite I-II transition. The release wave signatures of each transition differ dramatically.

The calcite I-II transition is known to be displacive [*Merrill and Bassett*, 1975], and, hence can be expected to proceed at a very rapid rate [*Buerger*, 1972]. Formation of a rarefaction shock wave on release due to the calcite I-II transition is consistent, then, with a rapid displacive transformation. This author is not aware of the nature (displacive or reconstructive) of the calcite II-III transition. However, comparison of the release wave shape due to the calcite II-III transition (point 4 in Figure 14) with the high pressure release wave shape in quartz (Figure 5) due to the reverse α-quartz-stishovite phase transition show considerable similarity in structure. The α-quartz-stishovite transition is a primary coordination transformation with a finite reaction rate. Rate dependence associated with the reverse phase change is expected to provide the profile shape observed in quartz (Figure 5) or as is indicated from the reverse calcite II-III transition in Figure 14. We tentatively suggest, therefore, that the calcite II-III transition is reconstructive and that the release wave signature of reconstructive and displacive transformations differ significantly.

Finally, it should be noted that there is no indication of elastic-plastic response in the release wave for marble (providing other features have been correctly identified).

Also, comparison of the initial release wave velocity with the calcite III phase static bulk modulus determined by *Singh and Kennedy* [1974] shows this velocity to be very close to the expected bulk value. This result is consistent with release wave behavior observed in quartz and feldspar [*Grady et al.*], both of which undergo a phase transition, but differ from behavior observed in corundum [*Bless and Ahrens*, 1976] and periclase [present work] both of which do not experience a phase transformation.

V. SUMMARY

Recent advances in both experimental techniques and methods for analyzing wave profile data have contributed to the study of shock processes in solids. There is, at present, a notable lack of time-resolved wave profile data on rocks and minerals approaching mantle pressures. Existing studies, however, have suggested a potential framework for a theory describing the shock processes of yielding and phase transition in silicate and oxide minerals, although further experimental studies are needed to confirm and to quantify some of the mechanisms proposed.

Our present understanding of the shock compression process in minerals is still not adequate to dictate the optimum experimental conditions necessary to prepare a thermodynamic Hugoniot state at pressures and temperatures compatible with mantle conditions. Results obtained to date, however, suggest some possibilities. Perhaps a starting material of gem quality single crystals in shock-wave experiments is not the optimum situation. A process of heterogeneous yielding during shock compression and the subsequent nonuniform temperature state may require more time for thermal equilibration than can be achieved in a standard shock-wave experiment. Perhaps a starting material of cold-pressed submicron sized mineral powder would force a more homogeneous yield process, and hence, more uniform energy dissipation. An accelerated thermal equilibration time should result in this case.

It is interesting to note that the release wave velocities in both corundum [*Bless and Ahrens*, 1976] and in periclase [*present work*, Figure 12] tend toward the expected value for the longitudinal wave speed at the higher stress levels. This would suggest that thermal equilibrium occurs more rapidly at the higher stress levels. The dramatic decrease in rise time of the deformation wave with increasing Hugoniot stress in periclase (Figure 10) and the reduction in the shear band separation with increasing strain rate predicted by the Taylor instability analysis (Equation 40) suggests that the energy dissipation process is more uniform at higher stress levels. Perhaps, then, at Hugoniot stresses far in excess of the elastic limit, the yield process will be sufficiently uniform to allow thermal equilibration within the required time period.

The experimental observations and analysis presented in the present paper point out the importance of technique and the complexity in the shock deformation of certain silicates and oxides. It appears that the dominant compressional features of shock-loaded minerals is strongly influenced by dynamic yielding. A model of heterogeneous yielding to describe the resulting adiabatic shear provides a plausible picture of the resulting shock compression behavior and indicates that material response under shock loading must be considerably more complex than has heretofore been viewed. Because of the importance of quartz, corundum, periclase, and certain other minerals to our understanding of mantle materials, it seems important to proceed with further work to develop a more detailed picture of the shock compression of these materials.

Acknowledgments. The author wishes to thank J. R. Asay, R. A. Graham, and T. J. Ahrens for critically reviewing the paper. This work was supported by the U.S. Energy Research and Development Administration, ERDA.

REFERENCES

Ahrens, T. J., High-pressure electrical behavior and equation of state of magnesium oxide from shock-wave measurements, *J. Appl. Phys., 37,* 2532-2541, 1966.

Ahrens, T. J., and G. E. Duvall, Stress relaxation behind elastic shock waves in rocks, *J. Geophys. Res., 71,* 4349-4360, 1966.

Ahrens, T. J., and J. T. Rosenberg, Shock metamorphism: experiments on quartz and plagioclase, in *Shock Metamorphism of Natural Materials,* edited by B. French and N. Short, pp. 59-81, Mono Press, Baltimore, Maryland, 1968.

Ahrens, T. J., W. H. Gust, and E. B. Royce, Material strength effect in the shock compression of alumina, *J. Appl. Phys., 39,* 4610-4616, 1968.

Ahrens, T. H., C. F. Petersen, and J. T. Rosenberg, Shock compression of feldspars, *J. Geophys. Res., 74,* 2727-2746, 1969.

Al'tshuler, L. V., S. V. Kormer, M. L. Brazhnik, L. A. Vladimirov, and M. P. Speranskaya, The isentropic compressibility of aluminum, copper, lead, and iron at high pressures, *Sov. Phys. JETP, 11,* 766-775, 1960.

Al'tshuler, L. V., M. N. Pavlovskii, and V. P. Drakin, Peculiarities of phase transitions in compression and rarefaction shock waves, *Soviet Phys. JETP, 25,* 260-265, 1965.

Anan'in, A. V., O. N. Brevsov, A. N. Dremin, S. V. Pershin, and V. F. Tatsii, The effect of shock waves on silicon dioxide. I. Quartz, *Combustion, Explosion, and Shock Waves, 10,* 426-436, 1974.

Andrews, D. J., Equation of state of the alpha and epsilon phases of iron, *J. Phys. Chem. Solids, 34,* 825-840, 1973.

Asay, J. R., G. R. Fowles, G. E. Duvall, M. H. Miles, and R. F. Tinder, Effects of point defects on elastic decay in LiF, *J. Appl. Phys., 43,* 2132-2145, 1972.

Asay, J. R., and D. B. Hayes, Shock-compression and release behavior near melt states in aluminum, *J. Appl. Phys., 46,* 4789-4800, 1975.

Barker, L. M., Fine structure of compression and release wave shapes in aluminum measured by the velocity interferometer technique, in *Behavior of Dense Media Under High Pressures,* edited by Gordon and Breach, p. 483, New York, 1968.

Barker, L. M., and R. E. Hollenbach, Interferometer technique for measuring the dynamic mechanical properties of materials, *Rev. Sci. Inst., 36,* 1617, 1965.

Barker, L. M., and R. E. Hollenbach, Shock-wave studies of PMMA, fused silica, and sapphire, *J. Appl. Phys., 41,* 4208-4226, 1970.

Barker, L. M., and R. E. Hollenbach, Laser interferometer for measuring high velocities of any reflecting surface, *J. Appl. Phys., 43,* 4669-4675, 1972.

Barker, L. M., C. D. Lundergan, and W. Herrmann, Dynamic response of aluminum, *J. Appl. Phys., 35,* 1203-1212, 1964.

Barker, L. M., and K. W. Schuler, Correction to the velocity-per-fringe relationship for the VISAR interferometer, *J. Appl. Phys., 45,* 3692-3693, 1974.

Bellman, R., and R. H. Pennington, Effects of surface tension and viscosity on Taylor instabilities, *Quart. Appl. Math., 12,* 151-162, 1955.

Bernstein, D., and D. D. Keough, Piezoresistivity of manganin, *J. Appl. Phys., 35,* 1471-1474, 1964.

Bless, S. J., and T. J. Ahrens, Measurement of release wave speed in shock-compressed polycrystalline alumina and aluminum, *11,* 1935-1942, 1976.

Boganov, A. G., S. A. Popov, and V. S. Rudenko, Location of the melting curve of quartz on the p,T-diagram of S_iO_2, (Translation) in *Acad. Sci. USSR, Proc. Chem. Section, 201,* 1011-1013, 1971.

Borg, I. Y., Some shock effects in granodiorite to 270 kbar at the pile-driver site, in *Flow and Fracture of Rock,* edited by H. Heard, I. Borg, N. Carter, and C. Raleigh, Amer. Geophys. Union, Monograph, 1974.

Bridgman, P. W., The high pressure behavior of miscellaneous minerals, *Amer. J. Sci., 237,* 7-18, 1939.

Bridgman, P. W., Physics above 20,000 kg/cm^2, *Proc. Roy. Soc. London, A203,* 1-17, 1950.

Buerger, M. J., Phase transformations, *Sov. Phys. Crystallography, 16*, 959-968, 1972.

Bunch, T. E., Some characteristics of selected minerals from craters, in *Shock Metamorphism of Natural Materials,* edited by B. French and N. Short, p. 509, Mono, Baltimore, Maryland, 1968.

Burke, J., *The Kinetics of Phase Transformations in Metals,* Pergamon Press, New York, 1965.

Chao, E. C. T., Pressure and temperature histories of impact metamorphosed rocks-based on petrographic observations, in *Shock Metamorphism of Natural Materials,* edited by B. French and N. Short, p. 149, Mono, Baltimore, Maryland, 1968.

Cowperthwaite, M., and R. F. Williams, Determination of constitutive relations with multiple gauges in non-divergent flow, *J. Appl. Phys., 42,* 456, 1971.

DeCarli, P. S., Observations of the effects of explosive shock on crystalline solids, in *Shock Metamorphism of Natural Materials,* edited by B. M. French and N. M. Short, p. 129, Mono, Baltimore, Maryland, 1968.

DeCarli, P. S., Lagrangian stress gauge study of the loading and unloading response of Raymond granite, *Trans. Am. Geophys. Union, 56,* 1061, 1975.

DeCarli, P. S., and D. J. Milton, Stishovite: synthesis by shock wave, *Science, 147* 144-145, 1965.

Dick, J. J., and D. L. Styris, Electrical resistivity of silver foils under uniaxial shock-wave compression, *J. Appl. Phys., 46,* 1602-1617, 1975.

Dremin, A. N., and G. A. Adadurov, The behavior of glass under dynamic loading, *Soviet Phys. Solid State, 6,* 1379-1384, 1964.

Dremin, A. N., and O. N. Breusov, Processes occurring in solids under the action of powerful shock waves, *Russian Chem. Rev., 37,* 396-402, 1968.

Dremin, A. N., and G. I. Kanel', Pressure dependence and electrical resistance of shock-compressed CuNiMn 3-12 manganin and CuNiMn 40-1.5 constantan, *Combustion, Explosion, and Shock Waves, 8.1,* 147-149, 1972.

Dremin, A. N., S. V. Pershin, and V. F. Pogorelov, Structure of shock waves in KCl and KBr under dynamic compression to 200,000 atm., *Combustion, Explosion, and Shock Waves, 1,* 1-4, 1965.

Duvall, G. E., and G. R. Fowles, *Shock Waves in High Pressure Physics and Chemistry II,* Academic Press, London and New York, 1963.

Duvall, G. E., and Y. Horie, Shock-induced phase transitions, in *Proc. Fourth Internat. Sympos. on Detonation*, pp. 248-257, Office of Naval Research, Washington, D.C., 1965.

Eden, G., and P. W. Wright, A technique for the precise measurement of the motion of a plane free surface, in *Proc. Fourth Internat. Sympos. on Detonation*, edited by S. J. Jacobs, p. 573, Office of Naval Research, Washington, D.C., 1966.

Engelhardt, W. von., and D. Stöffler, Stages of shock metamorphism in crystalline rocks in the Ries Basin, Germany, in *Shock Metamorphism of Natural Materials*, edited by B. French and N. Short, pp. 159-169, Mono Press, Baltimore, Maryland, 1968.

Fowles, G. R., Dynamic compression of quartz, *J. Geophys. Res.*, *72*, 5729-5742, 1967.

Fowles, G. R., Experimental technique and instrumentation, in *Dynamic Response of Materials to Intense Impulsive Loading*, edited by P. C. Chen and A. K. Hopkins, pp. 405-480, Air Force Materials Lab, Wright-Patterson AFB, Ohio, 1972.

Fowles, G. R., and R. F. Williams, Plane stress wave propagation in solids, *J. Appl. Phys.*, *41*, 460, 1970.

Fuller, P. J. A., and J. H. Price, Dynamic pressure measurements to 300 kilobars with a resistance transducer, *Brit. J. Appl. Phys.*, *15*, 751-758, 1964.

Gaffney, E. F., and T. J. Ahrens, Optical absorbtion spectra of ruby and periclase at high shock pressures, *J. Geophys. Res.*, *78*, 5942-5953, 1973.

Ginsberg, M. J., D. E. Grady, P. S. DeCarli, and J. T. Rosenberg, Effects of stress on the electrical resistance of ytterbium and calibration of ytterbium stress transducers, *Stanford Res. Inst. Final Rep.*, Menlo Park, California, 1973.

Grady, D. E., Experimental analysis of attenuating wave propagation, *Stanford Res. Inst. Rep. 002-72*, Menlo Park, California, 1972.

Grady, D. E., Experimental analysis of spherical wave propagation, *J. Geophys. Res.*, *78*, 1299, 1973.

Grady, D. E., R. E. Hollenbach, K. W. Schuler, and J. F. Callender, Strain rate dependence in dolomite inferred from impact and static compression studies, submitted for publication in *J. Geophys. Res.*, 1976.

Grady, D. E., and W. J. Murri, Dynamic unloading in shock compressed feldspar, *Geophys. Res. Let.*, *3*, 472-494, 1976.

Grady, D. E., W. J. Murri, and G. R. Fowles, Quartz to stishovite: wave propagation in the mixed phase region, *J. Geophys. Res.*, *79*, 332-338, 1974.

Grady, D. E., W. J. Murri, and P. DeCarli, Hugoniot sound velocities in two silicates, *Trans. Am. Geophys. Union*, *55*, 417, 1974*b*.

Grady, D. E., W. J. Murri, and P. DeCarli, Hugoniot sound velocities and phase transformations in two silicates, *J. Geophys. Res., 80,* 4857-4861, 1975.

Graham, R. A., Shock-wave compression of X-cut quartz as determined by electrical response measurements, *J. Phys. Chem. Solids, 35,* 355-372, 1974.

Graham, R. A., and J. R. Asay, Measurement of wave profiles in shock-loaded solids, *Appl. Mech. Rev.,* in press, 1976.

Graham, R. A., and W. P. Brooks, Shock-wave compression of sapphire from 15 to 420 kbar. The Effects of large anisotropic compression, *J. Phys. Chem. Solids, 32,* 2311-2330, 1971.

Graham, R. A., and R. D. Jacobson, Lithium niobate stress gauge for pulsed radiation deposition studies, *Appl. Phys. Lett., 23,* 584-586, 1973.

Graham, R. A., F. W. Neilson, and W. B. Benedick, Piezoelectric current from shock-loaded quartz-a submicrosecond stress gauge, *J. Appl. Phys., 36,* 1775-1783, 1965.

Gupta, Y. M., G. E. Duvall, and G. R. Fowles, Dislocation mechanisms for stress relaxation in shocked LiF, *J. Appl. Phys., 46,* 532-546, 1975.

Gust, W. H., and E. B. Royce, Dynamic yield strengths of light armor materials, Rep. UCRL-50901, Lawrence Livermore Laboratory, Livermore, California, 1970.

Hayes, D. B., Polymorphic phase transformation rates in shock-loaded potassium chloride, *J. Appl. Phys., 45,* 1208-1217, 1974.

Herrmann, W., On the evaluation of constitutive equations from experiment, in *Recent Advances in Engineering Science, Vol. 6,* pp. 297-307, 1976.

Holtzman, A. H., and G. R. Cowan, The strengthening of austenitic manganese steel by plane shock waves, in *Response of Metals to High Velocity Deformation,* Interscience Publishers, New York, 1961.

Ingram, G. E., and R. A. Graham, Quartz gauge technique for impact experiments, in *Proc. Fifth Internat. Sympos. on Detonation,* pp. 369-386, Office of Naval Research, ACR-184, Washington, D.C., 1970.

Ivanov, A. G., and S. A. Novikov, Capacitive data transmitter for recording the instantaneous velocity of moving surfaces, *Prib. Tekn. Eksp., 1,* 135-139, 1963.

Keeler, R. N., and E. B. Royce, Shock waves in condensed media, in *Physics of High Energy Density,* pp. 51-150, Academic Press, New York, 1971.

Keough, D. D., Procedure for fabrication and operation of manganin shock pressure gauges, *Stanford Res. Inst. Final Rep.,* Menlo Park, California, 1968.

Keough, D. D., and J. I. Wong, Variation of the shock piezoresistance coefficient of manganin as a function of deformation, *J. Appl. Phys., 41,* 3508-3515, 1970.

Keough, D. D., R. F. Williams, and D. Bernstein, in *Symposium on High Pressure Technology,* edited by Lloyd and Giardini, Amer. Soc. Mech. Eng., United Engineering Center, New York, 1964.

Klein, M. J., The structure of explosively shocked MgO crystals, *Phil. Mag., 12,* 735-739, 1965.

Lyle, J. W., R. L. Shriver, and A. R. McMillan, Dynamic piezoresistive coefficient of manganin to 392 kbar, *J. Appl. Phys., 46,* 4663-4664, 1969.

McQueen, R. G., S. P. Marsh, and J. N. Fritz, Hugoniot equation of state of twelve rocks, *J. Geophys. Res., 72,* 4999-5036, 1967.

Merrill, L., and W. A. Bassett, The crystal structure of $CaCO_3$ (II), a high-pressure metastable phase of calcium carbonate, *Acta Crystallogr. B* (Denmark), *B31,* 343-349, 1975.

Minshall, S., Properties of elastic and plastic waves determined by pin contractors and crystals, *J. Appl. Phys., 26,* 463-469, 1955.

Murri, W. J., D. E. Grady, and K. D. Mahrer, Equation of state of rocks, *Stanford Res. Inst. Final Rep.,* Menlo Park, California, 1975.

Nunziato, J. C., E. K. Walsh, K. W. Schuler, and L. M. Barker, Wave propagation in nonlinear viscoelastic solids, in *Handbuch der Physik, Vol. VI a/4,* edited by C. Truesdell, pp. 1-108, Springer-Verlag, Berlin, 1974.

Petersen, C. F., W. J. Murri, and M. Cowperthwaite, Hugoniot and release-adiabat measurements for selected geological materials, *J. Geophys. Res., 75,* 2063-2072, 1970.

Rice, M. H., Capacitor technique for measuring the velocity of a plane conducting surface, *Rev. Sci. Instr. 32,* 449-451, 1961.

Ringwood, A. E., A. F. Reid, and A. D. Wadsley, High-pressure $KAlSi_3O_8$, an aluminosilicate with six-fold coordination, *Acta Crystallogr., 23,* 1093-1095, 1967.

Schuler, K. W., and D. E. Grady, Compression wave studies in Solenhofen limestone, *Sandia Laboratories Rep.,* Albuquerque, New Mexico, 1976.

Seaman, L., Lagrangian analysis for multiple stress or velocity gauges, *J. Appl. Phys., 45,* 4303, 1974.

Shaner, J. W., and E. B. Royce, Shock-induced demagnetization of YIG, *J. Appl. Phys., 39,* 492-493, 1968.

Singh, A. K., and G. C. Kennedy, Compression of calcite to 40 kbar, *J. Geophys. Res., 79,* 2615-2622, 1974.

Spetzler, H., Equation of state of polycrystalline and single-crystal MgO to 8 kilobars and 800°K, *J. Geophys. Res., 75,* 2073-2087, 1970.

Stishov, S. M., and S. V. Popova, New dense polymorph modification of silica, *Geochimiya, 10,* 837-839, 1961.

Stöffler, D., Deformation and transformation of rock-forming minerals by natural and experimental shock processes,

Fortschr. Mineral., *49*, 50-113, 1972.
Teller, E., On the speed of reactions at high pressures, *J. Chem. Phys.*, *36*, 901-903, 1962.
Tyunyaev, Y. N., et al., *Combustion and Explosion*, Nauka, Moscow, 1972.
Wackerle, J., Shock-wave compression of quartz, *J. Appl. Phys.*, *33*, 922-937, 1962.
Walsh, J. M., and R. H. Christian, Equation of state of metals from shock-wave measurements, *Phys. Rev.*, *97*, 1544-1556, 1955.

PYROXENES AND OLIVINES: STRUCTURAL IMPLICATIONS OF SHOCK-WAVE DATA FOR HIGH PRESSURE PHASES

R. JEANLOZ and T. J. AHRENS
Seismological Laboratory
Division of Geological and Planetary Sciences
California Institute of Technology
Pasadena, California 91125

Abstract

A reexamination of Hugoniot equation of state data and three new release adiabat points indicates that results for enstatite-bronzite composition pyroxene are compatible with its transforming to a perovskite phase at high pressure (For En_{90}: ρ_0 = 4.20 g/cm^3, $K_0 \sim 2.6 \pm 0.35$ Mbar, $K'_0 \sim 3.5 \pm .65$). The release adiabat data, as well as results from porous samples, imply that the shock-wave data do not define an equilibrium, high-pressure phase Hugoniot below about 1.00 Mbar. These also suggest a further transformation to a phase (or assemblage) with density about 5% (or more) greater than that of orthorhombic perovskite. The data would allow such a transformation to occur at pressures as low as 0.60 Mbar under shock, representing an upper bound for the equilibrium transition pressure.

Hugoniot data on magnesian olivines also appear to represent states of thermodynamic disequilibrium or a mixed-phase region below about 0.80-1.00 Mbar. However, Hugoniot points for Mg-pyroxene and Mg-olivine coincide at pressures above 0.70 Mbar, suggesting that these minerals transform to high-pressure phases (or phase assemblages) of comparable density. Since MgO (presumably as periclase or in a closely related structure) attains relatively low densities at these pressures, the shock-wave data are in strong disagreement with the disproportionation of Mg_2SiO_4 to a $MgSiO_3$ (perovskite) + MgO assemblage above 0.80-1.00 Mbar. Conversely, the shock-wave data do not preclude a transformation of the type $Mg_2Si_2O_6 \rightarrow Mg_2SiO_4$ ("post-perovskite" phase) + SiO_2 (rutile or fluorite structure). Again, these results would imply polymorphism to very dense "post-perovskite" phases. Based on the arguments made for pyroxene, such a transformation could occur at pressures as low as 0.60 Mbar under equilibrium conditions.

We note that the combined results of high-pressure experiments allow Mg-pyroxene compositions to be as likely candidates for the lower mantle as olivine.

I. INTRODUCTION

Upon recognition that the features associated with plate tectonic processes on the earth (subduction zones, ridges, and plumes) either are minor surficial expressions or, alternatively, reflect an integral sampling of the mantle convective flow field, and with the perspectives attained from the last decade of terrestrial planetary exploration, experimental high-pressure geophysics has focused increasingly on studying candidate materials of the earth's lower mantle and core.

Current ideas [*e.g.*, *Schubert and Anderson*, 1974] suggest that active convection in the earth takes place to depths of at least ∿ 750 km. However, it is important to discern whether the upper mantle above ∿ 350 km depth, and possibly the phase transition region extending to depths of at least ∿ 670 km, represents a differentiate of the lower mantle and is thus atypical of the silicate zone of the earth as proposed by *D. L. Anderson et al.* [1972] and *D. L. Anderson* [in press], in contradistinction to *Ringwood* [1975], who has emphasized the essential homogeneity of the mantle.

Clearcut data specifying the composition of the upper mantle stem largely from inclusions in kimberlites [*e.g.*, *Boyd*, 1973] and nodules from volcanic rocks of deep-seated origin [*e.g.*, *Basu*, 1975; *MacGregor*, 1975] and demonstrate the upper mantle to be largely garnet lherzolite with minor amounts of eclogite. This composition is not unlike the "pyrolite" model originally proposed by *Clark and Ringwood* [1964], which, by means of variations in temperature, pressure, and degree of partial melting, accounts for the genesis of a large class of effusive and plutonic rocks that penetrate the crust [*Green*, 1972].

The lower mantle, extending from depths of 670 to 2900 km, represents 49% of the mass of the earth. Its mineralogy and thermal state can be inferred from comparison of shock-wave and seismological data [*Al'tschuler and Sharpidzhanov*, 1971*a*, 1971*b*; *D. L. Anderson et al.*, 1972; *D. L. Anderson*, in press], and by extrapolation of results based on quench products from static high-pressure, petrological experiments [*e.g.*, *Ringwood*, 1975; *Liu*, 1975*a*, 1975*b*, 1976; *Akimoto et al.*, 1976] using largely ultrasonic and static compression equations of state. The composition, structure, and thermal history of the earth must ultimately be related to the composition of the sun, the moon, meteorites and other terrestrial planets, and the processes of planetary accretion [*e.g.*, *Hanks and Anderson*, 1969; *Lewis*, 1972; *Ringwood*, 1975].

In the present study the nature of the shock-induced, high-pressure phases of olivine and pyroxene rocks is examined in the light of data for the densities of a new class of perovskite-related silicate structures synthesized by *Liu* [1975*a*, 1975*b*, 1976], and some new Hugoniot and release adiabat data for bronzite. Previous data for olivine-rich samples in the range Fo_{100} to Fo_{85} and for enstatitic samples of analogous composition are reported by *Trunin et al.* [1965] and *McQueen et al.* [1967]. *Trunin et al.* [1965] and recently *Simakov and Trunin* [1973] have reported single datum points for four different compositions, extending our knowledge of Hugoniot states for ferromagnesian silicates to pressures in excess of 2.40 Mbar.

II. EXPERIMENTAL METHODS

Discs (1 cm in diameter) of Bamle bronzite ($Mg_{.86}$, $Fe_{.14}$) SiO_3 were cored from the same aliquot previously studied to shock pressures of 0.48 Mbar [*Ahrens and Gaffney*, 1971] and machined to thickness tolerances of ∿ ± 0.05 mm. The density of each disc was determined at least twice by weighing in air and in reagent grade toluene using temperature corrections of *Berman* [1939]. The samples were mounted on 0.5 mm thick Cu or Ta driver plates and impacted by flyer-plate bearing projectiles at speeds ranging from 5.4 to 6.1 km/sec with a light-gas apparatus qualitatively similar to that described by *Jones et al.* [1966]. Pure (commercial grade) Cu and Ta flyer plates, 2.5 mm thick, with a minimum diameter of 17 mm (Figure 1) are hot-press welded into lexan projectile blanks using techniques developed by *A. Mitchell* [private communication]. After molding and stress relieving, a polyethylene, gas-sealing rear portion of the projectile was press fitted onto the 6 mm long lexan flyer plate (front) portion of the projectile. The final projectile diameter was machined so as to provide a ∿ 0.03 mm interference fit with the light-gas gun launch tube. The resulting projectiles had masses of 14 and 20 gm for Cu and Ta flyer plates, respectively. The light-gas gun used to accelerate these projectiles employs a 20 m long, 16 cm diameter pump tube in which a ∿ 20 kg high-density polyethylene piston is used to compress the propelling gas, H_2, initially at a pressure of 2 bar. The polyethylene piston achieves speeds of ∿ 0.4 km/sec using approximately 3 kg of double-based propellant. The projectile is launched upon bursting of a diaphram (at a gas pressure of ∿ 0.7 kbar) at the breech of the 25 mm diameter, 7 mm long launch tube. The total flight of the projectile from the muzzle of the launch tube to the target is ∿ 500 mm and occurs in a vacuum of 10 to 30 mtorr. Projectile velocities are measured using the flash X-ray method of *Jones et al.* [1966], except that interruption of a continuous X-ray beam [*Long and Mitchell*, 1972] is used to trigger the first flash X-ray unit (∿ 15 ns exposure), and the breakage of a 0.05 mm diameter copper wire intercepting the outer edge of the projectile is used to trigger the second flash X-ray unit. The center line of the two flash X-ray units are ∿ 350 mm apart, and the time interval between X-ray flashes is measured to within an uncertainty of ± 0.01 μsec. Since triggering in our initial experiments was not reliable, we have had to rely on the redundancy provided by measuring (1) the electronic delays of both flash X-ray units and (2) the time interval between the second X-ray flash and the closure of the redundant pin switch(es) mounted on the sample assembly (Figure 1) to infer impact velocity. The ∿ 3° projectile tilt observed in Figure 1, is highly reproducible, and the sample assembly was thus preoriented to achieve nearly normal impact.

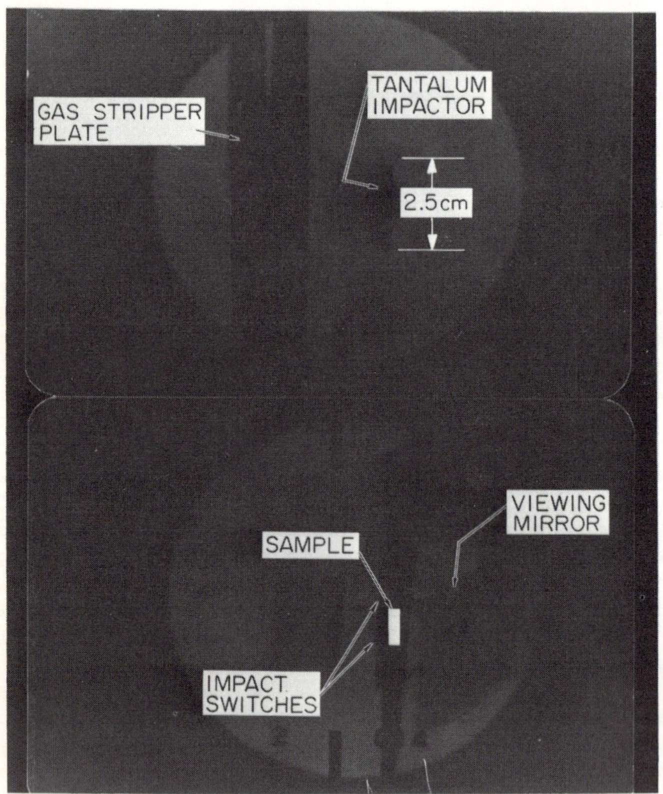

Fig. 1. Flash radiographs of projectile taken at stations 390 mm (upper) and 33 mm (lower) from target. Nominal exposure times are ∿ 15 nsec. Projectile speed is 5.5 km/sec. Shot LGG14.

As in the experiments of *Ahrens et al.* [1971], an electronically triggered, image-converter streak camera was used to record shock transit-time through the sample. However, in the present experiments, the onset of illumination on shock arrival at the mirrors (presumably arising from the high temperatures behind the shock in the glass) was used to record shock arrivals with the streak camera, and the arrival of the shock at the mirror-free surface was observed to result in the immediate loss of illumination (Figure 2). We infer this effect is caused by the disintegration of the specular mirror surface, probably upon incongruent vaporization of the glass and the accompanying sudden adiabatic expansion of the vaporized species. In the present experiments, the streak camera was writing at a speed of 25 mm/μsec for a duration of ∿ 2 μsec on 20 X 22.5 cm (4 X 5 inch) Tri-X film. Developing procedures were used to force the film to a sensitivity of ASA 3000. Time calibration was obtained using a Pockel-cell modulated Ar-ion laser beam modulated at 20 Mhz, thus providing time marks at 50-nsec intervals. The writing rate at any point on the streak camera trace was measurable to within ± 0.25%, and a practical time resolution of ∿ 2 nsec was achieved.

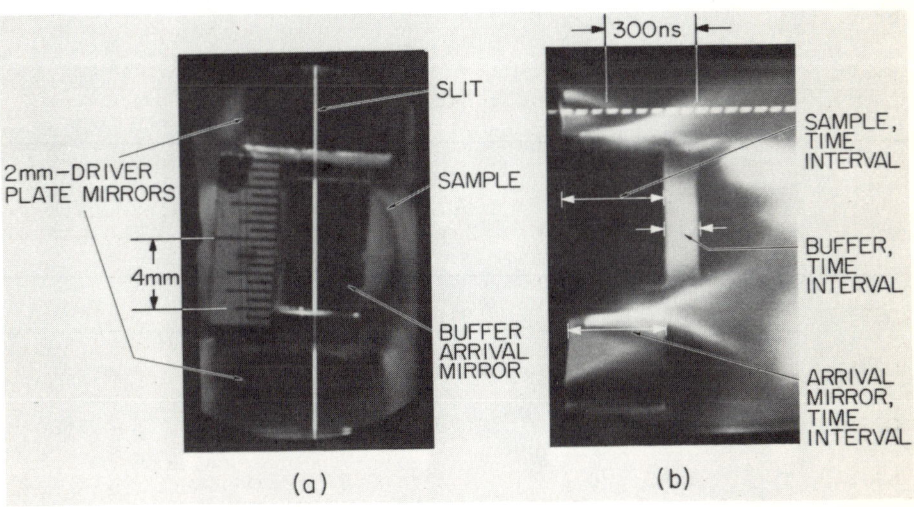

Fig. 2. Static (a) and dynamic (b) streak camera photograph of bronzite Hugoniot and release adiabat experiment. Shot LGG14.

Hugoniot states were determined using the impedance match method of *Rice et al.* [1958], and release adiabat data employ the buffer technique described by *Ahrens et al.* [1969*b*]. The equations of state for Ta and Cu employed in the impedance match solutions are those given by *McQueen et al.* [1970].

III. RESULTS FOR GLASS AND BAMLE BRONZITE

The unexpected self-illumination and prompt light decay upon passage of strong shocks through the soda-lime glass mirrors (microscope slide quality, plate glass) provided consistent and, we believe, reliable measurements of shock-wave velocities in this material. The resulting data are given in Table 1 and plotted in Figures 3 and 4. The shock-wave velocity (U_s) - particle velocity (u_p) relation in Figure 3 is concordant with data reported by *Dremin and Adadurov* [1964] to pressures of 0.41 Mbar for a glass of similar zero-pressure density (2.48 g/cm^3) but of slightly different composition. The linear U_s-u_p relation indicated in Figure 3 fits the present four data points, and *Dremin and Adadurov's* data points at 0.33 and 0.41 Mbar with a coefficient of correlation r^2 = 0.997. This fit was used to obtain the release adiabat points for the Bamle bronzite (Table 2, Figure 5).

TABLE 1. Hugoniot Data, Soda-Lime Glass[a]

Shot No.	Flyer Plate	Projectile Velocity, km/sec	Shock Velocity, km/sec	Pressure, Mbar	Density, g/cm^3
LGG9	Cu	5.76 ± 0.18[b]	8.90 ± 0.13	0.919	4.66
LGG11	Cu	5.85 ± 0.02[b]	8.91 ± 0.12	0.934	4.72
LGG14	Ta	5.541 ± 0.010[c]	9.094 ± 0.013	0.981	4.756
LGG12	Ta	5.66 ± 0.10[c]	9.11 ± 0.02	1.005	4.85

a. Na_2O, 0.1; MgO, 0.02; SiO_2, 0.75; CaO, 0.13

Initial Density, 2.49 ± 0.01 g/cm^3.

b. 1-mm-thick sample.

c. 2-mm-thick sample.

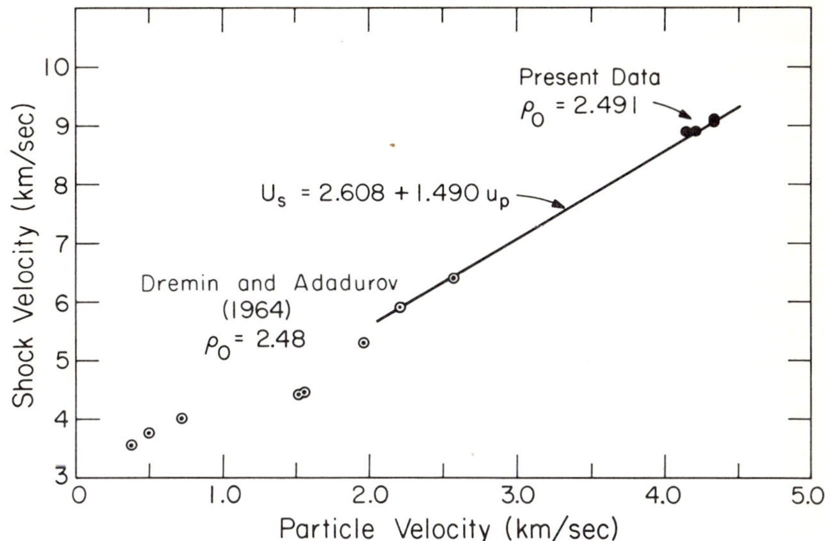

Fig. 3. Shock velocity versus particle velocity data for soda-lime glass.

Qualitatively, the behavior of Na-Ca glass is observed to be similar to that first reported by *Wackerle* [1962] for fused quartz. It appears that above ∿ 0.15 Mbar, the Si^{+4} ion begins to transform from tetrahedral to octahedral coordinations with the O^{2-} ion, as in the case of the quartz to stishovite transition [*McQueen et al.*, 1963]. This, and the possible coordination increases of the Na^+ and Ca^+ ions, account for the incompressible behavior indicated above 0.50 Mbar. The zero-pressure density (ρ_0) shown in Figure 4 is calculated using our microprobe analyses and standard oxide molar volumes assuming a molar volume of 14.014 cm^3 for SiO_2 (stishovite) [*Robie and Waldbaum*, 1968].

Although the uncertainties in the measurement of projectile and shock velocity can be obtained from the scatter of redundant projectile time-interval measurements and the uncertainties of discerning shock arrival times on the streak records, the estimation of uncertainties in the pressure-density plane (Figures 4 and 5) is less straightforward. The non-orthogonality of the uncertainty brackets shown in both figures for the Hugoniot data arise from the fact that the particle velocity and, hence, pressure and density, are not independently related to uncertainties in projectile and shock velocity via the impedance match solution. The uncertainties arising from both the density and pressure are calculated from the uncertainties in projectile velocity holding the shock velocity fixed (at its average value), and vice versa. The uncertainties in the three release points take into account only the uncertainties in shock velocity through the buffer (1-mm-thick glass).

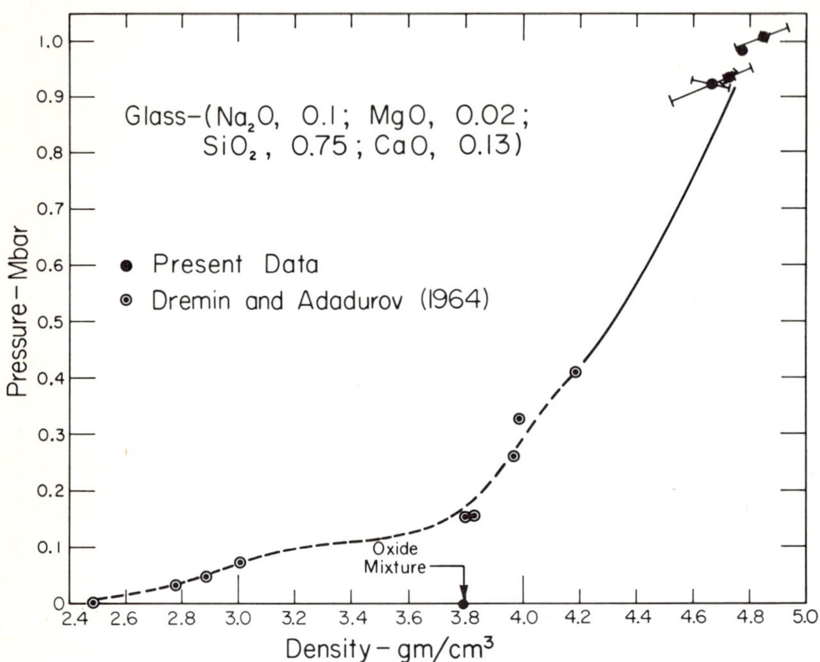

Fig. 4. Shock pressure versus density for soda-lime glass.

TABLE 2. Hugoniot and Release Adiabat Data, Bamle Bronzite

Shot No.	Flyer Plate	Initial Density, g/cm^3	Flyer Plate Velocity, km/sec	Shock Velocity, km/sec	Pressure, Mbar	Shock Density, g/cm^3	Buffer Shock Velocity, km/sec	Release Pressure, Mbar	Release Density, g/cm^3
LGG6	Cu	3.2923±0.0003	5.375[a]	10.08±0.13	1.126	4.963	9.14±0.11	0.803	4.950
LGG8	Cu	3.2943±0.0012	6.09±0.13	10.37±0.05	1.324	5.261	9.62±0.16	0.923	5.260
LGG14	Tu	3.3021±0.0007	5.54±0.01	10.07±0.03	1.335	5.491	8.83±0.15	0.730	5.258

a. Projectile velocity inferred from shock velocity through arrival mirrors.

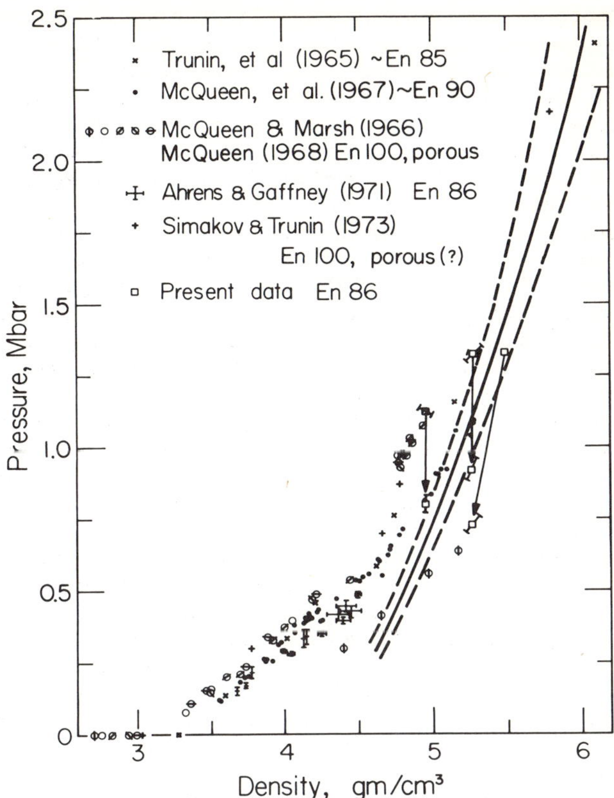

Fig. 5. Compilation of shock data of Mg-rich pyroxenes. Hugoniot points for samples of varying porosities (synthetic, polycrystalline aggregates) and pyroxene-rich (> 90%) rocks are included, as are newly determined release paths (arrows). Theoretical Hugoniot (solid curve) and envelope (dashed curve) are discussed in the text and Table 3. McQueen and Marsh [unpublished] are cited in *Birch* [1966].

IV. DISCUSSION

Shock-wave data for magnesian pyroxenes are summarized in Figure 5, including results for porous (synthetic aggregate) samples, nonporous samples, and our new data on Bamle bronzite (cf. Table 2). At pressures below about 0.70-1.00 Mbar, it is clear that the porous Hugoniot data bear an anomalous relation with respect to the nonporous data, since the former should lie on "hot compression" curves, systematically displaced to lower densities (at a given pressure) from the principal Hugoniot presumably defined by nonporous data. The energy associated with irreversible compaction of porosity represents a thermal pressure as a function of initial density, via Grüneisen's γ (V, T), which defines this displacement. For this reason, the porous data have, in general, been very scantily discussed, considered uninterpretable, or outright dismissed as wrong [*McQueen et al.*, 1967; *McQueen*, 1968; *Ahrens et al.*, 1969*a*; *Davies and Gaffney*, 1973; *Simakov and Trunin*, 1973].

The relatively high density attained by the porous materials on shocking can, nevertheless, be explained as a kinetic effect [*cf.*, *McQueen et al.*, 1967, 1970]. Assuming that pyroxene transforms to a dense phase or assemblage above a given pressure, this transformation, even once initiated under shock, may not have time to go to completion without substantial overdriving. Because porous samples become much hotter on shocking, their rate of transformation is expected to be significantly enhanced, thus yielding a larger proportion of high-pressure phase material for a given pressure compared with nonporous samples. Assuming this model applies, it is easy to explain the distribution of Hugoniot data in the figure.

The pressure-release points shown in Figure 5 provide independent support for such a model. Since the release path is adiabatic (isentropic to within the approximation of reversibility) while entropy increases monotonically up a material's Hugoniot, the P-V slope of the release path should be no steeper than the Hugoniot slope at a given pressure [*e.g.*, see *Duvall and Fowles*, 1966]. Because the measured release adiabat points lie below the Hugoniot data, it is indicated that a mixed-phase region extends to pressures around 0.70-1.00 Mbar with transformation to a high-density assemblage complete at pressures no lower than these. This high-pressure assemblage must have a metastable Hugoniot with a P-V slope at least as great as the release paths, which supports the hypothesis of an extended mixed-phase (or disequilibrium) region probably associated with the rate of phase transformations under shock.

At pressures above about 1.00 Mbar the nonporous samples achieve densities that are generally (and consistently) higher than the porous points. So, we believe that the shock-wave data begin to define an equilibrium, high-pressure assemblage Hugoniot at these pressures, below which one or more phase transformations result in a wide, mixed-phase region.

Similarly, Hugoniot data for Mg-rich olivines are summarized in Figure 6. These show the same anomalous relations between porous and nonporous samples, implying transformation to a high-density phase (or assemblage) subject to kinetic effects (though release adiabat data for such compositions are not yet available). This conclusion fits in with the discussion by *Ahrens and Petersen* [1969].

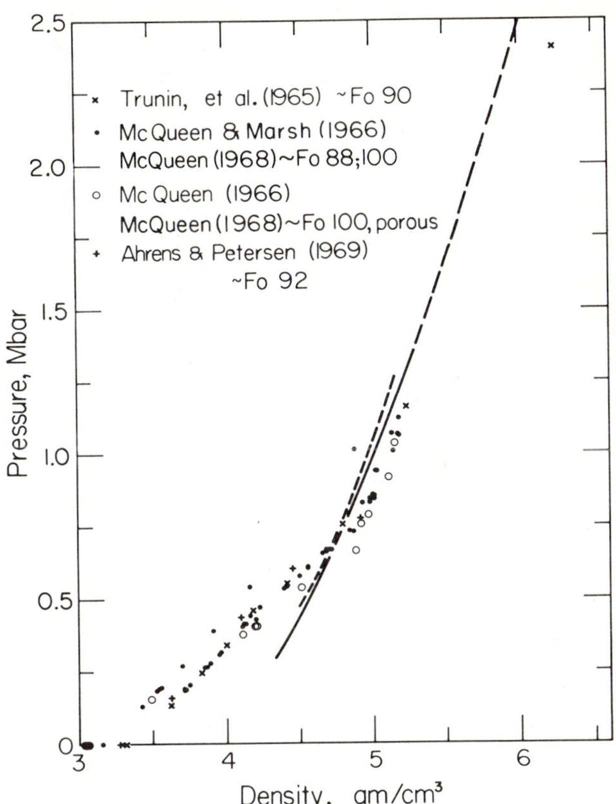

Fig. 6. Compilation of shock data for Mg-rich olivines, including results on porous samples (synthetic, polycrystalline aggregates) and olivine-rich (> 90%) rocks. Theoretical Hugoniot (solid curve) for a Mg-endmember and high-pressure assemblage are discussed in the text. Dashed curve corresponds to the left side of the envelope in Figure 5. McQueen and Marsh [unpublished] are cited in *Birch* [1966].

Aside from these results with porous data and steep, measured release paths [*cf.*, *Ahrens et al.*, 1969*b*; *Grady et al.*, 1974, 1977], a comparison of shock-recovery and static high-pressure experiments [*Ahrens and Graham*, 1972; *Liu*, 1974, 1975*c*; *Schneider and Hornemann*, 1976] also indicate that overdriving beyond equilibrium transition pressures is commonly needed in silicates to achieve a given phase transformation under shock, clearly implying control by kinetics of nucleation and growth of high-pressure phases [but note *Podurets and Trunin*, 1974]. Thus, porous samples may well indicate the presence and nature of high-pressure phases (or assemblages) more readily than nonporous samples under shock. Also, high-pressure phase Hugoniots will tend to be significantly steeper (and therefore centered at higher zero-pressure densities) than would be apparent from the distribution of the data.

Zero-pressure densities for high-pressure phases interpreted from pyroxene and olivine shock-wave data by several workers are summarized in Figure 7. The tendency has been to assign higher densities (for a given Mg:Fe ratio) to olivine (A_2BO_4) than to pyroxene (ABO_3) stoichiometries. One problem is apparent near the magnesian end, however, in that interpreted densities seem to actually decrease with increasing Fe content. Barring very complex phase relations, this is highly unlikely due to the enormous difference in the atomic masses of Fe and Mg. Rather, it appears that the density estimates for the magnesian endmembers have been biased upward relative to the Fe-bearing compositions by the high densities implied by the porous data (which only exist for the endmembers, forsterite and enstatite). Based on the arguments developed above, it is not appropriate to directly compare the porous data with the data for (nonporous) Fe-bearing compositions.

Liu's recent [1975*a*, 1975*b*, 1976; *Liu and Ringwood*, 1975] static high-pressure experiments have confirmed the previously suspected occurrence and significance of perovskite-related structures in silicates [*Reid and Ringwood*, 1969, 1974, 1975; *Shimizu et al.*, 1970; *Ringwood and Major*, 1971]. Distorted (noncubic), perovskite-like structures have been quenched from diamond-anvil experiments on both magnesian olivines and pyroxenes. These structures, apparently of the orthorhombic, rare-earth, orthoferrite type [*Marezio et al.*, 1970] will be loosely termed perovskites. Densities versus composition for the perovskite phases and assemblages of pyroxene (ABO_3) and olivine (ABO_3 + AO) derived from *Liu's* [1975*a*, 1975*b*] X-ray parameters are also shown in Figure 7. It is interesting to note that an olivine stoichiometry actually results in a lower density perovskite assemblage than a pyroxene stoichiometry. This is a direct consequence of the low density of (Mg, Fe)O in the ABO_3 + AO assemblage. The densities of such (Mg, Fe)O recovered from high pressure lie directly between the densities of MgO (periclase) and FeO (stoichiometric wüstite), as can be seen in Figure 7. This is compatible with the shock-wave data on MgO, which shows no evidence for phase transformations to pressures over 1.00 Mbar [*McQueen and Marsh* cited in *Birch*, 1966; *McQueen*, 1968; *Carter et al.*, 1971].

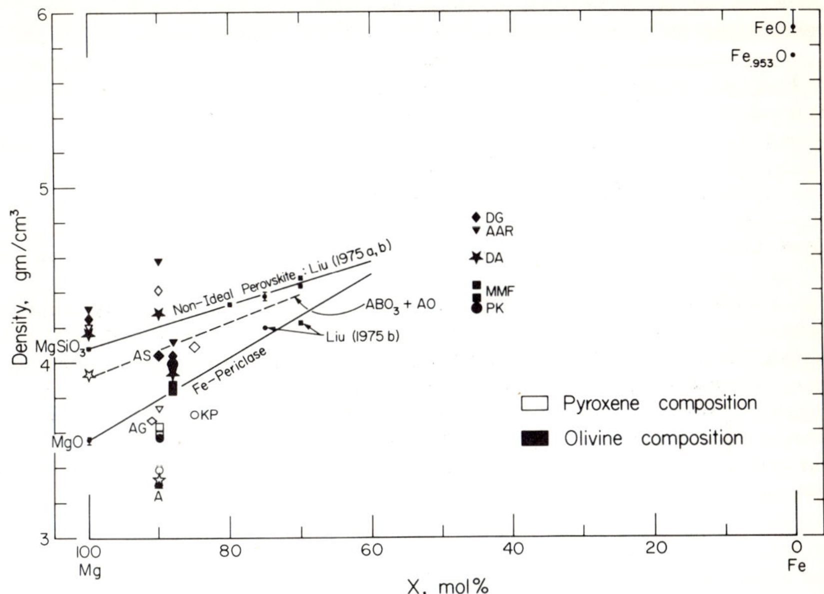

Fig. 7. Summary of zero-pressure densities for high-pressure phases of olivines and pyroxenes as a function of composition based on interpretation of shock-wave data (symbols), and calculated from X-ray parameters for ABO_3 (perovskite) and AO quenched from high-pressure diamond-anvil experiments (lines through points). MgO (periclase) and FeO (stoichiometric wüstite) values are from Bénard [1954] *Robie et al.* [1966] *Katsura et al.* [1967] *and Robie and Waldbaum* [1968] *with error bars showing variations in measurement and estimate, while error bars on the remaining points are based on Liu's* [1975a, 1975b] *estimates of error in the X-ray parameters. The dashed line (ABO_3 + AO) corresponds to the density of a perovskite assemblage with olivine stoichiometry (assuming $K_{D\ ol\text{-}perov.}^{Mg\text{-}Fe} = 1$). Symbols are keyed as follows:*

A	*Ahrens* [1971]
AAR	*Ahrens, Anderson, and Ringwood* [1969a]
AG	*Ahrens and Gaffney* [1971]
AS	*Al'tschuler and Sharpidzhanov* [1971a]
DA	*Davies and Anderson* [1971]
DG	*Davies and Gaffney* [1973]
KP	*Kalinin and Pan'kov* [1974]
MMF	*McQueen, Marsh, and Fritz* [1967]
PK	*Pan'kov and Kalinin* [1975]

In order to compare the static results with the shock-wave data, theoretical metastable Hugoniots were constructed for perovskite of appropriate composition ($Mg_{0.9}Fe_{0.1}SiO_3$) based on Birch-Murnaghan principal adiabats. This approach is philosophically and computationally similar to that of *Davies and Gaffney* [1973]. Necessary parameters are shown in Table 3 along with the values used here. Density and the energy of the pyroxene-perovskite transformation (at P = 0) were derived from *Liu's* work [1975*a*, 1975*b*, 1976], via Figure 7 and by using $\Delta E_{tr} \leq (-P\Delta V + T\Delta S)_{tr}$. Since the last term is not known, it is assumed to be negligibly small (i.e., $\frac{dP}{dT} = 0$ for the phase boundary). Because the experimental data upper bound the equilibrium pressure of transition, and most likely $T\Delta S \leq 0$, the resulting estimate for ΔE_{tr} is probably a strict upper bound (and rather high). Incidentally, though we discuss the direct transformation from pyroxene to perovskite structures under shock, this obviously represents a metastable phase boundary (based on our current knowledge from *Liu's* work). Intermediate structures may well be involved.

TABLE 3. Parameters for Theoretical Hugoniot of (Mg, Fe) SiO_3 Perovskite

Composition: $(Mg_{0.9}Fe_{0.1})SiO_3$	Source
$\rho_0 = 4.20$ g/cm^3	(1)
$K_0 = \sim 2.6\ (\pm .35)$ Mbar	(2)
$K'_0 = \sim 3.5\ (\pm 1.0)$	(2)
$\Delta E_{tr} < \sim 166$ *KJ*/mol	(1)
$\gamma = \gamma_0\ (V/V_0)^n$	
$\gamma_0 \cong 1.5\ (\pm 0.5)$	
$n \cong 1.0\ (\pm 0.5)$	

(1) From data of *Liu* [1975*a*,1975*b*].

(2) See *Davies* [1976], for example.

Estimates of the bulk modulus K_0 and $K'_0 = (dK/dP)_{P=0}$ were based on their empirical correlation with density, with particular emphasis given to data for perovskite-structured compounds [*Beattie and Samara*, 1971; *Davies*, 1976]. The wide range in values given easily overlap estimates based on the various K-ρ relations that have been proposed [*D. L. Anderson*, 1967, 1969; *O. L. Anderson et al.*, 1968; *D. L. Anderson and O. L. Anderson*, 1970; *O. L. Anderson*, 1972]. Similarly, Grüneisen's (bulk or thermodynamic) γ was estimated from typical values for various compounds, with a volume dependence that has been found to be not unreasonable for those cases where it has been studied. Temperature dependence was assumed to have a negligible effect, and, again, rather wide bounds were given for reasonable values. Finally, a family of theoretical perovskite Hugoniots was constructed for the best estimate values given in Table 3 and by varying each of the parameters within the stated bounds. An envelope containing this family of curves along with the Hugoniot based on the best values are shown in Figure 5.

It is immediately clear that the shock-wave data is compatible with Mg-pyroxene transforming to a perovskite structure at pressures above approximately 0.80 Mbar. This is contrary to the conclusions of *Simakov and Trunin* [1973]. In fact, by considering the most porous data (which achieve high densities at relatively low pressures) and the arguments presented above, it seems that perovskite-like densities can be achieved by about 0.50 Mbar. If anything, the left half of the envelope in Figure 5 agrees with the data as well as the best estimate, which could be consistent with an even lower pressure, for transformation to the perovskite structure. This, then, is an upper bound for the pressure of transformation based solely on Hugoniot data, yet it still represents a significant overdriving compared to the static experiments [*Liu*, 1975*b*, 1976].

Compared with the theoretical Hugoniots, the release paths still appear to have very steep slopes, and the most porous data cross over to yet higher densities. These latter data may be erroneous [*Simakov and Trunin*, 1973], but assuming that they are not, they suggest, along with the release adiabats, the possibility of a phase (or assemblage) with a zero-pressure density even higher than perovskite. Transformation to such a "post-perovskite" phase could occur at pressures as low as about 0.60 Mbar, according to the shock data; this would therefore be an upper limit for the equilibrium transition pressure. An initial density of 5% or so greater than for orthorhombic perovskite could be consistent with the data.

In similar fashion, a theoretical Hugoniot (Figure 6) was constructed for a perovskite assemblage corresponding to olivine stoichiometry ($MgSiO_3$ (perovskite) + MgO (periclase)) based on the best-estimate Hugoniot in Figure 5 and the rather well-determined Hugoniot of MgO [*McQueen and Marsh* cited in *Birch*, 1966; *McQueen*, 1968; *Carter et al.*, 1971] shown in Figure 8. Because the left half of the envelope in Figure 5 appeared most consistent with the pyroxene points, a corresponding band is shown in Figure 6.

In this case, perovskite-like densities are achieved in the 0.60-0.80 Mbar range. Once more, the shock-wave data are consistent with the static results that demonstrate olivine transforming to perovskite at high pressures (but again at pressures significantly lower than under shock). On the other hand, the olivine points are clearly shifted to higher densities than the perovskite-assemblage Hugoniot at pressures above about 0.70 Mbar. Because the theoretical curve for the perovskite assemblage cannot be shifted without causing a significant deviation from the pyroxene points, the olivine data are in strong disagreement with a perovskite assemblage at high pressure.

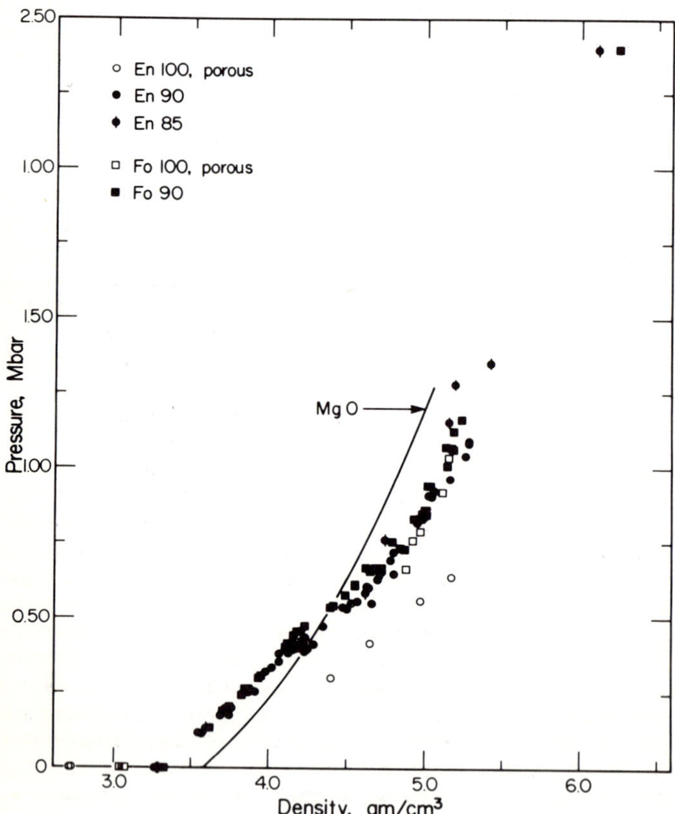

Fig. 8. Representative Hugoniot data from Figures 5 and 6 compared with the Hugoniot for MgO [*McQueen and Marsh,* unpublished, are cited in *Birch,* 1966; *McQueen,* 1968; *Carter et al.,* 1971].

If the olivine points did not deviate so systematically from the theoretical Hugoniot, which is consistent with the pyroxene shock data, this divergence might well be ascribed to be within experimental error. However, the same discrepancy emerges from Figure 8, quite apart from any theoretical curves. The point here is that above about 0.70 Mbar, the olivine and pyroxene data overlap, whereas a perovskite model would predict olivine densities to be lower than pyroxene densities (i.e., shifted toward the MgO Hugoniot) as was mentioned above and can be inferred from Figure 7. Although we have discussed kinetic effects and a mixed-phase region to the 1.00 Mbar range, the overlap continues to much higher pressures. Furthermore, the nature of the perovskite-forming transformations would suggest that pyroxene (which involves no disproportionation) would form perovskite at least as readily as olivine, whereas in the high-pressure data of Figure 8 one must either consider the olivine too dense or the pyroxene not dense enough to agree with a self-consistent perovskite model.

V. CONCLUSIONS

After reviewing experimental techniques, we presented new Hugoniot data for Bamle bronzite including simultaneously measured release paths, based on the mirror-buffer technique. Calculation of Hugoniot states for the mirror glass provides a satisfactory internal check, with the soda-lime glass behaving in a fashion similar to fused quartz under shock.

Reexamining available shock data for magnesian pyroxenes and olivines leads to the conclusion that these data define a mixed-phase (or disequilibrium) region to about the 1.00 Mbar range, related to the kinetics of phase transformation in these silicates. By recognizing this point, certain discrepancies in previous interpretations of shock data can be explained. A set of theoretical Hugoniots for pyroxene and olivine stoichiometry, perovskite-bearing assemblages was constructed based on the properties deduced from high-pressure work, showing that the shock data is compatible with transformations to perovskites in the 0.45-0.70 Mbar region (decidedly above the equilibrium pressure of transformation).

A perovskite model implies different densities for pyroxene and olivine stoichiometries, however. We note that by considering an extremely simplistic assemblage for the lower mantle, consisting of high-pressure phases of only olivine and pyroxene, varying the proportions of these components results in a significant variation in the overall density of the assemblage. In other words, any compelling evidence for density variations within the lower mantle (based, say, on seismological data) can be readily explained without resorting, for example, to variations in Mg:Fe ratios. This degree of freedom is likely to exist whatever the pertinent, multi-phase assemblage may be for the lower mantle. In fact, we consider our two-phase model simplistic if for no other reason than that we have ignored (for lack of data) the effects on phase relations of other likely components, such as Al or Ca.

In addition, the high-pressure shock data suggest the presence of a yet higher density phase (or assemblage) than perovskite. In particular, the results on olivine diverge from predicted estimates and from densities consistent with pyroxene Hugoniot points. Taken together, the data do not preclude transformations involving a "post-perovskite" phase of Mg_2SiO_4 at pressures above 0.80-1.00 Mbar, and including a reaction of the type $Mg_2Si_2O_6 \rightarrow Mg_2SiO_4$ ("post-perovskite" phase) + SiO_2 (rutile, α-PbO_2, or fluorite structure) [*cf.*, *German et al.*, 1974; *Jamieson*, 1977].

We reiterate that the static high-pressure results (and thus our analysis) involve an orthorhombic (i.e., non-ideal) modification of perovskite. Since this class of structures includes a wide variety of related modifications [succinctly summarized by *Salje*, 1976], it is quite possible that the higher pressure phase suggested by the shock data is, in fact, a perovskite such as the ideal cubic form. Indeed, it is quite possible that the phase quenched from high pressure is a modification of the actual structure attained under pressure. Bearing in mind the well-known difficulties of recovering these high-pressure polymorphs on quenching [*e.g.*, *Liu and Ringwood*, 1975] and the important effects of environmental conditions on the perovskite modifications [*e.g.*, see *Sis et al.*, 1973], it is not unlikely that they are involved in higher density assemblages than have yet been described. Recently discussed hexagonal forms may be an example of this [*cf.*, *Burbank and Evans*, 1948]. However, there are other structure types that could also be likely, some of which have only begun to be discussed as possible high-pressure polymorphs [*e.g.*, *Reid and Ringwood*, 1970; *Ringwood*, 1975; *Moore*, 1976].

Finally, the shock data indicate very similar properties for olivine and pyroxene at high pressures, making them both equally likely candidates for the lower mantle.

Acknowledgments. This is the first scientific report of results obtained with the light-gas gun apparatus housed in the Helen and Roland W. Lindhurst Laboratory of Experimental Geophysics. The physical facilities of this laboratory owe their existence to the generosity of Mrs. Helen W. Lindhurst. This research was supported under NSF Grant DES75-15006 and NASA Grant NGL05-002-105. We appreciate the skilled operation and maintenance of our apparatus by H. Richeson, D. Johnson, and V. Nenow. We also thank I. Jackson for helpful comments.

Contribution No. 2812, Division of Geological and Planetary Sciences, California Institute of Technology, Pasadena, California 91125.

REFERENCES

Ahrens, T. J., Shock-wave equations of state of minerals, in *Mantle and Core in Planetary Physics*, edited by L. Corso, pp. 157-187, Academic Press, New York, 1971.

Ahrens, T. J., and E. S. Gaffney, Dynamic compression of enstatite, *J. Geophys. Res.*, *76*, 5504-5513, 1971.

Ahrens, T. J., and E. K. Graham, Shock-induced phase change in iron-silicate garnet, *Earth Planet. Sci. Letters*, *14*, 87-90, 1972.

Ahrens, T. J., and C. F. Petersen, Shock wave data and the study of the earth, in *The Application of Modern Physics to the Earth and Planetary Interiors*, edited by S. K. Runcorn, pp. 449-461, Wiley-Interscience, New York, 1969.

Ahrens, T. J., D. L. Anderson, and A. E. Ringwood, Equations of state and crystal structures of high-pressure phases of shocked silicates and oxides, *Rev. Geophys.*, *7*, 667-707, 1969*a*.

Ahrens, T. J., J. H. Lower, and P. L. Lagus, Equation of state of forsterite, *J. Geophys. Res.*, *76*, 518-528, 1971.

Ahrens, T. J., C. F. Petersen, and J. T. Rosenberg, Shock compression of feldspars, *J. Geophys. Res.*, *74*, 2727-2746, 1969*b*.

Akimoto, S., M. Akuaogi, K. Kawada, and O. Nishizawa, Mineralogic distribution of iron in the upper half of the transition zone in the earth's mantle, in *The Geophysics of the Pacific Ocean Basin and Its Margin*, edited by G. H. Sutton, M. H. Manghnani, and R. Moberly, pp. 399-405, *Am. Geophys. Union Monograph 19*, 1976.

Al'tschuler, L., and I. Sharpidzhanov, Additive equation of state of silicates at high pressure, *Izv. Earth Physics*, *3*, 11-21, 1971*a*.

Al'tschuler, L., and I. Sharpidzhanov, Distribution of iron in the earth and its chemical differentiations, *Izv. Earth Physics*, *4*, 3-16, 1971*b*.

Anderson, D. L., A seismic equation of state, *Geophys. J. R. Astr. Soc.*, *13*, 9-30, 1967.

Anderson, D. L., Bulk modulus-density systematics, *J. Geophys. Res.*, *74*, 3857-3864, 1969.

Anderson, D. L., and O. L. Anderson, The bulk modulus-volume relationship for oxides, *J. Geophys. Res.*, *75*, 3494-3500, 1970.

Anderson, D. L., C. Sammis, and T. Jordan, Composition of the mantle and core, in *The Nature of the Solid Earth*, edited by E. C. Robertson, pp. 41-66, McGraw-Hill Co., New York, 1972.

Anderson, O. L., Patterns in elastic constants of minerals important to geophysics, in *The Nature of the Solid Earth*, edited by E. C. Robertson, pp. 557-613, McGraw-Hill Co., New York, 1972.

Anderson, O. L., E. Schreiber, R. C. Liebermann, and N. Soga, Some elastic constant data on minerals relevant to geophysics, *Rev. Geophys., 6,* 491-524, 1968.

Basu, A. R., Hot-spots, mantle plumes and a model for the origin of ultramafic xenoliths in alkali basalts, *Earth Planet. Sci. Letters, 28,* 261-274, 1975.

Beattie, A. G., and G. A. Samara, Pressure dependence of the elastic constants of $SrTiO_3$, *J. Appl. Phys., 42,* 2376-2381, 1971.

Bénard, J., Sur les paramètres limites de la phase FeO, *Acta Cryst., 7,* 214, 1954.

Berman, H., A torsion microbalance for the determination of specific gravities of minerals, *Am. Mineralogist, 24,* 434-440, 1939.

Birch, F., Compressibility; elastic constants, in *Handbook of Physical Constants,* edited by S. P. Clark, Jr., pp. 97-173, *Geol. Soc. Am. Mem. 97,* 1966.

Boyd, F. R., A pyroxene geotherm, *Geochim. Cosmochim. Acta, 37,* 2533-2546, 1973.

Burbank, R. D., and H. T. Evans, Jr., The crystal structure of hexagonal barium titanate, *Acta Cryst., 1,* 330-336, 1948.

Carter, W. J., S. P. Marsh, J. N. Fritz, and R. G. McQueen, The equation of state of selected materials for high-pressure reference, in *Accurate Characterization of the High-Pressure Environment,* edited by E. C. Lloyd, pp. 147-158, *Nat. Bur. Stds. Pub. 326,* 1971.

Clark, S. P., Jr., and A. E. Ringwood, Density distribution and constitution of the mantle, *Rev. Geophys., 2,* 35-88, 1964.

Davies, G. F., The estimation of elastic properties from analogue compounds, *Geophys. J. R. Astr. Soc., 44,* 625-647, 1976.

Davies, G. F., and D. L. Anderson, Revised shock-wave equations of state for high-pressure phases of rocks and minerals, *J. Geophys. Res., 76,* 2617-2627, 1971.

Davies, G. F., and E. S. Gaffney, Identification of high-pressure phases of rocks and minerals from Hugoniot data, *Geophys. J. Res. Astr. Soc., 33,* 165-183, 1973.

Dremin, A. N., and G. A. Adadurov, The behavior of glass under dynamic loading, *Soviet Physics - Solid State, 6,* 1379-1384, 1964.

Duvall, G. E., and G. R. Fowles, Shock waves, in *High Pressure Physics and Chemistry, 2,* edited by R. S. Bradley, pp. 209-299, Academic Press, New York, 1966.

German, V. N., N. N. Orlova, L. A. Tarasova, and R. F. Trunin, Synthesis of the orthorhombic phase of silica in dynamic compression, *Izv. Earth Physics, 7,* 50-56, 1974.

Grady, D. E., W. J. Murri, and G. R. Fowles, Quartz to stishovite: wave propagation in the mixed phase region, *J. Geophys. Res., 79,* 332-338, 1974.

Grady, D. E., Processes occurring in shock wave compression of rocks and minerals, in *High-Pressure Research: Applications to Geophysics*, edited by M. H. Manghnani and S. Akimoto, Academic Press, New York, 398-438, 1977

Green, D. H., Magmatic activity as the major process in the chemical evolution of the earth's crust and mantle, *Tectonophys.*, *13*, 47-71, 1972.

Hanks, T. C., and D. L. Anderson, The early thermal history of the Earth, *Phys. Earth Planet. Interiors*, *2*, 19-29, 1969.

Jamieson, J. C., Phase transitions in rutile-type structures, in *High-Pressure Research: Applications to Geophysics*, edited by M. H. Manghnani and S. Akimoto, Academic Press, New York, 209-218, 1977.

Jones, A. H., W. M. Isbell, and C. J. Maiden, Measurement of the very high pressure properties of materials using a light-gas gun, *J. Appl. Phys.*, *37*, 3493-3499, 1966.

Kalinin, V. A., and V. L. Pan'kov, Equations of state of rocks, *Izv. Earth Physics*, *7*, 10-20, 1974.

Katsura, T., B. Iwasaki, S. Kimura, and S. Akimoto, High-pressure synthesis of the stoichiometric compound FeO, *J. Chem. Phys.*, *47*, 4559-4560, 1967.

Lewis, J. S., Metal/silicate fractionation in the solar system, *Earth Planet. Sci. Letters*, *15*, 286-290, 1972.

Liu, L., Disproportionation of kyanite to corundum plus stishovite at high pressure and temperature, *Earth Planet. Sci. Letters*, *24*, 224-228, 1974.

Liu, L., Post-oxide phases of forsterite and enstatite, *Geophys. Res. Letters*, *2*, 417-419, 1975*a*.

Liu, L., Post-oxide phases of olivine and pyroxene and mineralogy of the mantle, *Nature*, *258*, 510-512, 1975*b*.

Liu, L., High-pressure reconnaissance investigation in the system $Mg_3Al_2Si_3O_{12}$-$Fe_3Al_2Si_3O_{12}$, *Earth Planet. Sci. Letters*, *26*, 425-433, 1975*c*.

Liu, L., The high pressure phases of $MgSiO_3$, *Earth Planet. Sci. Letters*, *31*, 200-208, 1976.

Liu, L., and A. E. Ringwood, Synthesis of a perovskite-type polymorph of $CaSiO_3$, *Earth Planet. Sci. Letters*, *28*, 209-211, 1975.

Long, J. R., and A. C. Mitchell, A dc X-ray fiducial pulse generator for light-gas gun work, *Rev. Scient. Instr.*, *43*, 914-917, 1972.

MacGregor, I. D., Petrologic and geophysical significance of mantle xenoliths in basalts, *Rev. Geophys. Space Phys.*, *13*, 90-93, 1975.

Marezio, M., J. P. Remeika, and P. D. Dernier, The crystal chemistry of the rare earth orthoferrites, *Acta Cryst.*, *B26*, 2008-2022, 1970.

McQueen, R. G., Shock-wave data and equations of state, in *Seismic Coupling*, edited by G. Simmons, pp. 53-150, Advanced Research Projects Agency Meeting, National Technical Information Service, Springfield, Virginia, 1960.

McQueen, R. G., J. N. Fritz, and S. P. Marsh, On the equation of state of stishovite, *J. Geophys. Res.*, *68*, 2319-2322, 1963.

McQueen, R. G., S. P. Marsh, and J. N. Fritz, Hugoniot equation of state of twelve rocks, *J. Geophys. Res.*, *72*, 4999-5036, 1967.

McQueen, R. G., S. P. Marsh, J. W. Taylor, J. N. Fritz, and W. J. Carter, The equation of state of solids from shock wave studies, in *High Velocity Impact Phenomena*, edited by R. Kinslow, pp. 294-419, Academic Press, New York, 1970.

Moore, P. B., The glaserite, $K_3Na[SO_4]_2$, structure type as a "super" dense-packed oxide: evidence for icosahedral geometry and cation-anion mixed layer packings, *N. Jb. Miner. Abh.*, *127*, 187-196, 1976.

Pan'kov, V. L., and V. A. Kalinin, Thermodynamic characteristics of rocks and minerals under the conditions obtaining in the earth's mantle, *Izv. Earth Physics*, *3*, 3-15, 1975.

Podurets, M. A., and R. F. Trunin, On the microstructure of the dense phase of shock-compressed quartz, *Izv. Earth Physics*, *7*, 21-24, 1974.

Reid, A. F., and A. E. Ringwood, High-pressure scandium oxide and its place in the molar volume relationships of dense structures of M_2X_3 and ABX_3 type, *J. Geophys. Res.*, *74*, 3238-3252, 1969.

Reid, A. F., and A. E. Ringwood, The crystal structure of dense M_3O_4 polymorphs: high pressure Ca_2GeO_4 of K_2NiF_4 structure type, *J. Solid State Chem.*, *1*, 557-565, 1970.

Reid, A. F., and A. E. Ringwood, New dense phases of geophysical significance, *Nature*, *252*, 681, 1974.

Reid, A. F., and A. E. Ringwood, High-pressure modification of $ScAlO_3$ and some geophysical implications, *J. Geophys. Res.*, *80*, 3363-3370, 1975.

Rice, M. H., R. G. McQueen, and J. M. Walsh, Compression of solids by strong shock waves, in *Solid State Physics*, *6*, edited by F. Seitz and D. Turnbull, pp. 1-63, Academic Press, New York, 1958.

Ringwood, A. E., *Composition and Petrology of the Earth's Mantle*, McGraw-Hill Co., New York, 1975.

Ringwood, A. E., and A. Major, Synthesis of majorite and other high pressure garnets and perovskites, *Earth Planet. Sci. Letters*, *12*, 411-418, 1971.

Robie, R. A., and D. R. Waldbaum, Thermodynamic properties of minerals and related substances at 298.15°K (25.0°C) and one atmosphere (1.013 bars) pressure and at high temperatures, *U.S. Geol. Survey Bull. 1259*, 256 pp., 1968.

Robie, R. A., P. M. Bethke, M. S. Toulmin, and J. L. Edwards, X-ray crystallographic data, densities, and molar volumes of minerals, in *Handbook of Physical Constants*, edited by S. P. Clark, Jr., pp. 27-73, *Geol. Soc. Am. Mem. 97*, 1966.

Salje, E., Symmetry and lattice dynamics of oxides with perovskite-like structures, *Acta Cryst.*, *A32*, 233-238, 1976.

Schneider, H., and U. Hornemann, Disproportionation of andalusite (Al_2SiO_5) into Al_2O_3 and SiO_2 at very high dynamic pressures, *Am. J. Sci.*, in press, 1976.

Schubert, G., and O. L. Anderson, The earth's thermal gradient, *Physics Today*, *27*, 28-34, 1974.

Shimizu, Y., Y. Syono, and S. Akimoto, High-pressure transformations in $SrGeO_3$, $SrSiO_3$, $BaGeO_3$, and $BaSiO_3$, *High Temp. - High Press.*, *2*, 113-120, 1970.

Simakov, G. V., and R. F. Trunin, On the existence of overdense perovskite structures in magnesium silicates under conditions of high pressure, *Izv. Earth Physics*, *9*, 80-81, 1973.

Sis, L. B., G. P. Wirtz, and S. C. Sorensen, Structure and properties of reduced $LaCoO_3$, *J. Appl. Phys.*, *44*, 5553-5559, 1973.

Trunin, R. F., V. I. Gon'shakova, G. V. Simakov, and N. E. Galdin, A study of rocks under the high pressures and temperatures created by shock compression, *Izv. Earth Physics*, *9*, 1-12, 1965.

Wackerle, J., Shock-wave compression of quartz, *J. Appl. Phys.*, *33*, 922-932, 1962.

PHASE-TRANSITION PRESSURES OF Fe_3O_4 AND GaAs DETERMINED FROM SHOCK-COMPRESSION EXPERIMENTS

Y. SYONO, T. GOTO, Y. NAKAGAWA
The Research Institute for Iron, Steel and Other Metals, Tohoku University Sendai 980, Japan

Abstract

Shock-wave technique is used in order to observe the phase transition of two iron oxides, magnetite (Fe_3O_4) and hematite (αFe_2O_3), and two semiconductors, GaAs and GaP. Shock data up to about 500 kbar are obtained by streak-camera photography.

Phase-transition pressure in magnetite is determined from two inclined mirror runs to be 216 ± 15 kbar, which is appreciably lower than the corresponding value at static pressures, 250 ± 15 kbar.

Two inclined mirror runs for GaAs reveal a three-wave structure, yielding a Hugoniot elastic limit and apparent phase-transition pressure of 84 ± 8 kbar and 203 ± 11 kbar. If a correction due to the shear strength effect on the phase transition is taken into consideration, this value is reduced to be 162 ± 11 kbar, which is again remarkably lower than the static value of 193 ± 5 kbar or 180 ± 1 kbar. The observed discrepancy between transition pressures determined from static and dynamic means suggests a partial loss of shear strength in shocked GaAs.

Preliminary experiments on αFe_2O_3 and GaP indicate phase transitions above about 500 kbar and at about 260 kbar respectively.

I. INTRODUCTION

The first discovery of shock-induced phase transition in iron by *Bancroft et al.* [1956] and the subsequent observation of this transition in static high-pressure experiments [*Balchan and Drickamer,* 1961] have stimulated interest in application of shock-compression techniques to observation of phase transition in various materials. *Jones and Graham* [1971] summarized the shock-induced phase transitions below about 150 kbar and gave a critical review of them from the viewpoint of establishment of pressure calibration points. The importance of shear strength effect, temperature effect, and several kinetic effects was pointed out in order to obtain a reliable equilibrium transition pressure by shock-wave techniques and to compare with the static observation of transition pressure.

Recent technical progress in the generation of static ultra-high pressure, however, has resulted in carrying out laboratory experiments covering even the lower mantle region of the earth's interior. This situation greatly enhances the need for new, appropriate, pressure-calibration points above about 150 kbar. This work contributes to the establishment of new pressure-calibration points in the relevant pressure range where the widely accepted NaCl scale still holds. For this purpose, two iron oxides, magnetite (Fe_3O_4) and hematite (αFe_2O_3), and two III-V semiconductors, GaAs and GaP, have been chosen. In this reports, shock-compression data up to about 500 kbar obtained by optical measurements are summarized. Experimental results have partly been reported elsewhere [*Syono et al.*, 1975; *Goto et al.*,1976].

II. EXPERIMENTAL METHODS

Detailed experimental techniques in shock-wave measurements have been elucidated elsewhere [*Syono et al.*, 1975; *Goto et al.*, 1975] and will only briefly be described here. High-speed, streak camera photography was used to determine the shock-wave velocity, U_s, through the specimen and the free-surface velocity, u_{fs}, using the inclined mirror technique or the argon flash gap technique. The particle velocity, u_p, was determined on the basis of the free-surface approximation, i.e., $2u_p \sim u_{fs}$. The Rankine-Hugoniot conservation equations,

$$\frac{V}{V_o} = \frac{U_s - u_p}{U_s} \qquad (1)$$

$$P - P_o = \frac{U_s u_p}{V_o} \qquad (2)$$

together with the experimental data (U_s, u_p), allow direct calculation of the pressure, P, and specific volume, V, along the shock-compression curve.

Plane shock waves were generated by the small size explosive lens system developed in our laboratory [*Goto et al.*,1975]. The shock wave was transmitted to the driver and specimen system either by an in-contact method, in which the high-explosive pad is directly contacted with the driver plate, or by a flyer run, in which a thin, metal flyer plate is accelerated by high-explosive pad to strike the driver plate. The high explosives used in the

TABLE 1. Characteristics of High Explosives

Explosive	Composition	ρ_o[a] g/cm^3	D[a] km/sec	γ	P_{C-J}[a] kbar
SEP	PETN/Paraffin	1.26	6.67	3.7	120
HABW	PETN/Pb_3O_4/ Paraffin	2.31	5.05		(147)[b]
Pentolite	PETN/TNT	1.59	7.50	3.3	216
Comp B	RDX/TNT	1.67	7.69		(247)[b]
Octol	HMX/TNT	1.77	7.81		(270)[b]

a. ρ_o = density, D = detonation velocity; and P_{C-J} = Chapman-Jouget pressure $\rho_o D^2/(\gamma+1)$.

b. γ = 3 is assumed.

present study are listed in Table 1.

For the purpose of observation of the multiple shock structure induced by phase transition, the inclined mirror technique was used to record the free-surface motion continuously. In the present study, the inclined mirror technique was applied only for in-contact runs, since good planarity of shock waves is required in this technique. Experimental arrangements are shown in Figure 1. Transparent mirrors silvered on their inside faces are placed on the surface of the driver or the specimen; the

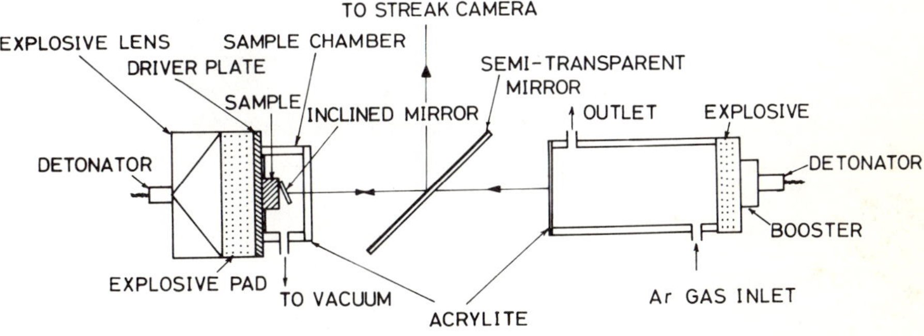

Fig. 1. Experimental arrangement of shock-wave experiments using the inclined mirror technique.

mirror on the specimen is inclined at a small angle, α. The whole assembly is illuminated by an explosive argon candle through a semi-transparent mirror, and the reflected light is introduced to a slit of the streak camera.

An example of photographic record of GaAs shocked to 220 kbar is shown in Figure 2. The slanted trace due to the inclined mirror shows a three-wave structure; an elastic precursor wave and a double plastic shock wave (plastic I and II waves) associated with shock-induced phase transition. The time of incidence of shock wave into the specimen and the times of arrival of three successive shock waves on the free surface of the specimen are indicated by dashed lines, t_o, t_1, t_2 and t_3, respectively. Assuming no wave interactions, the shock and particle velocities

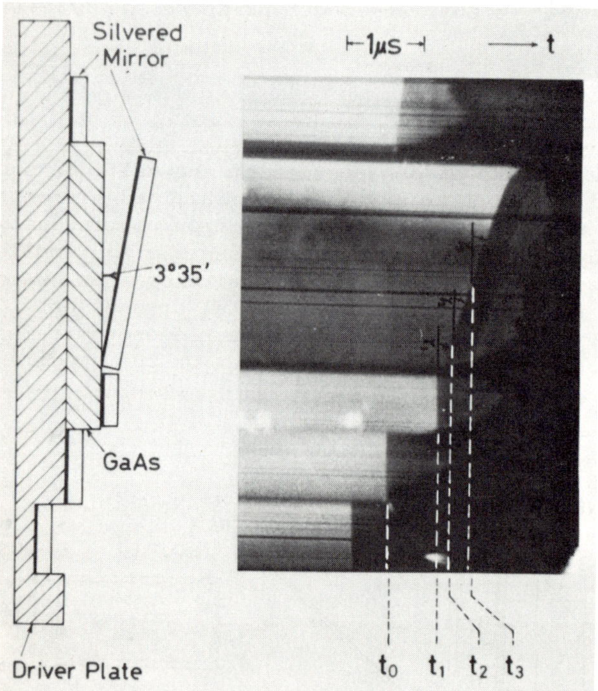

Fig. 2. Experimental arrangement and photographic record of an inclined mirror run for GaAs. The time of incidence of shock wave into the sample, and the times of arrival of three successive shock waves (elastic precursor wave, plastic I and II waves) on the free surface of the specimen are indicated by dashed lines, t_o, t_1, t_2, and t_3, respectively.

of the three waves are given by

$$U_n = \frac{d + 2\sum_{i=1}^{n} u_{i-1}(t_i - t_{i-1})}{t_n - t_o} \tag{3}$$

$$2u_n = \frac{W \tan \alpha}{M \tan \gamma_n} \qquad (n = 1, 2, 3) \tag{4}$$

where d is the thickness of sample, W the writing speed, M the space magnification of the streak-camera system, and γ_n the slanted angles of the inclined mirror trace as shown in Figure 2.

Use of the argon flash-gap technique is a simple means for observation of shock arrival. However, its use to measure u_{fs} can be subject to error, particularly when phase transitions are involved. Therefore, this technique was adopted as an auxiliary means except for the region where the shock wave is expected to be simple. Both the natural and synthetic crystals of Fe_3O_4, αFe_2O_3, GaAs, and GaP were used in this study. The sample sources are described in Table 2, together with unit cell para-

TABLE 2. Characterization of Specimen

Specimen	Source	Unit Cell		Density, g/cm^3	
		a(A)	c(A)	X-ray	bulk
Fe_3O_4	Synthetic[a]	8.395		5.133	5.20
Fe_3O_4	Natural[b]	8.397		5.194	5.16
αFe_2O_3	Natural[c]	5.032	13.78	5.264	5.26
GaAs	Synthetic[d]	5.652		5.320	5.326
GaP	Synthetic[d]	5.451		4.129	4.125

a. Aggregate of a few large crystals grown with Bridgman method by K. Chiba, Victor Co. of Japan. Chemical composition = FeO, 29.7 wt% and Fe_2O_3, 70.3 wt%.

b. Single crystal from Nisi-sonoki County, Nagasaki Prefecture, Japan.

c. Aggregate of a few large crystals from Minas Gerais, Brazil.

d. Undoped polycrystalline (coarse grained) ingot manufactured by Sumitomo Electric Ind., Ltd. Specific resistivity $\geq 10^6 \Omega \cdot cm$.

TABLE 3. Hugoniot Data for Fe_3O_4 and αFe_2O_3

Shot No.	Specimen	Assembly[a] (Tech.)[b]	ρ_o g/cm^3	d mm	U_s mm/µs	u_p mm/µs	P kbar	V/Vo	Remarks[c]
32-20	Fe_3O_4 Syn.	IC-A(AFG)	5.20	2.535	6.50	0.36	122	0.944	
33-19	Fe_3O_4 Syn.	IC-B(AFG)	5.20	3.035	6.40	0.74	247	0.884	
34-10	Fe_3O_4 Syn.	FL-P(AFG)	5.20	3.170	6.01	0.83	259	0.862	
34-19	Fe_3O_4 Nat.	IC-C(IM)	5.16	3.150	6.86	0.57	202	0.917	PW-I
					5.11	0.82	266	0.867	PW-II
34-17	Fe_3O_4 Syn.	IC-D(IM)	5.20	2.575	7.12	0.62	230	0.913	PW-I
					4.88	1.04	330	0.823	PW-II
33-21	Fe_3O_4 Syn.	IC-E(AFG)	5.20	3.588	6.79	0.87	306	0.872	
35-2	Fe_3O_4 Syn.	FL-Q(AFG)	5.20	2.980	6.50	1.55	522	0.762	
35-3	Fe_3O_4 Nat.	FL-R(AFG)	5.16	3.640	6.39	1.59	528	0.751	
34-7	αFe_2O_3	IC-A(AFG)	5.26	1.845	6.61	0.375	130	0.943	
35-15	αFe_2O_3	IC-B(AFG)	5.26	3.790	7.18	0.855	324	0.881	
35-7	αFe_2O_3	IC-D(IM)	5.26	2.810	7.77	1.03	418	0.869	

a. IC = in-contact run ; FL = flyer run .
A = 20 mm SEP + 5 mm brass driver ;
B = 20 mm Pentolite + 5 mm brass driver ;
C = 30 mm Composition B + 5 mm brass driver ;
D = 20 mm Octol + 5 mm brass driver ;
E = 20 mm Octol + 3.5 mm brass driver ;
P = 40 mm Pentolite + 2.5 mm brass flyer + 3 mm brass target;
Q = 40 mm Pentolite + 2 mm brass flyer + 3 mm brass target ;
R = 40 mm Pentolite + 1.5 mm brass flyer + 2.5 mm brass target.

b. AFG = argon flash gap technique ; IM = inclined mirror technique .

c. PW-I = first plastic wave (phase transition) ;
PW-II = second plastic wave (final state) .

TABLE 4. Hugoniot Data for GaAs and GaP

Shot No.	Specimen	Assembly[a] (Tech.[b])	ρ_o g/cm^3	d mm	U_s mm/μs	u_p mm/μs	P kbar	V/Vo	Remarks[c]
32-23	GaAs	IC-A(AFG)	5.326	2.01	3.53	0.415	78	0.882	
35-21	GaAs	IC-G(AFG)	5.326	2.75	3.95	0.93	196	0.765	
37-10	GaAs	IC-G(IM)	5.326	3.19	5.20	0.31	86	0.940	EPW
					4.27	0.83	203	0.817	PW-I
					3.55	0.92	220	0.788	PW-II
34-13	GaAs	IC-E(AFG)	5.326	1.96	4.04	1.11	239	0.725	
35-25	GaAs	FL-Q(AFG)	5.326	2.02	4.04	1.20	258	0.703	
37-9	GaAs	IC-H(IM)	5.326	3.16	5.36	0.29	83	0.946	EPW
					4.15	0.87	209	0.804	PW-I
					3.83	1.14	262	0.731	PW-II
34-12	GaAs	FL-S(AFG)	5.326	2.01	4.13	1.22	268	0.705	
34-14	GaAs	FL-U(AFG)	5.326	2.01	4.62	1.59	391	0.656	
37-5	GaP	IC-F(AFG)	4.125	2.07	5.13	0.64	135	0.875	
						0.62	131	0.879	
37-3	GaP	IC-G(AFG)	4.125	2.38	5.27	1.21	263	0.770	
						1.19	259	0.774	IMS
37-4	GaP	IC-H(AFG)	4.125	2.27	5.02	1.27	263	0.747	
						1.45	302	0.711	IMS
37-8	GaP	FL-T(AFG)	4.125	2.35	5.00	1.37	283	0.726	
						1.54	319	0.692	IMS
37-7	GaP	FL-V(AFG)	4.125	3.10	5.56	2.10	460	0.622	IMS

a. IC = in-contact run ; FL = flyer run .
A - E ; See footnotes on Table 3 .
F = 20 mm SEP + 5mm 2024 Al driver ;
G = 20 mm Pentolite + 5 mm 2024 Al driver ;
H = 20 mm Octol + 5 mm 2024 Al driver .
P - R ; See footnotes on Table 3 .
S = 40 mm Pentolite + 2.5 mm brass flyer + 5.5 mm brass target ;
T = 40 mm Pentolite + 2 mm brass flyer + 5 mm brass target ;
U = 40 mm Pentolite + 1.5 mm brass flyer + 3 mm brass target ;
V = 40 mm Pentolite + 1 mm brass flyer + 3 mm brass target .

b. AFG = argon flash gap technique ; IM = inclined mirror technique .

c. IMS = impedance match solution .
EPW = elastic precursor wave (Hugoniot elastic limit) ;
PW-I = first plastic wave (phase transition) ;
PW-II = second plastic wave (final state) .

meters, X-ray density, and bulk density which was determined by a pycnometer. The agreement between the X-ray and bulk densities is generally good, corroborating that the crystals are free from cracks and pores. The low value of bulk density of natural magnetite is presumably due to a small amount of impurity.

Thin plates, 2 - 4 mm in thickness and 15 × 20 mm in typical lateral dimension, are cut out from the crystals and carefully polished on two parallel sides. Both surfaces are finished with the alundum powder of #1500. The specimens are mounted on the driver plate. Experimental arrangement of each shot is shown in Tables 3 and 4.

Setting error in experimental arrangements is estimated to be ±0.005 mm for distance measurements and ±2' for angle measurements. Precision of film record reading is estimated to be ±0.2 mm for time interval measurements, corresponding to ±0.03 μsec in case of the writing speed of 7 mm/μsec, and $\pm 0.5^{\circ}$ for offset angle measurements of the inclined mirror trace. Summarizing, probable errors in shock- and free- surface velocities are estimated to be 5% and 2% for the inclined mirror run and 4% and 1% for the argon flash gap run, respectively.

III. RESULTS AND DISCUSSION

A. Magnetite Fe_3O_4

Shock compression data for Fe_3O_4 [*Syono et al.*, 1975] are shown in Figure 3 and Table 3. Also included in the figure are *McQueen's* [1966] and static compression X-ray data for making comparison.

In two inclined-mirror runs, a two-wave structure which gives an evidence for phase transformation, has been observed. The first wave corresponds to the phase transition and the second wave to the final deformational state. From the analysis of the first wave, transition pressures are estimated to be 202 and 230 kbar; the final pressures reached are 266 and 330 kbar, respectively. Considering uncertainties in velocity measurements, the transition pressure of magnetite is determined to be 216 ± 15 kbar. This value is appreciably lower than the static transition pressure of 250 ± 15 kbar determined from the X-ray data [*Mao et al.*, 1974]. Below the transition pressure, the shock compression curve is in good agreement with the static compression curve. Ultrasonic value of bulk modulus 1818.5 kbar [*Manghnani and Katahara*, unpublished data] also gives a reasonable slope to the shock compression curve at the origin. No elastic wave was observed in these runs, suggesting that the Hugoniot elastic limit of magnetite is not high. This is consistent with a relatively low Hugoniot elastic limit of 23 kbar observed in a manganese zinc ferrite [*Jones and Graham*, 1971].

In the mixed-phase region from the transition pressure to

about 800 kbar, the present shock data, *i.e.*, the velocity values determined from the argon flash gap technique, may be inaccurate (see full circles in Figure 3). Zero pressure density of a high-pressure phase of Fe_3O_4 estimated by extrapolation of the shock data above about 800 kbar is about 6.5 g/cm^3, considerably higher than the proposed density of the high-pressure phase of Fe_3O_4, 6.24 g/cm^3 at 250 kbar, estimated from X-ray diffraction measurements at static high pressure [*Mao et al.*, 1974]. This suggests that more than one transformation might take place in the mixed-phase region.

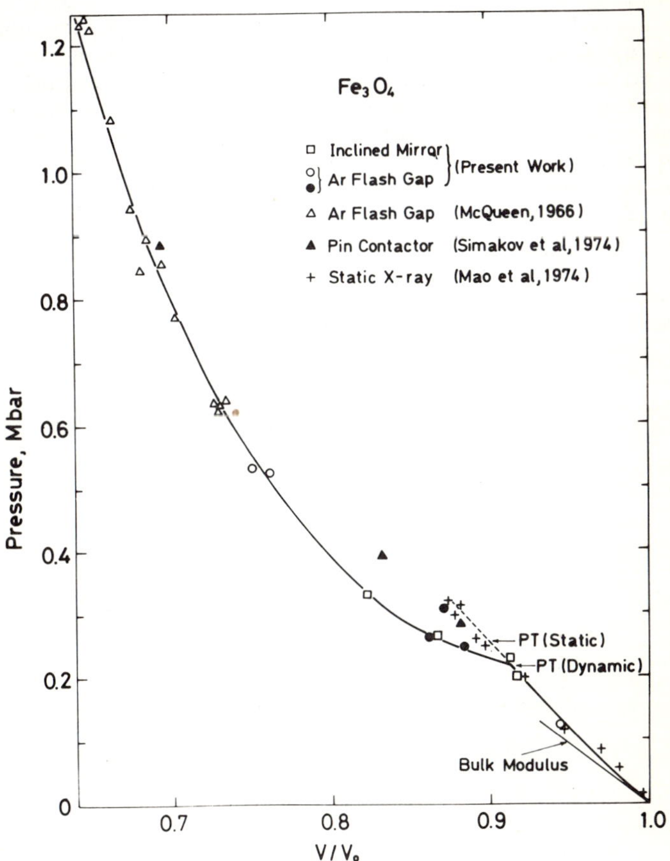

Fig. 3. Shock-compression curve of magnetite, Fe_3O_4. Shock data of McQueen[1966] and static compression X-ray data of Mao et al. [1974] are also included. The thin solid line indicates a slope corresponding to the bulk modulus at normal conditions. [Manghnani and Katahara, unpublished data].

B. Hematite, αFe_2O_3

Shock compression study of hematite still remains of preliminary nature. Only three shock data up to 418 kbar are included in this report, as shown in Figure 4 and Table 3. (Two more data previously reported [*Syono et al.*, 1975] were withheld in the present stage owing to internal inconsistencies). No evidence for phase transformation was detected in an inclined mirror run at 418 kbar. The shock-compression curve below this pressure agrees very well with the data by *McQueen* [*Davies and Gaffney*, 1973]. The present shock data in the low-pressure region are not reconciled with the shock data higher than 800 kbar [*McQueen*, 1966] by assuming a single compression curve. This suggests the existence of a phase transformation in the pressure region of 418 kbar to 800 kbar. A volume decrease of 14%, estimated from extrapolation of the shock-compression curve above 800 kbar, was suggested to be due to a high spin-low spin transition in Fe^{3+} [*Syono et al.*, 1971]. Recent conductivity measurements of hematite under shock compression seem to support this hypothesis [*A. Sawaoka*, private communication].

C. GaAs

Shock compression data [*Goto et al.*, 1975] are summarized in Figure 5 and Table 4. From the analysis of the three-wave

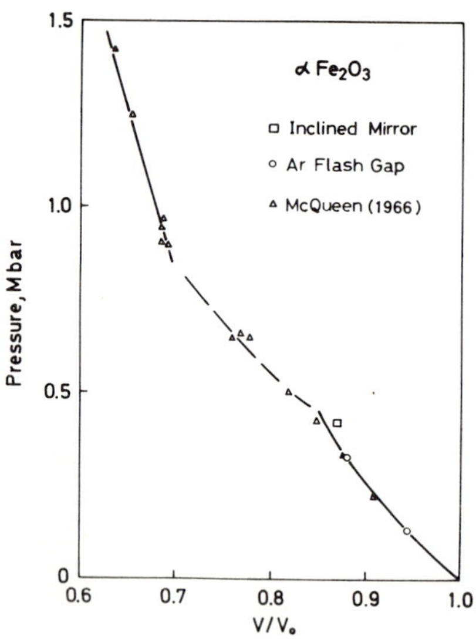

Fig. 4. Shock-compression curve of hematite, Fe_2O_3. Shock data by McQueen [1966, 1973] included.

structure observed in two inclined mirror runs, the Hugoniot elastic limit and the apparent phase transition pressure are determined to be 84 ± 8 kbar and 203 ± 11 kbar, respectively. The high shear strength observed in GaAs is similar to those in Ge [*Graham et al.*, 1966] and Si [*Gust and Royce*, 1971], and may be characteristic of such covalent crystals. Because of this high value of the Hugoniot elastic limit, the effect of the elastic-plastic transition on the phase-transition pressure must be considered, in order to make a direct comparison between transition pressures determined from dynamic and static conditions. According to the perfect elastic-plastic model, an offset between the Hugoniot pressure, P_{HUG} and the hydrostatic pressure, $\bar{P}$, above the Hugoniot elastic limit, P_{HEL}, is given by the following equation [*Jones and Graham*, 1971];

$$P_{HUG} - \bar{P} = P_{HEL} - K \frac{V_o - V_{HEL}}{V_o} , \qquad (5)$$

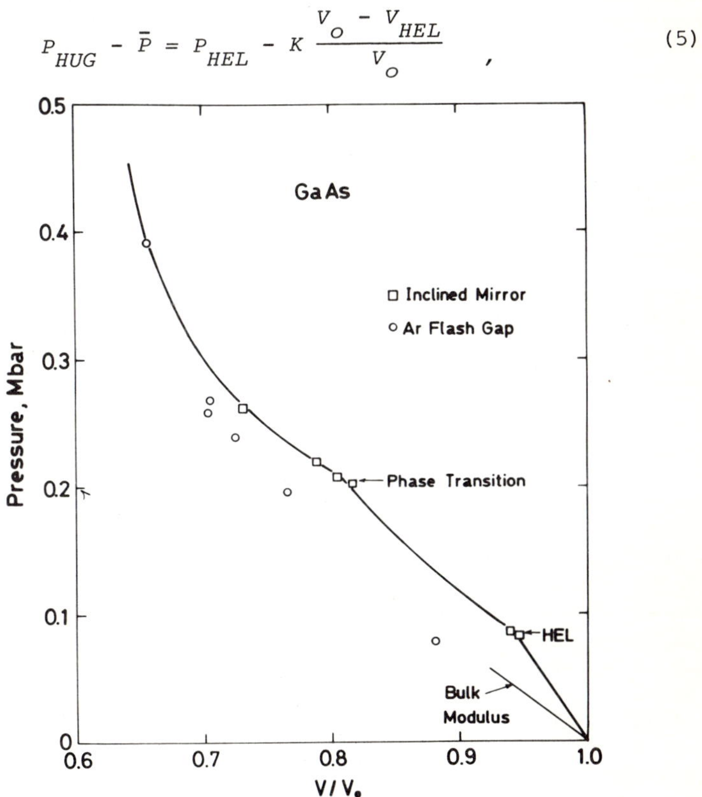

Fig. 5. Shock-compression curve of GaAs. No corrections were made on the argon flash-gap record [Goto et al., 1976]. The thin solid line shows a slope corresponding to the bulk modulus at normal conditions [Bateman et al., 1959].

where K and V_{HEL} are bulk modulus and the volume at the Hugoniot elastic limit. If this correction due to the shear strength effect is applied on the phase transition in GaAs, the transition pressure is reduced to 162 ± 11 kbar, which is remarkably lower than the static transition pressure of 175 - 185 kbar (based on *Drickamer's* [1970] new pressure scale) found by *Minomura and Drickamer* [1962], and those of 193 ± 5 kbar and 180 ± 10 kbar recently observed by X-ray diffraction study by *Yagi and Akimoto* [1976] and *Block et al.* [1977].

Since temperature increase at the phase-transition point is estimated to be 57°C based on the adiabatic approximation, the observed discrepancy between static and dynamic transition pressures may only partly be explained with a negative slope of the phase boundary. The discrepancy rather suggests a partial loss of the shear strength, as was observed in Al_2O_3 [*Ahrens et al.*, 1968; *Graham and Brooks*, 1971].

Shock wave data obtained using the argon flash-gap technique are also included in Figure 5 and Table 4. Since no corrections due to the existence of elastic precursor wave were made on the shock data, agreement between the inclined mirror data and argon flash-gap data is apparently poor. However, if such corrections are made on the flash data as have been done in an earlier report [*Goto et al.*, 1976], reasonable agreement is obtained.

D. GaP

Shock compression data taken by means of the argon flash-gap technique are shown in Figure 6 and Table 4. Although these data remain of preliminary nature and would be susceptible to the shear strength effect, a phase transition at around 260 kbar is clearly seen. Recent observation of a structural transition at 220 kbar in a diamond anvil cell reported by *Piermarini and Block* [1975] may correspond to this phase transition.

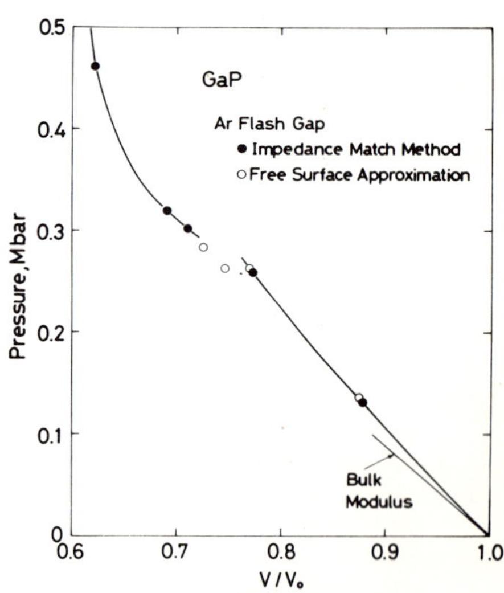

Fig. 6. Shock compression curve of GaP. The thin solid line shows a slope corresponding to the bulk modulus at normal conditions [Weil and Groves, 1968].

IV. Summary

Shock induced phase transition have been observed in Fe_3O_4 and GaAs and the dynamical transition pressures determined are found to be relatively lower than those corresponding transition, observed at static high pressures. The discrepancy observed in Fe_3O_4 may be explained by sluggish nature of its transition found by *Mao et al.* [1974]. On the other hand, the discrepancy found in GaAs suggests that the elastic-plastic model is not applicable to such covalent crystals. Further detailed investigations would be necessary to establish reliable pressure calibration points using these substances.

Acknowledgments. The authors express their most sincere thanks to the late J. Nakai for his invaluable contribution to to this work. They thank K. Chiba, Victor Co. of Japan, for supplying synthetic crystals of magnetite. They are also indebted to H. Moriya and Y. Fujita for their technical assistance.

REFERENCES

Ahrens, T. J., W. H. Gust, and E. B. Royce, Material strength effect in the shock compression of alumina, *J. Appl. Phys. 39*, 4610-4616, 1968.

Balchan, A. S., and Drickamer, H. G., High pressure electrical resistance cell, and calibration points above 100 kilobar *Rev. Sci. Instr., 32*, 308-313, 1961.

Bancroft, D., E. L. Peterson, and S. Minshall, Polymorphism of iron at high pressure, *J. Appl. Phys., 27*, 291-298, 1956.

Bateman, T. B., H. J. McSkimin, and J. M. Whelan, Elastic moduli of single-crystal gallium arsenide, *J. Appl. Phys., 30*, 544-545, 1959.

Block, S., R. A. Forman, and G. J. Piermarini, *High-Pressure Research Application in Geophysics*, edited by M. H. Manghnani and S. Akimoto, Academic Press, N. Y., 1977.

Drickamer, H. G., Revised calibration for high pressure resistance cell, *Rev. Sci. Instr., 41*, 1667-1669, 1970.

Goto, T., Y. Syono, J. Nakai, and Y. Nakagawa, Shock compression experiments in solids using high explosives, *Sci. Rept. RITU, A25*, 186-199, 1975.

Goto, T., Y. Syono, J. Nakai, and Y. Nakagawa, Pressure-induced phase transition in GaAs under shock compression, *Solid St. Commun., 18*, 1607-1609, 1976.

Graham R. A., and W. P. Brooks, Shock-wave compression of sapphire from 15 to 420 kbar. The effects of large anisotropic compressions, *J. Phys. Chem. Solids, 32*, 2311-2330, 1971.

Graham, R. S., O. E. Jones, and J. R. Holland, Physical behavior of germanium under shock wave compression, *J. Phys. Chem. Solids, 27,* 1519-1529, 1966.

Gust, W. H., and E. B. Royce, Axial yield strengths and two successive phase transition stresses for crystalline silicon, *J. Appl. Phys., 42,* 1897-1905, 1971.

Davies, G. F., and E. S. Gaffney, Identification of high-pressure phases of rocks and minerals from Hugoniot data, *Geophys. J. Roy. Astr. Soc. 33,* 165-183, 1973.

Jones, O. E., and R. A. Graham, Shear strength effects on phase transition "pressure" determined from shock-compression experiments, in *Accurate Characterization of the High Pressure Environment,* edited by E. C. Lloyd, N. B. S. Special Publ. 326, pp.229-242, 1971.

Mao, H. K., T. Takahashi, W. A. Bassett, G. L. Kinsland and L. Merrill, Isothermal compression of magnetite to 320 kbar and pressure-induced phase transformation, *J. Geophys. Res., 79,* 1165-1171, 1974.

McQueen, R. G., Unpublished data compiled by F. Birch, Compressibility; elastic constants, in *Handbook of Physical Constants,* edited by S. P. Clark, Jr., Sect. 7, pp.97-173, Geol. Soc. Amer., Memoir 97, N. Y., 1966.

Minomura, S., and H. G. Drickamer, Pressure induced phase transitions in silicon, germanium and some III-V compounds, *J. Phys. Chem. Solids, 23,* 451-456, 1962.

Piermarini, G. J., and S. Block, Ultrahigh pressure diamond-anvil cell and several semiconductor phase transition pressures in relation to the fixed point pressure scale, *Rev. Sci. Instrum., 46,* 973-979, 1975.

Simakov, G. V., M. N. Pavlovskii, N. G. Kalashnikov and R. F. Trunin, Shock compressibility of twelve minerals, *Bull. (Izv.) Acad. Sci. USSR, Earth Physics, No. 8,*11-17, 1974.

Syono, Y., S. Akimoto, and Y. Endoh, High pressure synthesis of ilmenite and perovskite type $MnVO_3$ and their magnetic properties, *J. Phys. Chem. Solids, 32,* 243-249, 1971.

Syono, Y., T. Goto, J. Nakai, and Y. Nakagawa, Shock compression study of transition metal oxides (Proc. 4th. Intern'l Conf. on High Pressure, Kyoto, 1974), *Rev. Phys. Chem. Japan,* Special Issue, 466-472, 1975.

Yagi, T., and S. Akimoto, Pressure fixed-points between 100 and 200 kbar based on the compression of NaCl, *J. Appl. Phys., 47,* 3350-3354, 1976.

Weil, R., and W. O. Groves, The elastic constants of gallium phosphide, *J. Appl. Phys., 39,* 4049-4051, 1968.

FORMATION OF DIAPLECTIC GLASS IN ANORTHITE BY SHOCK-LOADING EXPERIMENTS

Y. SYONO, T. GOTO, Y. NAKAGAWA
The Research Institute for Iron, Steel and Other Metals, Tohoku University Katahira, Sendai 980, Japan

M. KITAMURA
Institute of Mineralogy, Petrology and Economic Geology, Tohoku University, Aoba, Sendai 980, Japan

Abstract

Shock-recovery experiments on primitive anorthite have been carried out in which single crystals of anorthite are shocked to peak pressures of 150, 300, and 380 kbar, and the recovered specimen is examined by means of electron microscopy. Diaplectic glass is observed in specimens shocked above 300 kbar, and the specimen shocked to 380 kbar is found to be completely converted to diaplectic glass.

In fragments shocked to 300 kbar, the intergrowth textures of diaplectic glass and anorthite crystal are observed under an electron microscope. The diffuseness of *b* reflections in the electron diffraction pattern of intergrown crystals has no noticeable change in comparison with that of the starting specimen. The results indicate that the original configuration of antiphase domains is retained in the shocked crystal until an abrupt transition of primitive anorthite to diaplectic glass occurs. Diaplectic glass is supposed to be produced secondarily from an unquenchable high-pressure form induced under shock loading.

I. INTRODUCTION

Permanent change in the structure of crystals after shock loading has been one of the main subjects in shock-wave sciences and is termed "shock metamorphism." Investigation on natural materials from terrestrial and lunar impact craters and on experimentally shocked minerals in the laboratory both have been carried out extensively and a lot of interesting papers concerning shock metamorphism and crater formation have been reported so far [*e.g. French and Short*, 1968; *J. Geophys. Res., 76, No. 23*, 1971]. Most of these studies are concerned with the observation under an optical microscope of deformation structures, such as planar elements, twinning and kink bands. Characterization of these specific features and their relevance to the degree of shock metamorphism were first made by *Chao* [1967]. Formation of diaplectic glass in shock-loaded quartz [*DeCarli and Jamieson*, 1961; v. *Engerhardt and Bertsch*, 1969; *Müller*, 1969; *Stöffler and Hornemann*, 1972; *Kleeman and Ahrens*, 1973; *Hörz and Quaide*, 1973; *Kieffer et al.*, 1974] and feldspar [*Milton and DeCarli*, 1963; *Dworak*, 1969; *Kleeman*, 1971; *Stöffler and Hornemann*, 1972; *Hörz and Quaide*, 1973], is, among other things, one of the most attractive phenomena in shock metamorphism: various interesting features such as an increase in both density and refractive index and a change in infrared spectrum in comparison with ordinary fused glass have been observed [*Arndt et al.*, 1971; *Gibbons and Ahrens*, 1971]. However, application of electron microscopy, which will be a powerful tool in obtaining detailed informations about the shock disordering process, has been rather rare [*Müller*, 1969; *Kieffer et al.*, 1974].

In the present study, shock-recovery experiments to peak pressure of 380 kbar have been made on primitive anorthite, and shocked fragments are examined under electron microscope. Our goal in carrying out these experiments is to obtain microscopic information on the mechanism of the structural transformation from crystalline to glassy state under shock loading. For this purpose, among the plagioclases, primitive anorthite is particularly suitable, because it has a simple, integer-type supercell instead of the noninteger-type structure found in a low-temperature form of plagioclase with the chemical composition of diaplectic glass in meteorites.

II. EXPERIMENTAL METHODS

The recovery technique in shock-loading experiments is based on the momentum trap concept [*McQueen*, 1964], in which the specimen (or specimen container) is tightly surrounded by materials with shock impedance similar to that of the specimen. Tension waves originating from outside the specimen cannot enter because these interfaces will not support tension, and, hence, the specimen is protected against destructive effect from rarefaction waves.

The experimental arrangement in shock-recovery experiments is illustrated in Figure 1. Plane shock waves are produced by the 78-mm explosive lens system [*Syono and Goto* and associates, 1974, 1975] and transmitted to the specimen assembly either by amplifying the detonation wave via high explosive pad (in-contact

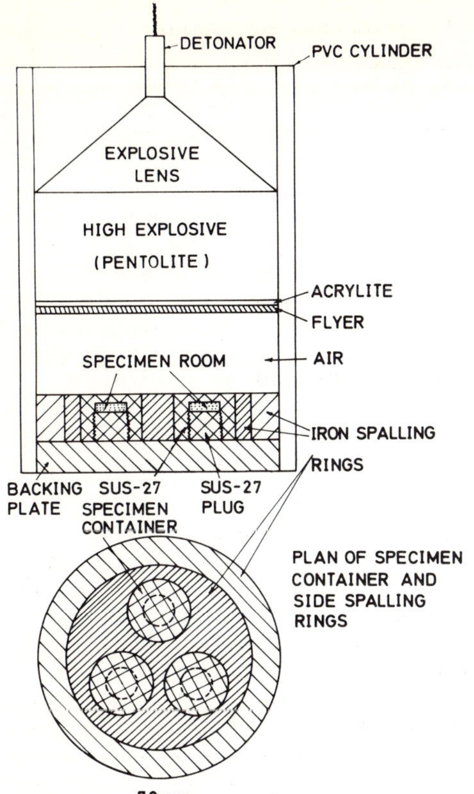

Fig. 1. Shock-wave recovery system (flyer run).

run) or by accelerating a thin metal plate to impact the specimen assembly (flyer run). The specimen assembly consists of three specimen containers, side spalling rings, and backing plates. The specimen container is made of stainless steel, 20 mm in diameter and 15 mm in height. The specimen room is 10 to 13 mm in diameter with a maximum height of 3 mm. The disc-shaped specimen is tightly fitted into the specimen room with a screwed plug so as not to be blown off upon impact. The whole assembly is shot down into a waterpool in order to assure recovery and to minimize the effect of post-shock temperatures. The specimen is usually recovered encased in the container (Figure 2).

Single crystals of primitive anorthite in a scoria fall, Fugoppe, Hokkaido, are used. An approximate composition is $An_{96}Ab_4$. Single crystals are cut perpendicular to the *b*-axis and shaped into a disc form (10 to 13 mm in diameter and 3 mm in thickness), and placed into a closely fitted sample room of the specimen container.

Three types of experimental assemblies (two in-contact runs and a flyer run) are used, as summarized in Table 1.

III. ESTIMATION OF PRESSURE

The pressure attained in the anorthite crystals is estimated graphically on the shock pressure versus particle-velocity plane based on the impedance-matching method. Published Hugoniot data of anorthosites [*McQueen*, 1966; *Ahrens and Rosenberg* and associates, 1968, 1969], stainless steel [*van Theil*, 1966] and brass [*van Thiel*, 1966] are shown in Figure 3. Hugoniot data for

TABLE 1. Types of Experimental Assemblies

Experimental Arrangement[a]	Pressure in Brass, kbar	Pressure in Anorthite, kbar
78ϕ PWG (HABW in-contact run)	150	150
78ϕ PWG + 30 mm Pentolite (in contact run)	300	300
78ϕ PWG + 60 Pentolite + 3 mm Brass flyer run	500	380

a. PWG = Plane wave generator [Syono et al., 1974]

anorthite have not been reported yet, but they would be expected to be very close to those of anorthosite. Crossed circles on brass Hugoniot have previously been determined by a series of shock-wave experiments using these shot assemblies [*Goto et al.*, 1975]. In the case of in-contact explosive runs, duration of shock pulse is estimated to be more than 8 μsec, therefore, the peak pressure in anorthite specimens is allowed to reach the shock pressure of the stainless steel container through successive shock reverberations as indicated by broken lines in Figure 3. Full circles at 150 kbar and 300 kbar on the stainless steel Hugoniot show the equilibrated pressure in the anorthite specimen sandwiched between stainless steel in the two in-contact runs. Since the effects of shock attenuation are neglected, these values are the upper limits of peak pressure realized in the

Fig. 2. Photograph of shock-wave recovery system before (left) and after (right) shock loading.

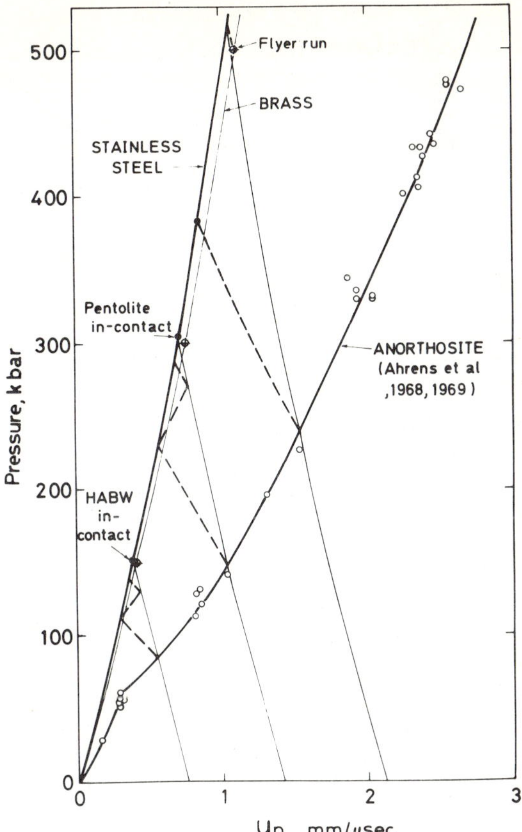

Fig. 3. Graphical representation of the impedance matching method of determining shock pressure. Hugoniot data of anorthosites are taken from Ahrens and associates [1968, 1969]. Crossed circles on brass Hugoniot show achieved shock states in brass using three kinds of shock-generating assemblies (Table I), and full circles on stainless steel Hugoniot represent estimated peak pressures in anorthite specimen. Broken lines represent shock reverberations in anorthite that progressively increase shock pressure.

anorthite specimen.

The duration of shock pulse in the flyer run, on the other hand, is short due to the thin flyer plate. Figure 4 shows a

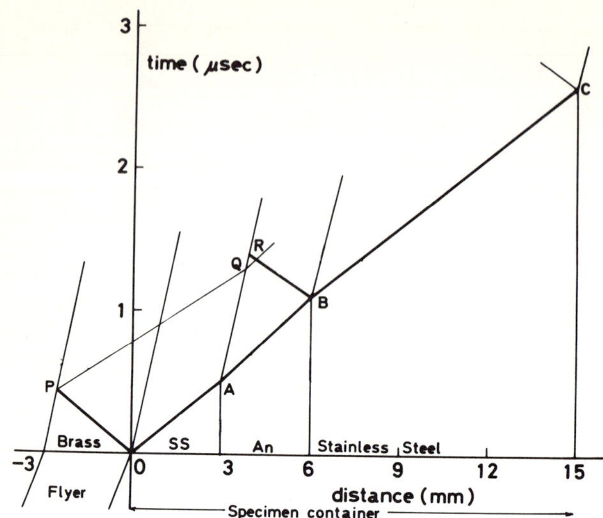

Fig. 4. A schematic distance-time plot of shock propagation in a specimen-container system of flyer run (380-kbar run). Thick and thin lines show trajectories of shock and rarefaction waves, respectively (An = Anorthite specimen).

schematic distance-time plot of shock propagation in the stainless steel container in the case of the flyer run. Upon impact of the flyer plate on the stainless-steel container at point O, shock wave OABC propagates within the specimen-container assembly, while the shock wave within the flyer plate terminates at the back face, point P, and is converted to a rarefaction wave. As shown in Figure 4, only one reflected shock wave, BR, is allowed to compress the specimen, before the rarefaction wave, PQ, originated from the back surface of the flyer plate, catches the shock reflection wave. Therefore, the shock pressure in anorthite in the flyer run is estimated to be about 380 kbar as shown in Figure 3 by a full circle on the Hugoniot of stainless steel.

The temperature along the Hugoniot is considered to be less than several hundred degrees centigrade even in the 380 kbar run [Ahrens et al., 1969], which is well below the melting point of anorthite (1553°C).

IV. POST-SHOCK OBSERVATION

The specimen recovered from shock-loading experiments were examined under an optical microscope and under JEOL 200 kV electron microscope at Tohoku University.

A. Observation under Optical Microscope

The fragments shocked to 150 kbar are aggregates of cleaved primitive anorthites, having a grain size of over ten microns. No diaplectic glass is observed in the 150 kbar run. On the other hand, all of the grains in the shocked product of the 380-kbar run are completely transformed to diaplectic glass. The results of the 300-kbar run show a mixture of primitive anorthite crystals and diaplectic glass. Correlation between the degree of glass formation and the pressure obtained in the present study is consistent with those for orthoclase by *Kleeman* [1971] and for oligoclase by *Hörz and Quaide* [1973], but quite different from the work on maximum microcline by *Robertson* [1973], in which almost no diaplectic glass was detected in the recovered products up to peak pressures of 417 kbar. Figure 5a is a microphotograph of a thin section of the 300 kbar specimen, cut parallel to the shock propagation direction. The textures of the shocked specimen are rather homogeneous within a plane parallel to the shock front, but heterogeneous along the shock propagation direction. Diaplectic glass is more abundant than crystalline anorthite. Planar deformation features, which are frequently found in naturally shocked quartz and feldspars [*Chao*, 1967; *Dworak*, 1969; *Engerhardt and Bertsch*, 1969], are not remarkable in the present recovered specimen. Figure 5b is an enlargement of a portion of Figure 5a, where such planar features are observed.

B. Fine Structures Observed in Electron Microscopy

Close examination of the shock-recovered products from the 300 kbar run was made using electron microscopy. Care was taken to pick up sample material from the central portion of the specimen disc to avoid edge effects of shock wave. An interesting texture in which glass and crystal forms a fine lamellae-like structure was found among the fragments examined. Since detailed description of these intergrowth textures observed under electron microscope has been reported elsewhere [*Kitamura et al.*, in press], special features of these intergrowth textures are briefly summarized in this report.

Figure 6a and 6b are electron micrographs of typical examples of intergrowth textures of diaplectic glass and untwinned anorthite. These intergrowth textures of glass and crystal is similar to those observed in shock-loaded quartz crystals by *Müller* [1969] and *Kieffer et al.* [1974].

Electron diffraction patterns of the intergrown particle were simultaneously taken in situ and indicated in Figure 6. The

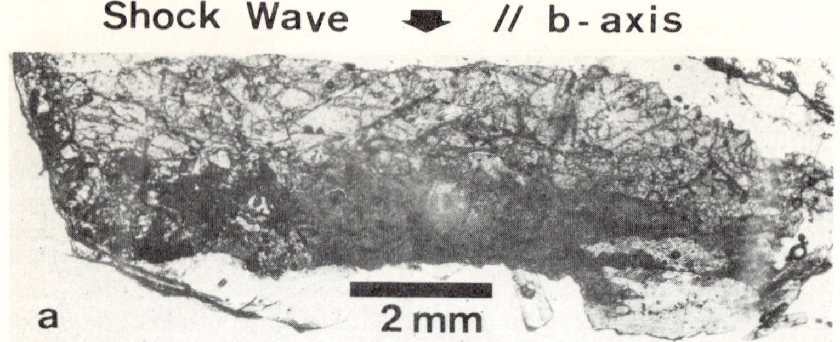

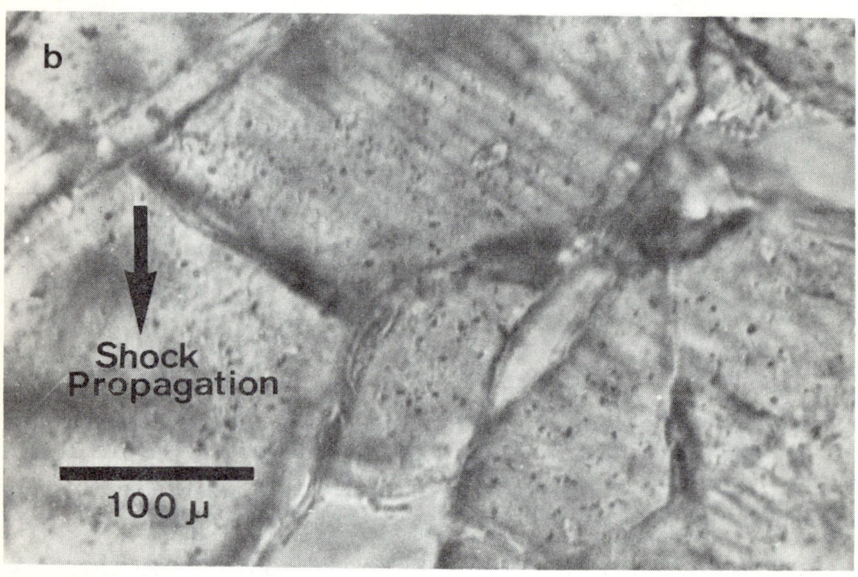

Fig. 5a (upper) Thin section microphotograph of the anorthite specimen shocked to 300 kbar (open nicols). Thin section is nearly perpendicular to the optical axis, X, and parallel to the direction of the shock propagation, [010], indicated by an arrow.

Fig. 5b (lower) An enlargement of a portion of Figure 5a. Note planar deformation lamellae and cracks running at about 45° with the direction of the shock propagation. Grains show wavy extinction under the crossed nicols.

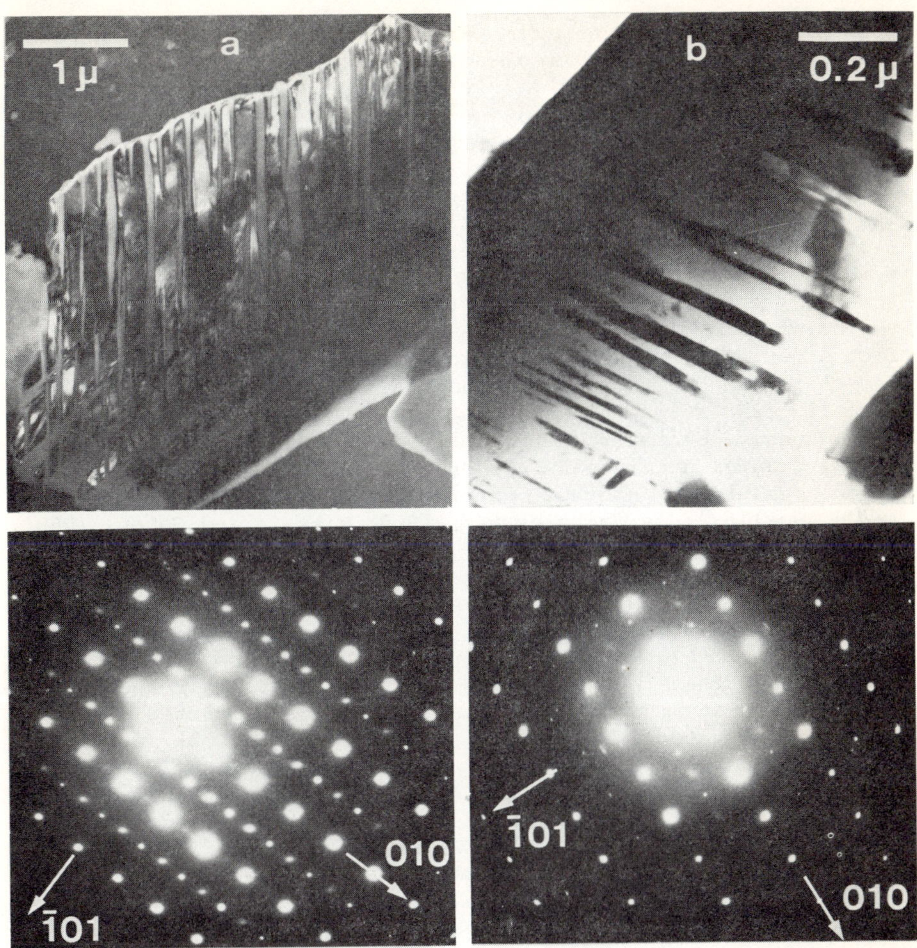

Fig. 6. Electron micrographs of intergrowth textures in shocked grains recovered from the 300-kbar run and their electron diffraction patterns. Strong spots are principal a-reflections of primitive anorthite; other weaker spots represent b-, c- and d-reflections due to the integer type of supercell.
(a) Two different orientations of the boundaries between diaplectic glass and crystal are observed. The boundary plane of the well developed lamellae is parallel to ($\bar{1}\bar{2}1$), while the other vague lamellae plane is nearly parallel to (010) to which shock waves propagate perpendicularly [Kitamura et al., submitted].
(b) With the increased width of diaplectic glass halo is more clearly observed in the diffraction pattern. The orientation of the boundary planes is nearly parallel to ($\bar{1}11$).

halo pattern due to the glass part of lamellae is superposed on the diffraction spots from the crystalline part of lamellae. With an increasing amount of glass, the halo pattern is more clearly observed (Figure 6b).

Special attention has been paid on the *b*-reflection ($h + k = 2n + 1$, $l = 2n + 1$), which is originated from an antiphase relation of distribution of Si and Al atoms [cf. *Smith*, 1974]. The diffuseness of the *b*-reflection corresponds to sizes of the *b*-domain. Since the diffuseness of the *b*-reflection of the intergrown particle shocked to 300 kbar does not differ much from that of the unshocked crystals, the domain configuration within the crystalline lamellae is conserved under shock loading. This fact directly corroborates previous suggestions [*Dworak*, 1969; *v. Engerhardt and Bertsch*, 1969] that the primitive anorthite abruptly transforms into the diaplectic glass, without involving a progressive disordering process. This is quite in contrast with a gradual decrease in the order parameter with increasing shock pressure, as has been observed in shock loading experiments in Cu_3Au alloy [*Beardmore et al.*, 1964; *Mikkola and Cohen*, 1966].

V. RELATION TO HIGH PRESSURE TRANSFORMATION

Ahrens et al. [1969] have shown from shock compression measurements that the high pressure transformation in anorthosites starts at around 130 kbar and almost is completed at around 400 kbar, as shown in Figure 6. It is noted that our 300-kbar run falls in the midst of the mixed phase region (Regime II), while the 150-kbar and 380-kbar runs may be considered to belong to the low pressure phase region (Regime I) and the high-pressure phase region (Regime III), respectively, if some uncertainties in pressure estimation are allowed. The observed facts in the present investigation can be compared directly with the above equation-of-state considerations, if we assume that the diaplectic glass is secondarily produced from an unquenchable high-pressure form induced by shock loading. The hollandite structure has generally been accepted as a possible high pressure form of feldspar [*Kume et al.*, 1966; *Ringwood*, 1970]. This type of high pressure transformation has not been proved for Ca-feldspar: phase transformation studies at static high pressure [*Hariya and Kennedy*, 1968; *Ringwood*, 1970] showed that anorthite breaks down to a mixture of grossularite, kayanite and stishovite. However, such disproportionate reactions, involving diffusion, are considered to be unfavorable under shock-loading condition. A hollandite-analog with Si and Al in sixfold coordination might remain a possible candidate of the high pressure form of Ca-feldspar. It may be interesting to interpret the observed lamellae-like structures as caused by partial reestablishment of the anorthite structure after complete phase transformation. However,

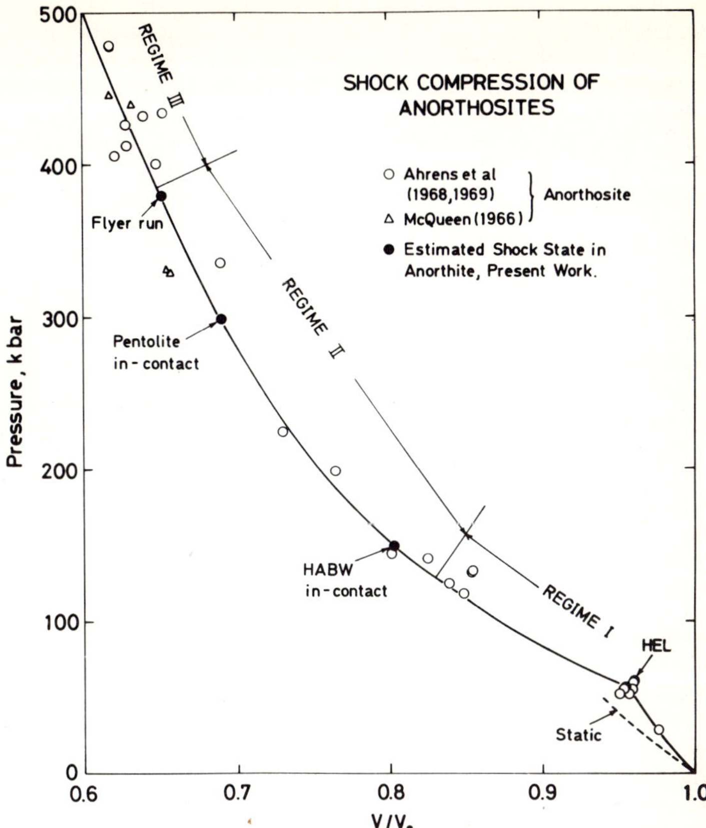

Fig. 7. Shock compression curve of anorthosite. Regime I, II, and III represent the low-pressure phase, mixed-phase, and high-pressure phase, respectively (after Ahrens and associates [1968, 1969]). Full circles on the curve show the estimated pressure-volume state achieved in the present shock recovery experiments for anorthite. The broken line is a compression curve at static high pressure.

the crystalline lamellae within a single grain have an identical crystallographic orientation as observed in electron diffraction. The present observation lends support to previous discussion on the origin of these lamellae [*Ahrens et al.*, 1969; *Dworak*,1969; *v. Engerhardt*, 1969]: these lamellae might be formed by periodical release of stored energy of the stressed state under high compression through structural transformation.

Aknowledgments. The authors wish to express their thanks to K. Yagi, Hokkaido University, for providing us with the crystal of anorthite. They are grateful to Ichiro Sunagawa, Tohoku University, and Hiroshi Takeda, University of Tokyo, for their kind advice and warm encouragement. Thanks are also due to Heishiro Ohta, Tohoku University, for his help in the experiments of electron microscopy. Critical comments on the manuscript by J. Arndt, Tübingen University, were most valuable and are cordially acknowledged.

REFERENCES

Ahrens, T. J., and J. T. Rosenberg, Shock metamorphism: Experiments on quartz and plagioclase, in *Shock Metamorphism of Natural Materials*, edited by B. M. French and N. M. Short, pp. 59-81, Mono Book, Baltimore, 1968.

Ahrens, T. J., C. F. Petersen, and J. T. Rosenberg, Shock compression of feldspar, *J. Geophys. Res., 74*, 2727-2746, 1969.

Arndt, J., U. Hornemann, and W. F. Müller, Shock wave densification of silica glass, *Phys. Chem. Glasses, 12*, 1-7, 1971.

Beardmore, P. B., A. H. Holtzman, and M. B. Bever, The effects of shock waves on the alloy Cu_3Au, *Trans. Metal. Soc. AIME, 230*, 725-730. 1964.

Chao, E. C. T., Shock effects in certain rock-forming minerals, *Science, 156*. 192-202, 1967.

DeCarli, P. S., and J. C. Jamieson, Formation of an amorphous form of quartz under shock condition, *J. Chem. Phys., 31*, 1675-1676, 1959.

Dworak, U., Stoßwellenmetamorphose des Anorthosits vom Manicouagan Krater, Québec, Canada, *Contrib. Mineral. Petrol., 24*, 306-347, 1969.

Engerhardt, W. v., and W. Bertsch, Shock induced planar deformation structures in quartz from the Ries Crater, Germany, *Contrib. Mineral. Petrol., 20*, 203-234, 1969.

French, B. M., and N. M. Short (eds.), *Shock Metamorphism of Natural Materials*, Mono Book, Baltimore, 1968.

Gibbons, R. V., and T. J. Ahrens, Shock metamorphism of silicate glasses, *J. Geophys. Res., 76*, 5489-5498, 1971.

Goto, T., Y. Syono, J. Nakai, and Y. Nakagawa, Shock compression experiments in solids using high explosives, *Sci. Rept. RITU, A25*, 186-199, 1975.

Hariya, Y., and G. C. Kennedy, Equilibrium study of anorthite under high pressure and high temperature, *Am. J. Sci., 266*, 193-203, 1968.

Hörz, F., and W. Quaide, Debye-Scherrer investigations of experimentally shocked silicates, *The Moon, 6*, 45-82, 1973.

Kieffer, S. W., P. P. Phakey and J. M. Christie, Microstructural relationships of high-pressure SiO_2 phase produced by shock waves in porous quartzite, *Proc. 8th Intn'l Congress on Electron Microscopy,* Canbera, Vol. 1, 484-485, 1974.

Kitamura, M., T. Goto, and Y. Syono, Intergrowth textures of diaplectic glass and crystal in shock-loaded P-anorthite, *Contrib. Petrol. Mineral.* (in press).

Kleeman, J. D., Formation of diaplectic glass by experimental shock loading of othoclase, *J. Geophys. Res., 76,* 5499-5503, 1971.

Kleeman, J. D., and T. J. Ahrens, Shock-induced transition of quartz to stishovite, *J. Geophys. Res., 78,* 5954-5960, 1973.

Kume, S., T. Matsumoto, and M. Koizumi, Dense form of germanate orthoclase ($KAlGe_3O_8$), *J. Geophys. Res., 71,* 4999-5000, 1966.

McQueen, R. G., Laboratory technique for very high pressures and the behavior of metals under dynamic loading, in *Conference on Metallurgy at High Pressures and Temperatures,* edited by K. A. Gshneider, Jr., M. T. Hepworth, and N. A. D. Purlee, Chap. 3, pp. 44-132, Gordon and Breach, N. Y., 1964.

McQueen, R. G., Unpublished data compiled by F. Birch, Compressibility; elastic constants in *Handbook of Physical Constants,* edited by S. P. Clark, Jr., Sect. 7, pp. 97-173, Geol. Soc. Amer., Memoir 97, N. Y., 1966.

Mikkola, D. E., and J. B. Cohen, The substructure of Cu_3Au after tensile deformation and shock loading, *Acta Metal., 14,* 105-122, 1966.

Milton, D. J., and P. S. DeCarli, Maskelynite: Formation by explosive shock, *Science, 140.* 670-671, 1963.

Müller, W. F., Elektronenmikroskopischer Nachweis amorpher Bereiche in Stoßwellenbeanspruchtem Quarz, *Naturwissenschaften, 5,* 279, 1969.

Ringwood, A. E., Phase transformations and constitution of the mantle, *Phys. Earth Planet. Interiors, 3,* 109-155, 1970.

Robertson, P. B., Experimental shock metamorphism of maximum microcline, *J. Geophys. Res., 80,* 1903-1910, 1975.

Smith, J. V., *Feldspar Minerals,* Vol. 1, Springer, Berlin, 1974.

Stöffler, D., and U. Hornemann, Quartz and feldspar glasses produced by natural and experimental shock, *Meteoritics, 7,* 371-394, 1972.

Syono, Y., T. Goto, J. Nakai, Y. Nakagawa, and H. Iwasaki, Shock compression of titanium monoxide up to 600 kbar, *J. Phys. Soc. Japan, 37,* 442-446, 1974.

van Thiel, M. (ed.), *Compendium of Shock Wave,* Vol. 1 and 2, UCRL-50108, Univ. of Calif., Livermore, 1966.

SESSION IV

INSTRUMENTATION, PRESSURE CALIBRATION, AND STANDARDIZATION

TECHNIQUES OF ELECTRICAL CONDUCTIVITY MEASUREMENT TO 300 KBAR

H. K. MAO and P. M. BELL
Geophysical Laboratory
Carnegie Institution of Washington
Washington, D. C. 20008

Abstract

Techniques of electrical conductivity measurement in the diamond pressure cell are developed for the pressure range 1 bar to 300 kbar and the temperature range -78° to 2500°C. Measurements of electrical conductivity demonstrate pressure and composition dependence of the parameters in the Boltzmann relationship of temperature and electrical conductivity. Electrical conductivity of the olivine and magnesiowüstite series was found to increase by orders of magnitude in the pressure range 1 bar to 300 kbar at room temperature.

I. INTRODUCTION

Electrical conductivity is a major parameter in modeling the temperature, composition, and magnetism of the solid earth. The Boltzmann equation

$$\sigma = \sum_{i} \sigma_{0i} \exp(-A_i/kT)$$

relates σ, electrical conductivity, to the absolute temperature, T, where i is a specific mechanism of conduction, k is the Boltzmann constant, σ_{0i} is the preexponential parameter, and A is the activation energy. The relationship has been used to estimate the temperature distribution of the earth's mantle [*Tozer*, 1959]. Electrical conduction contributes to electronic thermal conduction in the earth, and therefore is a controlling factor in the earth's thermal evolution [*Lawson and Jamieson*, 1958; *Lubimova*, 1967]. Geomagnetism also depends on the electrical conductivity of the mantle and core [*Bullard*, 1972]. These relations indicate the need for high-pressure electrical conductivity data of all materials thought to occur in the earth's mantle before a valid model of the earth can be constructed. In the absence of such measurements at pressure-temperature conditions equivalent to the lower mantle, all current models are necessarily based on parameters derived from general assumptions rather than on experimental determination. With diamond-anvil apparatus, however, one can now produce the *P-T* conditions of the mantle equivalent to depths of 2500 km [*Bell and Mao*, 1977].

Experimental difficulties of making electrical conductivity measurements at high pressure are mainly confined to the handling of microcircuits required for the minute sample size used in the diamond cell. *Mao and Bell* [1972] reported the first successful conductivity measurements in the diamond cell and found that the electrical conductivity of fayalite increases by six orders of magnitude from 1 bar to 300 kbar. The present paper reports experimental details of the technique including later improvements. Data on olivine, spinel, and magnesiowüstite are included as examples.

II. EXPERIMENTAL TECHNIQUES

A. Microcircuitry Template

Fabrication of the microcircuitry template is done under a ×40 magnification stereomicroscope. Metal wires (approximately 1 μm diameter) have been used as electrode leads for the conductivity measurements. Tungsten wire was found to have the lowest rate of breakage at low temperatures. Above 400°C, the portion of tungsten lead outside the contact area of diamond anvils tends to oxidize, and Pt, Pt-Rh, chromel, or alumel wires are used instead.

Wires are cemented to a Mylar template 0.25 mm thick to form the microcircuit arrangement. A hole 0.8 mm in diameter is drilled at the center of the template to accommodate diamond anvils with faces 0.5 mm in diameter and sample disc (Fig. 1A). Several configurations have been used with satisfactory results (Fig. 2). All the reported measurements are made with two leads, but four-lead techniques of electrical conductivity measurement (not shown in Fig. 2) are also practical.

B. Sample and Gasket

Powdered MgO, used as gasketing material, is first pressed between diamonds into a disc 0.5 mm in diameter and 20 μm thick. A 150-μm hole is drilled at the center, and the sample is pressed into the hole. When the gasketed sample and template are

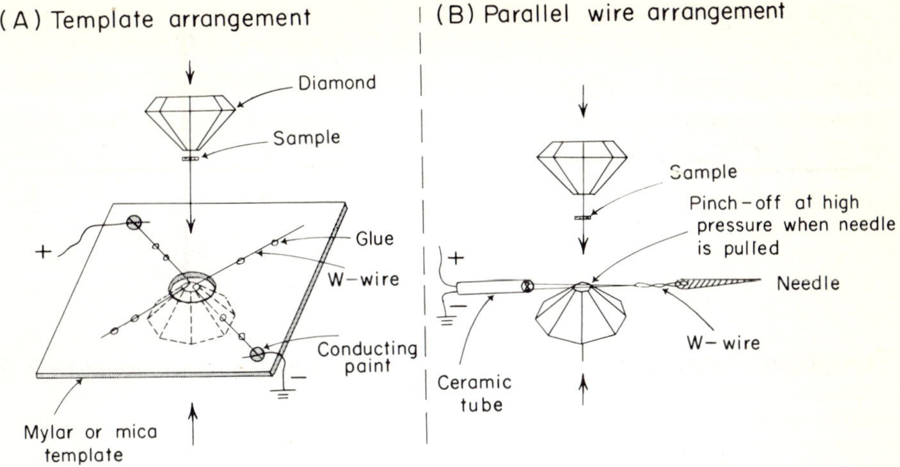

Fig. 1. Template and parallel wire arrangements for introducing electrical leads into the diamond-window pressure cell (the sample diameter is 500 μm).

introduced into the diamond cell, the metal wire cuts through the sample during initial pressurization (Fig. 3). At pressures higher than 50 kbar the sample and wires flow together to seal all fractures, whereas the wires still touch both diamond anvils. (Pressure of the sample is measured employing the shift of the R_1 line of a small amount of ruby powder embedded in the sample and gasket; see *Bell and Mao* [1977].)

C. Geometry for Resistivity Calculation

Resistance between the two leads is measured with an electrometer in the range 10^3 to 10^{13} ohm. Measurement of the conductivity of a solid MgO gasket without hole serves as the reference point of the open circuit. The measurement of resistance can be converted into resistivity because of the sample geometry. The distance between the two wires can be measured with the ocular graticule of a microscope viewing through the diamond window. The thickness of the wires, which is the same as that of the sample, is measured after the pressure is released and the sample preserved. Typically, the 13-μm wires flatten to 20 μm wide × 5 μm thick at pressures above 100 kbar.

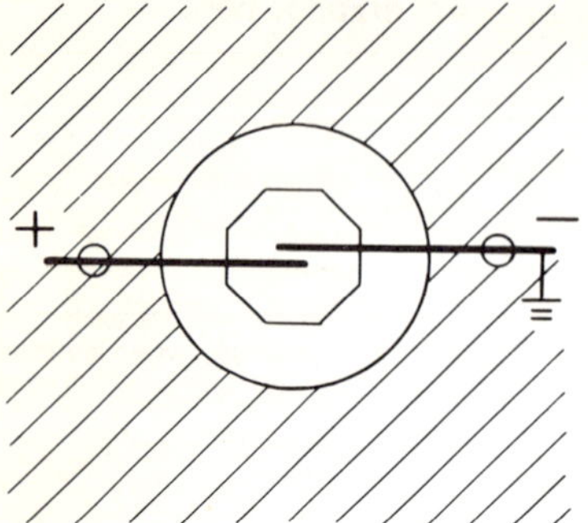

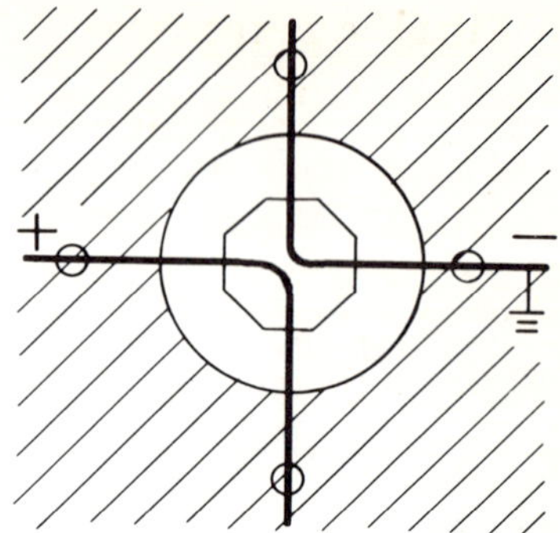

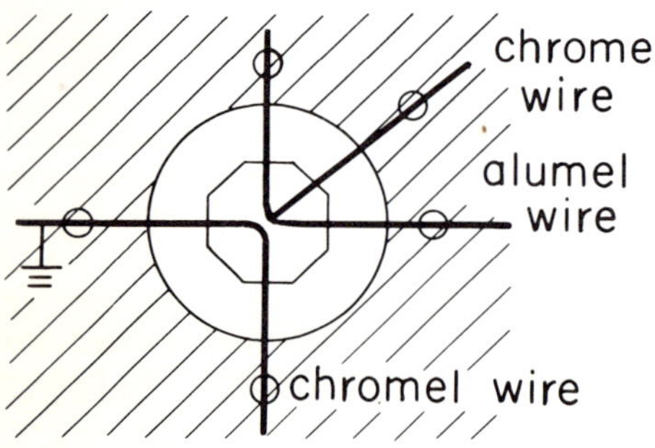

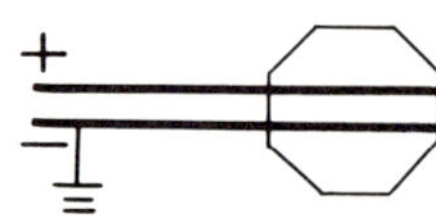

Fig. 2. Various configurations of electrical leads including thermocouples (octagonal outline of sample perimeter is approximately 500 μm across).

D. Temperature Other Than Ambient

The diamond-anvil cell can be heated with an external resistance furnace to 800°C or cooled with dry ice to -78°C. At high temperature, a mica template is used instead of Mylar, or a parallel wire arrangement with a two-hole ceramic tube is used instead of the template (Fig. 1B). Temperature is measured using an internal thermocouple with chromel and alumel wires mechanically pressed together (Fig. 2) or using an external thermocouple touching the sides of diamonds. No significant difference (±10°C) in temperature is found between the two types of measurement.

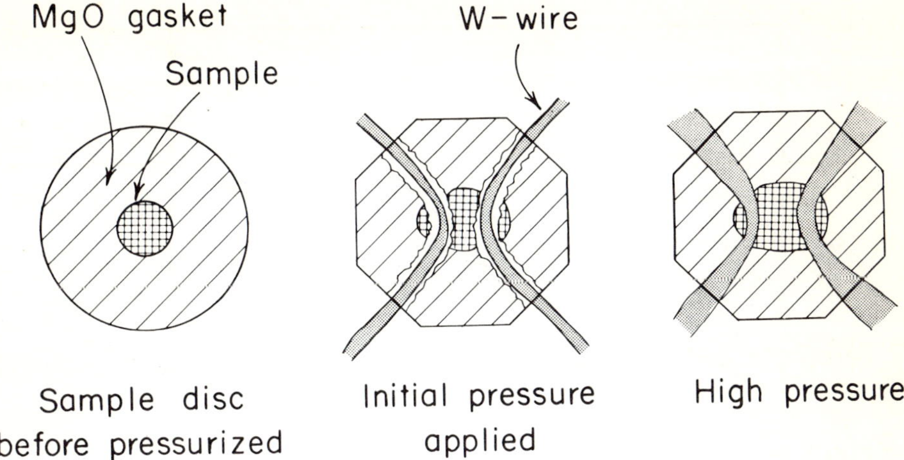

Fig. 3. Placement of electrical leads using a nonconducting gasket of MgO (octagonal diamond face is 500 μm across).

At pressures above 50 kbar, the sample is sealed by the MgO gasket and acts as a closed system. No redox change occurs, and conductivity measurements are reproducible on cycling the temperature to 600°C.

Heating of the sample in the diamond pressure cell with a Nd-doped YAG laser beam to 2500°C has also been used [*Bell and Mao*, 1975]. The electrical resistance measurements are surprisingly stable and reproducible. Because of a large temperature gradient, relatively small laser beam size (20 μm), and the different absorption properties of the sample and metal leads, however, a resistivity calculation is precluded.

III. EXAMPLES OF ELECTRICAL CONDUCTIVITY MEASUREMENTS AT HIGH PRESSURE

Figures 4 and 5 show the data for compositions in the olivine system fayalite (Fe_2SiO_4)—forsterite (Mg_2SiO_4) and in the magnesiowüstite system wüstite ($Fe_{1-x}O$)—periclase (MgO) [*Mao and Bell*, 1972; *Mao*, 1973a,b,c]. The electrical conductivity was measured stepwise as the pressure was cycled upward and downward. The changes in conductivity were reversible, and no hysteresis was observed. Evidently the conduction mechanism is related to the iron content in both systems because of the linear increase in conductivity observed with increasing iron (Figs. 4 and 5).

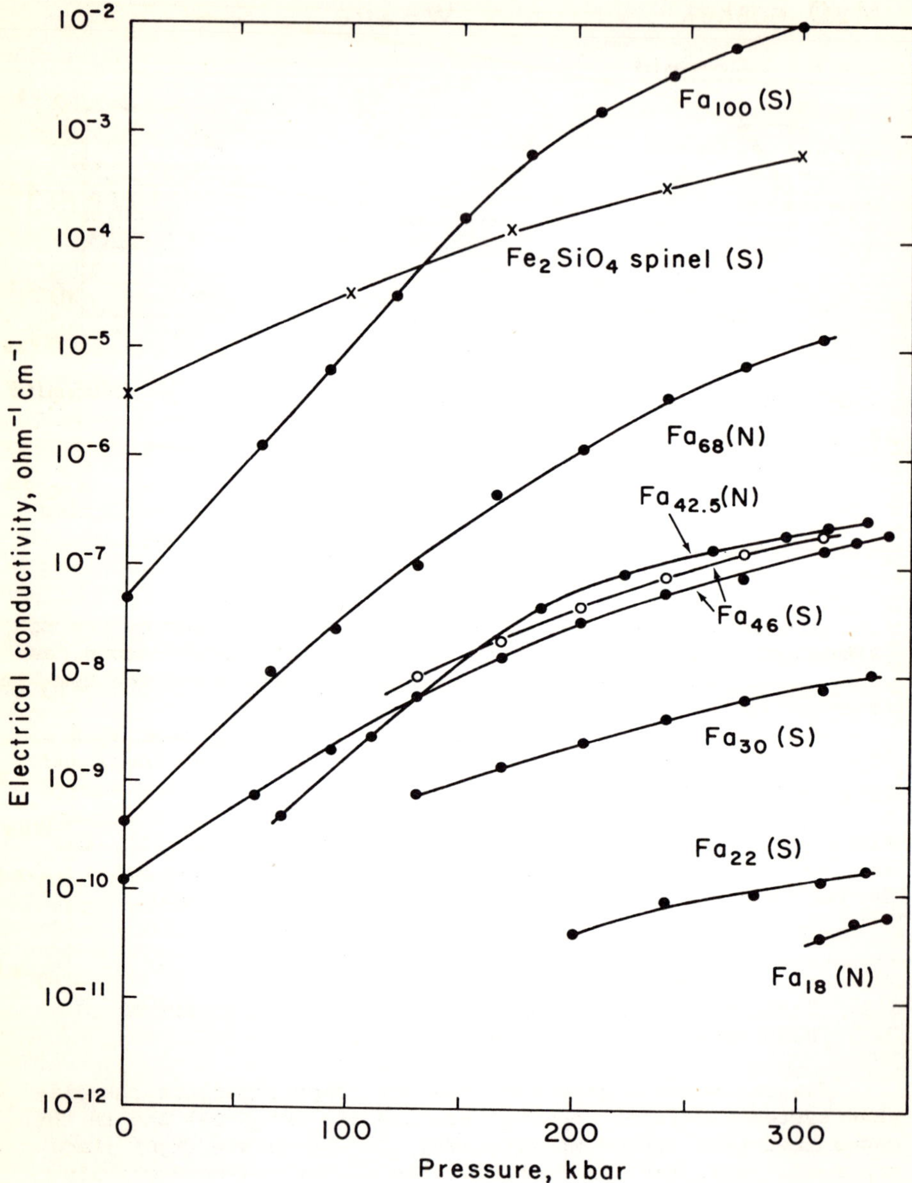

Fig. 4. Electrical conductivity of the olivine series. (S, synthetic sample; N, natural sample; see Mao [1973b] for origin of sample material.)

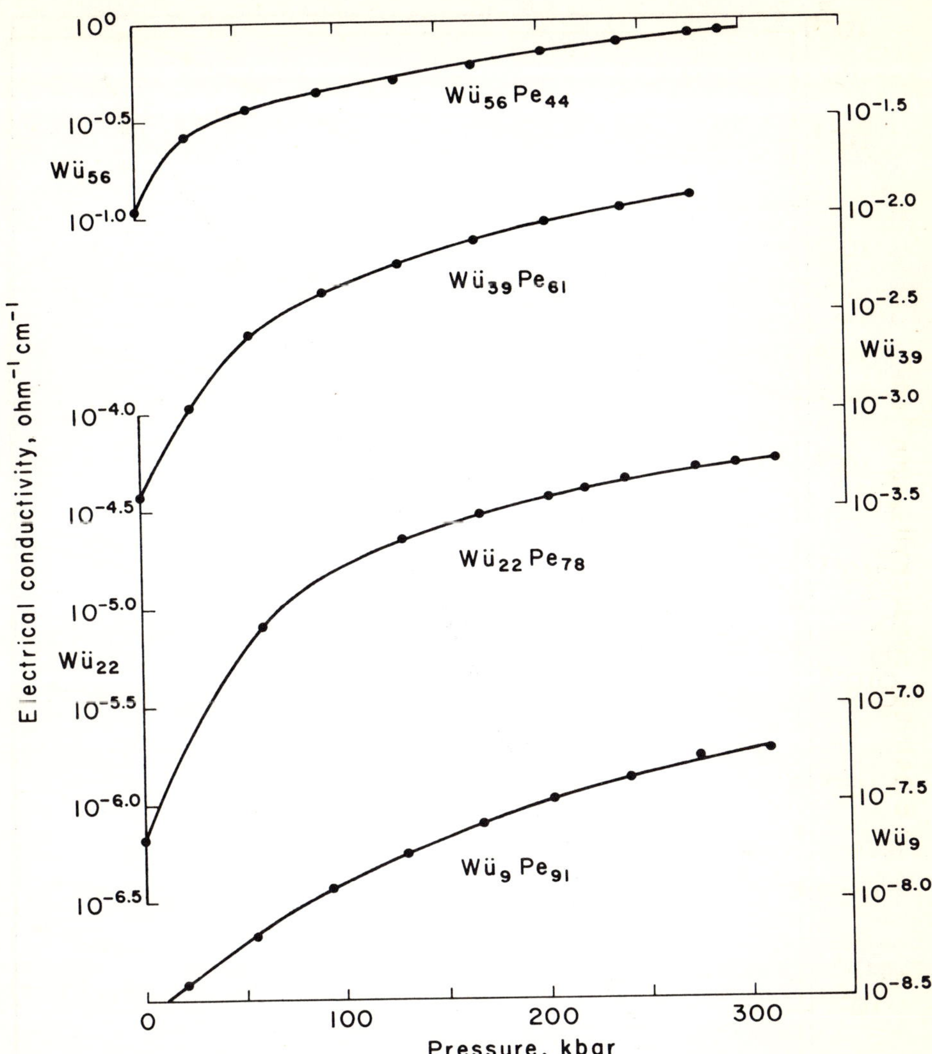

Fig. 5. Electrical conductivity of the magnesiowüstite series (see Mao [1973c] for origin of sample material).

IV. TEMPERATURE DEPENDENCE

It was found in experiments with two compositions, pure fayalite (Fig. 6) at 200 kbar and $Wü_{22}Pe_{78}$ (Fig. 7) at 150 kbar, that the temperature coefficients (Boltzmann) at high pressure

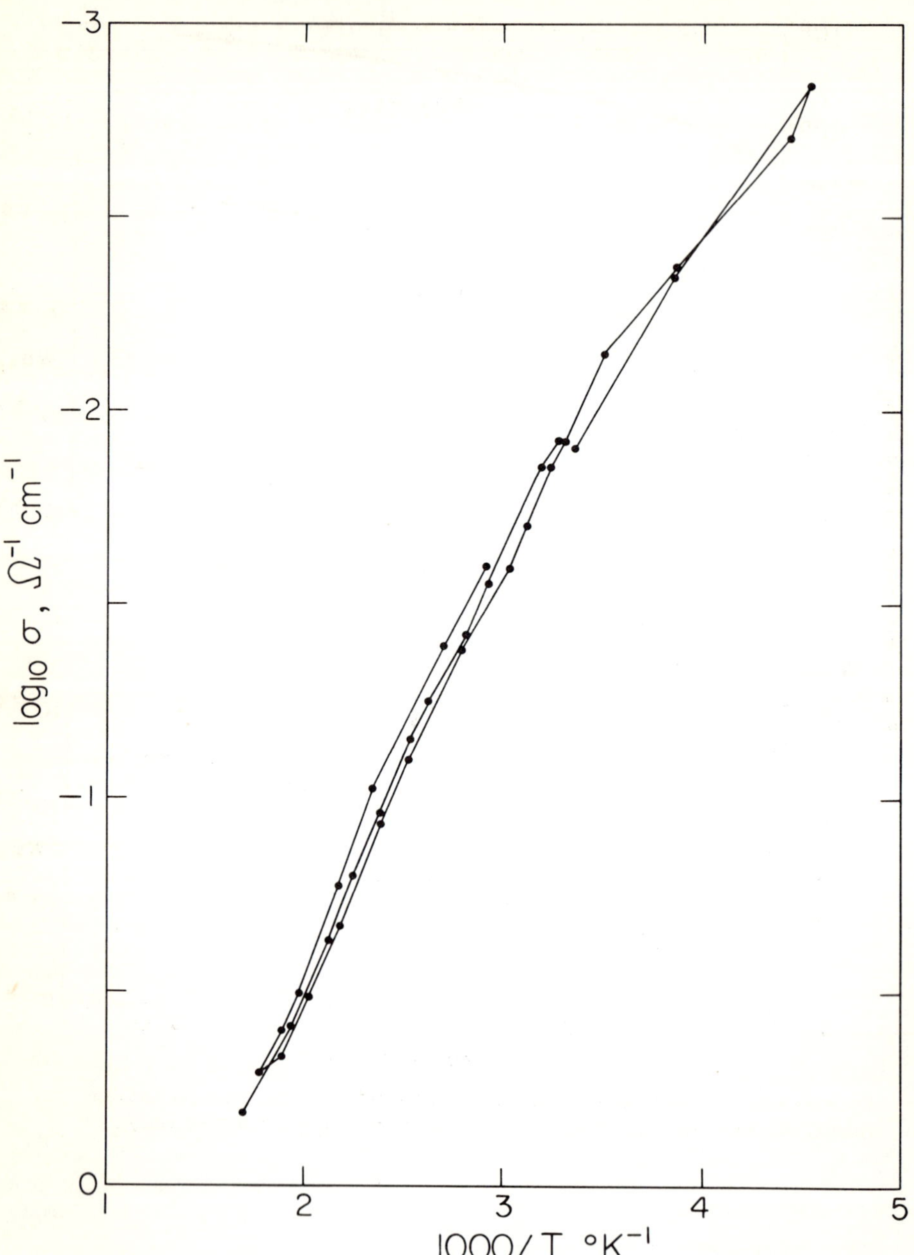

Fig. 6. Electrical conductivity of fayalite at 200 kbar as a function of temperature (two experimental runs are shown by connected solid dots).

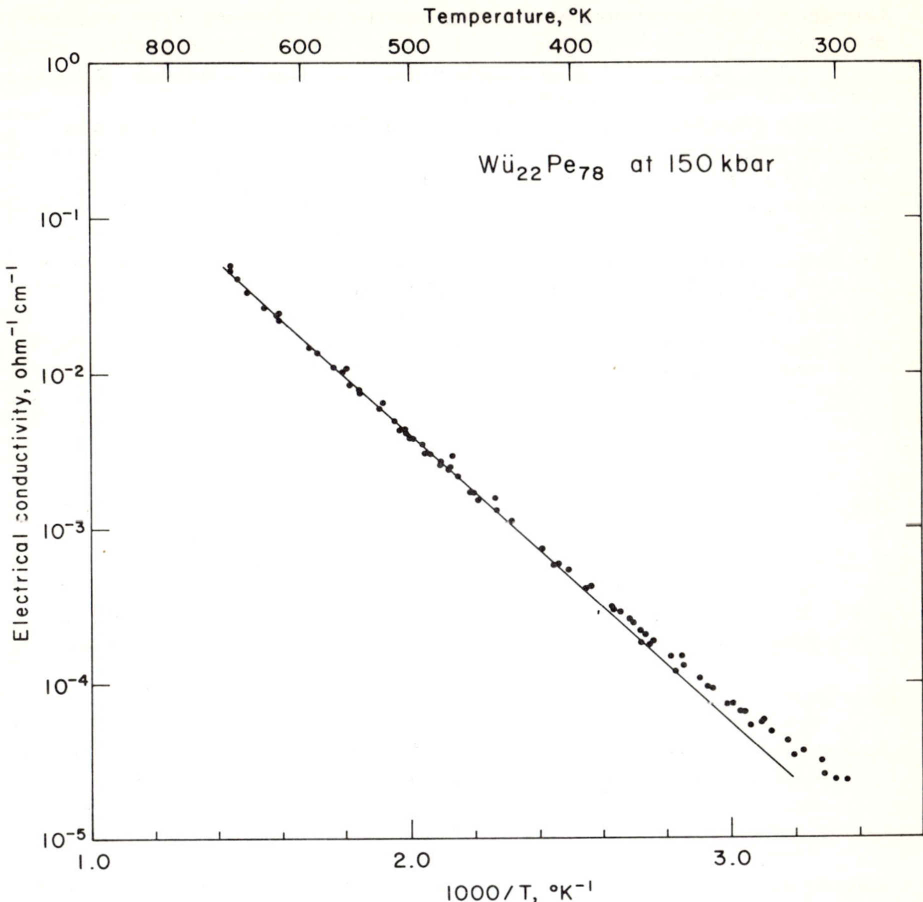

Fig. 7. Electrical conductivity of Wü$_{22}$Pe$_{78}$ at 150 kbar as a function of temperature (solid line is straight).

were different from those at 1 atm. The pressure cell was cooled and heated externally in these experiments by solid CO_2 and by use of a resistance heater.

V. CONCLUSIONS

The techniques described here for measuring electrical conductivity of minerals at high pressures are now routine, reproducible, and precise (average reproducibility in $\log_{10} \sigma$ (ohm-cm) = ± 0.05, Fig. 6). The introduction of electrical leads in the diamond-anvil pressure cell poses no errors arising from the geometric arrangement of the sample and leads, from the redox potential, or from the interaction with pressure media. The electrodes are in solid contact with the sample and are visible during all phases of measurement.

The changes in electrical resistance of the olivine and magnesiowüstite series with pressure are observed to be major, suggesting the existence of a large pressure coefficient of resistance for these minerals. This evidence for a relatively high electrical conductivity in materials thought to occur in the earth's mantle introduces new dimensions to the parameters for geophysical models of the earth. Conversely, the data given in Figures 4-7 can be used to relate chemical composition with depth, based on models of the radial distribution of pressure and electrical conductivity [*Mao*, 1973a].

REFERENCES

Bell, P. M., and H. K. Mao, Laser optical system for heating experiments and pressure calibration of the diamond-windowed, high-pressure cell, *Carnegie Inst. Washington Yearb.*, *74*, 399-402, 1975.

Bell, P. M., and H. K. Mao, Compression experiments on MgO and ruby with the diamond-window pressure cell to 1 megabar, in *High-Pressure Research: Applications to Geophysics*, edited by M. H. Manghnani and S. Akimoto, Academic Press, New York, 509-518, 1977.

Bullard, E., Geomagnetic dynamos, in *The Nature of the Solid Earth*, edited by E. C. Robertson, McGraw-Hill Book Co., New York, 232-244, 1972.

Lawson, A. W., and J. C. Jamieson, Energy transfer in the earth's mantle, *J. Geol.*, *66*, 540-551, 1958.

Lubimova, E. A., Theory of thermal state of the earth's mantle, in *The Earth's Mantle*, edited by T. F. Goshell, Academic Press, New York, 231-323, 1967.

Mao, H. K., Thermal and electrical properties of the earth's mantle, *Carnegie Inst. Washington Yearb.*, *72*, 557-564, 1973a.

Mao, H. K., Electrical and optical properties of the olivine series at high pressure, *Carnegie Inst. Washington Yearb.*, *72*, 552-554, 1973b.

Mao, H. K., Observations of optical absorption and electrical conductivity in magnesiowüstite at high pressures, *Carnegie Inst. Washington Yearb.*, *72*, 554-557, 1973c.

Mao, H. K., and P. M. Bell, Electrical conductivity and the red shift of absorption in olivine and spinel at high pressure, *Science*, *176*, 403-406, 1972.

Tozer, D. C., The electrical properties of the earth's interior, *Phys. Chem. Earth*, *3*, 414-436, 1959.

PRESSURE AND ELECTRICAL RESISTANCE MEASUREMENTS IN THE DIAMOND CELL

S. BLOCK, R. A. FORMAN, and G. J. PIERMARINI
Institute for Materials Research
National Bureau of Standards
Washington, D. C. 20234

Abstract

A method for measuring electrical resistance in the diamond anvil high pressure cell that employs gasketed samples and a fluid pressure transmitting medium has been developed. Measurements on the semiconductors, ZnS and GaP, both of which exhibit pressure-induced transitions to opaque phases have been made. A decrease in electrical resistance of several orders of magnitude has been measured with the appearance of the high pressure phase, thus indicating the formation of a metallic or semi-metallic state. These results are in agreement with our earlier work which indicated the need for downward revision of the fixed-point pressure scale.

I. INTRODUCTION

In previous papers, we reported that pressure-induced phase transitions in several semiconductors including GaP, ZnS, ZnSe, and Si produce high pressure modifications that are opaque at wavelengths less than 1.1μ [*Piermarini and Block,* 1975; *Weinstein and Piermarini,* 1975]. We assumed that these modifications had the properties of a metallic conductor since they were opaque in thin samples, as is typical of a metal, but we noted that verification required electrical resistance measurements in an optical cell, which we were unable to carry out at that time.

Semiconductor-to-metal transitions are ideal calibration markers on a fixed-point pressure scale because associated with the transition is a decrease in electrical resistance of several orders of magnitude that can be easily detected and measured. Thus, they provide convenient calibration points for high pressure apparatus designed for electrical resistance measurements. Numerous semiconductors have been studied in such apparatus by others, and transition pressures based on a fixed-point pressure scale have been reported. For these semiconductors we have measured transition pressures optically, which are based on the ruby scale [*Piermarini and Block*, 1975]. In some cases, our transition pressures do not agree with values determined electrically by others. The ruby and the previous fixed-point pressure scale were shown to diverge above about 135 kbar, and disagreement was by as much as a factor of two in the 500 kbar range with the ruby scale defining the lower pressures. To explain this discrepancy it had been suggested that perhaps we were observing different transitions in the same material not detectable in the electrical apparatus.

To prove that these transitions to opaque phases are the same ones that have been detected by electrical resistance measurements and used to define points on the fixed-point pressure scale, we have developed a method of measuring electrical resistance in the diamond anvil pressure cell. The method employs a gasketed system with the sample in a fluid environment.

A method of measuring electrical resistance in the ungasketed diamond cell was developed earlier by *Mao* [1972-3] who used two small diameter wires placed parallel across a diamond anvil flat with the powdered sample surrounding the entire wire arrangement. Because the system is ungasketed, the region of maximum pressure originates in the central area of the flat between the two parallel wires and decreases radially to 1 atm at the edge of the anvil. Thus, the high pressure phase appears first in the central area (region of maximum pressure) and increases in size until it makes contact with the parallel wire system as load is increased. By this method, *Mao* [1972-3] has obtained conductivity data on olivine and magnesiowustite. There is usually a large pressure distribution across the sample in the ungasketed method which, in more recent work, has been reduced considerably by the use of thin MgO washers [*Mao*, private communication, 1975].

The purpose of developing a gasketed technique for electrical measurements is to produce a hydrostatic environment for the sample by surrounding it with a fluid pressure transmitting medium. It is well known that in experiments conducted under nonhydrostatic conditions, the presence of stress gradients and the unknown magnitude of stress often casts serious doubt on the interpretation of the desired measurements. In this gasketed method these undesirable effects are eliminated in the hydrostatic range of a fluid pressure transmitting medium. Above the hydrostatic limit quasihydrostatic conditions apply and this fact must be taken into consideration in the interpretation of the desired measurement.

The gasketed approach has the added advantage of prolonging the useful lifetime of the diamond anvils and of allowing maximum pressures to be obtained without diamond failure.

II. EXPERIMENTAL METHODS

Two types of gaskets have been developed for electrical resistance measurements. One, which employs a sandwich arrangement depicted in Figure 1, consists of two Inconel sheets 0.15 mm thick separated by either a ceramic adhesive layer or an alumina - or silica-filled epoxy coating approximately 0.08 mm thick. It should be noted that these dimensions are initial values and may bear little resemblance to the final dimensions after pressures have been applied to the system and deformation of the gasket has occurred. The electrical connections are made to the two metal sheets which act as the electrodes. A high-sensitivity electrometer or a microvoltmeter with appropriate current sources can be used to measure the resistance. The powdered sample, which contains a small amount (10% by volume) of powdered ruby serving as the pressure sensor, is contained within the gasket hole with a small amount of fluid. Binary mixtures such as 1:1 pentane: isopentane and 4:1 methanol:ethanol (by volume) have been used.

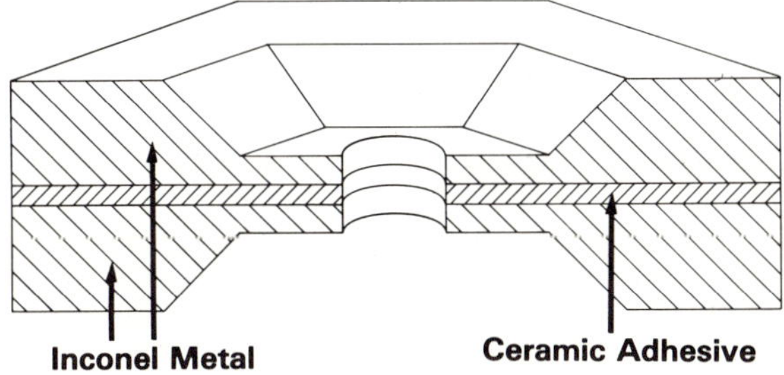

Fig. 1. The sandwich-type gasket in cross-section showing two Inconel alloy sheets, approximately 0.15mm in thickness separated by an insulating layer approximately 0.08mm thick. The two Inconel sheets are the electrodes.

The sandwich-type gasket has the tendency to allow the fluid to diffuse through the ceramic layer in cases where large gasket distortion has occurred under load. Nevertheless, even with this disadvantage, the approach is a simple and straightforward method of making electrical measurements in the diamond cell. It gives good measurements of the transition pressure since line broadening of the ruby R lines is minimal in the pressure range of interest.

Another gasket design is shown in Figure 2. An Inconel gasket, 0.2 mm thick, has an insulating ceramic layer, 0.1 mm thick, over which is placed a small diameter (0.013 mm) tungsten wire. This wire serves as one electrode and the gasket as the other. A fluid can be contained in this system quite easily without the difficulties experienced with the sandwich-type system. As in the previous case, a high sensitivity electrometer is used to measure resistance, and the shift of the ruby R_1 fluorescence line is used to measure pressure.

Frequently, a given gasket system can be recycled several times before an electrical short circuit or break develops. The short may occur when the wire penetrates the ceramic adhesive and makes electrical contact with the Inconel gasket. In the sandwich-type gasket a short circuit develops after much of the ceramic adhesive has been extruded from between the two Inconel sheets which then make electrical contact. As many as five cycles have been made in either type of gasket before a short develops.

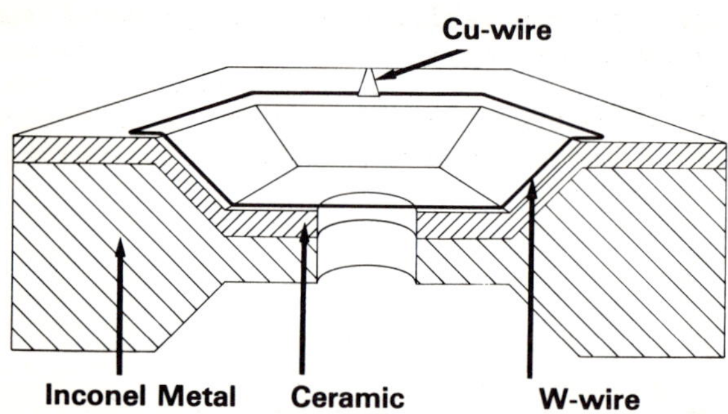

Fig. 2. The gasket-wire configuration in cross-section showing an Inconel alloy sheet approximately 0.2mm in thickness with an insulating layer about 0.1mm thick and a tungsten wire 0.013mm in diameter. In this case the wire serves as one electrode and the Inconel sheet as the other.

III. RESULTS

Preliminary results obtained by the gasket-wire method for predominantly cubic ZnS are shown in Figure 3. This graph represents the general features of the electrical resistance versus pressure behavior of ZnS, and is not intended to convey detailed information.

The graph shows that the resistance is initially in the 10^9 ohm range, and decreases to about 10^8 ohm with increase in pressure. At 140 kbar ZnS is still transparent with no indication of phase transition. As pressure is increased further, the resistance begins to decrease accompanied by the appearance of a small opaque region in the sample. Upon further increase in pressure, the opaque region rapidly increases in size, while simultaneously the resistance drops precipitously to about 10^2 ohm. No further decrease in resistance is observed above about 155 kbar. The precipitous resistance drop is definitely associated with the appearance of the opaque phase and is a reversible phenomenon as indicated by the dashed line in the graph, which indicates the approximate pressure for reconversion to the semiconductor phase. Similar results were obtained for GaP with the transition located in the region of 220 kbar in agreement with our earlier results [*Piermarini and Block*, 1975].

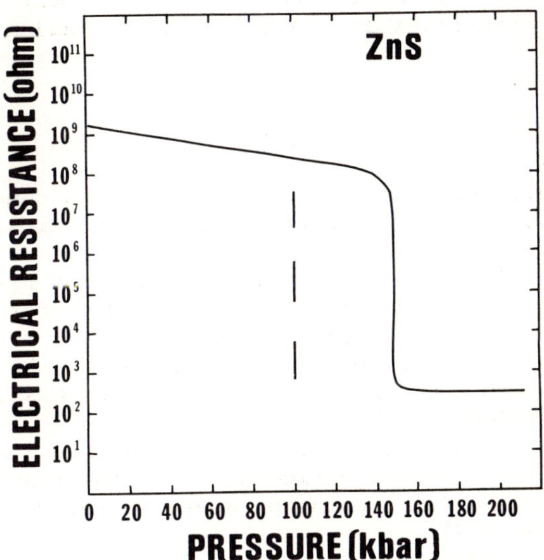

Fig. 3. Preliminary results for ZnS showing a large drop in electrical resistance (solid line) in the region of 145-150 kbar. The effect is reversible and a steep rise in resistance is observed at about 100 kbar (dashed line). The graph demonstrates the usefulness of the method in identifying transitions in semiconductor materials such as ZnS.

IV. CONCLUSIONS

We have demonstrated that electrical resistance measurements can be made in the diamond anvil high pressure cell using gasketed samples.

We have also established that the transitions to opaque phases that we had reported earlier for ZnS and GaP are indeed accompanied by large decreases in electrical resistance of several orders of magnitude. They are unequivocally the same transitions used to define markers on the fixed-point pressure scale. The discrepancy noted earlier between the fixed-point and ruby scales above 135 kbar is confirmed. Moreover, two very recent independent reports on the transition in GaP have appeared which support our transition pressure of 220±10 kbar. *Bundy* [1975] measured the electrical resistance of GaP using a newly developed press consisting of compacted sintered diamond tips as Bridgman anvils, and he reports a large drop in resistance in the 230-240 kbar range. *Homan et al.* [1975] have also measured the resistance of GaP using a variable lateral support Bridgman anvil press, and they place the transition at 220 kbar with a similar large drop in resistance.

REFERENCES

Bundy, F. P., Ultrahigh pressure apparatus using cemented tungsten carbide pistons with sintered diamond tips, *Rev. Sci. Instrum. 46*, 1318-1324, 1975.

Homan, C. G., Kendall, D. P. Davidson, T. E., and Frankel, J., GaP semiconductor-to-metal transition near 220 kbar and 298°K, *Solid State Comm. 17*, 831-832, 1975.

Mao, H. K., Electrical and optical properties of the olivine series at high pressure, *Carnegie Inst. of Washington Yearbook, 72*, 552-554, 1972-73.

Piermarini, G. J. and Block, S., An ultrahigh pressure diamond-anvil cell and several semiconductor phase transitions in relation to the fixed-point pressure scale, *Rev. Sci. Instrum. 46*, 973-979, 1975.

Weinstein, B. A. and Piermarini, G. J., Raman scattering and phonon dispersion in Si and GaP at very high pressure, *Phys. Rev. B12*, 1172-1186, 1975.

COMPRESSION EXPERIMENTS ON MgO AND RUBY WITH THE DIAMOND-WINDOW PRESSURE CELL TO 1 MEGABAR

P. M. BELL and H. K. MAO
Geophysical Laboratory
Carnegie Institution of Washington
Washington, D. C. 20008

Abstract

Experiments in which a sample of MgO contained crystals of ruby include data for the first calibration of the diamond-window pressure cell to approximately 1 Mbar. In the same experiments, pressures determined from volume measurements of MgO calibrated by shock-wave data and from spectral shift of the ruby R_1 fluorescent line are essentially identical.

I. INTRODUCTION

It has only been recently that means have been developed to extend the sustained experimental limits of research on the earth's mantle to conditions equivalent to depths greater than 600 km. *Piermarini and Block* [1975] and *Mao and Bell* [1975] conducted static experiments for the first time with internally calibrated pressures to 0.5 Mbar. Shortly thereafter, *Mao and Bell* [1976] reported calibrated pressures of approximately 1.12 Mbar. Equivalent conditions in the earth occur at 2500 km depth [*Clark and Ringwood*, 1964], close to the mantle-core boundary. New apparatus and techniques are being developed for routine use, and rapid advances are being made in obtaining experimental data under conditions to which 78% of the earth's volume is subjected.

Perhaps most important are the accuracy and reproducibility of the new techniques and the ease with which experiments are performed routinely to the highest limits of the pressure range. Although the technological increase in pressure range appears to have occurred over a relatively short period, it is calibration of pressure that has played a predominant role in the development. This paper is a discussion of the actual generation of pressures in the megabar range, and new data on the pressure calibration and its relation to shock-wave values are reported here.

The techniques give exceptionally consistent results, and therefore the experiments afford the first opportunity to determine the equation of state of materials believed to exist in the earth's interior.

II. APPARATUS

The diamond-window pressure cell originally devised by *Weir et al.* [1959] employed a spring-loaded lever to produce a clamping force on two opposed diamond anvils. The practical pressure limit was approximately 300 kbar because of yielding in support blocks and because torque from the lever caused displacement, resulting in misalignment of the diamonds at high loads. *Bassett et al.* [1967] avoided a lever in their design by spring-loading the diamonds axially, compressing a bellville-type spring. The axial design preserved alignment to 300 kbar, but lacked the mechanical advantage of a lever. Misalignment of the diamond anvils at high pressure ultimately results in their failure; *G. Piermarini and S. Block* (personal communication, 1975) noted that excellent alignment (to better than one interference fringe viewed on the contact surfaces of the diamonds) must be maintained if pressures above 300 kbar were to be reached. In the present modified diamond-window pressure cell, alignment is achieved by using a long cylindrical support for the diamonds, by utilizing carbide supporting seats, and by reconfiguring the lever fulcrum for linear motion [*Mao and Bell*, 1975]. Figure 1 shows the design of the piston (*A*), cylinder (*B*), carbide seats (*C, D*), and opposed diamond anvils. The carbide seats formed from half-cylinders are oriented axially 90^{0} to each other [*Bassett et al.*, 1967]. During alignment, the seats are translated and rotated. The present design employs a 5:1 lever and generates an approximate force of 1500 kg to sustain 1 Mbar pressure on the sample and gasket.

A. Compressibility of MgO

A series of experiments were performed to compare the ruby fluorescence (R_1) pressure scale with pressures independently measured by determining the equation of state of MgO. In these experiments, ruby crystals (10-30 μm diameter) were embedded in powdered, reagent-grade MgO, and the mixture was gasketed in the diamond-window pressure cell with thin iron foil (0.010 in. thick before pressure was applied). Ruby fluorescence was excited by epi-illumination of a He-Cd gas diffusion laser beam at 441 nm. Figure 2 shows the fluorescent crystal-field spectra of the *R* doublet of ruby recorded during a typical set of experiments. At high pressures, strain causes diffuseness of the bands, but with splitting of the bands consistent with the spin-forbidden transition ${}^4A_2 \rightarrow {}^2E$ in Cr^{3+}, it is possible to obtain satisfactory resolution.

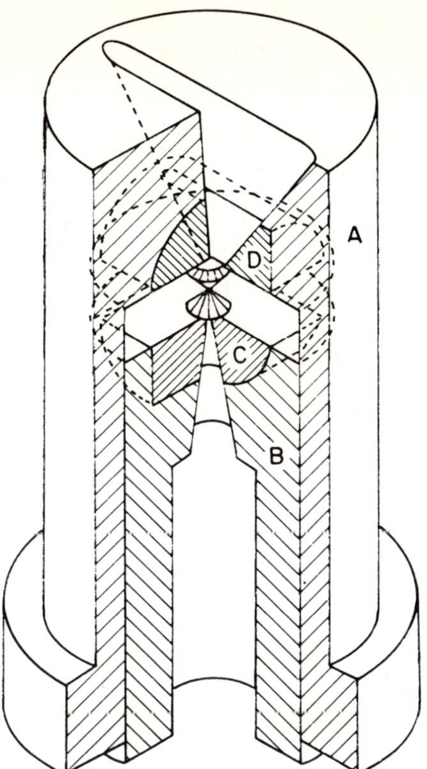

Fig. 1. Cut-away perspective drawing of the working mechanism of the modified diamond-window pressure cell used in this study. A, cylinder of 2 in. o.d.; B, piston; C, D, carbide seats.

Figure 3 shows the areas sampled during the experiments. The stippled areas are those sampled by the laser beam. The crosshatched area is the area sampled by a collimated $MoK\alpha_1$ X-ray beam. The five areas shown were sampled by the laser beam before and after X-ray diffraction measurements to ensure that no detectable pressure gradient existed.

For the purpose of comparing volume data for MgO with spectral shift of the ruby R_1 line, the equation of state of MgO determined by a direct dynamic method by *Carter et al.* [1971] was used. Figure 4 shows a plot of volume versus pressure for MgO based on these data. Figure 5 shows a plot of the present data (Table 1) for MgO and ruby. The quantity, $\Delta\lambda$ (nm), or spectral shift of the ruby R_1 line, is plotted against pressure in MgO determined simultaneously in the same experiment by X-ray

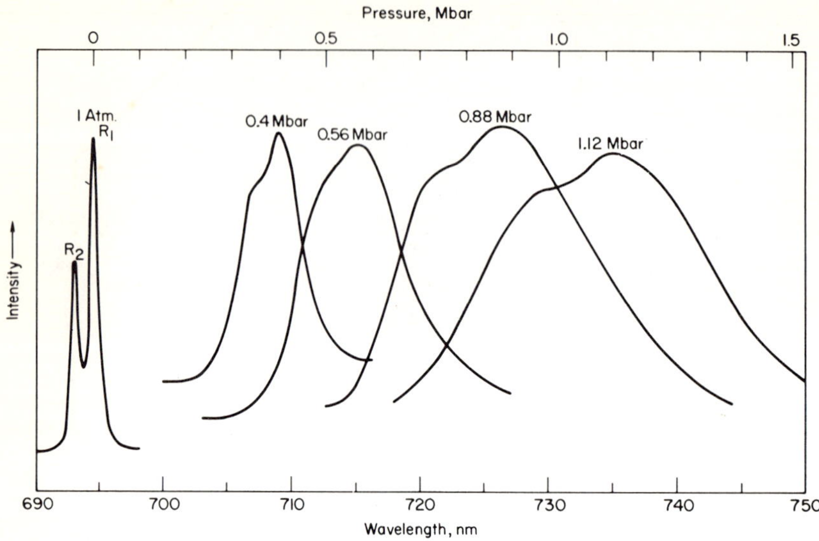

Fig. 2. Spectra of the ruby R fluorescence lines at various pressures from 1 atm to 1.12 Mbar (intensity in arbitrary units).

diffraction. Also plotted in Figure 5 is a linear extension of the ruby R_1 pressure curve of *Piermarini and Block* [1975]. A linear relationship exists to 0.88 Mbar.

B. Linearity of the Ruby R_1 Pressure Scale

As described above, the ruby R_1 pressure scale is linear to pressures approaching 1 Mbar, but the linearity observed was unexpected. It was not surprising that the data of *Piermarini et al.* [1975] were linear, because the total spectral shift to their maximum calibrated pressure (195 kbar) was small. Their data spanned a small, relatively linear section of the curve. Furthermore, if one plots wavelength (λ) versus frequency (ν), interatomic distance (r) versus frequency (ν), or the reciprocal fifth or ninth powers of interatomic distance (r^{-5}, r^{-9}) versus frequency (ν), a straight line to 195 kbar is obtained, within the 2% deviation of the NaCl equation of state that *Piermarini et al.* [1975] used. Although the relationship between $\Delta\nu$ versus

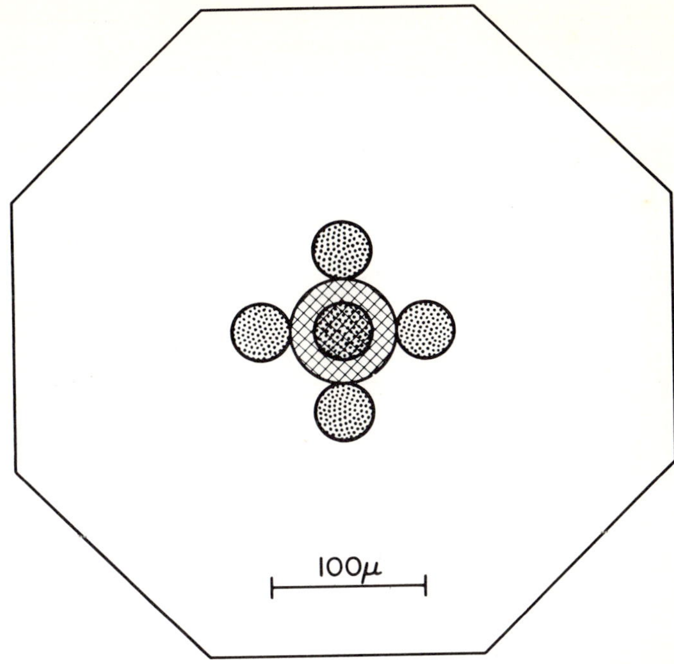

Fig. 3. Areas sampled during experiments by laser beam (stippled) and by X-ray beam (crosshatched). Octagonal shape is the outline of the pressurized sample viewed by microscope.

$\Delta\lambda$ is approximately linear to 1 Mbar (Figure 6), deviations of as much as 20% are observed in the relationship r/r_0 versus $\Delta\nu$ above 200 kbar (Figure 7), and thus linear extrapolation to 1 Mbar is not justified. Therefore, experiments such as those conducted in this study with MgO were required to calibrate the R_1 scale.

In attempting to clarify the observed linearity, one must examine the poorly known relations between crystal-field energy and Racah parameters with pressure. The relationship cannot be derived directly from energy-level diagrams until the pressure shift of other fluorescence lines or absorption bands is known. Even then, the local pressure field around Cr^{3+} ions may differ from that in the ruby structure itself.

If one assumes, however, that the Cr-O distance compresses identically with r of ruby, the Birch-Murnaghan equation of state can be applied. Using compressibility, K_{OT} - 2.901 Mbar, and the pressure derivative, K_{OT}' = 3.24 of *Ahrens et al.* [1969], the energy shift of the R_1 line was found to be proportional to $(r/r_0)^{-9}$ (Figure 8). If the recent values of *Sato* [1977] are used (K_{OT} = 2.38 Mbar, K_{OT}' = 1.6), the energy shift is proportional to $(r/r_0)^{-1}$. The data of *Ahrens et al.* [1969] are

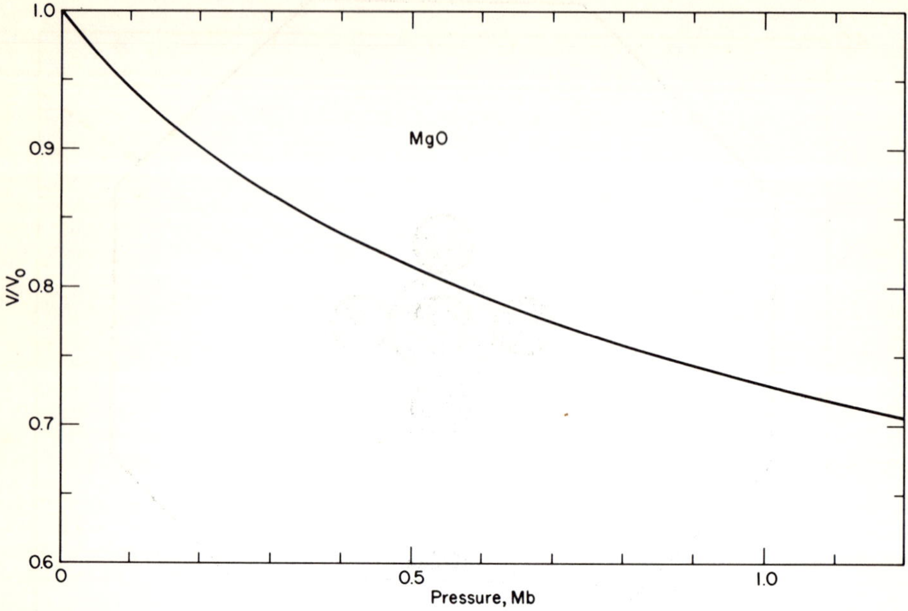

Fig. 4. Plot of specific volume (V) per specific volume at 1 atm (V_0) versus pressure used as primary calibration in the present experiments. Data from Carter et al. [1971].

TABLE 1. Pressure and Compressibility Measurements

Experiment No.	P_{ruby},[a] kbar			MgO	
	Center	Rim[b]	Average[c]	V/V_0	P, kbar[d]
4C1	528	495	520	0.8120	510
5C1	702	630	684	0.7802	670
5C2	806	759	794	0.7696	730
5C3	878	807	860	0.7473	865

a. Calculated on the basis of linear extrapolation of the calibration below 300 kbar [*Piermarini and Block*, 1975].

b. Average of four measurements outside the rim of the X-rayed area.

c. Weighed average of 1/4 P_{rim} + 3/4 P_{center}.

d. Based on the *P-V* relations determined by shock-wave technique [*Carter et al.*, 1971].

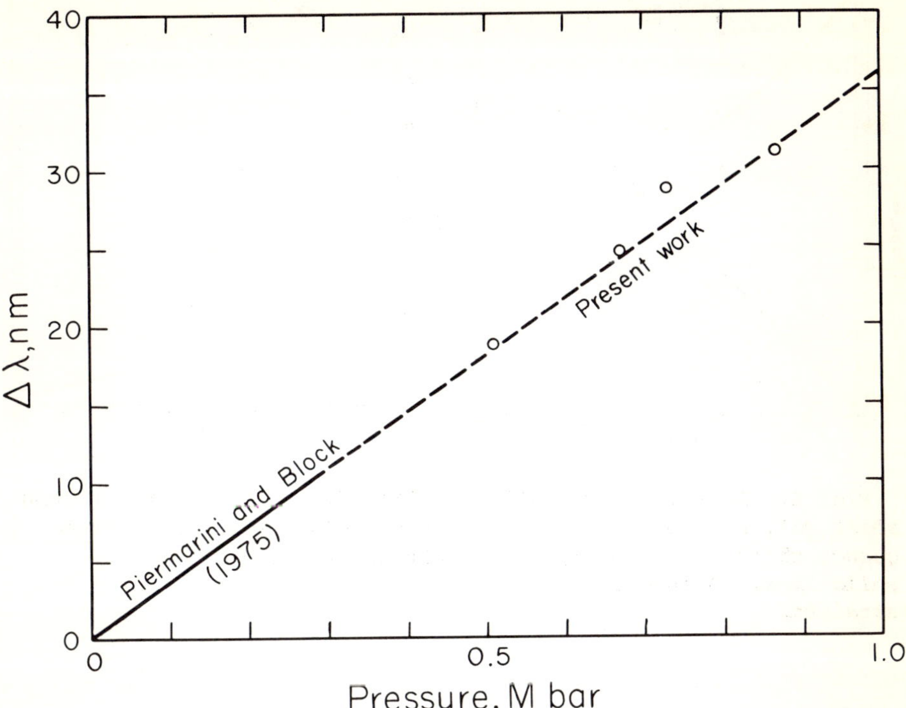

Fig. 5. Plot of spectral shift of the ruby R_1 fluorescent line (Δλ, nm) versus pressure (Mbar). Solid line, linear extrapolation of data by Piermarini and Block [1975]; dashed line, linear extension of solid line; circles, present data calibrated against volume of MgO. Volume of MgO and spectral shift of ruby R_1 line were measured simultaneously.

skewed in a plot of energy versus $(r/r_0)^{-5}$ toward the energy axis (Figure 9), and Sato's data, toward the $(r/r_0)^{-5}$ axis.

The deviation of $\Delta\nu$ from a straight line ($\delta\Delta\nu$) versus $(r/r_0)^q$ for various values of q is shown in Figure 10. This plot is relatively insensitive to K_{0T} but highly sensitive to K_{0T}'. It is difficult to judge the differences in the two sets of compression data, but if one assumes the true values of K_{0T} and K_{0T}' lie intermediate between the two sets, $\Delta\nu$ is proportional to r^{-5}. The proportionality contains a negative sign, and $\Delta\nu$ is not directly related to changes in the crystal-field energy, but is related to changes in the Racah parameters with pressure. At present, one cannot address the question of the linearity further than to note that no violations of existing theory, such as it is, have occurred.

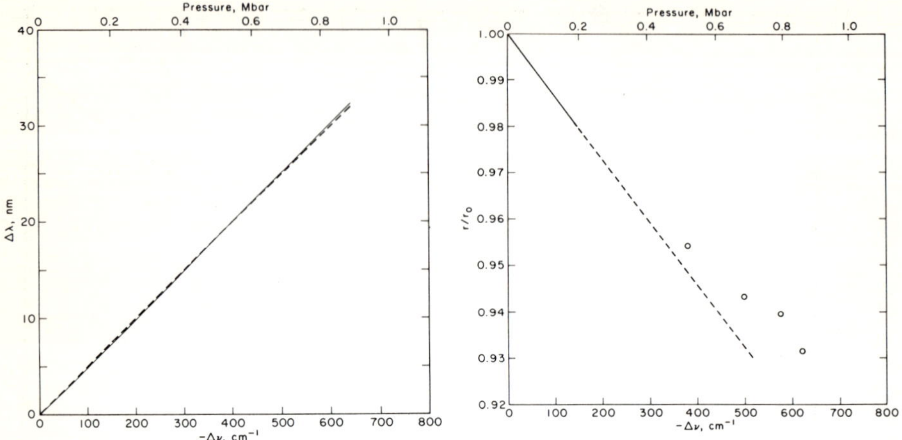

Fig. 6. Plot of wavelength shift (Δλ, nm) versus frequency shift (-Δν, cm^{-1}), solid line. Dashed line is straight.

Fig. 7. Plot of r/r_0 of ruby versus -Δν. Data for r from Ahrens et al. [*1969*].

III. CONCLUSIONS

Static pressures can be generated and sustained for several months up to approximately 1 Mbar, a range recently only accessible with explosions of nearly instantaneous duration. The pressures in the modified diamond-window cell can be sustained indefinitely as numerous measurements are made on the sample charge.

The data reported here form the basis for justification of use of the ruby R_1 linear pressure scale to 1 Mbar. Pressures determined from volume measurements of MgO and from the pressure shift of the ruby R_1 fluorescent line in the same experiment, do not deviate from each other, within the uncertainty in measurement (10% of the pressure, maximum). The MgO pressure standard is obtained from shock-wave data that are especially appropriate for comparison with ruby. Similar to ruby, MgO is a strong, close-packed oxide that undergoes no phase transition in the range of pressures spanned in the shock-wave experiments and in the present static experiments.

The reason for the linear spectral shift of the R_1 line cannot be predicted from existing incomplete theory, and therefore extrapolation must be supported by independent calibration.

Primary calibration of the diamond-window pressure cell is

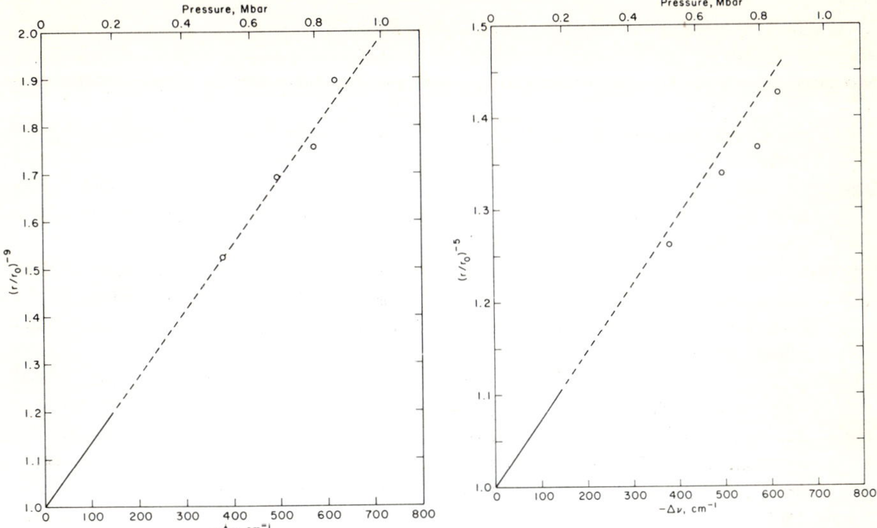

Fig. 8. Plot of $(r/r_0)^{-9}$ of ruby versus $-\Delta\nu$. Data for r from Ahrens et al. [1969].

Fig. 9. Plot of $(r/r_0)^{-5}$ of ruby versus $-\Delta\nu$. Data for r from Ahrens et al. [1969].

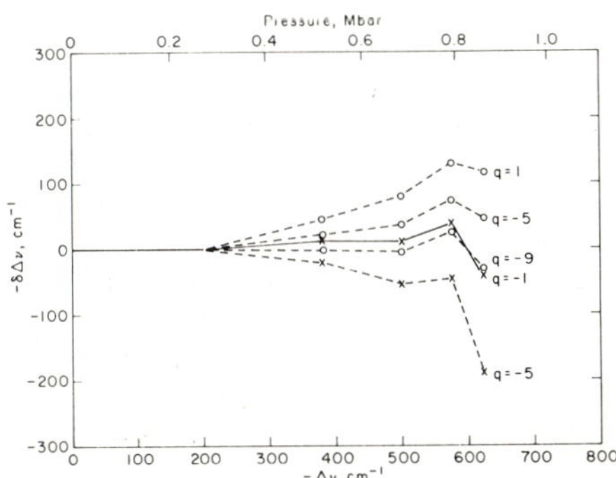

Fig. 10. Plot of the deviation of $\Delta\nu$ with pressure ($-\delta\Delta\nu$, cm^{-1}) versus $-\Delta\nu$ for various values of q, the exponent of r. Circles, data of Ahrens et al. [1969]; crosses, data of Sato [1977].

thus obtained from shock-wave data. The close agreement with the linear ruby R_1 scale suggests that the rube scale can be used safely as a technique for internal calibration to approximately 1 Mbar.

REFERENCES

Ahrens, T. J., D. L. Anderson, and A. E. Ringwood, Equations of state and crystal structures of high-pressure phases of shocked silicates and oxides, *Rev. Geophys.*, *7*, 667-707, 1969.

Bassett, W. A., T. Takahashi, and P. W. Stook, X-ray diffraction and optical observation on crystalline solids up to 300 kbar, *Rev. Sci. Instrum.*, *38*, 37-42, 1967.

Carter, W. J., S. P. Marsh, J. N. Fritz, and R. G. McQueen, The equation of state of selected materials for high pressure references, *Natl. Bur. Stand. (U.S.) Spec. Publ.*, *326*, 147-158, 1971.

Clark, S. P., and A. E. Ringwood, Density distribution and constitution of the mantle, *J. Geophys. Res.*, *2*, 35-88, 1974.

Mao, H. K., and P. M. Bell, Design of a diamond-window, high-pressure cell for hydrostatic pressures in the range 1 bar to 0.5 Mbar, *Carnegie Inst. Washington Yearb.*, *74*, 402-405, 1975.

Mao, H. K., and P. M. Bell, High pressure physics: the 1-megabar mark on the ruby R_1 static pressure scale, *Science*, *191*, 851-852, 1976.

Piermarini, G. J., and S. Block, An ultra high pressure diamond-anvil cell and several semiconductor phase transition pressures in relation to the fixed point pressure scale, *Rev. Sci. Instrum.*, *46*, 973-979, 1975.

Piermarini, G. J., S. Block, J. D. Barnett, and R. A. Forman, Calibration of the pressure dependence of the *R* ruby fluorescence line to 195 kbar, *J. Appl. Phys.*, *46*, 2774-2790, 1975.

Sato, Y., Equation of state of mantle minerals determined through high-pressure X-ray studies, in *High-Pressure Research: Applications to Geophysics*, edited by M. H. Manghnani and S. Akimoto, Academic Press, New York, 305-323, 1977.

Weir, C. E., E. R. Lippincott, and A. Van Valkenburg, Infrared studies in the 1- to 15-micron region to 30,000 atmospheres, *J. Res. Nat. Bur. Stand.*, *A*, *63*, 55-62, 1959.

BRILLOUIN SCATTERING: A NEW WAY YO MEASURE ELASTIC MODULI AT HIGH PRESSURES

W. A. BASSETT
Department of Geological Sciences
The University of Rochester
Rochester, New York 14627

E. M. BRODY
The Institute of Optics
The University of Rochester
Rochester, New York 14627

Abstract

Laser light scattered from thermal phonons in a transparent sample undergoes frequency shifts from which phonon velocities in the sample can be calculated. Elastic moduli of the sample can then be calculated from the phonon velocities. Since no mechanical or electrical coupling to the sample is necessary, very small (0.15 mm), single-crystal samples can be used. It is possible to place these samples under hydrostatic pressure in a gasketed diamond-anvil cell and to pass the laser light through the transparent diamond anvils. Measurements on NaCl have yielded data that are in good agreement with measurements by ultrasonic techniques, and with the Hildebrand and Decker equations of state.

I. INTRODUCTION

Ultrasonic measurements have yielded a great deal of information about the effect of pressure on the elastic moduli of geophysically important materials. Extension of the ultrasonic method to very high pressures has been hindered by two aspects of the method: (1) transit time is measured, and, therefore, a fairly large sample of accurately known dimensions is required, and (2) sound waves must be introduced into the sample either by bonding a transducer to the sample or by sending sound waves through an anvil that is applying pressure to the sample. Although the most significant data have been collected on single-crystal samples under truly hydrostatic conditions, investigators have been able to obtain valuable data using polycrystalline samples under nonhydrostatic conditions. Measurements have been made up to 270 kbars under these conditions [*Frankel et al.*, in preparation].

The Brillouin scattering method is ideal for measuring elastic moduli in very small samples [*Anderson et al.*, 1969; *Weidner et al.*, 1975]. It permits the measurement of all of the elastic moduli because it utilizes a single-crystal sample, and it avoids error introduced by strain resulting from nonhydrostatic conditions. It has been shown [*Piermarini et al.*, 1973] that a 4:1 mixture of methonol:ethanol remains a liquid in a gasketed diamond-anvil cell up to a pressure of 100 kbars. Therefore, placing a single-crystal sample under hydrostatic pressure in a diamond-anvil cell at pressures up to 100 kbars and introducing a laser beam through one of the diamond anvils will enable analysis of the light that emerges through the other diamond anvil.

The ruby-pressure-calibration method [*Barnett et al.*, 1973; *Piermarini et al.*, 1975] provides an ideal way to measure the pressure in the gasketed diamond-anvil cell. It is rapid; it is sensitive to the degree of hydrostaticity; it requires only a very small chip of ruby that is easily accommodated along with the single crystal; and, since it is an optical method, it requires merely that there be access for visible light. Thus, the combination of the ruby method and the Brillouin scattering method allows measurements of elastic constants as a function of pressure by purely optical methods.

II. THE THEORY OF BRILLOUIN SCATTERING

When visible light from a laser traverses transparent samples, scattered light contains components whose frequencies have been shifted due to interactions of the photons with thermal phonons that are present in the sample. The frequency shift in the scattered light can be considered to result from a Doppler shift when light is Bragg-reflected off thermal phonons traveling through the sample. The frequency shift, then, is directly related to the velocities of the phonons. The Bragg reflection can be stated by the equation

$$\lambda_{o,n} = 2\Lambda \sin \phi \tag{1}$$

where $\lambda_{o,n}$ is the wavelength of the impinging light in the sample, Λ is the wavelength of the sound wave in the sample, and ϕ is the angle between the impinging light ray and the phonon wave front (Figures 1 and 2). In this relationship it is assumed that the wavelength of the impinging light is the same as the wavelength of the scattered light since the Brillouin frequency shift is very small. It is important to keep the surfaces of the diamond anvils as well as the surfaces of the sample all parallel to each other. This allows the use of Snell's law

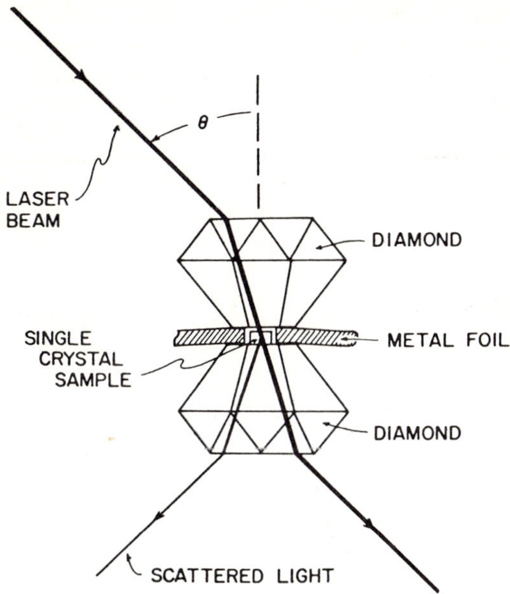

Fig. 1. Arrangement for making Brillouin scattering experiments in a diamond-anvil, high-pressure cell.

$$n \sin \phi = (c_o \sin \phi)/c_n = \sin \theta \tag{2}$$

where n is the refractive index, c_o is the velocity of light in air, c_n is the velocity of light in the sample, and θ is the angle of incidence between the laser beam and the outer surface of the diamond anvil. Doppler shift is given by the relationship

$$\nu/\nu_o = 1 \pm (2\ v \sin \phi)/c_n \qquad \text{or} \tag{3a}$$

$$\pm\ (\nu_o - \nu)/\nu_o = (2\ v \sin \phi)/c_n \tag{3b}$$

where ν is the frequency of the light and v is the velocity of the phonon. When combined with Snell's law, this relationship becomes

$$\pm\ (\nu_o - \nu)/\nu_o = (2\ v \sin \theta)/c_o \qquad \text{or} \tag{4a}$$

$$\pm\ \Delta\nu = (2\ v \sin \theta)/\lambda_o \tag{4b}$$

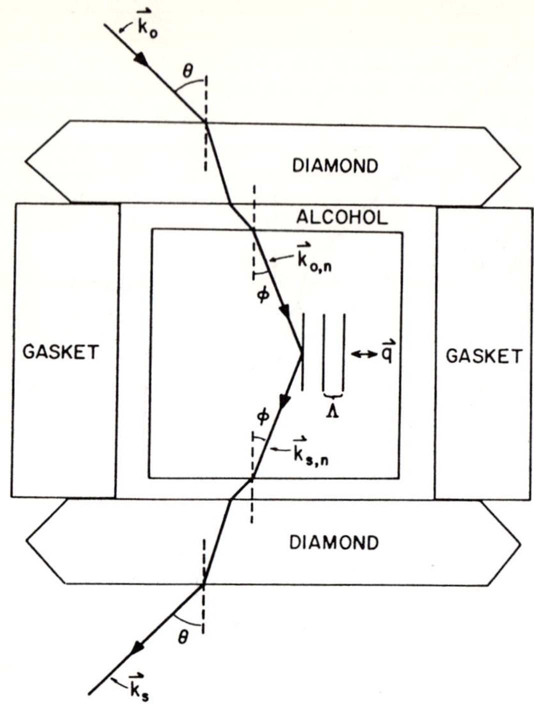

Fig. 2. Schematic diagram of the diamond-anvil cell, but with dimensions distorted so that the angular relationships can be illustrated. $|\vec{k}| = 2\pi/\lambda$, $|\vec{q}| = 2\pi/\Lambda$.

If the angle θ is 45°, the equation becomes

$$v = (\Delta\nu\lambda_o)/\sqrt{2} \tag{5}$$

When a parallel derivation in terms of the conservation of momentum and energy is carried out, it becomes apparent that the frequency shift of the photon is equal to the frequency of the phonon.

It is important to note that the geometry used in the experimental setup permits the substitution of the external angle and wavelengths without any knowledge of the refractive index of the diamond, the pressure transmitting fluid, or the sample.

The technique is easily extended to optically anisotropic crystals. However, there is a weak dependence on the magnitude of the birefringence which can be measured independently in the diamond cell.

III. EXPERIMENTAL METHODS

The diamond-anvil pressure cell used in the Brillouin scattering experiment is designed to permit light to enter one diamond anvil at an angle of 45° to the axis of the cell and to emerge from the other at 45° to the axis of the cell. This arrangement makes it possible to completely avoid reflection from any of the surfaces through which the light passes. A single crystal, about 0.15 mm across, is placed in a mixture of alcohol (4:1 methanol: ethanol) inside a hole 0.30 mm in diameter in a metal foil 0.25 mm thick. A sample, such as NaCl, is prepared by cleaving a very small cube. One of the cleavage faces comes to rest against one of the diamond-anvil faces, thus partially fixing the orientation. The other cleavage faces provide a visual means for completely orienting the crystal. A small ruby chip is placed inside the gasket to one side of the crystal. With this arrangement, it has been possible to collect data for all the modes up to 35 kbars.

Once the single crystal and ruby have been loaded in the alcohol mixture inside the gasketed diamond-anvil cell, the cell is transferred to a microscope and spectrometer for the measurement of pressure by the ruby method. A 0.25-watt beam of green laser light (5145 Å) from an argon laser is used to excite fluorescence in the ruby chip. Tests have shown that the heating due to the laser beam is negligible. The laser beam enters a standard reflecting microscope, is reflected downward by a beam splitter, passes down through an objective lens, and enters the diamond cell through the upper diamond anvil. Fluorescent emission of R lines (6928 Å and 6942 Å) is excited in the ruby chip by the laser light. This emitted light is collected by the same microscope objective and is imaged on the slit of a Hitatchi-Perkin-Elmer spectrometer (model 139). A red-sensitive RCA 4463 photomultiplier tube is used to collect the light analyzed by the spectrometer. Pressure on the ruby causes a wavelength shift of 0.365Å/kbar, which can then be used to determine the pressure. In this system, pressures have been measured with a reproducibility of ± 0.75 kbars. Detailed descriptions of the ruby method are given by *Barnett et al.* (1973) and *Piermarini et al.* (1975).

IV. BRILLOUIN SPECTROMETER

Once the pressure has been determined, the diamond cell is transferred to the Brillouin spectrometer. A 0.15-watt laser beam with a wavelength of 4880 Å or 5145 Å from an Ar laser enters the sample (Figure 3). The light that scatters from the sample at 90° to the beam is imaged onto pinhole P_1, which serves as a spatial filter and collimator for the spectrometer, making it possible to select a volume 0.05 mm across within the sample crystal inside the diamond cell. Scattered light from everything else in the path of the beam is eliminated.

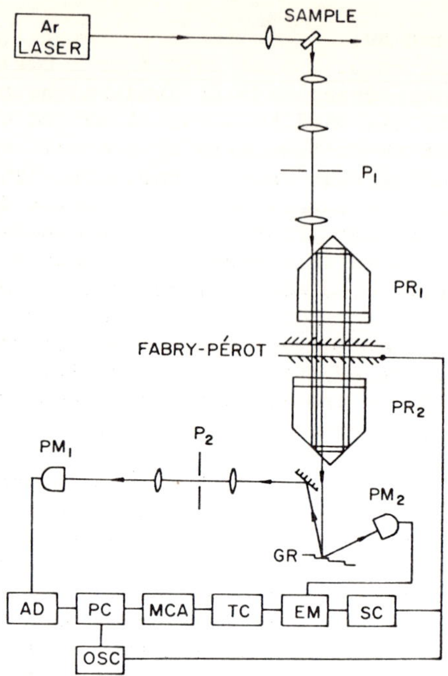

Fig. 3. The Brillouin spectrometer. A beam of monochromatic light (λ = 5145 Å) from the Ar laser is focused onto the sample. Light scattered at 90° by the sample is imaged on the pinhole, P_1. It then passes through the Fabry-Pérot plates five times as it is reflected back and forth by the corner cube prisms, PR_1 and PR_2. The grating, GR, and the pinhole, P_2, reject extraneous light shifted more than 10 cm^{-1} by fluorescence. The signal received by the photomultiplier tube, PM_1, and processed by an amplitude discriminator, AD, and photon counter, PC, is displayed on an oscilloscope, OSC, and stored in a 400-channel analyzer, MCA. A weakly diffracted first-order signal is received by photomultiplier tube, PM_2, and is used to stabilize the interferometer and to trigger the sweep of the multiscaling mode of the 400-channel analyzer by means of the electrometer, EM, stabilizing circuit, SC, and trigger circuit, TC. The Fabry-Pérot plates are piezoelectrically scanned by a ramp voltage from the oscilloscope.

The light that passes through the pinhole enters a piezoelectrically scanned Fabry-Pérot interferometer, which is used in a five-pass configuration--i.e., the light is routed through the parallel plates of the Fabry-Pérot interferometer five times by corner prisms PR_1 and PR_2. This interferometer has a contrast of 10^{10} and a transmission of 45% at 4880 Å or a contrast of 10^{13} and a transmission of 20% at 5145 Å. This high contrast makes it possible to suppress the light that is elastically scattered and therefore has the original frequency ν_o.

After leaving the Fabry-Pérot interferometer, the light is reflected from a grating and is imaged onto pinhole P_2 in order to eliminate light having frequencies beyond 10 cm^{-1} from ν_o. Fluorescence is the most serious source of such light. The Brillouin-shifted light passes through the system and is collected by photomultiplier PM_1. Another photomultiplier PM_2, collects light that is weakly scattered in another direction by the diffraction grating for the purpose of electronically stabilizing the interferometer and triggering the sweep of the 400-channel analyzer.

The interferometer is scanned through two orders of frequency every 5 sec. The signal from photomultiplier PM_1 is converted to a number of standardized pulses by the photon-counting electronics, which are then fed into the multichannel analyzer. Each channel corresponds to a discrete frequency. The frequency range of interest is repeatedly scanned until the channels of the analyzer show satisfactory intensities--usually a total of 10 min is sufficient.

Figure 4 shows a typical spectrum of scattered light. The initial frequency, ν_o, is responsible for the very high peaks at the extreme left and extreme right of the spectrum. The smoothed curve was computer-generated, using the method of cubic splines (FORTRAN Scientific Subroutine ICSSMU). This smoothing process may distort the intensities, but it does not distort the frequencies. The right-hand portion of the spectrum is a mirror image of the left-hand portion of the spectrum becuase of the up-and-down-shifted frequencies resulting from phonons moving toward and away from the spectrometer. The large peaks near the center are due to longitudinal phonons, and the weaker peaks are due to the transverse phonons. The transverse phonons are weaker because of the much smaller light-sound coupling (Pockels) coefficients.

V. DATA REDUCTION

A cubic crystal, such as NaCl, has three elastic moduli, C_{11}, C_{12}, and C_{44}, which can be determined from longitudinal and transverse phonon velocities measured in two crystal orientations. In the first orientation, the [100] direction bisects the angle between the entering and the scattered rays, thus orienting the [100] direction parallel to the direction of propagation, $\vec{q}$, of the phonon. In the second orientation, the [110] direction bisects the angle between the entering and scattered light ray, thus orienting it parallel to the direction of propagation, $\vec{q}$, of the phonon. The following two equations show the relationship of C_{11} and C_{44} to the longitudinal and transverse phonon velocities

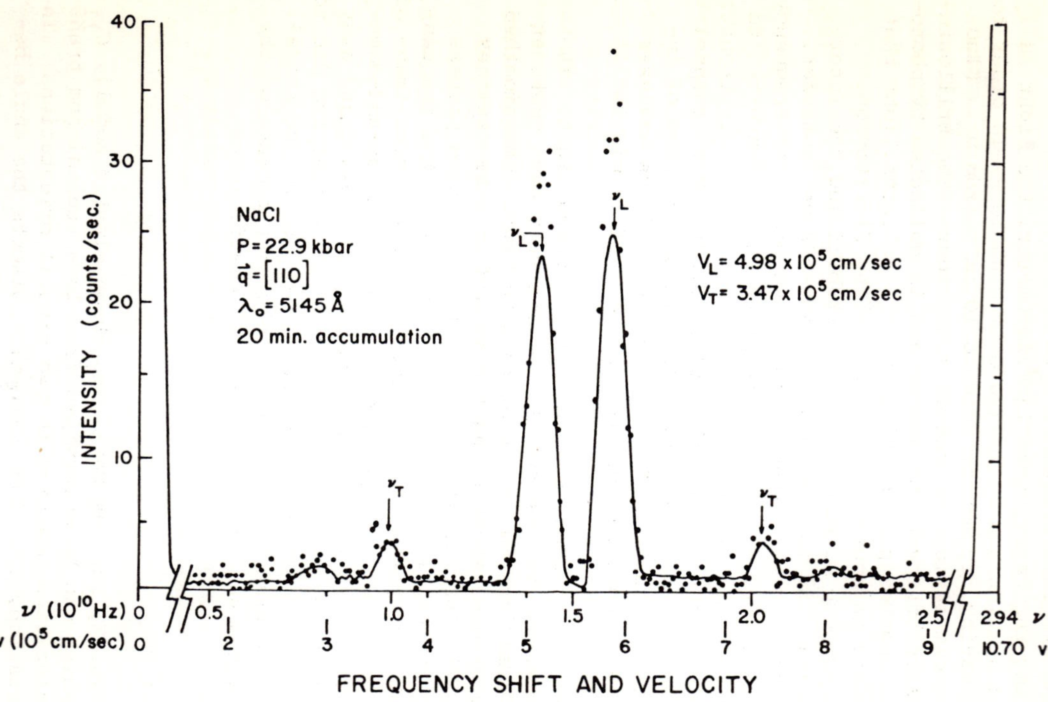

Fig. 4. Example of typical experimental data. Tne actual photoelectron counts per channel are shown by the black dots. The smooth curve was generated by a cubic-splines, computer-smoothing program.

in the first orientation:

$$\rho v_L^{2} = C_{11} \quad \text{for} \quad \vec{u}\ [100] \tag{6}$$

$$\rho v_T^{2} = C_{44} \quad \text{for} \quad \vec{u}\ [100] \tag{7}$$

where ρ is the density, v_L is the velocity of longitudinal sound wave, v_T is the velocity of the transverse sound wave, and $\vec{u}$ is the displacement vector. The following relationships can be employed in orientation $\vec{q}$ [100]:

$$\rho v_L^{2} = \tfrac{1}{3}(C_{11} + 2C_{44} + C_{12}) \quad \text{for} \quad \vec{u}\ [100] \tag{8}$$

$$\rho v_T^{2} = \tfrac{1}{2}(C_{11} - C_{12}) \quad \text{for} \quad \vec{u}\ [1\bar{1}0] \tag{9}$$

Selection rules prohibit scattering from the other transverse sound waves in each of the orientations [*Cummins and Schoen*, 1972]. The density is calculated by an internally consistent procedure. The Hildebrand equation of state [*Weaver et al.*, 1971] was used to determine the initial expression for the density versus pressure. A least-squares routine was then used to fit the elastic moduli versus pressure data to a straight line. New densities were then calculated by integration, employing the following equations. The isothermal compressibility, β_T, is given by

$$\beta_T = \frac{1}{\rho}\frac{\partial \rho}{\partial P} = \frac{3}{C_{11}^{T} + 2C_{12}^{T}} \tag{10}$$

When this is integrated it yields

$$\int_{\rho_o}^{\rho} \frac{\partial \rho}{\rho} = \int_{P=0}^{P} \frac{3}{C_{11}^{T} + 2C_{12}^{T}}\, dP \tag{11}$$

Using the linear pressure dependence of the elastic moduli C_{11} and C_{12} and of the adiabatic bulk modulus K_S, we have

$$C_{11} = C_{11}^{o} + aP$$

$$C_{12} = C_{12}^{o} + bP$$

$$K_S = \frac{1}{3}(C_{11} + 2C_{12})$$

$$K_S^{o} = \frac{1}{3}(C_{11} + 2C_{12}^{o})$$

$$K_S' = \frac{\partial K_S}{\partial P} = \frac{a + 2b}{3} \tag{12}$$

and converting from the adiabatic to isothermal values by the relationship

$$K_S = K_T (1 + \Delta) \tag{13}$$

with

$$\Delta = \frac{9\alpha^2 T K_S}{\rho c_p} \tag{14}$$

where α is the linear thermal expansion coefficient and c_p is the specific heat at constant pressure, we obtain

$$\ln \frac{\rho}{\rho_o} = \frac{1 + \Delta}{K'_S} \ln \left(1 + \frac{K'_S}{K^o_S} P\right) \tag{15}$$

The new densities were then used to compute new elastic moduli, from Eq 6-9, which were again fit with straight lines. New densities were then calculated. This iterative process was continued until further cycling produced no further change. At 35 kbars, the final density value is in agreement with that obtained from the Hildebrand equation of state. This approach is similar to Cook's method [*Cook*, 1957] except that our technique does not involve the dimensions of the sample. Decker's equation of state [*Decker*, 1971] differs negligibly from the Hildebrand equation [*Weaver*, 1971]. Virtually the same results would have been obtained with either equation.

The estimated error in C_{11} and C_{44} is ± 3%. Each is determined from a single velocity using either equation (6) or (7). In the case of C_{12}, the error is ± 6%, since it is derived from two separate velocity measurements using both equations (6) and (9).

VI. DISCUSSION

NaCl was selected as a trial sample for the Brillouin scattering method because of its simplicity and because of the extensive studies that have been made on its elastic properties previously. Figure 5 shows graphically the pressure dependence of the elastic moduli calculated from the phonon velocities, which were measured by the Brillouin scattering method compared with elastic constants calculated from ultrasonic measurements up to 10 kbars. These results are also summarized in Table 1. The 1-bar pressure values of the elastic moduli determined by the two methods agree within the stated experimental error of the Brillouin scattering method. The pressure derivatives of the elastic constants have been calculated from straight lines fit to the plots shown in Figure 5 and are given in Table 1 along with

TABLE 1. Adiabatic Elastic Moduli and Their Pressure Derivatives for NaCl

Source	C_{11}	$\partial C_{11}/\partial P$	C_{12}	$\partial C_{12}/\partial P$	C_{44}	$\partial C_{44}/\partial P$
This work	482 ± 014[a]	11.62 ± 0.70[b]	128 ± 008[a]	3.05 ± 0.37[b]	127 ± 004[a]	0.759 ± 0.046[b]
Spetzler et al. [1972]	496	11.65	130	2.06	128	0.369
Barsch and Chang [1967]	490	11.82	126	2.18	127	0.372
Bartels and Schuele [1965]	490	11.74	126	2.16	127	0.37

a. Units are in kbars and evaluated at $P = 0$ and $T = 300°K$.
b. Evaluated at $P = 0$ and $T = 300°K$.

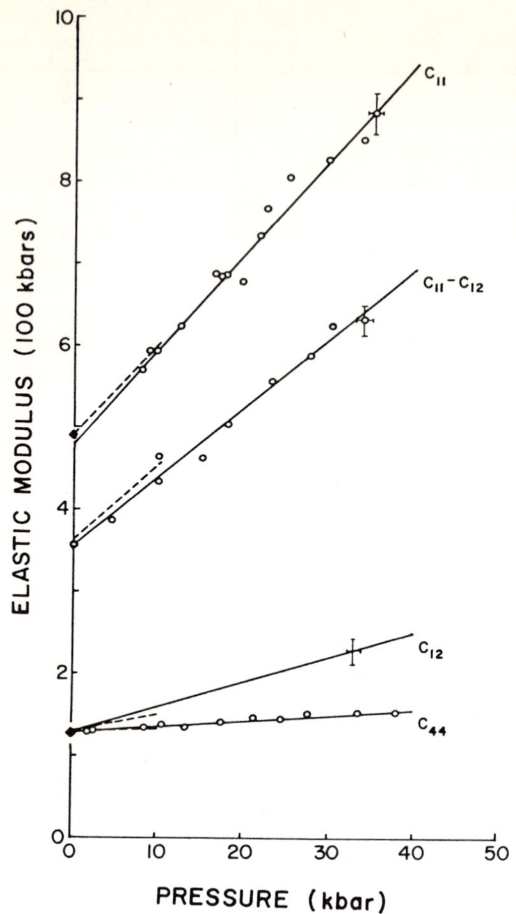

Fig. 5. The three elastic moduli of NaCl, C_{11}, C_{12}, and C_{44}, as a function of pressure. Data points are not shown for C_{12} since the curve must be calculated indirectly from other measurements. Solid lines are straight line fits to the data. Dotted lines are ultrasonic measurements by Spetzler et al [1972].

the zero pressure values reported by other investigators employing ultrasonic techniques [*Spetzler et al.*, 1972; *Barsch and Chang*, 1972; *Bartels and Schuele*, 1965]. Although the zero pressure slope of the C_{11} is in excellent agreement with the ultrasonic measurements, the ultrasonic data show a downward curvature, which we do not observe up to 35 kbars. A small discrepancy exists in the zero pressure slope of the C_{12}, and since the ultrasonic data show a slight negative curvature with pressure, the discrepancy becomes greater with increasing pressure. The zero pressure slope of C_{44} is considerably higher than the zero pressure slope based on ultrasonic measurements. The difference between the two slopes, however, is well within the experimental error for the Brillouin scattering method. Taken point by point, the data derived from Brillouin scattering agrees within the experimental error with the data derived from ultrasonic measurements. However, extrapolation of the data from the ultrasonic measurements leads to values significantly lower than ours at pressures of 35 kbars.

Although the Brillouin scattering method cannot rival the ultrasonic method in the precision of the measurements, its ability to measure sound velocities in single crystals under hydrostatic conditions at pressures up to 100 kbars should provide an important means of testing the validity of extrapolation procedures that have been used to extrapolate low-pressure ultrasonic measurements to high pressures.

Acknowledgments. The authors wish to express their appreciation to Charles Whitfield for making most of the measurements reported in this paper. The research was supported by NSF grant DES 74-22501.

REFERENCES

Anderson, D. L., C. G. Sammis, and R. A. Phinney, Brillouin scattering--a new geophysical tool, in *Applications of Modern Physics to the Earth and Planetary Interiors*, edited by S. K. Runcorn, pp. 465-477, Wiley Interscience, London and New York, 1969.

Barnett, J. D., S. Block, and G. J. Piermarini, An optical fluorescence system for quantitative pressure measurement in the diamond-anvil cell, *Rev. Sci. Instrum. 44*, 1-9, 1973.

Barsch, G. R., and Z. P. Chang, Adiabatic, isothermal and intermediate pressure derivatives of the elastic constants for cubic symmetry, *Phys. Stat. Sol. 19*, 139-151, 1967.

Bartels, R. A., and D. E. Schuele, Pressure derivatives of the elastic constants of NaCl and KCl at 295°K and 195°K, *J. Phys. Chem. Solids 26*, 537-549, 1965.

Cook, R. K., Variation of elastic constants and static strains with hydrostatic pressure: a method for calculation from ultrasonic measurements, *J. Acoust. Soc. Amer. 29*, 445-449, 1957.

Cummins, H. Z., and P. E. Schoen, Linear scattering from thermal fluctuations, in *Laser Handbook*, pp. 1029-1075, edited by F. T. Arecchi and E. O. Schuly-Dubois, North-Holland, Amsterdam, 1972.

Decker, D. L., High pressure equation of state for NaCl, KCl, and CsCl, *J. Appl. Phys. 42*, 3239-3244, 1971.

Piermarini, G. J., S. Block, and J. D. Barnett, Hydrostatic limits in liquids and solids to 100 kbar, *J. Appl. Phys. 44*, 5377-5382, 1973.

Piermarini, G. J., S. Block, J. D. Barnett, and R. A. Forman, Calibration of the pressure dependence of the R_1 ruby fluorescence line to 195 kbar, *J. Appl. Phys. 46*, 2774-2780, 1975.

Spetzler, H., C. G. Sammis, and R. J. O'Connell, Equation of state of NaCl: ultrasonic measurements to 8 kbar and 800°C and static lattice theory, *J. Phys. Chem. Solids 33*, 1727-1750, 1972.

Weaver, J. S., W. A. Bassett, and T. Takahashi, Calculation of P-V relation for sodium chloride up to 300 kilobars at 25°C, in *Accurate Characterization of the High Pressure Environment*, edited by E. C. Lloyd, pp. 325-330, National Bureau of Standards Publ. 326, 1971.

Weidner, D. J., K. Swyler, and H. R. Carleton, Elasticity of microcrystals, *Geophys. Res. Lett. 2*, 189-192, 1975.

TRANSDUCER AND BOND PHASE SHIFTS IN ULTRASONICS, AND THEIR EFFECTS ON MEASURED PRESSURE DERIVATIVES OF ELASTIC MODULI

G. F. Davies
Department of Geological Sciences
University of Rochester
Rochester, New York 14627

R. J. O'Connell
Department of Geological Sciences
Harvard University
Cambridge, Massachusetts 02138

Abstract

Phase shifts introduced into ultrasonic signals by the presence of transducers and bonds at sample surfaces have been measured using an automated variable frequency ultrasonic interferometer. At zero pressure, phase shifts have been resolved due to transducers, bonds between transducers and samples, and bonds between buffer rods and samples. Observed transducer-bond-sample phase shifts are in good accord with theoretical estimates, and bond thicknesses of about 0.3 μ are inferred. Measurements to 7 kbar are consistent with theoretical estimates of the effect of pressure on transducer-bond-phase shifts. Providing the frequency of the ultrasonic signal is within a few percent of the resonance frequency of the transducer, and the effect of pressure on the transducer resonance frequency is accounted for (as recommended by *McSkimin*, [1961]), the effect of the bond phase shift on the measured pressure derivative of the elastic modulus should amount to less than about 0.02. If the frequency deviates substantially from the transducer resonance frequency, especially at zero pressure, errors of the order of 0.25 could be incurred in the pressure derivatives. The nonlinearity of transducer-bond phase shifts could cause significant errors in second-pressure derivatives, even under favorable conditions. For shear waves at zero pressure, the observed buffer-bond-sample phase shifts are consistent with those estimated theoretically for a bond of about 1 μ thickness. For compressional waves at zero pressure, phase shifts are very sensitive to the buffer-sample contact: large differences in phase are observed between dry lapped, "wetted," immersed, and resin-bonded contacts. The sources of these differences are not fully understood, but they may be due to variations in contact area produced by the ultrasonic wave. "Normal" buffer-sample bonds are estimated to be capable of affecting measured pressure derivatives by about 0.25. The behavior of the anomalous buffer-sample phase shifts under pressure is unknown, but the shifts could easily give rise to substantial errors in measured pressure derivatives.

I. INTRODUCTION

The geophysical objective of ultrasonic measurements is to be able to extrapolate laboratory measurements of elastic properties and density to the pressure and temperature conditions of the earth's deep interior. For these extrapolations to be useful, considerable accuracy is required of the measurements. In principle, the widely used pulsed ultrasonic techniques [*McSkimin*, 1957; 1961], or variations thereof, are capable of measuring elastic velocities to an accuracy of at least 0.05%, and of detecting relative changes in elastic velocity of even smaller magnitude [*McSkimin*, 1956]. In practice, however, this accuracy apparently has not usually been achieved.

Table 1 compares the results of different measurements of the same material for several materials: the spread in results for elastic moduli is commonly in the range 0.2-1%, and greater in some cases. The velocity measurements taken to pressures of the order of 10 kbar would show changes of the order of a few percent. Thus, in principle, the measured pressure derivatives of elastic modulus should be accurate to a few percent at least. It can be seen in Table 1, however, that discrepancies of the order of 10% commonly occur. For materials of geophysical interest, an uncertainty of this magnitude produces an uncertainty of the order of 10% in elastic velocities extrapolated to 1 Mbar, and this uncertainty is much too large for the extrapolation to be useful in interpreting the seismically observed elastic velocities in the deep mantle. It is thus of interest to determine the source of the discrepancies.

This paper reports on the effects of phase shifts in ultrasonic waves caused by the presence of transducers, buffer rods, and bonding material at the surface of a sample. These phase shifts are one possible source of the discrepancies discussed above. Theoretical and experimental determination of buffer-bond phase shifts at zero pressure were considered by *McSkimin* [1950; 1957]. Transducer-bond phase shifts were considered by *Redwood and Lamb* [1956] and *McSkimin* [1957; 1961]; the effects of pressure and temperature were considered by *McSkimin* [1961] and *McSkimin and Andreatch* [1962]. In this study the phase shifts were observed over a wide frequency range using ultrasonic interferometry, and the effects of pressure are considered in more detail.

TABLE 1. Comparison of Different Measurements of Elastic Moduli and Their Pressure Derivatives (Denoted by a Prime) for Several Materials[a]

Material	*Elastic Moduli*, Mbar				*Pressure Derivatives*				Reference
	C_{11}	C_{12}	C_{44}	K_S	C'_{11}	C'_{12}	C'_{44}	K'_S	
MgO	2.974	0.958	1.562	1.628	8.70	1.42	1.09	3.85	*Spetzler* [1970]
	2.966	0.951	1.558	1.623	9.16	1.82	1.12	4.27	*Chang and Barsch* [1969]
	2.967	0.951	1.560	1.623	9.48	1.99	1.16	4.49	*Anderson and Andreatch* [1969]
Range (%)	0.2	0.5	0.3	0.3	8	30	7	10	
NaCl	.4956	.1303	.1280	.251	11.65	2.06	0.37	5.26	*Spetzler et al.* [1972]
	.4942	.1269	.1281	.249	11.62	1.58	0.10	4.93	*Drabble and Strathen* [1967]
	.4958	.1306	.1279	.252	11.89	2.13	0.37	5.38	*Swartz* [1967]
	.4899	.1257	.1272	.247	11.66	2.08	0.37	5.27	*Bartels and Schuele* [1965]
Range (%)	1	3	0.7	2	2.5	25	25	6	
TiO_2	2.701	1.766	1.239	2.10	6.29	9.02	1.08	6.9	*Fritz* [1974]
	2.714	1.780	1.244	2.15	6.47	9.10	1.10	6.8	*Manghnani* [1969]
Range (%)	0.5	0.2	0.4	2	3	1	2	1.5	
Mg_2SiO_4	3.284	0.639	0.652	1.28	8.47	4.67	2.12	5.36	*Kumazawa and Anderson* [1969]
	3.291	0.663	0.672	1.29	8.32	4.30	2.12	4.83	*Graham and Barsch* [1969]
Range (%)	0.2	4	3	1	2	9	0	10	

a. Not all of the single-crystal moduli are shown for TiO_2 (tetragonal) and Mg_2SiO_4 (orthorhombic).

II. EXPERIMENTAL METHODS

The ultrasonic system, illustrated schematically in Figure 1, is described in detail elsewhere [*O'Connell et al.*, in preparation] and is based on the phase comparison technique [*McSkimin*, 1950].

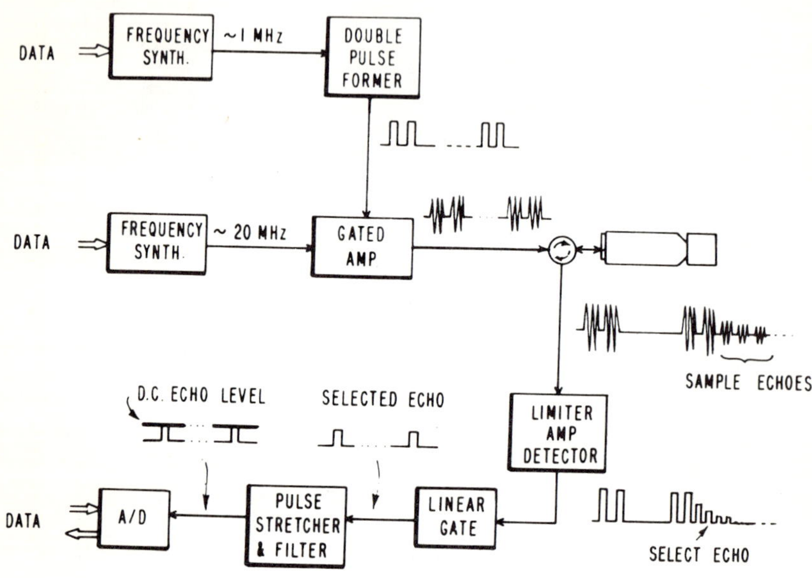

Fig. 1. Schematic diagram of the ultrasonic interferometer. Lines marked DATA indicate connection to minicomputer.

The instrumentation is similar to that described by *Spetzler* [1970]. An RF (carrier) wavetrain is gated to produce two RF pulses, which are phase coherent--i.e., the phase relationship of the pulses is the same as in the original wavetrain, independent of the spacing of the pulses. The electrical pulses are converted to acoustic pulses in the sample via a quartz transducer and the spacing of the pulses is adjusted so that the second pulse is superimposed on an echo of the first pulse in the sample. Alternate constructive and destructive interference of the pulses can be obtained by varying the carrier frequency. One echo from the train of echoes is selected with a linear gate, and its peak amplitude is converted to a DC output. The carrier frequency and pulse spacing are digitally controlled by an on-line minicomputer, which also monitors echo amplitude. The computer is programmed to sweep through a specified range of carrier frequency and record the echo amplitude as a function of frequency, and it is also used to process the data as described below.

High pressures were generated with a Bridgman piston-cylinder apparatus with kerosene pressure medium. Pressure was determined to within 1% with a Heise gauge.

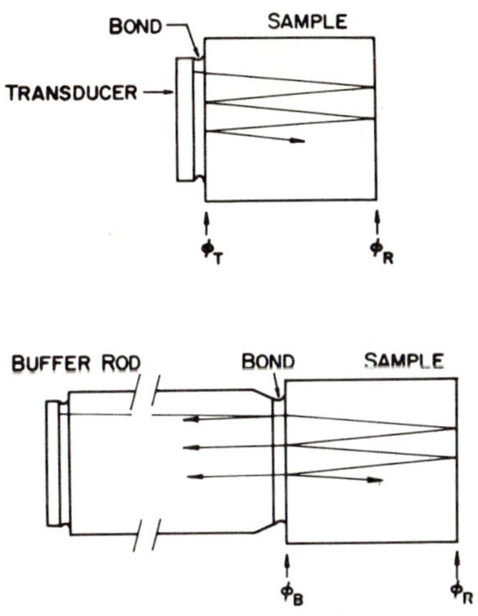

Fig. 2. Schematic diagram of the transducer-bond-sample and buffer rod-bond-sample assemblies in this study.

The transducer-buffer-bond-sample configurations used are illustrated in Figure 2. Samples used in this study--both about 1 cm in size--were a synthetic stoichiometric spinel ($MgAl_2O_4$) crystal obtained from Union Carbide and a synthetic MgF_2 crystal obtained from Optovac, Inc. Opposite faces on the samples were polished flat and parallel to within about one μ. A fused quartz buffer rod was used. Transducers were bonded by Dow Corning resin 276-V9 or, for some zero-pressure runs, phenyl salicylate. Coupling between the buffer rod and samples was achieved in several different ways: with dry lapped contact; with lapped contact immersed in a fluid pressure medium, such as isopentane; with a lapped contact wetted with the fluid (but not immersed); and with a V9 resin bond.

III. DATA PROCESSING

An example of amplitude-frequency data is illustrated in Figure 3(a). The broad amplitude envelope results from the combined response of the electronics and the transducer resonance (in this case the fundamental resonance frequency of the transducer

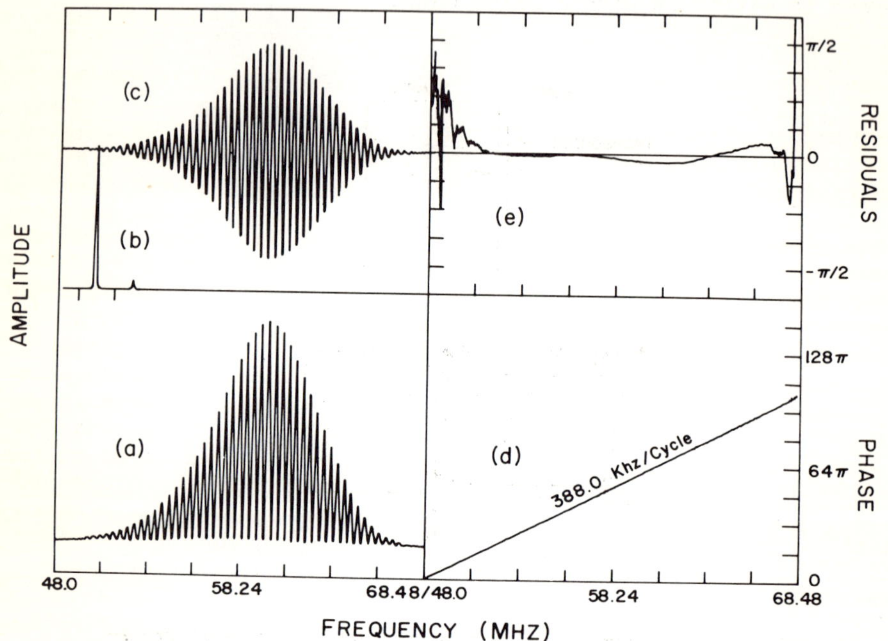

Fig. 3. (a) Amplitude-frequency interference pattern obtained from the ultrasonic interferometer. (b) Spectrum of a. Filter window is indicated by tick marks below spectrum. (c) Filtered amplitude-frequency data with envelope, harmonics, and noise removed. (d) Instantaneous phase of c versus frequency. (e) Residuals of d from a bestfit line.

was 20 Mhz). The narrow peaks result from successive constructive interferences of the superposed echoes in the sample. The condition for constructive interference is that the phase difference between successive echoes be equal to $2n\pi$, where n is an integer, i.e.:

$$\frac{4\pi f L}{v} + \phi_r = 2n\pi \tag{1}$$

where f is the carrier frequency, L is the sample length, v is the sound velocity in the sample, and ϕ_r is the total phase shift produced by reflections at the sample faces. If the carrier frequency is increased until the next constructive interference is achieved, the change in frequency, Δf, is given by

$$\frac{4\pi L}{v} \Delta f + \Delta\phi_r = 2\pi \tag{2}$$

where $\Delta\phi_r$ is the accompanying change in ϕ_r. Thus, if $\Delta\phi_r$ is known or can be neglected, the velocity, v, can be determined by measuring Δf; Δf can be determined by directly measuring the frequencies at successive peaks and troughs [*Spetzler*, 1970], but their positions may be affected by the amplitude envelope and by noise. Instead, the phase of the sinusoid in Figure 3(a) was calculated continuously as a function of carrier frequency [*O'Connell et al.*, in preparation]. The phase of this sinusoid is directly related to the phase difference between superposed echoes in the sample; thus pressure derivatives can also be determined by measuring the phase as a function of pressure.

The first step in calculating the phase is to filter out the amplitude envelope and any harmonics and noise from the data. This is done via the spectrum (Figure 3b) of the data calculated by the Fast Fourier Transform (FFT) algorithm, and the result is illustrated in Figure 3(c). The filtered data are phase shifted by 90°, via a Hilbert transform in the frequency domain, and the "instantaneous" phase is then given by the inverse tangent of the ratio of the transformed to the original filtered data. The process is described in more detail by *O'Connell et al.* [in preparation]. The calculated variation of phase with carrier frequency is illustrated in Figure 3(d), and residuals from a best-fit straight line are shown in Figure 3(e). If reflection phase shifts in the sample could be neglected, the slope of this line would be a measure of the sound velocity in the sample. Small systematic deviations from linearity can be seen in Figure 3(e), which are interpreted below as arising from reflection phase shifts in the sample.

IV. TRANSDUCER-BOND PHASE SHIFTS

A sound wave reflected from a sample face to which a transducer is bonded is actually the superposition of waves reflected from three faces: the sample-bond and bond-transducer interfaces and the outer face of the transducer. The phase and amplitude of the resultant wave depend in a complicated way on the thickness and relative acoustic properties of the bond, transducer, and sample. This dependence, for the case of plane, parallel waves and interfaces, has been given by *Redwood and Lamb* [1956], who used the analogous theory of transmission lines [see also *McSkimin*, 1957; *Williams and Lamb*, 1958]; identical results can be obtained by considering plane elastic waves directly. The total transducer phase shift is

$$\phi_{tf} = 2\psi_{tf} - \pi \tag{3}$$

where

$$\tan \psi_{tf} = \frac{Z_f}{Z_s} \frac{Z_t \tan \theta_t + Z_f \tan \theta_f}{Z_f - Z_t \tan \theta_t \tan \theta_f} \tag{4a}$$

$$= \frac{Z_f}{Z_s} \tan (\eta_t + \theta_f) \tag{4b}$$

and

$$\tan \eta_t = \frac{Z_t}{Z_f} \tan \theta_t \tag{5}$$

Here, Z_s, Z_t, and Z_f are the acoustic impedance of the sample transducer, and bond ("film"), respectively, and θ is the thickness, in terms of the phase of the sound wave, of the transducer or bond

$$\theta = kl = 2\pi l/\lambda = 2\pi lf/v \tag{6}$$

where k is the wave number, λ is the wavelength, and l is the thickness. Subscript t or f in (6) would indicate transducer or bond properties, respectively. The acoustic impedance, Z, of a medium is the product of the density, ρ, and sound velocity, v, of the medium

$$Z = \rho v \tag{7}$$

The total reflection phase shift, θ_r, between successive echoes in the sample is the sum of (3) and a phase shift of π occurring on reflection at the other end of the sample. An example of calculated phase shifts as a function of carrier frequency is illustrated in Figure 4 for the case of a 10 Mhz quartz P-transducer (X-cut) bonded to a (100) face of MgF_2. The relevant acoustical properties are listed in Table 2, along with those of other materials discussed in this paper. Different bond thicknesses are represented by the parameter

$$\tau_f = \frac{l_f}{v_f} \tag{8}$$

which is the transit time (in nsec) of the wave through the bond. Wave velocities in typical bond materials are 1 to 2 km/sec. For v_f = 1 km/sec, τ_f is the bond thickness in μ.

For zero bond thickness, the phase shift is symmetrical about multiples of the transducer resonance frequency (in this case, the third), and equation (4) simplifies to

TABLE 2. Densities, Velocities, and Acoustic Impedances of Various Materials

Material	ρ, g/cm^3	v_p, km/s	v_s, km/s	Z_p	Z_s	Reference
α-quartz	2.65	5.75[a]	3.92[b]	15.2	10.3	*McSkimin et al.* [1965]
MgF_2 [100]	3.18	6.70	4.22	21.3	13.4	*Davies* [in preparation]
Fused quartz	2.20	5.9	3.8	13.0	8.4	*Peselnick et al.* [1967]
Spinel [100]	3.58	8.87	6.57	31.8	23.5	*Chang and Barsch* [1973]
276-V9 resin (20°C) (Dow-Corning)	-	-	-	2.2	1.0	*McSkimin and Andreatch* [1962]

a. For [100] (X-cut).

b. For [010] (Y-cut).

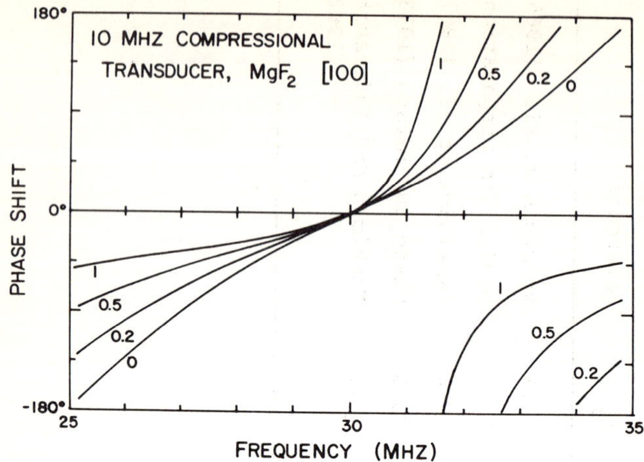

Fig. 4. Calculated transducer-bond phase shifts for compressional [100] waves in MgF_2 with a 10-Mhz quartz transducer bonded to the sample with V9 resin. Curves are labelled with bond transit time τ_f (nsec). Note that +180° is equivalent to -180° so that the curves in the lower right are continuations of the curves intersecting the upper margin.

$$\tan \psi_t = \frac{Z_t}{Z_s} \tan \theta_t . \qquad (9)$$

For non-zero bond thickness, the calculated phase shift is not symmetrical. At multiples of the transducer resonance frequency, the effect of the bond is very small, as pointed out by *McSkimin* [1961], but the bond effect can increase quite rapidly away from resonance multiples, especially on the high-frequency side (Figure 4).

Figure 5 compares some measured and calculated phase shifts as functions of carrier frequency. A reference measurement was made using a fused quartz buffer rod bonded to the sample. As described in the next section, the slope of this phase line should be very little affected by the buffer-sample bond. Accordingly, Figure 5 also shows the measured phase residual, with transducer bonded to sample, relative to a line with a slope measured with the buffer rod and coinciding with the measurement of 30 Mhz. The measured residual is compared with calculated transducer-bond phase shifts with τ_f = 0, 0.1, and 0.2 nsec. The form of the measured residual agrees well with that of the calculated phase shifts, and a value of τ_f between 0.1 and 0.2 nsec can be inferred from Figure 5. This corresponds to a bond thickness of the order of 0.2 to 0.4 μ, which is perhaps thinner than might have been

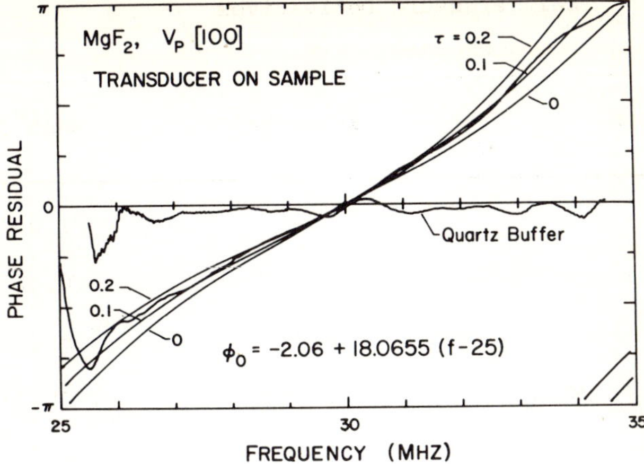

Fig. 5. Comparison of measured transducer-bond phase residuals (relative to phase obtained with buffer rod) with calculated phase shifts assuming τ_f = 0, 0.1, and 0.2 nsec.

expected. *McSkimin* [1961] reports θ_f = 5 to 15°, which, at 20 Mhz, corresponds to about τ_f = 0.7 to 2 nsec.

V. PRESSURE EFFECTS ON TRANSDUCER-BOND PHASE SHIFTS

Pressure affects, of course, all three components of the mechanical system: sample, bond, and transducer, but it is desired to isolate the effect on the sample from the effects on the other two. The effect of pressure on the transducer was considered by *McSkimin and Andreatch* [1962] who noted that the resonance frequency would change. The pressure derivatives of the resonance frequencies of X- and Y-cut quartz transducers (Table 3) were determined by *McSkimin and Andreatch* [1962] and *McSkimin et al.* [1965]. The effect of pressure on the bond was not discussed in detail by *McSkimin and Andreatch* [1962], their point being to establish that the bond phase shift is very small at the transducer resonance frequency and that it should change very little with pressure, providing the resonance frequency is followed. The effect of pressure on the bond phase shift will be discussed in more detail here, since it appears that the latter condition has not been strictly adhered to by many workers (including *McSkimin and Andreatch* [1962]). (*McSkimin and Andreatch* [1962] actually argued that the bond phase shift depends only on the total mass per unit area of the bond, and that the latter would not change with pressure. However, this conclusion depends on the assumption

TABLE 3. Transducer (α-quartz) and Bond Properties and Their Pressure Derivatives

Parameter	α-quartz[a]	Bond[b]
ρ, g/cm^3	2.65	1.0
K, Mbar	0.38	0.02
$\partial K/\partial P$	6.4	5.0
$M_p = \rho v_p^2$, Mbar	0.868	0.033
$\partial M_p/\partial P$	3.28	7.7
$M_s = \rho v_s^2$, Mbar	0.399	0.010
$\partial M_s/\partial P$	-2.69	2.0
$(\partial \ln f_r/\partial P)_p$, Mbar^{-1}	1.51	-
$(\partial \ln f_r/\partial P)_s$, Mbar^{-1}	-3.68	-

a. *McSkimin et al.* [1965].
b. Assumed.

that the bond compresses uniaxially, which is unlikely to be true, since it would require a large shear strength of the bond. Even with the assumption that the bond compresses isotropically, the phase shift changes very little with pressure, because it is so small to begin with, so that their main conclusion is still justified.)

A. Transducer Phase Shifts

A fairly common procedure has been to hold the carrier frequency constant at the zero-pressure resonance frequency, since the transducer resonance frequency varies by only a few percent over a pressure range of 10 kbar [e.g., *McSkimin et al.*, 1965]. A correction can be applied to allow for the changing transducer resonance frequency, and it is usually assumed that the bond effect remains negligible. Close to a transducer resonance frequency, where θ_t is close to an integral multiple of π, and for a vanishingly thin bond (negligible θ_f), equations (3) and (9) yield approximately, and to within an additive factor of 2π,

$$\phi_t = 2\pi \frac{Z_t}{Z_s} \left(\frac{f}{f_r} - \frac{f}{f_{ro}}\right) \tag{10}$$

where f_{ro} is the value of f_r at zero pressure. Differentiating (10), and evaluating the result at zero pressure gives

$$\frac{\partial \phi_t}{\partial P} = -2\pi \frac{Z_t}{Z_s} \frac{f}{f_{ro}} \frac{\partial \ln f_r}{\partial P} \tag{11}$$

(Although Z_t and Z_s vary with pressure, they produce only a second-order effect.) A convenient expression can be derived for the corresponding correction to the derivative of the appropriate combination of elastic moduli, $M = \rho v^2$. The desired phase derivative is

$$\frac{\partial \phi_s}{\partial P} = \frac{\partial \phi_m}{\partial P} - \frac{\partial \phi_t}{\partial P} \tag{12}$$

where ϕ_m is the measured phase. The sample phase is $\phi_s = 4\pi f L/v$, so that

$$\frac{1}{\phi_s} \frac{\partial \phi_s}{\partial P} = -\frac{1}{2K} \left[\frac{K}{M} \frac{\partial M}{\partial P} - (1 - 2\ K\ \beta)\right] \tag{13}$$

where $\beta = -\partial \ln L/\partial P$ is the linear compressibility of the sample. The correction is, using (12),

$$\left[\frac{\partial M}{\partial P}\right]_{\text{corr.}} = -\frac{\frac{M}{v}}{2\pi f L} \left(\frac{\partial \phi_t}{\partial P}\right). \tag{14}$$

Now combining (14) with (11)

$$\left[\frac{\partial M}{\partial P}\right]_{\text{corr.}} = \frac{v^2}{L} \left(\frac{Z_t}{f_r} \frac{\partial \ln f_r}{\partial P}\right). \tag{15}$$

Thus, apart from the transducer properties, the correction to $\partial M/\partial P$ depends only on the sample velocity and length. Some representative values are given in Table 4, for quartz transducers on MgF_2 and spinel. The corrections to the pressure derivatives of the moduli are of the order of 0.1 for compressional waves and -0.05 for shear waves.

TABLE 4. Representative Corrections to Pressure Derivatives of Elastic Moduli Due to Transducer Phase Shifts

Mode	v,km/s	L,mm	f_r, Mhz	$(\frac{\partial M}{\partial P})_{corr.}$	$\frac{\partial M}{\partial P}$
MgF_2[a]					
[001]P	8.01	10.80	20	.069	5.66
[110]P	8.15	10.80	10	.141	8.43
[001]S	4.22	9.64	20	-.035	0.79
[110]S[c]	2.82	10.80	20	-.014	-0.68
$MgAl_2O_4$[b]					
[001]P	8.88	10[d]	20	.091	5.15
[110]P	10.22	10	20	.120	5.85
[001]S	6.57	10	20	-.082	0.89
[110]S[c]	4.23	10	20	-.034	0.19

a. *Davies* [in preparation].

b. *Chang and Barsch* [1973].

c. Polarization [1$\bar{1}$0].

d. Assume lengths for spinel.

Note that this difference in sign (arising from the opposite signs of the derivatives of the relevant quartz transducer frequencies [Table 3]). causes the corrections for some derived moduli to be compounded. This is most easily seen for an isometric material. If C_s denotes the modulus for a shear wave in the [110] direction, polarized in the [1$\bar{1}$0] direction, then the bulk modulus is $K = C_{11} - 4C_s/3$, and $C_{12} = C_{11} - 2C_s$. The effect is illustrated in Table 5 for spinel (isometric) and MgF_2 (tetragonal).

TABLE 5. Corrections to Measured and Derived Modulus Derivatives Due to Transducer Phase Shifts, Illustrating the Compounding of Corrections in Derived Quantities

	Modulus, M.	$(\frac{\partial M}{\partial P})_{corr.}$	$\frac{\partial M}{\partial P}$
MgF_2	C_{11}	.10	5.01
	C_{33}	.07	5.66
	C_{12}	.15	6.37
	C_{13}	.24	4.18
	C_{44}	-.03	0.79
	C_{66}	.03	2.90
	K_S [a]	.17	5.05
	μ [a]	-.03	0.55
$MgAl_2O_4$	C_{11}	.09	5.15
	C_{12}	.16	4.76
	C_{44}	-.08	0.89
	K_S	.14	4.89

a. Isotropic aggregate modulus calculated from the Hashin-Shtrikman bounds [*Davies*, in preparation].

B. Bond Phase Shifts

Pressure affects the bond phase shift in two ways: directly, by changing the thickness and wave velocity of the bond material, and indirectly, if the carrier frequency is not continuously matched to the transducer resonance frequency to keep the bond phase shift minimal.

The pressure dependence of the properties of bond materials are not well known. For the purpose of assessing the importance of the effect, some representative properties have been assumed. These are shown in Table 3, and the resultant variations in density, velocities, and impedances to 40 kbar are given in Table 6. The variation in density was calculated from the Murnaghan equation

TABLE 6. Density, Velocities, and Impedances of a Hypothetical Bond Material as a Function of Pressure[a]

P,kbar	ρ,g/cm^3	v_p,km/s	v_s,km/s	Z_p	Z_s
0	1.00	1.82	1.00	1.82	1.00
10	1.28	2.92	1.53	3.75	1.96
20	1.43	3.61	1.87	5.16	2.67
30	1.53	4.14	2.14	6.35	3.28
40	1.62	4.59	2.36	7.41	3.81

a. Assumed bond properties are given in Table 3.

$$\rho = \rho_o \left(1 + \frac{K'}{K} P\right)^{1/K'} \tag{16}$$

(where $K' = \partial K/\partial P$), which is derived from the assumption that the bulk modulus varies linearly with pressure. Consistent with this, the longitudinal and shear moduli were also assumed to vary linearly with pressure. The bond was assumed to compress isotropically, so that ℓ is proportional to $\rho^{-1/3}$. The velocities and impedances increase by factors of 2 to 4 in this pressure range, and θ_f (equation 6) decreases by a comparable amount, especially in the first few kbars. The effect of changing thickness on θ_f is relatively minor. This rapid effective thinning of the bond (relative to the sound wavelength in the bond) is important because it means that even if the frequency deviates from the transducer resonance frequency at high pressures, the bond phase shift may still not be large.

The total transducer-bond phase shift, including the indirect effect on the bond phase shift due to the variation of transducer resonance frequency (assuming fixed carrier frequency), has been calculated from equations (3-5), assuming bond properties similar to those in Table 6 and neglecting the variation of sample and transducer impedances. Results for a 10 Mhz quartz transducer on a [100] MgF_2 sample (see Tables 2 and 3) are illustrated in Figures 6 and 7 for fixed carrier frequencies f = 29, 30, and 31 Mhz. For $\theta_f = 0$, the variation of phase shift with pressure arises from the variation of the transducer resonance frequency. At f = 30 Mhz, the additional bond phase shift is fairly small at all pressures. The curves for f = 29 and 31 Mhz are included to illustrate more clearly the effects of pressure on bond phase

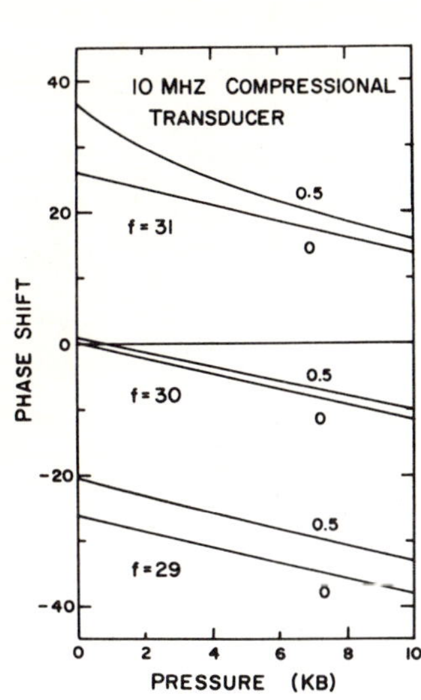

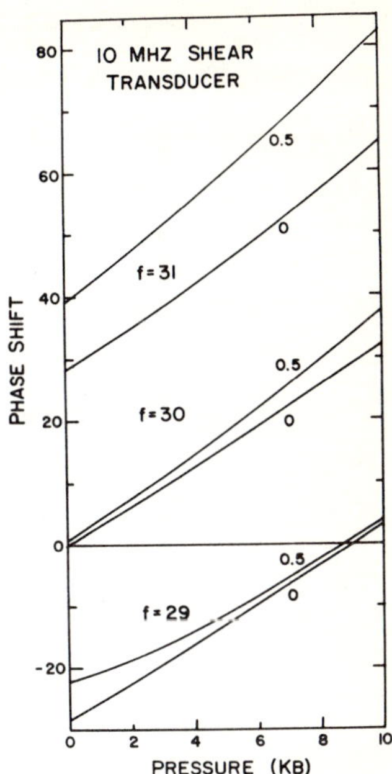

Fig. 6. (Left) Estimated effect of pressure on transducer-bond phase shifts for compressional [100] waves in MgF_2 with a 10-Mhz quartz transducer. Curves are for τ_f = 0 and 0.5 nsec and carrier frequency f = 29, 30, and 31 Mhz. Bond properties are estimated (Table 3). Fig. 7. (Right) As in Figure 6 for shear waves.

shifts. In Figure 6, at $f = 31$ Mhz, the bond phase shift decreases rapidly with increasing pressure, both because of the decreasing effective thickness of the bond and because f_r increases toward f. On the other hand, at f = 29 Mhz, f_r increases away from f, and the effects roughly cancel, so that the bond phase shift is fairly constant. These situations are reversed in Figure 7 because f_r decreases with increasing pressure for shear waves.

The reality of these effects was tested by comparison with some measurements on MgF_2 [100] compressional waves. Measurements of phase versus f at pressures up to 7 kbar were converted to phase versus pressure at frequencies f = 30, 31, and 32 Mhz. The results are shown in Figure 8(a). A systematic difference in slopes at the different frequencies, amounting to about 5%, can be discerned. Residuals relative to a line through the 30 Mhz data

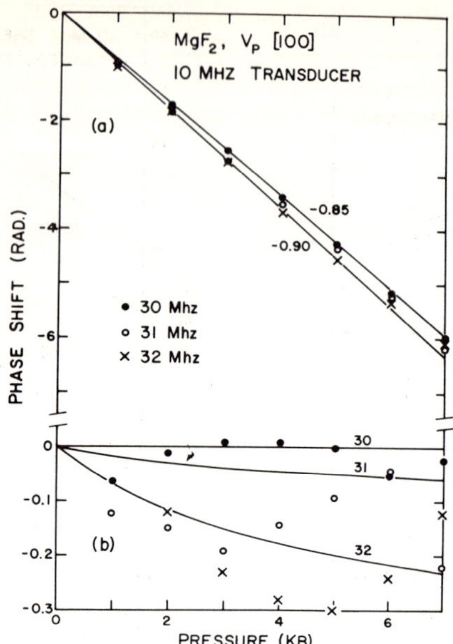

Fig. 8. (a) Measured phase shifts versus pressure (normalized to 30 Mhz) at 30, 31, and 32 Mhz for compressional [100] waves in MgF_2. Lines are labelled with slopes (rad/kbar). (b) Residuals of measured phase shifts relative to line through 30 Mhz data, compared with calculated transducer bond phase shifts, assuming bond properties (Table 3) and τ_f = 0.2 nsec.

(shown in Figure 8(b)) are compared to calculated transducer-bond phase shifts, assuming τ_f = 0.2 nsec (see previous section). Although there is considerable scatter in the residuals, their ordering and magnitudes are quite comparable to the calculated phase shifts, which is evidence that the phase shifts are being correctly described and that there are no other important effects on the measurements.

If the carrier frequency is accurately fixed at the transducer resonance frequency, the effect of the bond phase shift should be quite small. For instance, in Figure 7 the effect of the bond at 30 Mhz is about 15% of the effect of the transducer on the slope. Since the latter changes the value of $\partial M/\partial P$ by about -0.05, the effect of the bond phase shift on $\partial M/\partial P$ would be less than -0.01. If f deviates from f_{ro}, then of course the bond effect can be considerably larger. For instance, in Figure 6, at 31 Mhz the bond effect is comparable to that of the transducer.

Measurements of second pressure derivatives of elastic moduli have been reported by several workers [e.g., *Chang and Barsch*, 1967, 1973; *Frisillo and Barsch*, 1972]. Bond phase shifts could be significant sources of error in such measurements. Some curvature is evident in Figures 6 and 7, even for f = 30 Mhz. In Figure 7 at f = 30 Mhz, the slope increases by about 20% over the 10 kbar pressure range. This would correspond to a change in $M' = \partial M/\partial P$ of about -0.01. Assuming a bulk modulus K = 2 Mbar then gives a change in KM'' of about -2. This is comparable to the accuracy claimed in some of the measurements referred to above. Of course, any deviation of f from f_{ro} could considerably increase this effect.

C. Other Measurement Procedures

Spetzler [1970] determined pressure derivatives by measuring the pressure dependence of the peaks and troughs in the interference pattern (Figure 3a), i.e. by measuring the derivative $(\partial f/\partial P)_\phi$ rather than $(\partial \phi/\partial P)_f$. (*Spetzler* used a buffer rod, but the same procedure might be used without a buffer rod.) The transducer-bond phase shifts in this case will depend on the difference between the transducer and sample derivatives, $\partial(f_s - f_r)/\partial P$. The expressions above can be generalized to allow for the variation of the carrier frequency f. For compressional waves, both derivatives will be positive and will tend to cancel, so that the bond phase shifts should be no greater than when f is constant. For shear waves, however, the sample derivative will most commonly be positive, while the transducer derivative is negative (for quartz), so that f_s may deviate considerably from f_r and the bond phase shift become significant. Also, *Spetzler* [1970] averaged measurements on a series of peaks and troughs near the transducer resonance frequency. If this was done without a buffer rod, significant bond phase shifts might affect those peaks and troughs not initially close to the transducer resonance frequency.

Another procedure that might be followed is to use the maximum amplitude response of the transducer to attempt to follow the resonance frequency of the transducer under pressure. However, when a transducer is bonded to a sample, the frequency at which maximum amplitude is obtained is not the free resonance frequency of the transducer, owing to the effect of the bond between the transducer and the sample. The expression for the response of the bonded transducer can be found in *Mason* [1958].

Figure 9 shows the relative amplitudes calculated for an X-cut quartz transducer bonded with V9 resin to a sample of olivine; curves are shown for several bond thicknesses ranging from 0 to 2 μm. The shift of the point of maximum amplitude to higher frequencies, owing to the presence of the bond, is apparent. The variation of the point of maximum amplitude with pressure is shown in Figure 10. The pressure dependence arises primarily from the change of the properties of the bond with pressure.

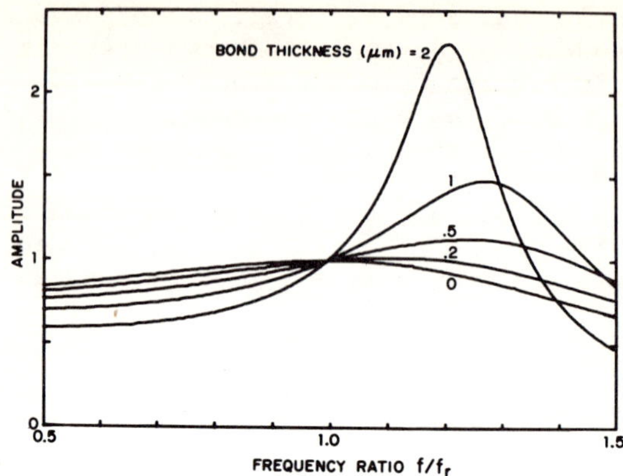

Fig. 9. Calculated relative amplitude versus frequency of X-cut quartz transducer bonded to olivine with various thicknesses of V9 resin.

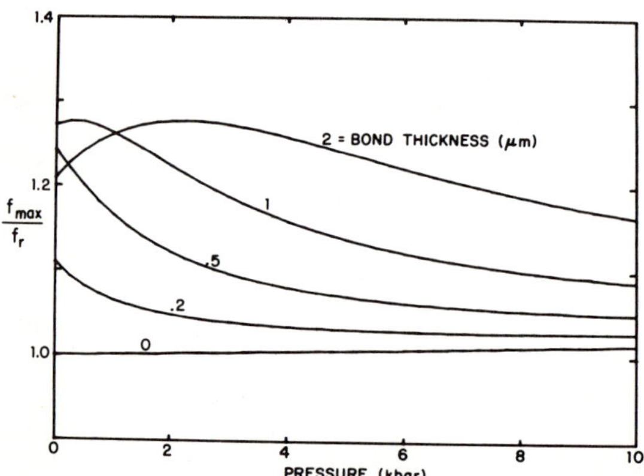

Fig. 10. Calculated frequencies of maximum amplitude in Figure 9 as a function of pressure.

If during an experiment the point of maximum amplitude was (wrongly) taken to be the resonant frequency of the transducer, then substantial errors could occur. It can be seen in Figure 4 that the phase shift due to the bond can be quite large, 1 or 2 Mhz above the free resonance frequency. Figure 11 shows the bond phase shift as a function of pressure when the operating frequency is kept at the point of maximum amplitude as shown in Figure 10. The phase shifts can easily be a large fraction of π, and vary significantly with pressure. A rough calculation shows that the

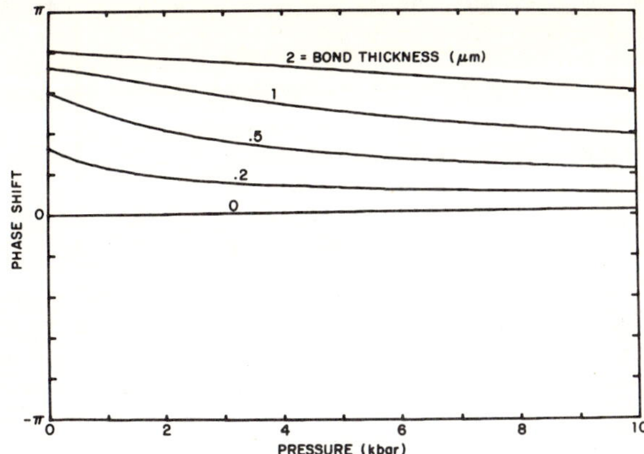

Fig. 11. Calculated transducer-bond phase shift versus pressure for the case where the operating frequency is kept at the frequencies of maximum amplitude of Figure 10.

magnitude of the phase shifts could cause errors of the order of ½% in determining absolute velocities, and errors of several tens of percent in pressure derivatives of velocities for typical minerals.

These potential errors are large, and would be best avoided. This is easily done by operating the transducer at its free resonant frequency in experiments using the pulse superposition technique. It is not clear from published reports of experiments if this procedure has been always used in pulse superposition measurements, but in at least some experiments the transducer has been operated at the maximum amplitude frequency rather than at the free resonance frequency [*E. Schreiber*, personal communication].

VI. BUFFER ROD-BOND PHASE SHIFTS

The phase shift which occurs upon reflection in the sample at a buffer rod-sample interface (Figure 2) is [*McSkimin*, 1950; 1957]

$$\phi_r^- = \tan^{-1} \frac{2Z_f Z_s\,(Z_b^2 - Z_f^2)\tan\theta_f}{Z_f^2\,(Z_b^2 - Z_s^2) + (Z_f^4 - Z_s^2 Z_b^2)\tan^2\theta_f} \quad (17)$$

where subscript b denotes buffer rod properties, and superscript "-" denotes the reflection of a wave traveling in the negative (leftward) direction (Figure 2). This phase shift (plus the phase shift of π occurring at the free end of the sample) applies to superpositions of all but the first and second reflections received by the transducer. Since the first reflection is from the buffer side of the interface, and the wave does not penetrate the sample, whereas the second and succeeding reflections are transmitted twice through the interface, the appropriate phase difference due to the interface is

$$\phi_1 = 2\phi_t - \phi^+_r \tag{18}$$

where ϕ_t, the phase shift on transmission, is

$$\phi_t = \tan^{-1} \frac{Z_s Z_b + Z_f^2}{Z_f (Z_s + Z_b)} \tan \theta_f \tag{19}$$

and ϕ^+_r has the same form as (17), but with Z_b and Z_s interchanged. The superposition of the first and second reflections will be referred to as echo 1, and succeeding combinations as echoes 2, 3, and so on.

An example of calculated buffer-bond phase shifts is shown in Figure 12 for the case of shear waves in the [100] direction in spinel, with a fused quartz buffer rod and a V9 bond. The

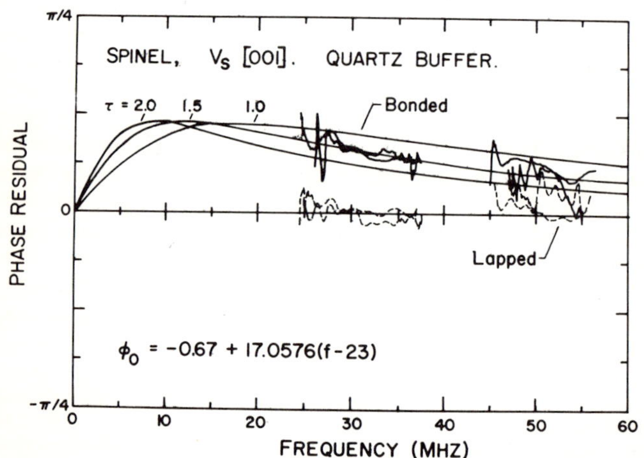

Fig. 12. Measured phase residuals (solid traces) relative to phase with dry lapped buffer-sample contact (dashed traces) for shear [100] waves in spinel, with fused quartz buffer rod bonded to sample, compared with calculated buffer-bond phase shifts (solid curves) for τ_f = 1.0, 1.5, and 2.0 nsec.

relevant properties are given in Table 2. Curves are shown for three different values of τ_f. It can be seen that the effect of thinning the bond is to increase the frequency at which the maximum phase shift occurs. Compared to these in Figure 12 are some measured phase residuals in two frequency intervals (around the third and fifth harmonics of a 10 Mhz transducer). Measurements with a dry lapped contact between the buffer and sample (i.e., no bond) were taken as the standard, and the other measurements referred to a best-fit line through these. Another series of measurements was made with a V9 bond between the buffer and the sample. All runs were duplicated. It can be seen in Figure 12 that the results are consistent with a bond having a value of τ_f between 1 and 2 nsec. For the wave velocity of 1 km/sec measured for V9 (Table 2), this corresponds to a bond thickness between 1 and 2 μ. These results indicate that the phase shift due to the bond has been resolved and that it is described by equation (17) derived from plane wave theory.

Results of similar measurements with compressional waves are not so easily interpreted, however. Figures 13 and 14 compare calculated phase shifts with measured residuals for the cases of

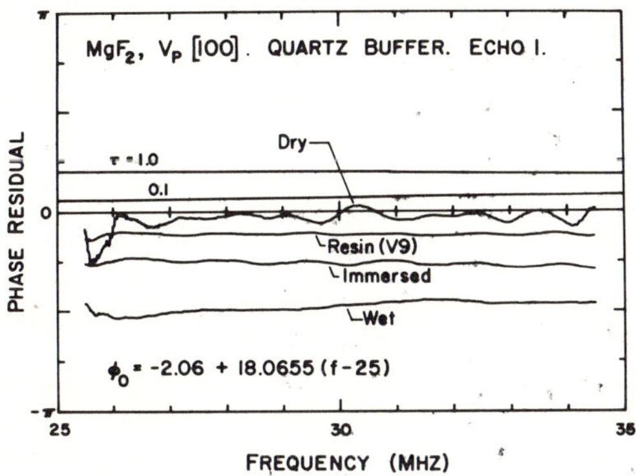

Fig. 13. As in Figure 12, for echo 1 of compressional [100] waves in MgF_2, τ_f = 0.1 and 1.0 nsec, and various buffer-sample contacts (see text).

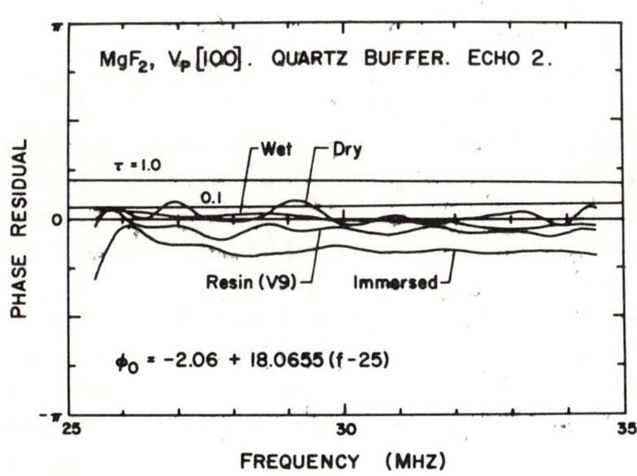

Fig. 14. As in Figure 13, for echo 2.

[100] compressional waves in MgF_2, echoes 1 and 2, with a fused quartz buffer rod. The measurements with the dry lapped contact have again been taken as references for the calculation of residuals. It can be seen that the measured effect of the V9 bond relative to the dry lapped contact is smaller and of opposite sign than expected.

Measurements were made in two other situations. First, in an attempt to get a very thin bond, a dry lapped contact was first achieved and the contact was then squirted with isopentane (used in other experiments as a pressure medium). The result was an immediate enhancement of the signal, indicating that the liquid quickly penetrated the contact. Although isopentane is quite volatile, the enhancement would persist for some time (5 to 15 min) before the signal would drop to its original level relatively rapidly (the order of one minute), indicating that the isopentane had evaporated. The phase shift measured for this "wet" contact was very small for echo 2, but about 90° for echo 1 (Figures 14 and 13)! To test for the possibility that the above procedure had not completely wetted the contact, a dry lapped contact was again established, and the part of the assembly (the sample and part of the buffer rod) was immersed in isopentane. The resulting phase shifts are different again (Figures 13 and 14): they are more negative than with the V9 bond for both echoes 1 and 2. This suggests that the previous procedure had indeed only partially wet the interface.

The explanation for these results is not clear at this stage, but one interpretation will be suggested below. There is an indication that the assumptions of simple plane wave theory may not be appropriate. The maximum calculated phase shift (see Figure 12) is controlled by the impedance ratio of the buffer and sample, Z_b/Z_s (at least, when Z_b is less than Z_s; when Z_b is greater than Z_s, the phase shift increases to π at zero frequency). It is about 40° in Figures 13 and 14 ($Z_b/Z_s = 0.61$) and about 20° in Figure 12 ($Z_b/Z_s = 0.36$). It approaches 90° only as Z_b/Z_s approaches 1.0. Thus, even if the wrong reference were used in calculating the residuals, or if their signs were wrong, the plane wave theory cannot explain the range of measured residuals of about 90° for echo 1 (Figure 13).

It may be seen in Figures 13 and 14 that the measured phase shifts with the V9 bond and the immersed contact are mutually consistent if it is assumed that the "dry lapped" measurements are not the correct reference (i.e., do not correspond to zero phase shift). The V9 bond and immersed results can be matched by the calculated curves by assuming τ_f is about 1 nsec for the V9 bond (comparable to the results found with shear waves, Figure 12) and about 0.1 nsec with the contact immersed in isopentane (consistent with the expectation of getting a much thinner bond in that case). The results for the wet and dry contacts then require some other explanation.

Even with accurately lapped surfaces, only partial contact between the buffer and the sample can be expected (considerable manual force is required to achieve a measurable signal), and presumably this contact is at a series of points scattered across the surfaces. The area of contact would be expected to vary with the normal force across the interface. A passing shear wave would not affect this force, but a compressional wave would. Thus, parts of the surfaces may be in contact, and thus transmitting the stress wave, only near peaks in a compressional wave; further, that part of the wave might be phase delayed because of the time required to bring the surfaces into contact, which might be a considerable fraction of the period of the wave. It is not at all clear at this stage that this kind of mechanism can explain the observations, but it serves to illustrate the possibility that the mechanical coupling at lapped contacts may be complicated and nonlinear. Partial wetting of the interface might produce additional effects, such as "squirting" of the liquid during compressional cycles, which could considerably modify the coupling. Presumably, complete wetting with even a weak couplant, such as isopentane, could eliminate or considerably reduce these effects. On the other hand, even the "dry" contact might be affected by condensation of vapor or by surface films in the interface. Although the surfaces were carefully cleaned with solvent before assembly, no extraordinary precautions were taken to keep them clean and dry. (Some measurements were made under modest vacuum, about 0.01 atm, with no observable effect. We also had to contend with curious colleagues, who wondered if we intended extrapolating to mantle pressures from measurements at 0 and 1 atm.) Until a better understanding of these phenomena is achieved, it seems clear that the results of measurements with buffer rods should be treated with some caution, especially in view of the large phase shifts observed with the dry or "wet" contacts.

These cautionary remarks also apply to measurements of pressure derivatives. It is premature to try to estimate the effect of pressure on the lapped contacts, except to note that, unless they are remarkably unaffected, they could be significant sources of error.

Spetzler et al. [1972] reported a large negative second pressure derivative of the bulk modulus of NaCl, which it is difficult to reconcile with shock compression data to 250 kbar [*Fritz et al.*, 1971], as was discussed by *Spetzler et al.* [1972]. They used a buffer rod, in dry lapped contact with the sample, and a gas (Ar) pressure medium. The quality of the contact may have been somewhat better than in the present experiments because the surfaces were polished flat to within 1/10 wavelength of light, NaCl is relatively soft, the surfaces were very carefully wrung together, and the contact area was only 5 mm in diameter [*Spetzler et al.* 1969a, 1969b, 1972; *Spetzler*, personal communication]. However, their measured zero-pressure velocities differ significantly from the extrapolation of the higher-pressure velocities to zero pressure. This may indicate some problem with the zero-pressure contact which was reduced as the Ar acoustic impedance increased at higher pressures.

Assuming that a bonded contact between the buffer and sample is "well behaved," as suggested above, the effect of pressure on the bond phase shift can be estimated. Bond properties similar to those in Table 3 were again assumed. Some results are illustrated in Figure 15 for compressional [100] waves in MgF_2, a fused quartz buffer, and several carrier frequencies and bond thicknesses.

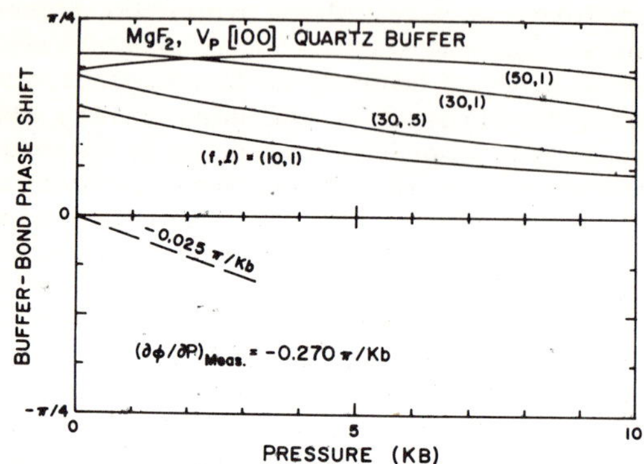

Fig. 15. Estimated effect of pressure on buffer-bond phase shifts for compressional [100] waves in MgF_2, fused quartz buffer rod, τ_f = 0.5 and 1 nsec and carrier frequency f = 10, 30, and 50 Mhz.

Again, the effective thickness of the bond decreases rapidly with increasing pressure. As mentioned earlier, this causes the position of the maximum phase shift to move to higher frequencies. Thus, in some cases, the phase shift first increases with increasing pressure, then decreases; in other cases, it decreases montonically. The magnitude of the effects are significant: for the curves in Figure 15, the slopes are up to 3° kbar. This could amount to 5% of the total phase shift; for the example in Figure 15, the measured phase shift was 49.2° kbar.

The buffer-bond phase shift can be reduced by having a thinner bond (as may be the case for the "immersed" contact discussed above), and by keeping the buffer to sample impedance ratio small. Values of Z_b/Z_s close to or greater than 1 allow phase shifts of up to 90°. (This may be another source of error in the measurements of *Spetzler et al.* [1972], since the relevant impedances of NaCl are probably all less than those of the fused quartz buffer rod used, although it is difficult to evaluate because of the uncertainty in the nature of the bonding.)

VII. CONCLUSIONS

No common source of errors in measurements of pressure

derivatives of elastic properties has been clearly identified in this study, but some possibilities may be suggested. Most measurements seem to have been done with the transducer bonded directly to the sample, and many of those measurements were done by the method described by *McSkimin* [1961] and *McSkimin and Andreatch* [1962]. If their procedure is closely followed, in particular, if the carrier frequency is fixed at the zero pressure (free) resonance frequency of the transducer and the correction made for the change of the resonance frequency with pressure, then the effect of the bond on the pressure derivatives of elastic moduli should amount to less than 0.02. If the transducer correction is neglected, an error of the order of 0.1 might be incurred in the modulus derivative. If, in any experimental procedure in which the transducer is bonded to the sample, the carrier frequency deviates more than a few percent from the transducer resonance frequency, then the bond phase shift might cause significant error, especially if the deviation is at zero pressure, where the bond effects are largest. Measurements reported here demonstrate an error of 5% in the slope (i.e., about 0.25 in the pressure derivative of modulus) incurred by operating a 10 Mhz transducer at 32 Mhz rather than 30 Mhz.

On the other hand, it seems that measurements made with buffer rods between transducers and samples may be subject to significant errors unless special precautions are taken. It has been estimated here that a conventional bond between the buffer and sample could change the measured modulus derivative by the order of 0.25, a very thin bond (such as was apparently achieved here by immersing a lapped contact in a liquid pressure medium) would produce a somewhat smaller effect, perhaps of the order of 0.05. The ratio of buffer to sample acoustic impedance should be significantly different from (preferably less than) unity, to reduce the bond phase shift.

It seems that the (not unreasonable) hope of *Spetzler et al.* [1969*a,b*; also, *Spetzler*, 1970; *Spetzler et al.*, 1972] of avoiding the bond (or interface) phase shift by using a dry lapped contact has not been completely borne out by the results reported here, at least for compressional waves. Dry lapped contact seems to produce substantial phase shifts in compressional waves. Since the mechanism producing the phase shift is not understood at present, its pressure dependence cannot be estimated, but it could easily produce substantial errors in measured pressure derivatives. With what is inferred to have been a partially wetted lapped contact, even larger phase shifts were observed at zero pressure. As indicated above, no anomalous buffer-sample interface phase shifts were observed for shear waves.

Further study, especially with buffer rods, should clarify these problems. In particular, a careful comparison at zero pressure (which was not achieved in the present study) of the relative phases measured with the transducer on the sample and with the various buffer rod-sample contacts should identify unambiguously which of the buffer-sample contacts are producing anomalous phase shifts. Then, more measurements under pressure are required, both to confirm the expected effects of "normal" buffer-sample bonds and to determine the pressure dependence of the "anomalous" interface phase shifts.

Finally, we note again that the effects of transducer bonds can account for significant systematic errors, especially in the measurement of first and higher order pressure derivatives of elastic constants. The results of this paper show that such effects are in accord with theoretical predictions; consequently it is possible to correct for them. But in order to correct for such effects one must know the details of the experimental procedure, such as the operating frequency of the transducer. We strongly recommend that such details be routinely included in published reports, so that the effects of these and possibly other yet unknown systematic errors can be corrected for without repeating the experiment.

Acknowledgments. We are grateful to J. Peter Watt for assistance in computer program development, and to H. Spetzler for useful discussions. This research was supported by the Committee on Experimental Geology and Geophysics, Harvard University, and by National Science Foundation Grant GA38899 (Earth Sciences).

REFERENCES

Anderson, O. L., and P. Andreatch, Jr., Pressure derivatives of elastic constants of single-crystal MgO at 23° and -195.8°C, *J. Amer. Ceram. Soc.*, *49*, 404-409, 1966.

Bartels, R. A., and D. E. Schuele, Pressure derivatives of the elastic constants of NaCl and KCl at 295°K and 195°K, *J. Phys. Chem. Solids*, *26*, 537-594, 1965.

Chang, Z. P., and G. R. Barsch, Non-linear pressure dependence of elastic constants and fourth-order elastic constants of cesium halides, *Phys. Rev. Lett.*, *19*, 1381-1383, 1967.

Chang, Z. P., and G. R. Barsch, Pressure dependence of the elastic constants of single-crystalline magnesium oxide, *J. Geophys. Res.*, *74*, 3291-3294, 1969.

Chang, Z. P., and G. R. Barsch, Pressure dependence of single-crystal constants and anharmonic properties of spinel, *J. Geophys. Res.*, *78*, 2418-2433, 1973.

Drabble, J. R., and R. E. B. Strathen, The third-order elastic constants of potassium chloride, sodium chloride, and lithium fluoride, *Proc. Phys. Soc., London*, *92*, 1090-1095, 1967.

Frisillo, A. L., and G. R. Barsch, Measurement of single-crystal elastic constants of bronzite as a function of pressure and temperature, *J. Geophys. Res.*, *77*, 6360-6384, 1972.

Fritz, I. J., Pressure and temperature dependences of the elastic properties of rutile (TiO_2), *J. Phys. Chem. Solids*, *35*, 817-826, 1974.

Fritz, J. N., S. P. Marsh, W. J. Carter, and R. G. McQueen, The Hugoniot equation of state of sodium chloride in the sodium chloride structure, in *Accurate Characterization of the High Pressure Environment*, edited by E. C. Lloyd, pp. 201-208, NBS Special Publication 326, Washington, D. C., 1971.

Graham, E. K., Jr., and G. R. Barsch, Elastic constants of single-crystal forsterite as a function of temperature and pressure, *J. Geophys. Res., 74*, 5949-5960, 1969.

Kumazawa, M., and O. L. Anderson, Elastic moduli, pressure derivatives, and temperature derivatives of single-crystal olivine and single-crystal forsterite, *J. Geophys. Res., 74*, 5961-5972, 1969.

Manghnani, M. H., Elastic constants of single-crystal rutile under pressure to 7.5 kilobars, *J. Geophys. Res., 74*, 4317-4328, 1969.

Mason, W. P., *Piezoelectric Crystals and Their Application to Ultrasonics*, Van Nostrand, Princeton, New Jersey, 1958.

McSkimin, H. J., Ultrasonic measurement techniques applicable to small solid specimens, *J. Acoust. Soc. Amer., 22*, 413-418, 1950.

McSkimin, H. J., Use of high frequency ultrasound for determining the elastic moduli of small specimens, *IRE Trans. on Ultrasonic Eng., PGUE-5*, 25, 1957.

McSkimin, H. J., Pulse superposition method for measuring ultrasonic wave velocities in solids, *J. Acoust. Soc. Amer., 33*, 12-16, 1961.

McSkimin, H. J., Variations on the ultrasonic pulse-superposition method for increasing the sensitivity of delay-time measurements, *J. Acoust. Soc. Amer., 37*, 864-871, 1965.

McSkimin, H. J., and P. Andreatch, Jr., Analysis of the pulse superposition method for measuring ultrasonic wave velocities as a function of temperature and pressure, *J. Acoust. Soc. Amer., 34*, 609-615, 1962.

McSkimin, H. J., P. Andreatch, Jr., and R. N. Thurston, Elastic moduli of quartz versus hydrostatic pressure at 25° and -195.8°C, *J. Appl. Phys., 36*, 1624-1632, 1965.

Peselnick, L., R. Meister, and W. H. Wilson, Pressure derivatives of elastic moduli of fused quartz to 10 kb, *J. Phys. Chem. Solids, 28*, 635-639, 1967.

Redwood, M., and J. Lamb, On the measurement of attenuation in ultrasonic delay lines, *Proc. Inst. Elec. Engrs., Pt. B., 103*, 773-780, 1956.

Spetzler, H., Equation of state of polycrystalline and single-crystal MgO to 8 kilobars and 800°K, *J. Geophys. Res., 75*, 2073-2087, 1970.

Spetzler, H., C. G. Sammis, and R. J. O'Connell, Equation of state of NaCl: ultrasonic measurements to 8 kbar and 800°C and static lattice theory, *J. Phys. Chem. Solids, 33*, 1727-1750, 1972.

Spetzler, H., E. Schreiber and D. Newbigging, Coupling of ultrasonic energy through lapped surfaces at high temperature and pressure, *J. Acoust. Soc. Amer.*, *45*, 1057-1058, 1969*a*.

Spetzler, H., E. Schreiber, and L. Peselnick, Coupling of ultrasonic energy through lapped surfaces: application to high temperatures, *J. Acoust. Soc. Amer.*, *45*, 520, 1969*b*.

Swartz, K. D., Anharmonicity of sodium chloride, *J. Acoust. Soc. Amer.*, *41*, 1083-1092, 1967.

Williams, J., and J. Lamb, On the measurement of ultrasonic velocity in solids, *J. Acoust. Soc. Amer.*, *30*, 308-313, 1958.

A NOVEL DEVICE TO REACH HIGHER PRESSURE IN LARGER VOLUME

M. KUMAZAWA
Department of Earth Science
Nagoya University
Nagoya, Japan

Abstract

The present stage of the development in MASS-type, high-pressure generating apparatus is reported, and the technical prospects of higher pressure generation in a larger volume are discussed.

I. INTRODUCTION

MASS (Multiple-Anvil Sliding System) is a group of novel mechanisms of reducing the volume of a closed space. The basic idea of this mechanism appeared earlier in some literature [*e.g.*, *Hall*, 1960; *Epain et al.*, 1967]. Independently of these earlier contributions, the present author found this mechanism in 1969 and judged it quite invaluable in generating very high pressure for geophysical research.

However, the practical apparatus utilizing the new mechanism had not been in successful operation until a systematic and comprehensive analysis was made on the mechanisms in three dimension [*Kumazawa*, 1971]; the practical ways of driving the mechanism [*Kumazawa, et al.*, 1972]; the mechanics of a pressure-generating system driven by an external force [*Kumazawa*, 1973]; and also a number of test experiments on the different types of MASS mechanisms and their driving systems [*Masaki*, et *al.*, 1974].

The purpose of the present paper is to describe the present MASS apparatus in our laboratory and to discuss the future development of pressure-generating devices in connection with the geophysical requirements in high-pressure experimentation.

II. MASS MECHANISMS

The basic principle of MASS is shown by two-dimensional schematic illustrations in comparison with the other mechanisms

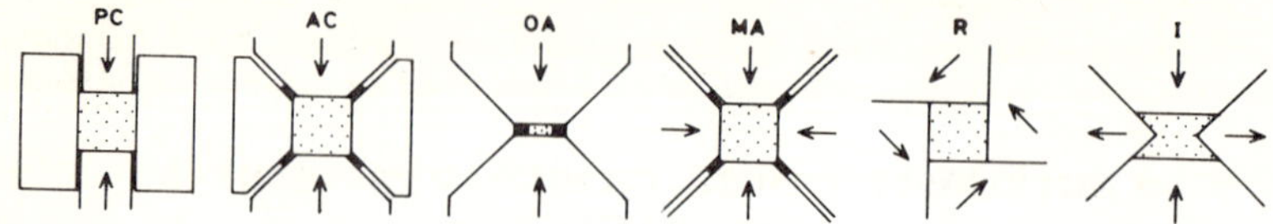

Fig. 1. Schematic, two-dimensional illustration of pressure-generating mechanisms. PC = piston-cylinder type; AC = anvil-cylinder type (belt and girdle); OA = opposed anvil type; MA = multiple anvil type (tetrahedral, cubic, octahedral, etc.); I = irrotational type of MASS; R = rotational type of MASS.

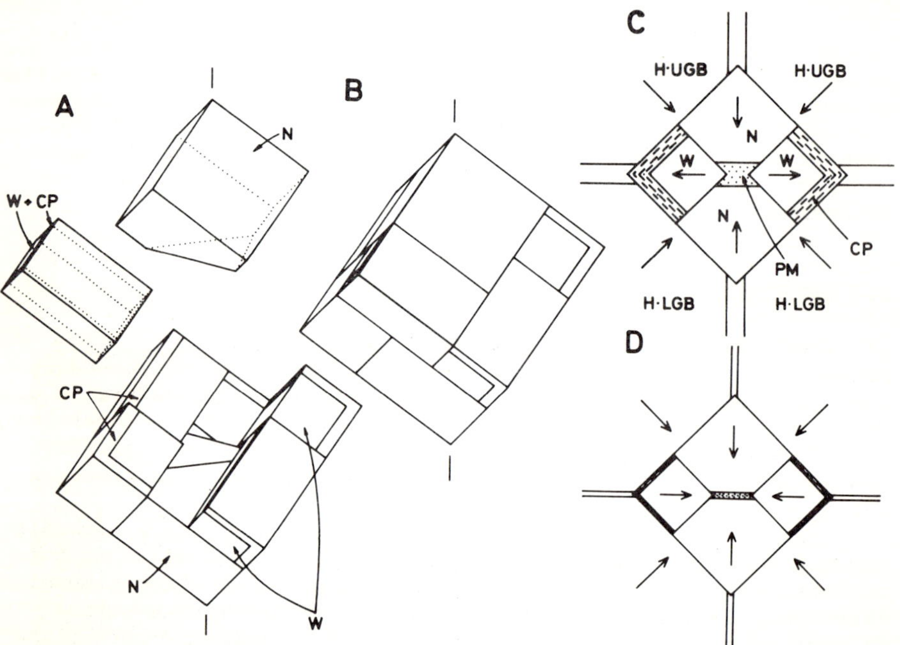

Fig. 2. Geometry of MASS 3I8-90 type anvil assembly and the schematic illustration of operational mechanism. A and B; Perspective view of anvil assembly. In A, a normal anvil (N), a wedge anvil (W) combined with two compressible pads (CP) and pressure medium are removed from the cube-shaped assembly B. Pressure medium is set between the two truncated faces of normal anvils. C and D; Operational mechanism of cube-shaped assembly of 3I8 surrounded by heads of upper guide block (H-UGB) and of lower guide block (H-LGB). Arrows in C show the relative displacements of components during the pressurization. Arrows in D show the force at the pressurized state with a reduction of thickness in pressure medium (PM) and compressible pads.

known so far (Figure 1). There are more than a dozen different types of three-dimensional MASS mechanisms, which seem to have practical use. A two-dimensional illustration of three-dimensional illustration of three-dimensional mechanisms (both in geometry and operation) is difficult and too technical for this paper. Therefore, a simple illustration of 3I8-90 type anvil assembly alone is shown in Figure 2. Readers interested in the actual perspective view and the nature of the other types of MASS mechanisms in three dimension are urged to make plastic models by referring to *Kumazawa* [1971] and *Kumazawa et al.* [1972] and to manipulate them by themselves.

The advantages of MASS mechanisms over any other pressure-generating mechanism reported hitherto are (1) each component (anvil) has a geometry quite favorable for massive support and lateral support to utilize the strength of material to the maximum extent, and (2) the volume of a closed space is subjected to reduction in three dimension without any spatial limitation. Previously the MASS mechanism was considered to have a disadvantage in that too many anvils, having complicated geometry, had to be driven synchronously by many hydraulic rams. However, an introduction of guide blocks and compressible pads has resolved this difficulty. The MASS mechanism is, the present author believes, the best one among all mechanisms ever known for generating very high static pressure in a larger volume by means of external force.

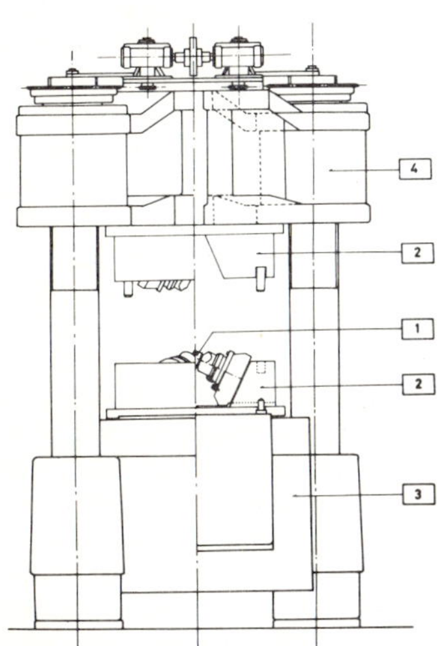

Fig. 3. Drawing of a press system for MASS apparatus. 1 = MASS anvil assembly; 2 = guide block; 3 = hydraulic ram; 4 = movable upper platen of uniaxial press.

Fig. 4. Photograph of lower guide block (800 mm in diameter). 1 = head of guide block (60 mm in edge length); 2 = power lead; 3 = electric lead for thermocouple.

III. THE ACTUAL APPARATUS IN OPERATION

The illustration of the presently working MASS apparatus for routine use in our laboratory is shown in Figure 3, and its photograph is shown in Figure 4.

A couple of RH3-type guide blocks (monoblock construction) are in the daylight of the 2,000-ton uniaxial press. This system works as a cubic press in a similar operation to that of the SLIDE press [*Ichinose et al.*, 1975]. The three heads of cubic anvils in each of the guide blocks (upper and lower) are replaceable if any suffers from failure or if the MASS anvil assembly of a different size is to be driven. The MASS 3I8-90-type anvil assembly [*Masaki et al.*, 1975] is contained in a couple of guide blocks and compressed. The anvil assembly is a cube in shape and is set [1 1 1] axis vertical (Figure 2). Therefore, the whole system works as a two-stage cascading compressor. The first stage (cube; 20, 40, 60, or 80mm in edge length) just inside the guide blocks is compressed up to 20 kbar, and the second stage by a 3I8-90 goes up to several hundred kbar.

In the present system, the initial volume of the pressure medium can be made to be from 5 to 50mm^3, depending on the required condition of the experiments. The electric furnace, thermocouple wires, and specimen of 1mm^3 are easily set in a volume of 20mm^3, and the high-pressure and high-temperature experiments are made successfully.

The pressure values reached by the present apparatus is calibrated with the uniaxial force of the ram at room temperature by means of detecting the change in electric resistance associated with the phase transitions in some metals and semiconductors. The force-pressure relation thus obtained is not linear, and pressure-generating efficiency is lowered with the increase of press load, however, an appropriate extrapolation to higher pressure gives 500 kbar in a volume of 20mm^3 at the press load of 1,000 tons. The highest reliable calibration point determined by this method is 220 kbar (metallic transition in gallium phosphide [*Piermarini and Block*, 1975]).

An attempt to detect metallic transition in BP (predicted to occur at 400 kbar by *Van Vechten* [1973] has been made in a reduced volume (5 mm^3) of the pressure medium [*Fukao et al.*, unpublished manuscript, 1976]. A half dozen runs invariably showed that the electric resistivity of BP decreased gradually with pressure up to 150-300 kbar, increased after the minimum, and then showed a sharp drop by several orders of magnitude at a pressure estimated to be above 500 kbar. However, the sharp drop in the electric resistivity is not convincingly identified with the metallic transition in BP since there remained a possibility of a short circuit in the pressure cell. According to personal communication from *Piermarini*, BP does not transform to a metallic state below 575 kbar.

Although there is a maximum limit of the pressure attainable with cemented WC anvils in a given volume, the technical improvements in the constitution of compressible pads and other factors are still raising the pressure-generating efficiency.

High temperature in the pressure cell is attained by electrical resistance heating. The metal-heating elements proved unsuccessful, probably because of large shear deformation and the small size of the pressure cell. Therefore, a carbon heater is being used for the present. However, the attainable temperature in the cell is limited to 1,500°C as a result of very fast conversion of carbon to diamond above this temperature. The use of appropriate material for a heating element will resolve the present temperature limit.

The thermocouple wires for temperature measurement are taken out through the anvils, which are electric conductors. However, as many electric leads as needed can be taken out through the small gaps (filled with a very thin electric insulator, such as mica sheet) between the anvils in contact.

The present apparatus and its earlier primitive versions have been used to carry out the phase change experiments on the geophysically interesting materials (for Mg_2SiO_4, *Kumazawa,et al.*, 1974; for Fe_2SiO_4, *Sawamoto et al.*, 1974; for $MgAl_2O_4$, *Ohtani et al.*, 1974; and for $(Mg,Fe)SiO_3$, *Sawamoto*, 1977) above 150 kbar.

IV. PRESSURE AND VOLUME OF PRESSURE-GENERATING APPARATUS

The attainable pressure and volume of the pressure cell are the important characteristics of a pressure-generating device. The static force, F, to be used to generate pressure, P, in a volume, V, by means of material with strength, τ (sustainable maximum stress difference), are theoretically related to each other [*Ruoff*, 1972; and *Kumazawa*, 1973]. Using the theory of conditional pressurization [*Kumazawa*, 1973], the functional relation $V(P, F, \tau)$ in a spherical pressure-generating system is calculated and shown in Figure 5 for two representative cases. The material strength is assumed to increase linearly with an equivalent hydrostatic pressure $\bar{P}$;

$$\tau = \tau_0 + a\bar{P} \tag{1}$$

The present author does not believe that material strength increases linearly with pressure up to very high pressure. However, the strength is a property related directly to rigidity, which has usually a positive value for pressure-derivatives. Therefore, we can expect the material strength increases with pressure. A simple linear relation (1) is useful in knowing the effect of pressure-dependent strength. The result of simple arithmetic manipulation of force balance equation is represented by

$$V^{2a}P^{3+2a}F^{-3a} \cong (2^2\ 3^5\pi)^{-a}\ (3/a + 2)^{2a+3}\ (3-a)^{a-3}\ \tau_0^{3-a} \qquad (2)$$

at very high pressure $P > \tau > \tau_0$. Here F is a total force $4\pi P_0 R^2$ to compress a spherical system of outer radius, R, by means of external pressure, P_0. The possible maximum value of P_0 is $3\tau_0/(3-a)$, which is the limit of geometric pressurization. Since we can employ the reaction of static force to compress the system in an actual apparatus, the effective value of F is larger than the force actually used up to a factor of 6.

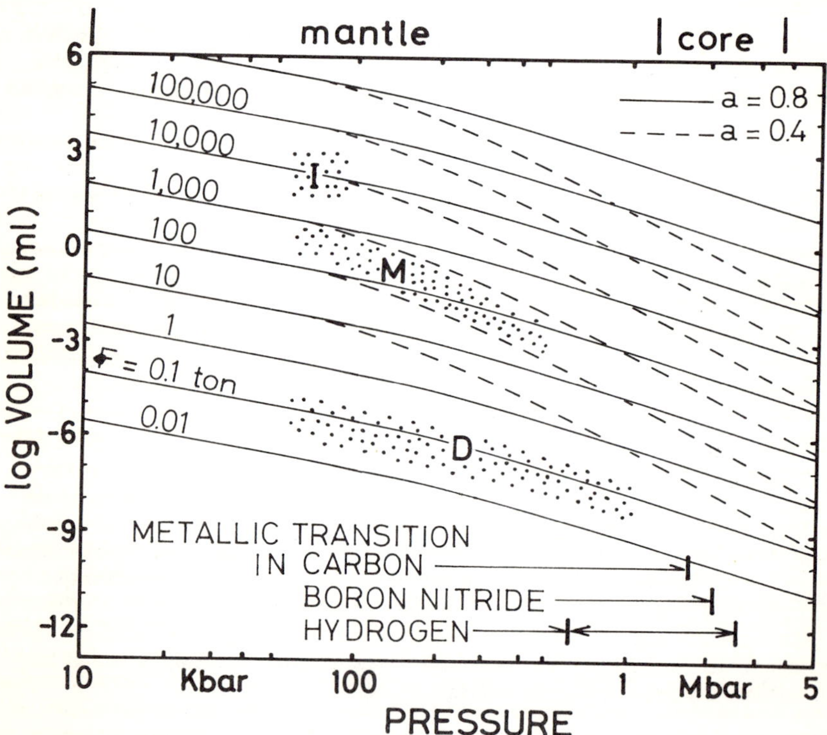

Fig. 5. Pressure-volume relation of several different types of high-pressure apparatus. I = device for industrial use; M = anvil cylinder, multiple anvil, and multiple anvil sliding system with tungsten carbide anvil; D = opposed anvil-type diamond cell. The theoretical relation of pressure and volume to the static force used for generating pressure is also shown for two cases of pressure-dependency of material strength. See text for details.

As is inferred from equation(2) and also from Figure 5, the three important factors to attain higher pressure in larger volume are the large force utilized, high strength of material, and also the large pressure derivative of strength.

By referring to a number of papers describing high-pressure devices, the reported values of pressure and volume are critically evaluated and shown by dotted areas for three representative groups of devices.

The apparatus for industrial use (region I in Figure 5 *e.g.*, for diamond synthesis) has very large volume (10^2 - $10^3 cm^3$) for efficiency. Usually, the anvil-cylinder type (belt or girdle) made of tungsten carbide or hardened steel is used; however, large presses (10^3 - 10^4tons or more) are needed as a basic facility as a result of volume requirement.

An extreme case, in contrast to the above, is the opposed diamond-anvil apparatus, which employs very hard material and very small force (0.1 - 1 tons) to compress a very small volume (10^{-4} - $10^{-8} cm^3$). The diamond-anvil device has almost a 30-year history of development, and it has been reported recently to have reached pressures exceding 1 Mbar [*Mao and Bell*, 1976]. The diamond-anvil cell is essentially an opposed anvil-type apparatus and there is a geometric limitation in volume despite the high strength of material. However, it has an invaluable advantage in that small size and the transparency of diamond enable it to be used for a number of important measurements for basic research.

The intermediate devices between the two groups mentioned above are the multiple anvil (tetrahedral, cubic, and octahedral), anvil-cylinder (belt and girdle), and MASS. This group of apparatus has been used for most of the high-pressure and high-temperature phase-change experiments ever published (except the piston-cylinder type, used below several 10 kbar) in the basic research fields. Although the attained maximum pressure is not confirmed with reliability, it is estimated to be 300-500 kbar for the present. In this category of apparatus, the material is usually cemented WC, simply because carbide is the hardest material available commercially with a sufficient size. The pressure limit of these types of apparatus is partly due to the smaller strength of anvil material. However, the most improtant point originates from the volume requirement, since larger volume is an essential specification of high-pressure apparatus for a number of experiments. If we do not insist on larger volumes of the pressure cell and the pressure-generating device is driven by a sufficiently large force for lateral support at the flank of anvils, the high pressure as 1 Mbar could be generated as expected from Figure 5 even with WC anvils. Another limitation of pressure generation in some apparatus in this group is a geometric restriction of the amount of compression (that is, the anvil-cylinder and multiple-anvil type devices). It is noted again that MASS is free from this geometric limitation.

By comparison of the theoretically calculated relation of

$V(P,F,\tau)$ with those attained in different groups of apparatus, the pressure derivatives of strength of diamond and of tungsten carbide are respectively estimated to be about 0.8 and 0.4. It is noted that the usefulness of diamond for anvil material lies not only in the high strength at atmospheric condition but also in its large pressure-derivative. The search for a material with large pressure derivative of strength is considered to be very important.

Figure 5 shows that the very large force has to be used in order to reach a given target of high pressure in a very large volume. Then the whole size of high-pressure apparatus would be necessarily huge.

V. SUMMARY AND CONCLUSION

When we compare the present capabilities of diamond-anvil device and MASS apparatus currently used, the diamond-anvil generates higher pressure. The usefulness of the diamond-anvil to make exploratory research on new materials and new phenomena is convincingly accepted. On the other hand, much larger volumes--several orders of magnitude -- can be pressurized by means of MASS apparatus, even with tungsten carbide anvils. Therefore, both types of apparatus are complementary in geophysical research.

However, if MASS apparatus is made of diamond, the potentiality of MASS mechanism is undoubtedly larger than the simple opposed-anvil-type diamond cell. MASS was invented very recently and has been developed in a few laboratories only for the past several years. Further development of MASS apparatus is anticipated.

A crucial problem in the present MASS-3I8-90 apparatus is that the generated pressure is not satisfactorily determined. In the case of MASS-OB8 or OC8 mechanisms, however, it is quite possible to devise a sufficient window for *in situ* X-ray measurement of a compressed specimen. Our present effort is being directed to development of MASS apparatus equipped with an X-ray system.

The predicted pressures of metallic transition in hydrogen (considered in many theoretical predictions), diamond, and cubic boron nitride [*Van Vechten*, 1973] are also shown in Figure 5. If the higher estimation of the metallic transition in hydrogen is right, and the metallic phase of carbon and boron nitride is mechanically soft as expected, the laboratory production of metallic hydrogen by static methods is quite a difficult task, unless some other anvil material is found. However, the author is quite optimistic about the appropriate approach in the future and in realizing the state of Jupiter's mantle in MASS apparatus. We are now constructing a larger MASS apparatus (with a press of 10^4 ton capability) in order to reach higher pressure in larger volume.

Acknowledgments. I would like to thank my colleague Hiroshi Sawamoto for his help in every respect of the high-pressure research program in my laboratory and F. P. Bundy for his valuable comments on this paper. I am grateful to Y. Shimazu, S. Akimoto and N. Kawai for their support, and to The Mitsubishi Foundation and the Japanese Ministry of Education for financial support of this work.

REFERENCES

Epain, R., C. Susse, and B. Vodar, Appareils étanches, a coins mobiles, pour production de trés hautes pressions statiques en milieu solide, *C. R. Acad. Sc. Paris, Serie A, 265*, 323-326, 1967.

Hall, H. T., High pressure methods, in *High Temperature Technology*, pp. 145-156, 335-336, McGraw-Hill, New York, 1960.

Ichinose, K., M. Wakatsuki, and T. Aoki, A new sliding type cubic anvil high pressure apparatus, *Japan High Pressure Institute 13*, 244-253,1975.

Kumazawa, M., Multiple-anvil sliding system - a new mechanism of producing very high pressure in a large volume, *High Temperatures-High Pressures, 3*, 243-260, 1971.

Kumazawa, M., Theory of generation of very high static pressure by an external force, *High Temperatures-High Pressures, 5*, 599-610, 1973.

Kumazawa, M., K. Masaki, H. Sawamoto, and M. Kato, Guide blocks and compressible pads for the practical operation of multiple-anvil sliding system for the production of high pressure, *High Temperatures-High Pressures, 4*, 293-310,1972.

Kumazawa, M., H. Sawamoto, E. Ohtani, and K. Masaki, Postspinel phase of forsterite and evolution of the earth's mantle, *Nature, 247*, 356-358, 1974.

Mao, H. K., and P. M. Bell, High-pressure physics, the 1-megabar mark on the ruby R_1 static pressure scale, *Science, 191*, 851-852, 1976.

Masaki, K., H. Sawamoto and M. Kumazawa, High pressure generation by MASS apparatus, *Proc. 4th Internat. Conf. High Pressure, Kyoto*, 832-839, 1974.

Masaki, K., H. Sawamoto, E. Ohtani, M. Kumazawa, M. Machida, S. Mizukusa and M. Nakayama, High pressure generation by MASS 3I8-90 type apparatus, *Rev. Sci. Instr., 46*, 84-88, 1975.

Ohtani, E., H. Sawamoto, K. Masaki and M. Kumazawa, Decomposition of spinel $MgAl_2O_4$ at extremely high pressure, *Proc. 4th Internat. Conf. High Pressure, Kyoto*, 185-189, 1974.

Piermarini, G. J., and S. Block, Ultrahigh pressure diamond-anvil cell and several semiconductor phase transition pressures in relation to fixed point pressure scale, *Rev. Sci. Instr. 46*, 973-980, 1975.

Ruoff, A. L., Penultimate static pressure containment considerations and possible applications to metallic hydrogen preparation, *Advances Cryogenic Eng.*, *18*, 435-440, 1972.

Sawamoto, H., Orthorhombic perovskite $(Mg,Fe)SiO_3$ and constitution of the lower mantle, *High Pressure Research; Applications to Geophysics*, edited by M. H. Manghnani and S. Akimoto, Academic Press, New York, 219-244, 1977.

Sawamoto, H., E. Ohtani, and M. Kumazawa, High pressure decomposition of γ-Fe_2SiO_4, *Proc. 4th Internat. Conf. High Pressure, Kyoto*, 194-201, 1974.

Van Vechten, J. A., Quantum dielectric theory of electronegativity in covalent system. III. Pressure-temperature phase diagrams, heats of mixing, and distribution coefficients, *Phys. Rev.*, *B*, *7*, 1479-1507, 1973.

PRESSURE CALIBRATION ABOVE 100 KBAR BASED ON THE NaCl INTERNAL STANDARD

T. YAGI and S. AKIMOTO
Institute for Solid State Physics, University of Tokyo
Roppongi, Minato-ku, Tokyo 106, Japan

Abstract

Transformation pressures of seven materials -- Ba, Pb, ZnSe, ZnS, GaAs, and GaP -- were studied, employing NaCl as the internal pressure standard. A cubic-anvil-type, high-pressure electrical-resistance apparatus was used in connection with an X-ray diffraction system. Based on Decker's equation of state for NaCl, transition pressures (in units of kbar) are:

Material	Transformation Pressure, kbar	
	Increasing Pressure Cycle	Decreasing Pressure Cycle
Ba	126	112
Pb	142	111
Si	133	-
ZnSe	139	105
ZnS	162	105
GaAs	193	107
GaP	>233	-

Present results are approximately 7% higher than those based on the ruby scale. The relation between present results and the results based on the new Drickamer scale is complicated. The effect of nonhydrostatic component of pressure is discussed.

I. INTRODUCTION

The establishment of an accurate pressure scale in the hundreds of kbar range is increasingly essential for the further refinement of ultrahigh-pressure research. Recently *Piermarini and Block* [1975] proposed a new pressure scale based on the shift of the R_1 ruby fluorescence line, calibrated against the compression of NaCl with pressures derived from Decker's equation of state. The absolute accuracy of that scale is, of course, open to question since it is based on an equation of state and not on an absolute measurement. They determined the transition pressure of several materials such as Si, ZnSe, ZnS, and GaP, and they found that a large discrepancy exists between their ruby scale and the new Drickamer scale [*Drickamer*, 1970] above 135 kbar. The disagreement may be as much as a factor of 2 in the 500-kbar range. *Bundy* [1975] studied the pressure-fixed points with a Drickamer-type electrical-resistance apparatus, and he also obtained the result that the new Drickamer scale needs further revision downward for the range above 150 kbar.

In this paper, we report a set of new determinations of the pressure-fixed points between 100 and 230 kbar based on the NaCl internal standard technique. In previous papers we have determined the transformation pressures of Ba [*Akimoto et al.*, 1975], Pb, ZnS, and GaAs [*Yagi and Akimoto*, 1976] by means of an X-ray diffraction technique using NaCl as an internal pressure standard and an electrical-resistance measurement for detecting the transformation. In this paper, we present additional data on Si, ZnSe and GaP, and compare all the results obtained by X-ray diffraction technique with other results obtained by different techniques.

II. EXPERIMENTAL METHODS

Experiments were made with a cubic-anvil type, high-pressure electrical-resistance apparatus connected with an X-ray diffraction system to measure the change in lattice parameter of NaCl. We have made various improvements on the pressure-transmitting media and the sample assembly and successfully extended the pressure limits to 230 kbar. The experimental procedure used for determining the transformation pressure is, in principle, the same as developed by *Jeffery et al.* [1966] for pressure calibration to 100 kbar. The details have been described in an earlier paper [*Yagi and Akimoto*, 1976].

Two types of cemented WC anvils with square faces of 2 mm and a 1-mm edge length were used for the present study. A cube-shaped pressure-transmitting medium was made of a mixture of amorphous boron and epoxy resin in the ratio 2:1 by weight. The edge length was 3.75 mm and 2.50 mm for 2 mm and 1-mm anvils, respectively. Although this material improved the efficiency of producing very high pressure, the net amount of self-formed gaskets under compression is rather reduced owing to the material hardness. In order to avoid the stress concentration at the top of the anvils and to keep away from the probable shift of the specimen during an experiment, pre-formed gaskets with softer material were used. A mixture of amorphous boron and epoxy resin in the ratio 1:2 by weight was used for the pre-formed gaskets.

In the sample assembly shown in Figure 1, the NaCl internal-pressure standard was inserted in the center hole of the boron/epoxy resin cube. Calibrant materials for pressure-fixed points were sandwiched between two columns of NaCl to ensure equal pressure in both the calibrant and NaCl. A ribbon-shaped specimen was used for metals such as Ba and Pb, and single crystal was used for semiconductors such as ZnS, GaAs, and GaP.

The pressure-induced phase transformation of the calibrant material was detected at room temperature by observing the change in the electrical resistance with pressure. In the case of Ba and Pb, a four-probe technique was developed to eliminate contact resistances and to achieve high sensitivity on resistance measurements. The arrangement of the gold current lead, the gold potential lead, and the metal specimen is schematically illustrated in Figure 1. A two-wire method was adopted in resistance measurements of semiconductors, because the change in the resistance amounts to several orders of magnitude in these semiconductor-to-metal transitions. The pressure was increased at the rate of approximately 1 kbar/min near the phase transformation. When the transition was observed by the resistance measurement, the ram load of the cubic-anvil press was held constant, and the lattice parameter of the NaCl internal pressure standard was then determined.

A Mo-Kα radiation generated by a rotary-target-type, high-power X-ray source (operated at 50 kV and 160 mA) entered the sample chamber through a gap in the anvils. Diffracted X-rays passed through the opposite gap and were measured by a fixed time step scanning method using a scintillation counter. In order to reduce the error caused by a probable shift of the zero point, 4θ angle instead of 2θ were measured for each diffraction line by

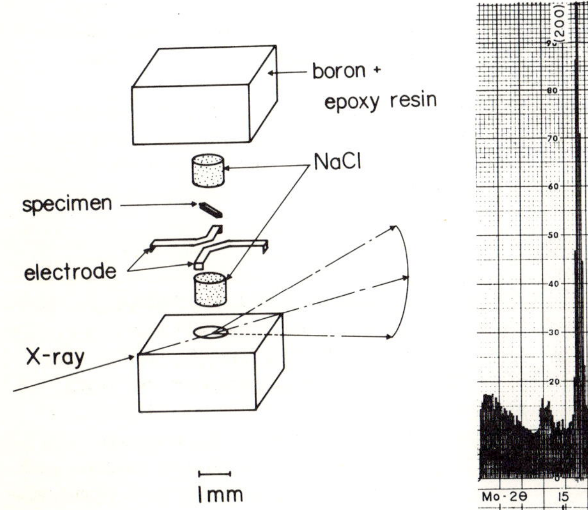

Fig. 1. Sample assembly used for the present study.

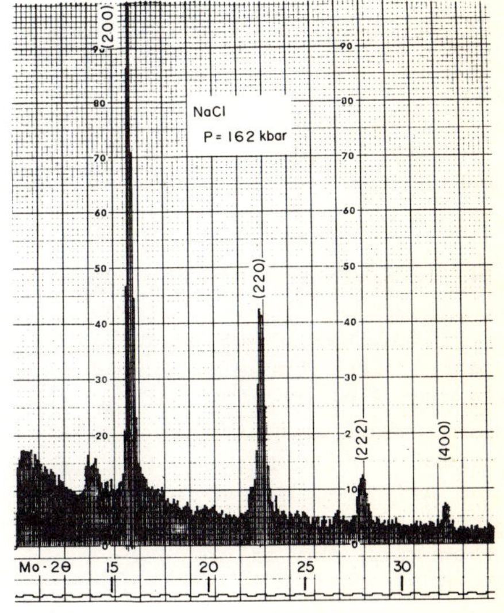

Fig. 2. X-ray diffraction profile of NaCl observed at 162 kbar.

TABLE 1. Experimental Results of the Transition Pressures Based on the NaCl Internal Pressure Standard

Material	Increasing Pressure Cycle		Decreasing Pressure Cycle	
	NaCl-$\Delta a/a_o$	P, kbar	NaCl-$\Delta a/a_o$	P, kbar
Ba	0.0871(6)[a]	126(2)	0.0810(1)	112(1)
Pb	0.0931(8)	142(2)	0.0807(8)	111(2)
Si	0.0898(9)	133(2)	-	-
ZnSe	0.0921(4)	139(1)		110-100
ZnS	0.0999(10)	162(3)		110-100
GaAs	0.1096(6)	193(2)		115-100
GaP		>233	-	-

a. The standard deviations in parentheses represent the scatter of the measurements and are expressed in units of the last digit.

scanning the counter to both sides of the direct beam.

Three diffraction lines of NaCl -- (200), (220), and (222) -- were used in determining the lattice parameter, and the pressure value was calculated according to Decker's NaCl pressure scale at 25°C [*Decker*, 1971]. The angle width of a step was adopted as 0.05°, and the fixed time of a step was 8-20 sec above 150 kbar. The diffraction profile of NaCl was displayed on a pen-recorder chart (Figure 2) simultaneously punched out on a paper tape, and analyzed by a computer program developed by *Ida* [*Yagi et al.*, 1975]. The 2θ-angle of the diffracted lines of NaCl was calculated by the least-squares fit of the Gaussian curve to the diffraction profile.

The precision of this X-ray diffraction system is approximately ± 0.04% for lattice parameter determination, corresponding to ± 0.7% for pressure values above 100 kbar.

III. RESULTS AND DISCUSSION

All the experimental results are summarized in Table 1, together with the previous data for the high-Ba transition [*Akimoto et al.*, 1975] and for the transitions in Pb, ZnS, and GaAs [*Yagi and Akimoto*, 1976]. The lattice parameter of NaCl at normal condition, a_o, was determined from the observation at 1 bar after each compression-decompression cycle. It was in precise agreement of ± 0.04% from measurement to measurement and also with the value given on the ASTM card. The lattice compression of NaCl, $-\Delta a/a_o$, at the transition listed in Table 1 is calculated from the arithmetical mean of several measurements, and the standard deviation in parenthesis represents the scatter of the values from measurement to measurement. In each measurement the lattice parameter was determined by averaging three values from d_{200}, d_{220}, and d_{222}.

In Table 2, we summarized the transition pressure data reported by various investigators. Detailed comparison of the experimental results for Ba, Pb, ZnS, and GaAs was made in an earlier paper [*Yagi and Akimoto*, 1976]. We will now make a brief comparison of the present data for Si, ZnSe, and GaP with previous investigator's data.

As shown in Tables 1 and 2, the most probable value for the transition pressure of Si is 133 ± 3 kbar for the increasing pressure cycle. In the increasing pressure cycle, electrical resistance of the specimen drops sharply from 10^5 Ω to 1 Ω; therefore, we can easily determine the transition pressure precisely. On the other hand, in the pressure-decreasing cycle, the reverse reaction is very sluggish, and it is very difficult to determine the transition pressure.

A pressure-induced phase transition in Si was first reported by *Minomura and Drickamer* [1962] at 195-200 kbar. When the new Drickamer scale is used, this value is lowered to about 150 kbar. *Piermarini and Block* [1975] reported the transition pressure of Si to be 125 ± 5 kbar based on the ruby scale and determined in a diamond-anvil apparatus. It is remarkable that in the gasketed diamond-anvil apparatus and the cubic-anvil apparatus, the metallic transition in Si takes place at a lower pressure than the transition in Pb. On the other hand, in a Drickamer anvil-type apparatus, which is considered to generate less hydrostatic pressure, the transition pressure in Si is higher than that of Pb. This fact indicates that, as *Minomura and Drickamer* [1962] already pointed out, the transition pressure in a covalent crystal depends largely on the stress state of the crystal. Therefore, such a covalent crystal can be used as a pressure-calibration material only when special caution is taken with respect to the pressure environments. On the contrary, a covalent crystal might be available as a sensor of the nonhydrostatic component of pressure.

The metallic transition in ZnSe was found at 139 kbar in the pressure increasing cycle and 100-110 kbar in the pressure decreasing cycle. This result for ZnSe is, however, only a preliminary one, because the purity of the specimen is not clear, and we made only one measurement for this material. The transition in ZnSe to a conducting state was first found by *Samara and Drickamer* [1962], at 165 kbar, and this value is lowered to 134 kbar when we use the new Drickamer scale. *Piermarini and Block* [1975] observed the same transition at 137 ± 4 kbar based on the ruby scale. The present result is in agreement with these values.

Although we have made several measurements for GaP to 233 kbar, we could not observe the metallic transition at all up to this pressure. The phase transition in GaP was first found by *Onodera et al.* [1974] using a split-sphere apparatus. The transition pressure was determined to be 220 ± 10 kbar by *Piermarini and Block* [1975] based on the ruby scale, and this value is in

TABLE 2. Summary of Transition Pressure Data

Material	Transition Pressure, kbar Increasing	Decreasing	Pressure Device	Source
Ba	118-122	-	Drickamer cell	*Drickamer* [1970]
	126±2	112±2	cubic press	*Akimoto et al.* [1975]
Pb	128-132	-	Drickamer cell	*Drickamer* [1970]
	142±3	111±2	cubic press	*Yagi and Akimoto* [1976]
Si	195-200	-	Drickamer cell	*Minomura and Drickamer* [1962]
	150-155	-	Drickamer cell	*Drickamer* [1970]
	125±5	-	diamond cell	*Piermarini and Block* [1975]
	133±3	-	cubic press	This work
ZnSe	165	-	Drickamer cell	*Samara and Drickamer* [1962]
	134	-	Drickamer cell	*Drickamer* [1970]
	137±4	-	diamond cell	*Piermarini and Block* [1975]
	139±4	110-100	cubic press	This work
ZnS	240-245	-	Drickamer cell	*Samara and Drickamer* [1962]
	180-185	-	Drickamer cell	*Drickamer* [1970]
	150±5	-	diamond cell	*Piermarini and Block* [1975]
	162±4	110-100	cubic press	*Yagi and Akimoto* [1976]
GaAs	240-250	-	Drickamer cell	*Minomura and Drickamer* [1962]
	180-190	-	Drickamer cell	*Drickamer* [1970]
	180±8	-	diamond anvil	*Piermarini and Block*[a]
	193±5	115-100	cubic press	*Yagi and Akimoto* [1976]
GaP	220±10	-	diamond cell	*Piermarini and Block* [1975]
	230-240	-	Drickamer type	*Bundy* [1975]
	>233	-	cubic press	This work

a. Private communication.

good agreement with the theoretical estimate of 216 kbar by *Van Vechten* [1973]. *Bundy* [1975] has also determined the transition in GaP using compacted sintered diamonds as Bridgman anvils in an apparatus designed for electrical resistance measurements. He places the transition in the 230-240 kbar range. *Homan et al.* [1975] has also measured the transition in GaP electrically with a variable lateral support Bridgman anvil type press and reports the transition at 220 kbar. Based on the present results, however, it is higher than 233 kbar. If we extrapolate the systematic discrepancy between present results and ruby scale, as will be discussed later, we can estimate the transition pressure to be approximately 240 kbar.

In Figure 3, we compare the present results with results based on the ruby scale and the new Drickamer scale. The horizontal axis is the transition pressure of calibrant materials determined in the present study, and the vertical axis is the difference between the present result and other results. It is clear from Figure 3 that the relationship between the present results and the new Drickamer scale is complicated. For example, the present results for the metals are 5-10% higher than those from the new Drickamer scale, while some of the results for

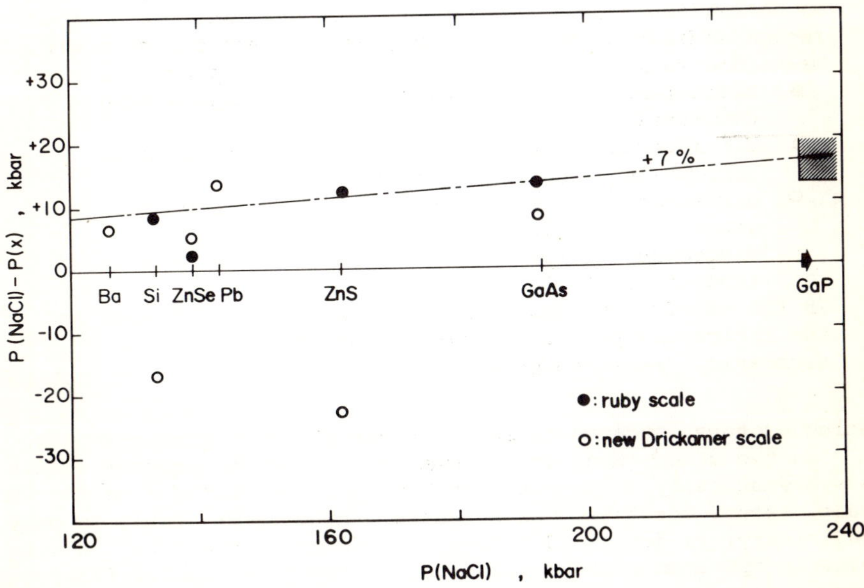

Fig. 3. Comparison of the present results with ruby scale (solid circle) and new Drickamer scale (open circle). Horizontal axis is the transition pressure of calibrant materials determined in the present study, and vertical axis is the difference between present result and other result.

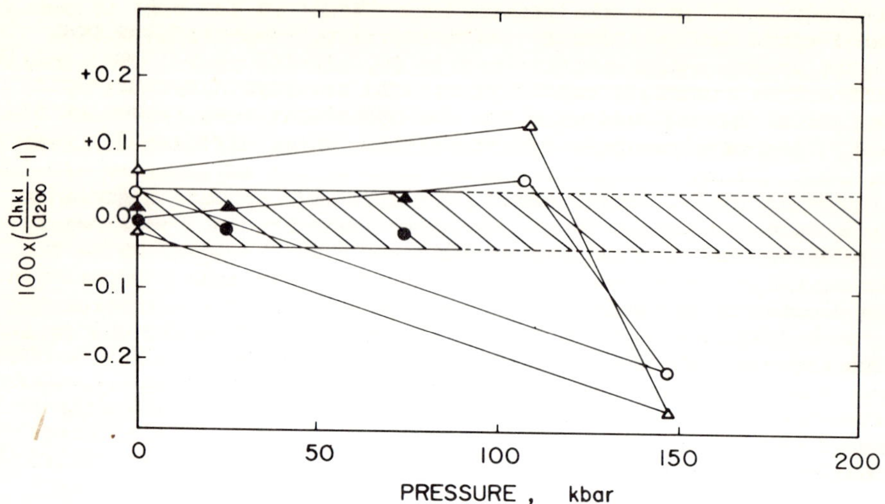

Fig. 4. Discrepancies of the lattice parameter of NaCl as determined from different diffraction lines. A hatched region indicates the uncertainty of the lattice parameter measurements in the present X-ray diffraction system.

●: discrepancy between a_{200} and a_{220} in hydrostatic compression (LQ.LIF-101 run)

▲: discrepancy between a_{200} and a_{222} in hydrostatic compression (LQ.LIF-101 run)

○: discrepancy between a_{200} and a_{220} in the present experiment (PB-103 run)

△: discrepancy between a_{200} and a_{222} in the present experiment (PB-103 run)

In the PB-103 run, measurements were made at 147 kbar in the increasing pressure cycle and at 108 kbar in the decreasing pressure cycle.

semiconductors, such as Si and ZnS, are lower by more than 10%.

On the other hand, it is remarkable that the present result is systematically 7% higher than the result obtained from the ruby scale, except for ZnSe. As for ZnSe, the present result is only preliminary, as described before, and further investigation with a reliable specimen will be required. Although we cannot fully account for this discrepancy, we found an interesting phenomena related to this problem.

Under nonhydrostatic conditions, a systematic discrepancy exist among the value of the NaCl lattice parameters determined from different diffraction lines. In Figure 4, the discrepancy of the lattice parameter of NaCl determined from different diffraction lines in the present experiment (PB-103 run) is compared with that obtained in the hydrostatic compression with a liquid pressure-transmitting medium (LQ.LIF-101 run). The X-ray diffraction technique under hydrostatic pressure developed by *Yagi et al.* [1975] was used for the LQ.LIF-101 run. For both runs, lattice parameters determined from three diffraction lines -- (200), (220), and (222) -- agree well within an accuracy of ± 0.04% at 1 bar. At elevated pressure, however, the difference in lattice parameters in the PB-103 run amounts to approximately ± 0.14% in the increasing order of $a_{222} < a_{220} < a_{200}$, while in the LQ.LIF-101 run, the difference remains less than ± 0.04%. In the decreasing pressure cycle, the discrepancy of the lattice parameter still exists in the PB-103 run, and, further, the relative magnitude of the lattice parameter is reversed to $a_{200} < a_{220} < a_{222}$. Mechnical distortion of the apparatus with the applied load cannot account for these phenomena. They would be caused by the existance of a nonhydrostatic component of pressure.

The diffraction profile of a cubic crystal under nonhydrostatic pressure with uniaxial stress component was analyzed by *Singh and Kennedy* [1974] , and the present results are qualitatively in good agreement with their analysis. However, various assumptions for stress distribution and a precise knowledge of the pressure dependence of the elastic parameters of NaCl are required for a detailed quantitative discussion. Therefore, we cannot accurately evaluate the amount of uniaxial stress component in the present experiments. It is clear, however, that some amount of uniaxial stress component exists even in a cubic-anvil type high-pressure apparatus.

The present values of pressure were calculated by averaging the three values a_{200}, a_{220}, and a_{222}. Since these values differ in the increasing order of $a_{222} < a_{220} < a_{200}$ in the pressure increasing cycle, the transition pressure will be lowered systematically by 2-3% if we use only a_{200} for the calculation of pressure. This fact might account for a part of discrepancy between present results and those of the ruby scale, because the calibration of the ruby scale against the compression of NaCl is based primarily on the observation of the (200) diffraction line of NaCl [*Piermarini et al.*, 1975].

The complicated relationship between the present results and the new Drickamer scale might be also caused by the existance of a nonhydrostatic component of pressure, because the sensitivity of transition pressure to the stress will differ with the nature of the force acting in the crystal.

Stress distribution within the high-pressure vessel will differ significantly with the type of high-pressure apparatus. Therefore, an experiment under hydrostatic conditions will be required for the correct understanding of these problems.

IV. CONCLUSIONS

A new, consistent set of pressure-fixed points between 100 and 193 kbar was determined based on the compression of NaCl. Although a slight systematic discrepancy exists between the present results and the transition pressures determined by the ruby scale, we can say that no drastic revision of the pressure will be made in this pressure region.

For the detailed discussion, however, it was clarified that the effect of nonhydrastatic stress becomes important. Therefore, futher investigations under hydrostatic conditions will be required to make the pressure scale more reliable.

Acknowledgments. The authors wish to express their gratitude to G. J. Piermarini for critically reviewing the paper.

REFERENCES

Akimoto, S., T. Yagi, Y. Ida, K. Inoue, and Y. Sato, High-pressure X-ray diffraction study on barium up to 130 kbar, *High Temp.-High Pressures*, *7*, 287-294, 1975.

Bundy, F. P., Ultrahigh pressure apparatus using cemented tangsten carbide pistons with sintered diamond tips, *Rev. Sci. Instrum.*, *46*, 1318-1324, 1975.

Decker, D. L., High-pressure equation of state for NaCl, KCl, and CsCl, *J. Appl. Phys.*, *42*, 3239-3244, 1971.

Drickamer, H. G., Revised calibration for high pressure electrical resistance cell, *Rev. Sci. Instrum.*, *41*, 1667-1668, 1970.

Homan, C. G., D. P. Kendall, T. E. Davidson, and J. Frankel, GaP semiconducting-to-metal transition near 220 kbar and 298°K, *Solid State Commun.*, *17*, 831-832, 1975.

Jeffery, R. N., J. D. Barnett, H. B. Vanfleet, and H. T. Hall, Pressure calibration to 100 kbar based on the compression of NaCl, *J. Appl. Phys.*, *37*, 3172-3180, 1966.

Minomura, S., and H. G. Drickamer, Pressure induced phase transitions in silicon, germanium and some III-V compounds, *J. Phys. Chem. Solids*, *23*, 451-456, 1962.

Onodera, A., N. Kawai, K. Ishizaki, and I. L. Spain, Semiconductor-to-metal transition in GaP under high pressure, *Solid State Commun.*, *14*, 803-806, 1974.

Piermarini G. J., and S. Block, Ultrahigh pressure diamond-anvil cell and several semiconductor phase transition pressure in relation to the fixed point pressure scale, *Rev. Sci. Instrum.*, *46*, 973-979, 1975.

Piermarini, G. J., S. Block, J. D. Barnett, and R. A. Forman, Calibration of the pressure dependence of the R_1 ruby fluorescence line to 195 kbar, *J. Appl. Phys.*, *46*, 2774-2780, 1975.

Samara, G. A., and H. G. Drickamer, Pressure induced phase transitions in some II - VI compounds, *J. Phys. Chem. Solids*, *23*, 457-461, 1962.

Singh, A. K., and G. C. Kennedy, Uniaxial stress component in tungsten carbide anvil high-pressure X-ray cameras, *J. Appl. Phys.*, *45*, 4686-4691, 1974.

Van Vechten, J. A., Quantum dielectric theory of electronegativity in covalent systems. III. Pressure-temperature phase diagrams, heat of mixing, and distribution coefficients, *Phys. Rev.*, *B7*, 1479-1507, 1973.

Yagi, T., Y. Ida, Y. Sato, and S. Akimoto, Effect of hydrostatic pressure on the lattice parameters of Fe_2SiO_4 olivine up to 70 kbar, *Phys. Earth Planet. Interiors*, *10*, 348-354, 1975.

Yagi, T., and S. Akimoto, Pressure fixed points between 100 and 200 kbar based on the compression of NaCl, *J. Appl. Phys.*, *47*, 3350-3354, 1976.

HIGH TEMPERATURE-PRESSURE PHASE BOUNDARIES IN SILICATE SYSTEMS USING IN SITU X-RAY DIFFRACTION

S. AKIMOTO, T. YAGI
Institute for Solid State Physics, University of Tokyo
Roppongi, Minato-ku, Tokyo 106, Japan

K. INOUE
Asada Fundamental Research Laboratory, Kobe Steel Ltd.
Nada-ku, Kobe 657, Japan

Abstract

Phase boundaries between olivine and spinel in Fe_2SiO_4, between coesite and stishovite in SiO_2, and between pyroxene and ilmenite in $ZnSiO_3$ were determined by *in situ* X-ray diffraction measurements using cubic-anvil-type high pressure apparatus. Pressures at high temperatures were determined by X-ray diffraction from NaCl, the internal pressure standard. The usual Debye-Scherrer configuration was used in the investigation of SiO_2 and $ZnSiO_3$, while an energy dispersive X-ray diffraction technique was used for preliminary investigation of the olivine-spinel phase boundary of Fe_2SiO_4. The following equations for the phase boundaries were determined:
olivine-spinel transition in Fe_2SiO_4, $P(\text{kbar}) = 34.6 + 0.025T(°\text{C})$;
coesite-stishovite transition in SiO_2, $P(\text{kbar}) = 80 + 0.011T(°\text{C})$;
pyroxene-ilmenite transition in $ZnSiO_3$, $P(\text{kbar}) = 91 + 0.02T(°\text{C})$.
Present results form a set of pressure fixed-points at high temperatures in the 50 to 130 kbar range.

I. INTRODUCTION

Phase transformations in silicates at high pressures and temperatures have usually been detected by the quenching method. In most high-pressure and high-temperature apparatus with solid pressure media and compressible gaskets, pressure values are conventionally calibrated at room temperature by a number of resistance-jump transitions in standard materials, such as Bi, Ba, Pb, ZnS, GaAs, and GaP. This method is not satisfactory for high-temperature work. Depending upon the type of high-pressure apparatus and the temperature-pressure conditions, the change in shear strength of the solid pressure media, thermal expansion of the samples and pressure media, and phase change or chemical alteration of the samples and pressure media are all expected to affect the pressure values considerably. It is likely that the phase boundaries between various silicate phases determined by the quenching method contain uncertainties in the values of pressure at high temperatures. Thus, the refinement of the pressure calibration points at high temperatures is urgently needed for the finer analyses of the mineralogical models of the earth's mantle.

The high-pressure *in situ* X-ray diffraction technique is a promising method for measurement of the pressure acting on a sample at elevated temperatures. Pressure values at high temperatures can be determined by measuring the lattice parameter of an internal pressure standard material, such as NaCl, and, consequently, they are absolutely accurate to within the uncertainty of the equation of state of the standard material used.

Reliable phase boundaries, established by *in situ* X-ray measurements, offer a convenient basis for pressure calibration at high temperatures. Usually, the high-pressure investigator has no specialized X-ray equipment, and can readily make use of boundary curves for calibrating his devices. Preliminary experiments with the objective of observing the phase boundary by *in situ* X-ray measurements have recently been reported in a few high-pressure laboratories. *Böhler and Arndt*[1974] first succeeded in determining the quartz-coesite boundary using the belt-type high-pressure X-ray apparatus. Using cubic-anvil press and X-ray diffraction technique, the phase relations at high pressures and high temperatures are currently being investigated in our laboratory. Both the angle-dispersive [*Yagi and Akimoto*, 1976] and the energy-dispersive methods in X-ray analysis [*Inoue*, 1975] have been attempted. In the present paper, phase boundaries determined by the angle-dispersive technique are shown for the coesite-stishovite transformation in SiO_2 and for the pyroxene-ilmenite transformation in $ZnSiO_3$. Preliminary investigation of the phase boundary by means of an energy-dispersive X-ray diffraction technique is also carried out for the olivine-spinel transformation in Fe_2SiO_4.

II. EXPERIMENTAL METHODS

A. Angle-dispersive Type X-ray Diffraction

A cubic-anvil high-pressure apparatus was combined with a Debye-Scherrer X-ray diffraction configuration and a rotary-target type high power X-ray source operated at 50 kV and 160 mA. The Zr-filtered Mo Kα radiation was used. Details of the apparatus have been described in a previous paper [*Yagi et al.*, 1975].

1. SiO_2

The sample assembly used for the study of coesite-stishovite transformation [*Yagi and Akimoto*, 1976] is shown in Figure 1. A pressure-transmitting medium consisting of 2:1 mixture of amorphous boron and epoxy resin shaped into cube with edge length of 6 mm and was compressed by six WC anvils with square faces of 4-mm edge length. The pressure cell was successfully used up to 1100°C at 100 kbar. In the center hole of the pressure cell were stacked the pyrophyllite end plugs, graphite heater disks,

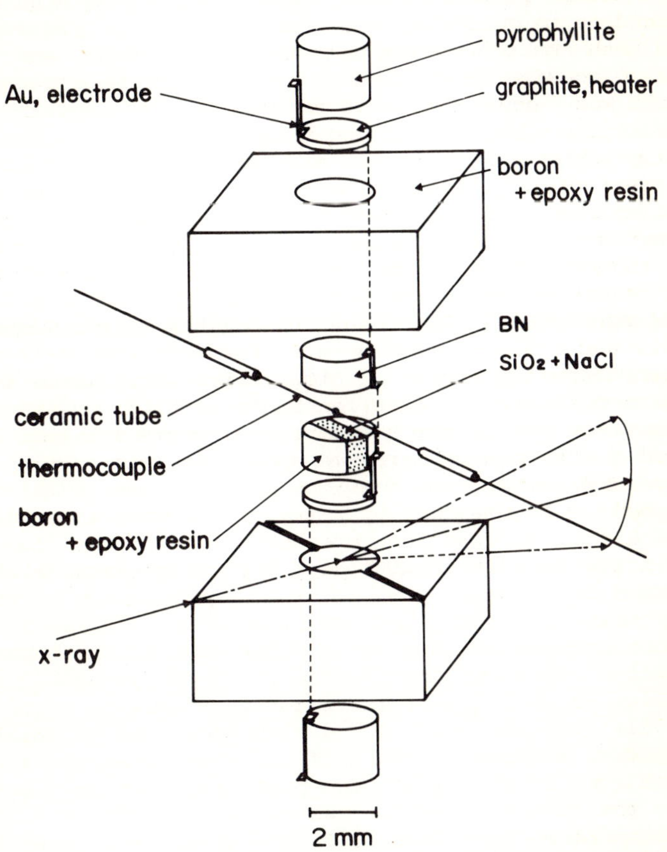

Fig. 1. Sample assembly used for the direct determination of coesite-stishovite transformation by in situ X-ray measurements.

and gold-electrodes. An intimate mixture of amorphous hydrous SiO_2 (Mallinckrodt silicic acid) and NaCl was embedded in the central portion of the pressure cell. NaCl served as an internal pressure standard and also as a pressure-transmitting medium. Temperature was measured with a chromel/alumel thermocouple, and its leads were taken out of the high-pressure cell through the gaskets and gaps of the anvils. No correction was made for the effect of pressure on the emf of the thermocouple throughout the present study. The position of the incident X-ray beam was adjusted very close to the hot junction of the thermocouple by monitoring its shadow in the direct path of the X-ray beam. The accuracy of the temperature measurement in this sample assembly was estimated to be ± 10°C.

Since both reactions from coesite to stishovite and from stishovite to coesite were found to be sluggish near the phase boundary, it was rather difficult to determine the boundary position accurately along the isotherm or the isobar. For this reason, an experimental procedure similar to the quenching technique was used in conjunction with the *in situ* X-ray measurements. First, the pressure was raised to the desired value at room temperature, and then the temperature was rapidly raised to the desired value. Phase change in the SiO_2 sample was detected by *in situ* X-ray measurements. Two diffraction lines of coesite -- (130)+(111) and (040)+(200) -- and three lines of stishovite -- (110), (101) and (111) -- were used to identify the high-pressure and high-temperature products of SiO_2. Pressure values were determined from the two diffraction lines of NaCl -- (200) and (220) -- based on the Decker's equation of state for NaCl [*Decker*, 1971]. After the *in situ* X-ray measurements, the sample was quenched under pressure, and then the pressure was released. The quench products were examined by X-ray diffraction at ambient conditions. No discrepancies were found between the phases detected under the *in situ* conditions and those present in the quench products.

2. *$ZnSiO_3$*

The sample assembly used for the study of pyroxene-ilmenite transformation in $ZnSiO_3$ is illustrated in Figure 2. The 3-mm-edge WC anvils were used with the 4.5-mm-edge cube of boron/epoxy resin mixture. Because of the difficulty in obtaining clear diffraction patterns of $ZnSiO_3$ at high-pressure and high-temperature conditions, the quenching method was generally favored. Since the strongest diffraction line of $ZnSiO_3$ (ilmenite) interferes with that of NaCl at the 110-kbar range, the $ZnSiO_3$ sample and NaCl were placed separately in the sample chamber space. In order to prevent the grain growth of NaCl at high temperature, powdered hexagonal boron nitride was mixed with it. Temperature was measured with a Pt/Pt-13%Rh thermocouple, and its leads were taken out to the surface of the anvils. Necessary corrections were made for the temperature rise of the anvil surface. Uncertainty

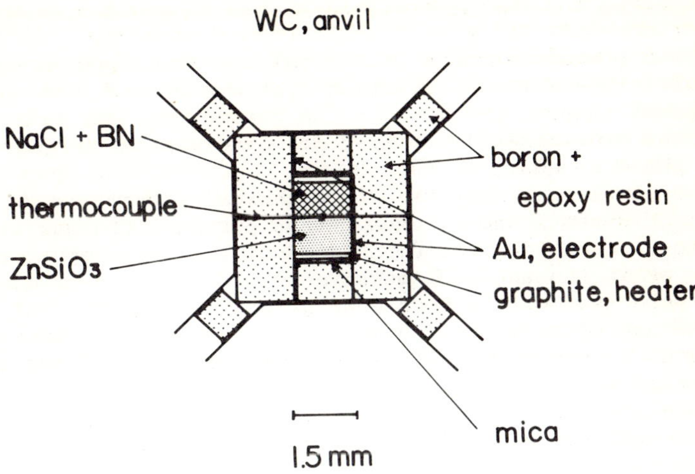

Fig. 2. Sample assembly used for determining the pyroxene-ilmenite transformation boundary of $ZnSiO_3$ by in situ X-ray measurements.

of the temperature measurement in this assembly was estimated to be ± 20°C.

Clinopyroxene-type $ZnSiO_3$, which was prepared in advance at 72 kbar and 1000°C using the tetrahedral-anvil press, was used as a starting material. Experimental procedures for determining the pressure values at high temperatures were the same as in the case of coesite-stishovite transformation. After the nominal pressure and temperature conditions were fixed for each run, the lattice parameter of NaCl was measured and converted to absolute pressure according to Decker's equation of state for NaCl. Duration of runs was varied from 30 to 60 min, depending upon the reaction temperature. After being maintained at a desired pressure and temperature for a given time, the $ZnSiO_3$ sample was quenched isobarically. The quench products were examined by the X-ray diffraction method at ambient conditions.

B. Energy-dispersive Type X-ray Diffraction

The energy dispersive X-ray diffraction technique was first proposed by *Giessen and Gordon* [1968]. In this method, white X-ray radiation is used as the incident beam in contrast to the characteristic X-ray for the usual Debye-Scherrer method. The X-rays diffracted from a powder sample to a fixed direction are analysed by a system consisting of a solid state detector and a multichannel pulse-height analyzer. *Freud and LaMori* [1969] first applied the energy-analysis system to a high-pressure X-ray diffraction study using split-belt type apparatus. Recently, one of the present authors (K. Inoue) has developed an energy-dispersive type X-ray diffraction system combined with a cubic-anvil press [*Inoue*, 1975]. In the present investigation, the system was preliminarily applied for determining the olivine-spinel phase boundary of Fe_2SiO_4.

The sample assembly used is principally the same as already shown in Figures 1 and 2, except that pyrophyllite was used as a pressure-transmitting medium instead of the boron/epoxy resin mixture as in the angle-dispersive method. The 4-mm-edge WC anvils were used with a 6-mm-edge pyrophyllite cube. It is noted that the use of relatively high-energy X-rays (20 - 50 keV) enables us to use pyrophyllite as a pressure medium for X-ray diffraction work.

Fayalite, Fe_2SiO_4(olivine), was prepared by sintering a 1:1 mixture of αFe_2O_3 and SiO_2(α-quartz) under a controlled partial pressure of oxygen and was used as the starting material. A plate of intimate powder mixture of Fe_2SiO_4 olivine and hexagonal boron nitride (2:1 in volume) was set in the boron nitride sample holder. The holder was placed in the center of the pyrophyllite cell. A pair of graphite disk heaters were used for heating the samples. Temperature was measured with chromel/alumel or Pt/Pt-13%Rh thermocouples embedded in the sample plate. The thermocouple leads were taken out through the gaskets and anvil gaps. Temperatures thus measured were estimated to be accurate to ± 15°C.

As a source of white X-ray radiation a W-fine-focus X-ray tube operated at 50 kV x 20 mA was used. The diffracted X-rays from the sample were detected by an ORTEC 8113 Ge(Li) solid state detector fixed at a Bragg angle of 12.2°. A data processing system was also developed, and its detail was reported by *Inoue* [1975].

Both isobaric and isothermal measurements were carried out in traversing the olivine-spinel phase boundary at pressure-temperature conditions extending to 70 kbar and to 1200°C. Figures 3 and 4 represent typical energy-dispersive, X-ray-diffraction patterns for the olivine and spinel phases of Fe_2SiO_4, respectively. An isothermal series of diffraction patterns of Fe_2SiO_4 at 800°C are shown in Figure 5. Each pattern was obtained by accumulating the data in a multichannel analyzer for only 400 sec. It is seen that both olivine and spinel phases coexist at a pressure corresponding to the press load of 380 kg/cm^2.

Diffraction patterns of NaCl powder mixed with BN powder were observed in separate runs at the same press load and temperature conditions where the olivine-spinel phase boundary was detected in the phase identification runs. In the present sample assembly, only the (220) diffraction line of NaCl could

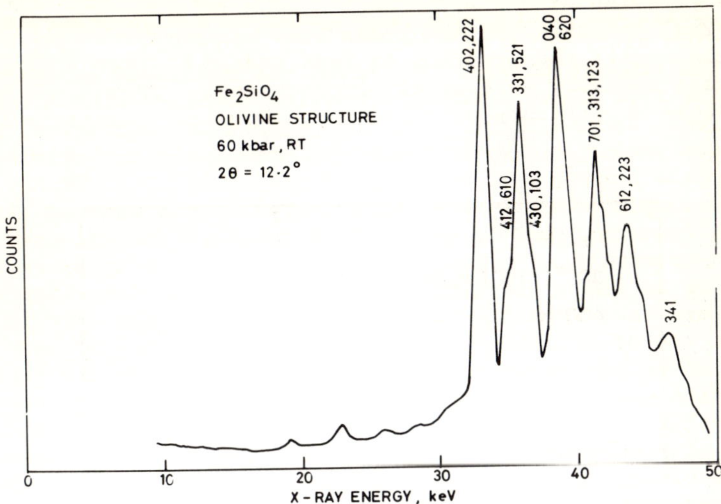

Fig. 3. Energy-dispersive X-ray diffraction pattern of Fe_2SiO_4 olivine at 60 kbar and room temperature.

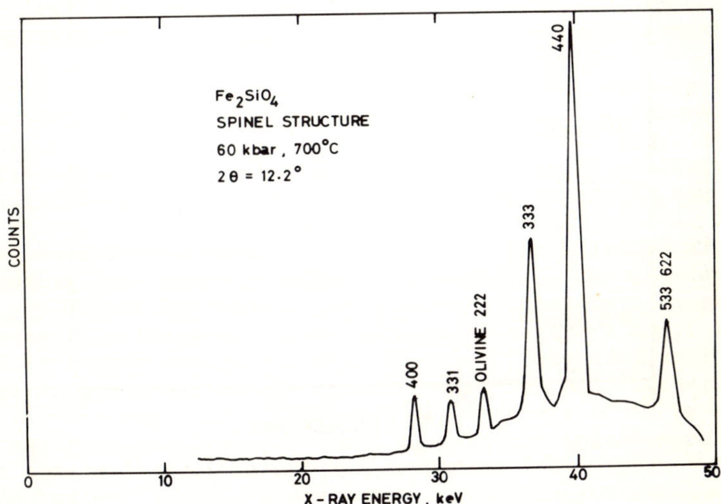

Fig. 4. Energy-dispersive X-ray diffraction pattern of Fe_2SiO_4 spinel at 60 kbar and 700°C. The Fe_2SiO_4 olivine sample shown in Figure 3 was heated to 700°C isobarically.

be resolved from the diffraction lines of the BN sample holder, and this line was used to determine the value of pressure using Decker's equation of state for NaCl. The relationship between the internal pressure and the press load was confirmed to be reproducible with the accuracy of less than 3 % when the parts in the cell assembly were carefully machined. This leads to the pressure uncertainty of ± 3 % in the present measurements.

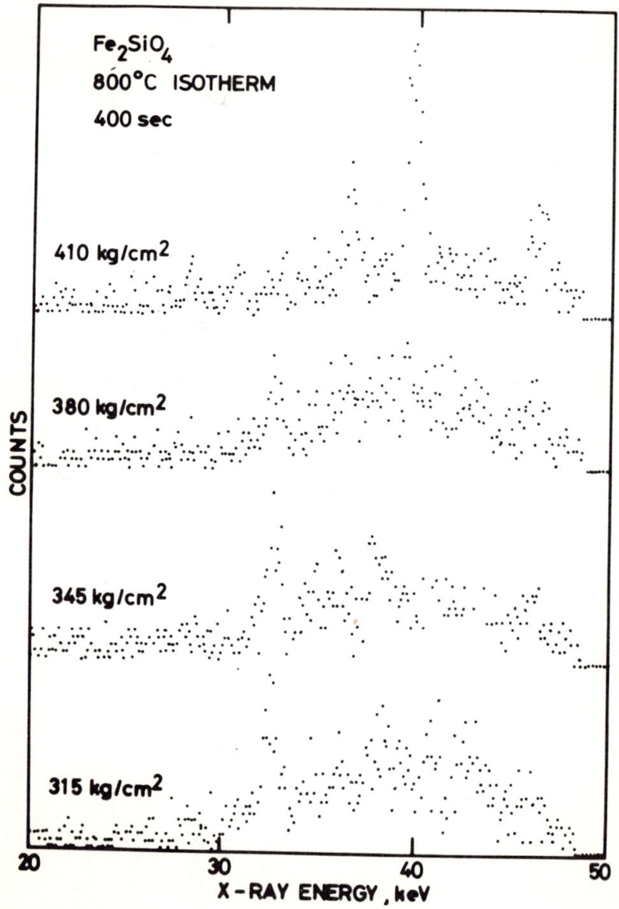

Fig. 5. Isothermal series of energy-dispersive X-ray diffraction patterns of Fe_2SiO_4 at 800°C. The coexistence of both olivine and spinel phases is recognized at the oil pressure of 380 kg/cm^2.

III. RESULTS AND DISCUSSION

A. Coesite-Stishovite Transition

Experimental results of 21 runs are shown in Figure 6. Detailed tables indicating pressure-temperature conditions and phases present in individual runs are given elsewhere [*Yagi and*

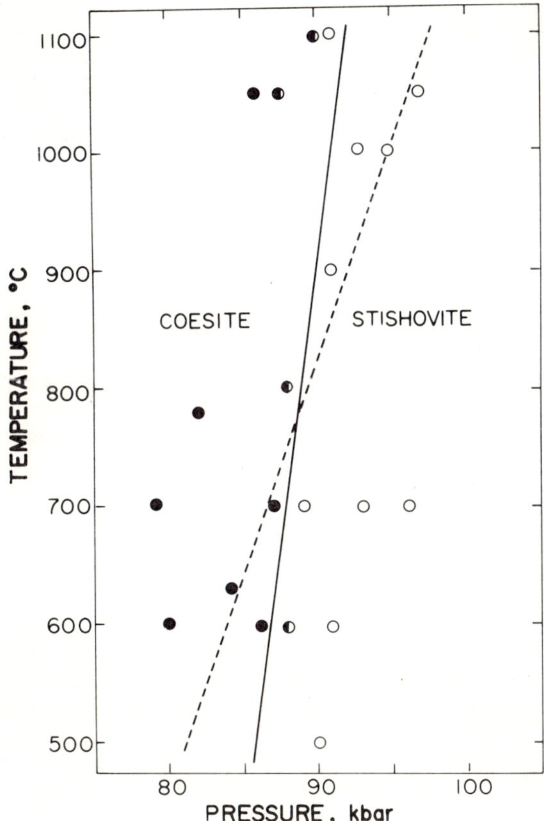

Fig. 6. Pressure-temperature stability diagram determined by in situ X-ray measurements for the coesite-stishovite transformation. Boundary curve determined by the quenching experiments [Akimoto and Syono, 1969; Akimoto, 1972] is shown by a broken line.

Akimoto, 1976]. As described before [*Yagi and Akimoto*,1976], in the present experiments coesite in the high-temperature region above 1000°C always formed via stishovite, and at those conditions the phase transformation rate from stishovite to coesite was found to be extremely slow near the phase boundary. Considering these facts and the uncertainty of the observed values of pressure and temperature, we finally obtained the following equation for the coesite-stishovite phase boundary, assuming it to be linear:

$$P(\text{kbar}) = (80 \pm 2) + (0.011 \pm 0.003)T(°\text{C})$$

In Figure 6, a previous determination of the boundary curve, $P(\text{kbar}) = 67 + 0.028T(°\text{C})$[1], by quenching experiments using the pyrophyllite pressure cell [*Akimoto and Syono*, 1969; *Akimoto*, 1972] is shown for comparison. It is found that the slope of the present curve is a little steeper than the previous determination; at temperatures below approximately 800°C, the present curve is at higher pressures, while at the higher temperatures, the present curve is at lower pressures. This suggests that a systematic increase and decrease of pressure with increasing temperature take place in the pyrophyllite pressure cell.

B. Pyroxene-Ilmenite Transformation in $ZnSiO_3$

Experimental results of seven runs are summarized in Table 1 and Figure 7. Present data confirmed the previous report by *Ito and Matsui* [1974] that $ZnSiO_3$(clinopyroxene) transforms directly to the ilmenite structure without disproportionating into an intermediate mixed-phase of stishovite plus high-pressure polymorphs of Zn_2SiO_4. Although the number of data is still insufficient to determine the precise location of the pyroxene-ilmenite boundary of $ZnSiO_3$, the present results indicate definitely that the slope of the phase boundary dP/dT is positive. We obtained the following tentative linear boundary equation:

$$P(\text{kbar}) = (91 \pm 8) + (0.02 \mp 0.01)T(°\text{C})$$

Ito and Matsui [1974] claimed that the transition pressure from pyroxene to ilmenite was higher than 120 kbar at 1000°C, while in the present experiments the transition pressure is 111 ± 2 kbar at that temperature. Since *Ito and Matsui* used pyrophyllite octahedron as pressure-transmitting material in their experiments, the situation is quite similar to the case of the coesite-stishovite transformation.

C. Olivine-Spinel Transformation in Fe_2SiO_4

The results of phase relationships between olivine and spinel polymorphs of Fe_2SiO_4 are summarized in Figure 8 as a function of

[1] The boundary equation given by *Akimoto and Syono* [1969] was later revised to this form by *Akimoto* [1972] based on the NBS pressure scale.

TABLE 1. Experimental Results of Pyroxene-Ilmenite Transformation in $ZnSiO_3$

Run No.	Temperature, °C	NaCl Compression, $-\Delta a/a_o$	Pressure, kbar	Phases Present[a]
ZN-102	800	0.0684	107.5	il + cpx
103	800	0.0626	96.5	cpx
104	900	0.0684	110.5	il
105	800	0.0622	96.0	cpx
107	1000	0.0695	115.5	il
108	850	0.0718	115.5	il
109	1000	0.0656	108.0	cpx

a. cpx = clinopyroxene; il = ilmenite.

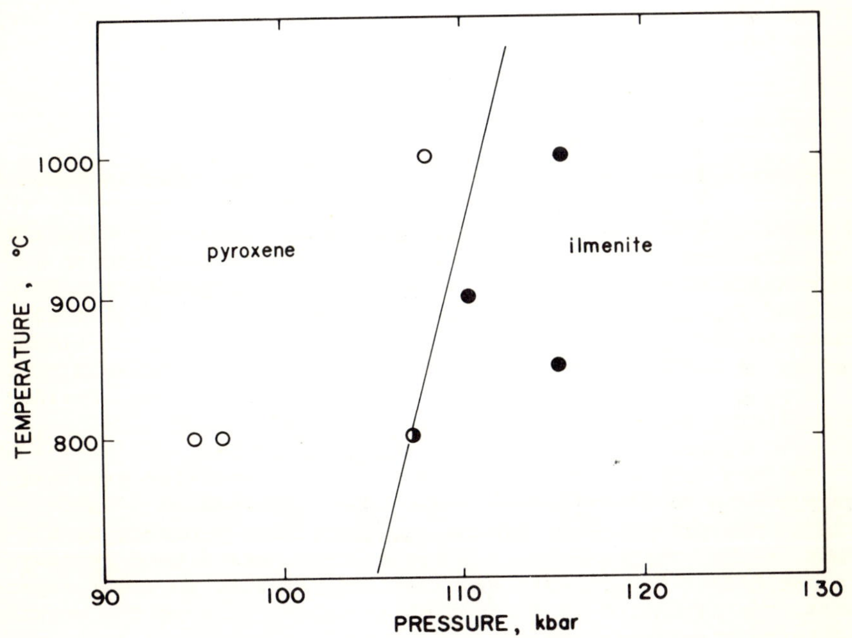

Fig. 7. Pressure-temperature stability diagram determined by in situ X-ray measurements for the pyroxene-ilmenite transformation of $ZnSiO_3$.

the press load and the heating power. The phase boundary curve shows remarkable inflection from a straight line above the heating power of 180 watts (corresponding to approximately 920°C). Since the linearity between the heating power and the sample temperature holds well in the present temperature range, this deviation is chiefly due to the fall of the pressure-generation efficiency in the pyrophyllite cell at high temperatures.

In Figure 9 the olivine-spinel phase boundary in Fe_2SiO_4 is shown on the conventional pressure-temperature diagram. Data

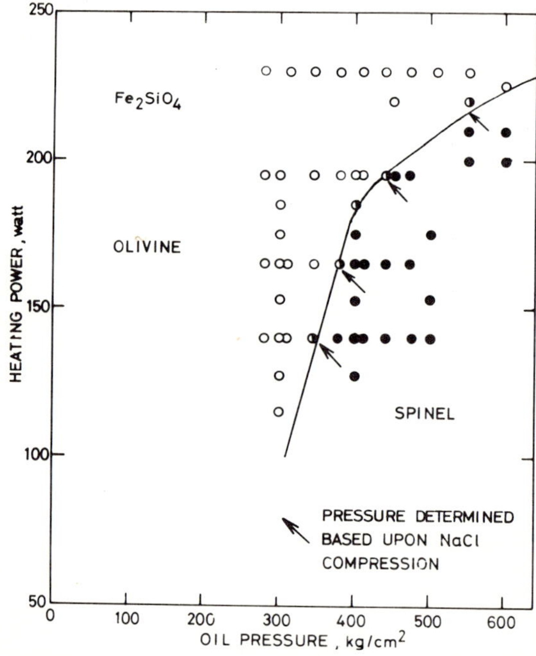

Fig. 8. Summary of the phase identification runs for the olivine-spinel transformation of Fe_2SiO_4. Both isobaric and isothermal measurements were done for traversing the phase boundary. Pressure values were determined on the points marked by arrows based on the NaCl pressure scale.

TABLE 2. Olivine-Spinel Transition Pressure of Fe_2SiO_4 Based on the NaCl Internal Pressure Standard

Temperature, °C	NaCl Compression, $-\Delta a/a_o$	Pressure, kbar
700	0.03417	51.5
800	0.03415	55.8
1000	0.03457	60.4
1120	0.03356	62.8

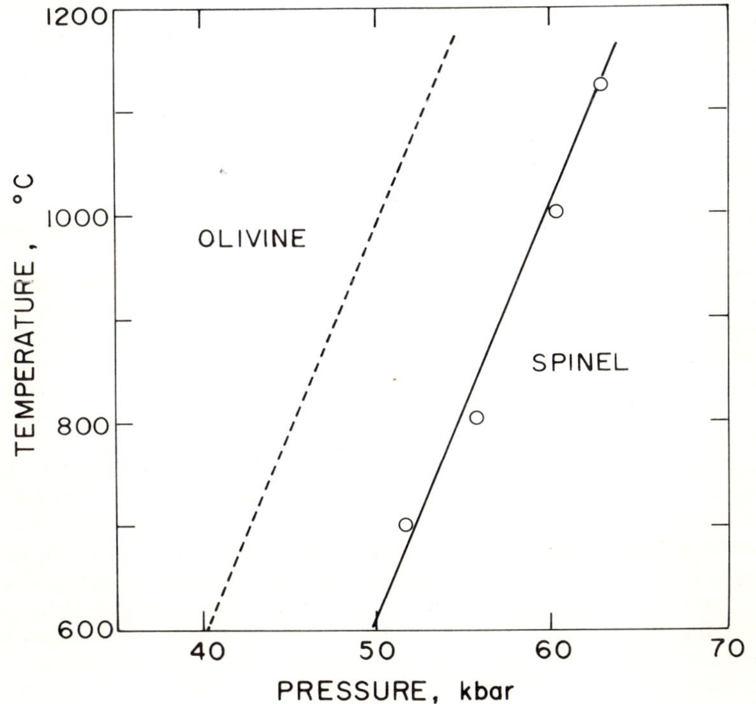

Fig. 9. Olivine-spinel transformation curve based on the NaCl pressure scale. Data points represent the runs marked by arrows in Figure 8. Boundary curve determined by the quenching experiment [Akimoto et al., 1967] is also shown by a broken line.

points in the figure represent the runs marked by arrows in Figure 8. Pressure values for these points were determined on the basis of the NaCl pressure scale and listed in Table 2. The boundary curve thus determined is expressed by a linear equation

$$P(\text{kbar}) = 34.6 + 0.025T(°\text{C})$$

The results of the quenching experiments by *Akimoto et al.* [1967], $P(\text{kbar}) = 25.3 + 0.025T(°\text{C})$ [2], are also compared in Figure 9. No significant discrepancy was found for the slope of the boundary curve between these two independent measurements. For the absolute value of the transformation pressure, however, there exists a fairly large discrepancy amounting to 9 kbar. This suggests that the pressure values were underestimated in the previous quenching experiments using the tetrahedral-anvil press and pyrophyllite pressure cell.

D. Variation of Internal Pressure of the Pyrophyllite Cell with Increasing Temperature

It is evident from the experimental data mentioned above that pressures on the sample within the pyrophyllite cell are remarkably affected by the process of internal heating. The following experiments reported by *Inoue* [1975] indicate more clearly the high-temperature behaviors of the pyrophyllite pressure cell. As shown in Figure 10, *Inoue* has successfully observed the variation of internal pressure during heating and cooling cycles up to 1400°C under constant press load by means of the energy-dispersive type *in situ* X-ray diffraction technique. In the numerical sequence shown in the figure, temperature was successively changed, and pressure values were determined by measuring the compression of NaCl. It was found that at temperatures below approximately 600°C, pressure increases slightly with increasing temperature, possibly due to the thermal expansion of the sample and sample container. On further heating, however, gradual decrease in pressure was revealed. This may be chiefly attributable to the decrease in the shear strength of pyrophyllite gaskets as well as the flow of the pyrophyllite pressure cell. *Inoue* [1975] also found that, when temperature is elevated above approximately 1000°C, disproportionation of pyrophyllite to the denser minerals takes place around the furnace space. It is very likely that the marked pressure decrease observed at 1200 to 1400°C is partly due to this disproportionation effect. It must be noted that, in the press load condition in Figure 10, coesite and kyanite are the possible disproportionation products. If pyrophyllite cell is used in the higher pressure range, internal heating results in the formation of stishovite, and much larger pressure decrease is expected.

---

[2] The equation given by *Akimoto et al.* [1967] was later revised to this form by *Mao et al.* [1969] based on the NBS pressure scale.

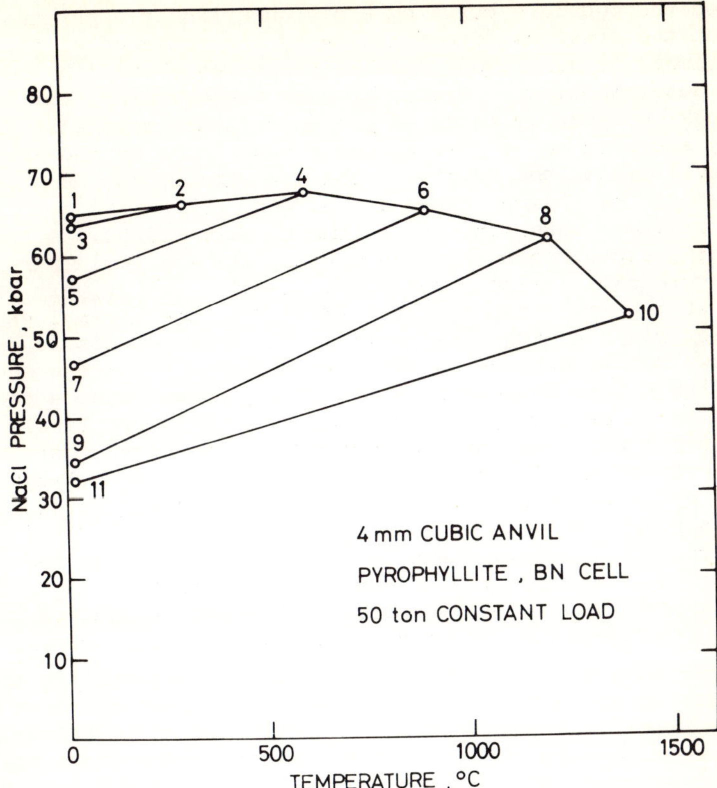

Fig. 10. Variation of internal pressure of the pyrophyllite pressure cell at constant press load during heating and cooling cycles up to 1400°C. Temperature was changed successively along the sequence number attached to the data points.

In the following, the discrepancies between phase boundaries determined by the quenching experiments and by the *in situ* X-ray measurements are successfully interpreted based on Figure 10. In the case of olivine-spinel transformation in Fe_2SiO_4, the quenching experiment [*Akimoto et al.*, 1967] was done using the tetrahedral-anvil press. In that experiment, volume ratio of heater space to total pyrophyllite cell space is much smaller than the case shown in Figure 10. Consequently, the effect of thermal expansion will be more dominant than the other mechanical or chemical effects. This may be the reason why previous determination by the quenching method shifts to the lower pressure side. Recently, a similar shift was also pointed out by

Ma [1974] on the olivine-spinel transformation in Ni_2SiO_4. He determined the boundary curve by a piston-cylinder apparatus and found that the previous determination by the tetrahedral-anvil press [*Akimoto et al.*, 1965] lies at pressures about 12% lower than his determination. Since, at high temperatures, the pressure values obtained by means of piston-cylinder apparatus are generally accepted to be more accurate than those by the tetrahedral anvil press, the origin of the discrepancy may be chiefly attributed to the thermal expansion effect.

The previous determination of the coesite-stishovite boundary by quenching experiments was made with relatively small pyrophyllite pressure cell [*Akimoto and Syono*, 1969]. It is very probable that at the lower temperatures (T<700°C), the thermal expansion effect plays an important role in the observed discrepancies between two determinations as shown in Figure 6. On the other hand, at the higher temperatures (T>700°C), phase changes and the decreases in shear strength of pyrophyllite cells are likely to be responsible for these discrepancies. It is noted that the pressure corrections due to the thermal expansion are always positive, while those due to phase changes and shear-strength change are negative.

The present interpretation is also consistent with the investigation of the pyroxene-ilmenite transformation in $ZnSiO_3$: a little higher transformation pressure has been suggested in the quenching experiments by *Ito and Matsui* [1974].

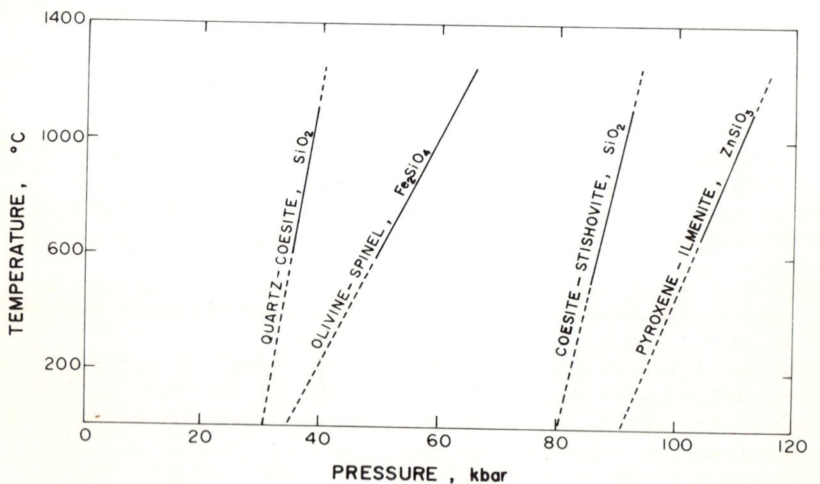

Fig. 11. Phase boundaries determined by in situ X-ray measurements.

E. Pressure Calibration at High Temperature

Several phase transformation diagrams of silicates determined by *in situ* X-ray measurements are summarized in Figure 11. There still exist many sources of error in determination of these boundaries, such as (1) the error in measured lattice parameters of NaCl by *in situ* X-ray diffraction technique, (2) the validity of Decker's equation of state of NaCl for high temperature use, (3) the effect of pressure on the emf of the thermocouple used, (4) temperature distribution within the sample space, and (5) problems in the equilibrium relation of the appeared phases. More detailed analyses of these experimental uncertainties are necessary for the further refinements of the present results. Since all the boundaries shown in Figure 11 are determined with the common NaCl internal pressure standard, however, they can be used as a consistent set of high-pressure calibration points at high temperatures.

Acknowledgments. This work was supported by the Ministry of Education under research grant no. 042014. The authors wish to thank B. Olinger for his critical review of the manuscript.

REFERENCES

Akimoto, S., The system $MgO-FeO-SiO_2$ at high pressures and temperatures — phase equilibria and elastic properties, *Tectonophys.*, *13*, 161-187, 1972.

Akimoto, S., and Y. Syono, Coesite-stishovite transition, *J. Geophys. Res.*, *74*, 1653-1659, 1969.

Akimoto, S., H. Fujisawa, and T. Katsura, The olivine-spinel transition in Fe_2SiO_4 and Ni_2SiO_4, *J. Geophys. Res.*, *70*, 1969-1977, 1965.

Akimoto, S., E. Komada, and I. Kushiro, Effect of pressure on the melting of olivine and spinel polymorph of Fe_2SiO_4, *J. Geophys. Res.*, *72*, 679-686, 1967.

Böhler, R., and J. Arndt, Direct determination of the quartz-coesite transition by *in situ* X-ray measurements, *Contr. Mineral. Petrol.*, *48*, 149-152, 1974.

Decker, D. L., High pressure equation of state for NaCl, KCl, and CsCl, *J. Appl. Phys.*, *42*, 3239-3244, 1971.

Freud, P. J., and P. N. LaMori, Non-dispersive high pressure high temperature X-ray diffraction analysis, *Trans. Am. Cryst. Assoc.*, *5*, 155-162, 1969.

Giessen, B. C., and G. E. Gordon, X-ray diffraction: New high-speed technique based on X-ray spectrography, *Science*, *159*, 973-975, 1968.

Inoue, K., Development of high temperature and high pressure X-ray diffraction apparatus with energy dispersive technique and its geophysical applications, Ph.D. dissertation, University of Tokyo, 1975.

Ito, E., and Y. Matsui, High-pressure synthesis of $ZnSiO_3$ ilmenite, *Phys. Earth Planet. Interiors, 9*, 344-352, 1974.

Ma, Che-Bao, Reinvestigation of the olivine-spinel transformation in Ni_2SiO_4 and the incongruent melting of Ni_2SiO_4 olivine, *J. Geophys. Res., 79*, 3321-3324, 1974.

Mao, H. K., T. Takahashi, W. A. Bassett, J. S. Weaver, and S. Akimoto, Effect of pressure on the molar volumes of wüstite and three $(Fe,Mg)_2SiO_4$ spinel solid solutions, *J. Geophys. Res., 74*, 1061-1069, 1969.

Yagi, T., and S. Akimoto, Direct determination of coesite-stishovite transition by *in situ* X-ray measurements, *Tectonophys.*, in press, 1976.

Yagi, T., Y. Ida, Y. Sato, and S. Akimoto, Effect of hydrostatic pressure on the lattice parameters of Fe_2SiO_4 olivine up to 70 kbar, *Phys. Earth Planet. Interiors, 10*, 348-354, 1975.

ULTRASONIC WAVE VELOCITY MEASUREMENTS IN SOLIDS UNDER HIGH PRESSURE USING SOLID PRESSURE MEDIA

H. ITO[1]
Geological Survey of Japan
Hisamoto, Kawasaki 213, Japan

H. MIZUTANI[2]
Geophysical Institute
University of Tokyo
Hongo, Bunkyo-ku
Tokyo 113, Japan

K. ICHINOSE
Toshiba Research and Development Center
Komukai, Kawasaki 210, Japan

S. AKIMOTO
Institute for Solid State Physics
University of Tokyo
Roppongi, Minato-ku
Tokyo 106, Japan

Abstract

An experimental technique to measure ultrasonic wave velocities of solids at pressures up to 50 kbar and temperatures up to 600°C using a new sliding-type, cubic-anvil, high-pressure apparatus is described. The sample assembly is designed to make the pressure distribution around a sample as hydrostatic as possible. Good agreement of the present data on a single crystal of MgO at room temperature with *Spetzler's* [1970] data indicates that the present technique can considerably extend the pressure range of ultrasonic wave velocity measurements. Compressional and shear wave velocities of a polycrystalline enstatite ($MgSiO_3$)

---

[1]Now at Department of Geophysics, Stanford University, Stanford, Ca. 94305.

[2]Now at Cooperative Institute for Research in Environmental Sciences, University of Colorado, Boulder, Co. 80302

were also measured to 40 kbar at room temperature. The results are in good agreement with those for a bronzite single crystal by *Frisillo and Barsch* [1972], but not with those for a bronzitite or pyroxenite by *Christensen* [1974] and by *Wang* [1974]. Shear wave velocities in Twin Sisters dunite have been determined to 50 kbar at room temperature and from 25-600°C at P = 48.5 kbar. These results encourage us to make further refinements of the present technique because of the potential for enabling us to determine seismic velocities of mantle minerals or rocks within a reasonable accuracy under upper mantle P-T conditions without any extrapolation.

I. INTRODUCTION

Since seismic wave velocities in the earth are the most accurately known quantities in geophysics, comparison of laboratory data of elastic wave velocities of mantle minerals with observed seismic data offers the most direct information about the internal structure and composition of the earth.

At present, we have zero-pressure data for the elastic wave velocities of most geophysically important materials, including high-pressure minerals, which are stable only at high pressure but metastable at zero pressure. However, we have few determinations of the pressure and temperature dependence of the elastic wave velocities of mantle minerals. These data have been obtained by ultrasonic measurements at relatively low pressure (usually below 10 kbar) using liquid or gas pressure media. Judging from large discrepancies among various data on MgO single crystals and polycrystals [*Spetzler and Anderson*, 1971], the existing data for the pressure dependence of the elastic constants for mantle minerals may contain fairly large experimental errors; some elastic constants (especially shear constants) may show nonlinear pressure dependence. Therefore, extrapolation of the experimental data obtained at low pressures to mantle conditions may lead to substantial errors.

High-pressure measurements are essential to determine intrinsic elastic properties of natural rocks and hot-pressed samples because such specimens usually contain cracks and pores whose effect on the elastic properties may persist up to 10 to 20 kbar. Measurements of elastic wave velocities on high-pressure minerals at their equilibrium condition are also of great importance to geophysics because quenched metastable minerals may show abnormal mechanical behavior.

Some attempts have been made to extend the pressure range of ultrasonic wave velocity measurements. *Christensen* [1974], *Fritz* [1974], and *Wang* [1974] succeeded in measuring ultrasonic wave velocities of rocks and oxide minerals at pressures up to 30 kbar using a Birch-Bridgman type apparatus with a liquid pressure medium. Several authors tried to make ultrasonic measurement using a piston-cylinder type high-pressure apparatus with a liquid pressure cell [*Heydemann and Houck*, 1971; *Fujisawa and Kinoshita*, 1974]. However, it seems quite difficult to extend the pressure range above 30 kbar with a liquid pressure system. High-temperature measurement is also very difficult in the liquid pressure system. Use of solid pressure media minimizes these difficulties [*Montgomery et al.*, 1967; *Matsushima et al.*, 1974]. *Mizutani et al.* [1974] have measured the elastic wave velocities of solids up to 50 kbar, using a tetrahedral-anvil type high-pressure apparatus. In the present paper, a new sliding-type, cubic-anvil, high-pressure apparatus is used; the advantage of using this solid media apparatus over a piston-cylinder type is that larger sample size and more nearly hydrostatic pressure are obtained.

II. EXPERIMENTAL METHODS

The sliding-type, cubic-anvil, high-pressure apparatus was designed and developed by *Ichinose et al.* [1975]. This apparatus is basically a modified improvement of the National Bureau of Standards multi-anvil wedge-type high-pressure device [*Lloyd et al.*, 1959; *Houck and Hutton*, 1963]. The modification is made on its sliding mechanism, which realizes perfect synchronization in anvil movements and improved reproducibility in generated pressure. With a combination of 18.6 mm edge tungsten-carbide anvils and a 24-mm edge cubic pyrophillite coated with α-Fe_2O_3 powder, pressures above 50 kbar can be obtained. Figure 1 shows a part of the apparatus with a sample assembly.

Cylindrical specimens 4 mm in diameter and 4 mm in length are prepared for ultrasonic measurements by grinding the ends parallel within 5 μm and polishing the surfaces. The surfaces of the sample are nickel-plated (1 μm thickness) with an electroless plating solution (SOMER, Japan Kanigen Co., Ltd.). $LiNbO_3$ transducers, 5 MHz frequency, are then bonded to the Ni-coated sample with Dow Chemical grease 276-V9.

The sample assembly used in the present experiment is shown in Figure 2. A 24-mm edge cube pyrophyllite, which was fired at 500°C in air condition for about 1 hour, was divided into three square disks, one of which has the same thickness as the sample. Each disk has an 8-mm diameter hole. A Teflon cylinder with the same length as the sample and Teflon disks are inserted into these holes in order to make the pressure distribution around the sample as hydrostatic as possible. The pyrophyllite and Teflon cylinders were coated with conductive Ag paste, which serves as the electrode for ground connection. The three disks were then glued with a dental cement. Thin strips of Bi and Tl are embedded in the Teflon near the sample to determine pressure values. Electrodes from the transducers and Bi and Tl exit through the edges of the pyrophyllite cube.

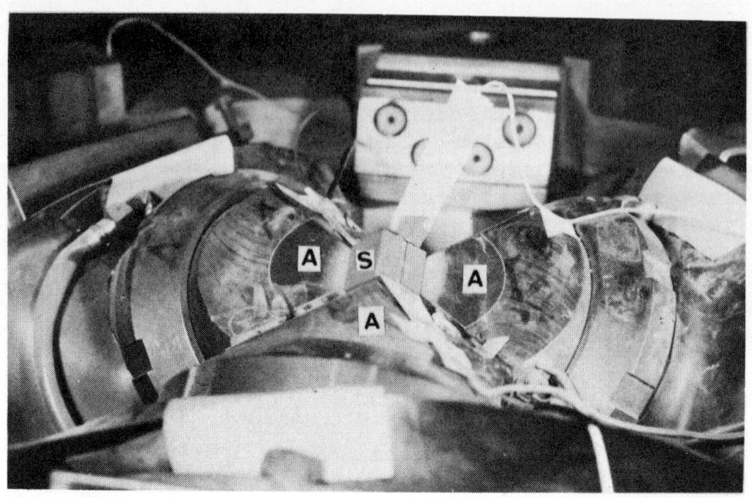

Fig. 1. Photograph of the high-pressure apparatus used in this study. Three of the six anvils (designated by A) and the 24-mm edge pyrophillite cube coated with α-Fe_2O_3 (designated by S) are shown.

In addition to the high-pressure, room temperature measurements, high-pressure, high-temperature measurements were also made, and the sample assembly is shown in Figure 3. Graphite disks 0.5 mm thick, placed close to both ends of the sample, served as heater. Temperature was measured by a chromel-alumel thermocouple placed at the center of the sample. Teflon used in the room temperature measurements was replaced by BN for the high-temperature measurements.

The ultrasonic wave velocities were determined by a modified pulse transmission method [*Mizutani et al.*, 1972]. When a PZT transducer was used, transmitted signals were weak above 40 kbar (Figure 4). $LiNbO_3$ transducers of 5-MHz natural frequency were used to generate acoustic signals: a 36°Y-cut single crystal was used for compressional waves (designated P-wave hereafter), and a 163°Y-cut single crystal was used for shear waves (designated S-wave hereafter) [*Warner et al.*, 1967]. Both compressional wave and shear wave transducers were found to generate very strong signals at pressures up to 50 kbar. A typical signal from a shear wave transducer is shown in Figure 5. Although there is a small compressional wave forerunner, the onset of shear wave can be easily observed within 0.002 μsec.

Observed travel time consists of the true travel time through the sample, and the zero length time delay. The zero

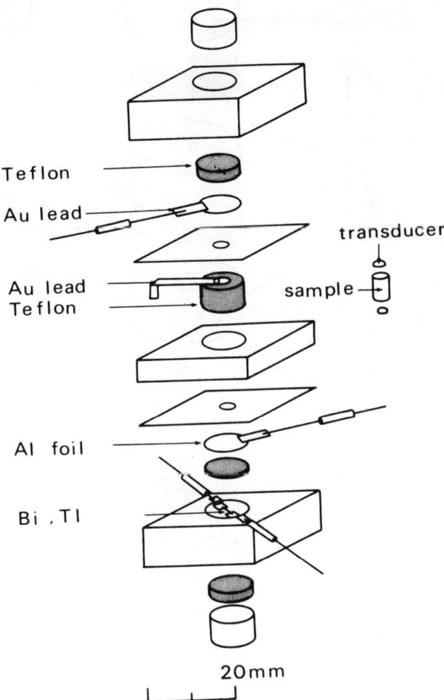

Fig. 2. Sample assembly for measurement at room temperature and high pressure.

length time delay includes delays originating in transducers and bonding grease between the sample and the transducers. Using a series of cylindrical aluminum tubes 3-15 mm in length, the zero length time delay was determined to be 0.01 and 0.02 μsec for the compressional and shear wave measurements, respectively. It was assumed that this delay is independent of pressure [*Liebermann*, 1972] although *Peselnick and Stewart* [1975] observed a small dependence of the zero length delay on pressure and temperature.

Elastic wave velocity in the sample is determined from the travel time of the elastic wave through the sample and the length of the sample. The sample length changes in two ways: one is elastic, the other is plastic. The elastic shortening of the sample can be easily estimated, by *Cook's* [1957] method, for

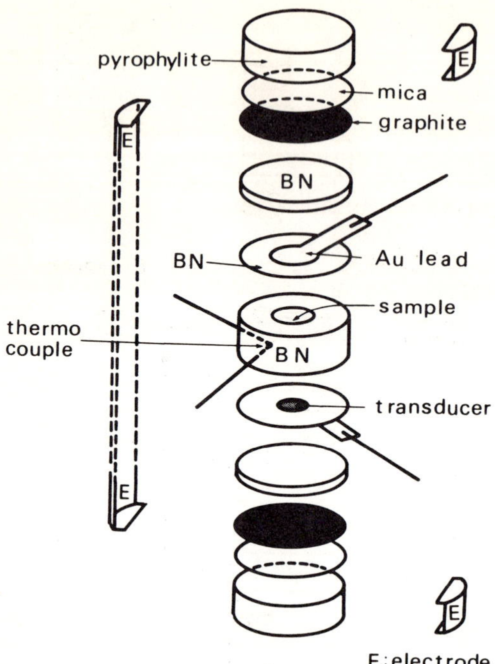

Fig. 3. Sample assembly for high-temperature measurements.

instance. In the present study, correction of the elastic shortening of the sample was made by the following equation, which is found to be accurate enough for the present purpose,

$$\frac{\ell_o}{\ell} = 1 + \frac{P}{3K_o^T} \tag{1}$$

where ℓ and ℓ_o are the lengths of the sample at pressure P and at zero pressure, respectively. K_o^T is the isothermal bulk modulus at zero pressure.

Length change due to plastic deformation, if it occurs, is difficult to estimate. Since the cubic high-pressure apparatus, the sample assemblage, and the large volume ratio of the pressure medium to that of the sample (125:1) are all intended to produce a pressure distribution around the sample as hydrostatic as possible, we can reasonably expect plastic deformation of the sample to be very small. In fact, the thin $LiNbO_3$ single-crystal transducers and the sample have never been broken during

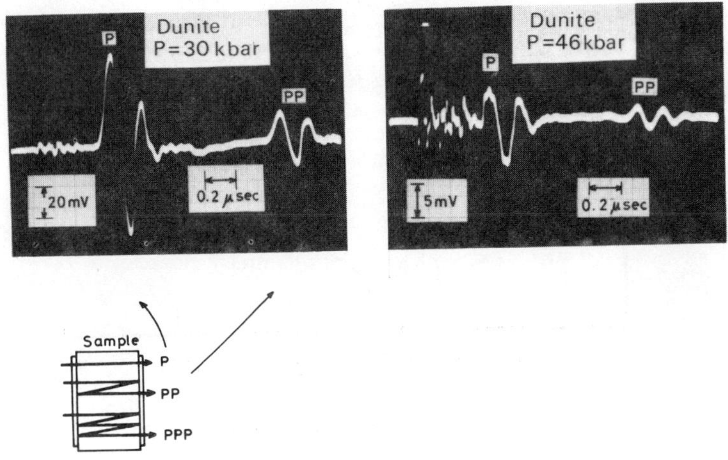

Fig. 4. Compressional wave signals obtained with PZT transducer at P = 30 and 46 kbar. The amplitude of the signal at P = 46 kbar is about one-tenth that at P = 30 kbar.

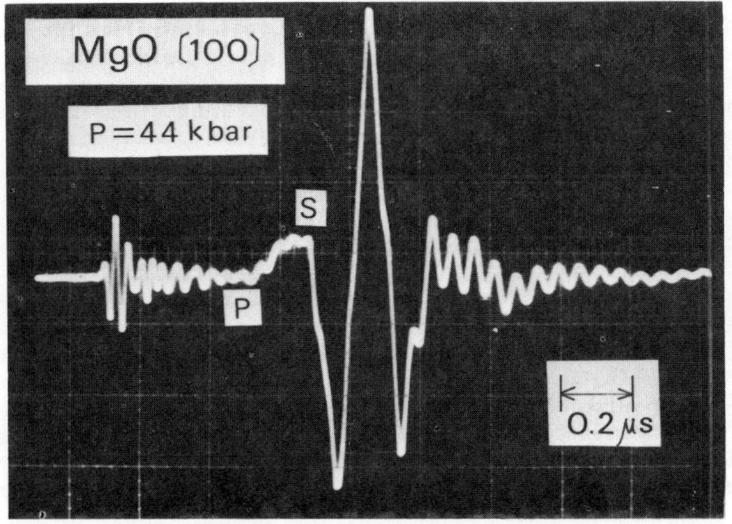

Fig. 5. Shear wave signal obtained at 44 kbar for MgO single crystal. The amplitude scale is 10 mV/div.

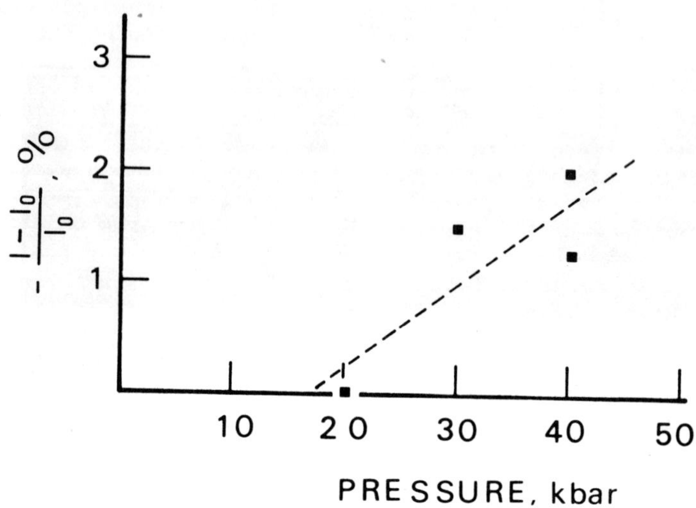

Fig. 6. The amount of plastic deformation as a function of pressure. ℓ_o is initial length and ℓ is the sample length after the experiment. The dotted line indicates correction.

the experiments. However, we have observed permanent (plastic) deformation of MgO single crystals in several preliminary experiments. The permanent length change of the sample observed in a series of experiments is shown as a function of applied pressure in Figure 6. The data indicate that the plastic deformation of MgO does not occur at pressures below 20 kbar, and that it amounts to 1 ∿ 2% at a pressure of 40 kbar. Although plastic deformation of the sample is dependent on the stress (*i.e.*, loading) rate, as well as pressure (nonhydrostatic part), we assume that the plastic deformation shown in Figure 6 is typical of other samples: we assume no plastic deformation below $P = 20$ kbar and a linear increase of plastic deformation up to the highest pressure from $P = 20$ kbar. This assumption is partly justified because the loading rate was kept almost the same for every sample in the present study. The sample length measured after an experiment gives a degree of the plastic deformation at the highest pressure. Correction of the sample length change due to plastic deformation is only required for the shear wave velocity measurement of MgO in the [100] direction: the amount of correction is shown in Figure 6 by the dotted line.

After the experiment on MgO [100] for shear waves, we learned that the degree of plastic deformation is very sensitive, depending on how the sample assemblage is prepared. If clearance between each part in Figure 2 is kept to a minimum, and only a very small amount of dental cement is used in gluing three pyrophillite disks, the plastic deformation is reduced to less than 0.05% in length. Therefore, we do not need to make any correction for plastic deformation in the other experiments.

In the case of room temperature measurements, the pressure calibration was made simultaneously with velocity measurement by utilizing the resistance transitions of Bi and Tl. In the high-temperature and high-pressure experiments, pressure calibration was made at room temperature using the Bi I-II (25.5 kbar), Tl II-III (36.7 kbar) and Ba I-II (55 kbar) transitions as defined by the new NBS scale [*Hall et al.*, 1971]. Typical relations between oil pressure and generated pressure are shown in Figure 7. The high-temperature sample assembly (Fig. 3) was somehow a little more effective in generating high pressure than the room temperature sample assembly (Fig. 2), as shown in Figure 7.

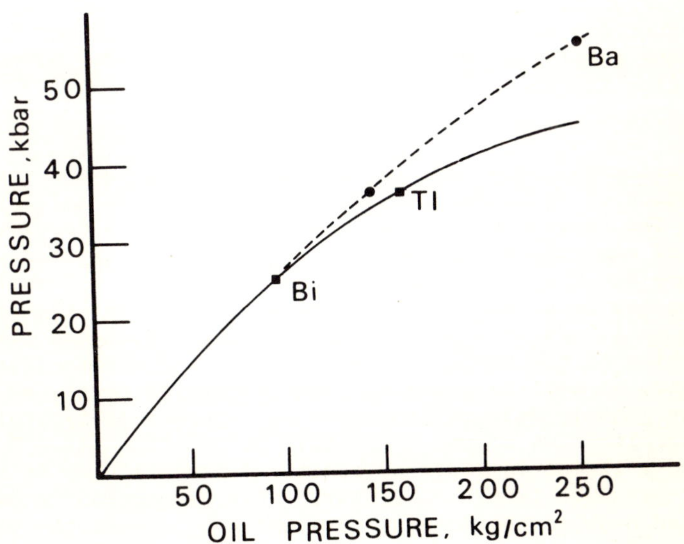

Fig. 7. Calibration curve between oil pressure and generated pressure. The solid line is for the measurements at room temperature; the broken line is obtained with the sample assembly for high-temperature measurement.

III. RESULTS

A. MgO Single Crystal

The single-crystal sample of MgO was purchased from Tateho Chemical Industry Co. (Japan). The purity of the sample is better than 99.9%. Velocity measurements were made at room temperature.

In the first experiment, shear velocities along [100] direction were measured. After the run, the sample was found to be shorter by about 1.25%. Correction of the permanent deformation was made in the way described in section II. The corrected data are shown in Figure 8 and listed in Table 1. A least-squares fitting of a straight line to the data between P = 20-40 kbar yields

$$V_s[100]\,(\text{km/sec}) = \sqrt{C_{44}/\rho} = 6.671 - 0.0013\,P\;(\text{kbar}) \tag{2}$$

which is shown by a solid line in Figure 8. The data obtained below 20 kbar and above 40 kbar were not used in the least squares calculations, because the pressure values are less reliable in this pressure range. Although the present result gives a small negative pressure derivative, it is not inconsistent with the small positive value in the low pressure range obtained by *Spetzler* [1970]. The difference between the present slope and *Spetzler's* (-0.0013 to + 0.0003 in $\partial V_S/\partial P$) is certainly within the experimental error, but the present data may indicate that C_{44} in MgO has a maximum around P = 25 kbar if the data at P = 11 and 16.5 kbar are included. This possible nonlinear behavior of C_{44} in NgO, however, must await further verification.

The data on the compressional wave velocity along [110] direction are presented in Figure 9 and Table 2. A least-squares fitting to the data between P = 20 - 40 kbar yields

$$V_p[110]\,(\text{km/sec} = \sqrt{(C_{11} + C_{12} + 2C_{44})/2\rho}$$

$$= 9.934 + 0.0046\,P\;(\text{kbar}) \tag{3}$$

This is in excellent agreement with *Spetzler's* [1970] data, which are given by

$$V_p[110]\,(\text{km/sec}) = 9.921 + 0.0053\,P\;(\text{kbar}) \tag{4}$$

Overall agreement of the present data with Spetzler's low-pressure data, as shown in Figures 8 and 9, proves the present technique to be very valuable for determination of elastic properties of solids under high pressure.

TABLE 1. Shear Wave Velocities of MgO Single Crystal along [100] Direction

Pressure, kbar	Velocity, km/sec
11.0	6.607[a]
16.5	6.624[a]
21.0	6.640
23.5	6.648
25.5	6.625
29.0	6.636
32.0	6.631
34.5	6.633
36.7	6.627
38.0	6.619
39.0	6.616
40.0	6.612
40.0	6.615

a. Not used in the calculation of pressure derivatives.

TABLE 2. Compressional Velocities of MgO Single Crystal along [110] Direction

Pressure, kbar	Velocity, km/sec
11.5	9.927[a]
16.5	9.966[a]
21.5	10.030
26.5	10.054
30.5	10.086
34.0	10.084
37.0	10.098
39.5	10.118
41.5	10.114[a]
43.0	10.111[a]
44.0	10.109[a]
44.0	10.109[a]

a. Not used in the calculation of pressure derivatives.

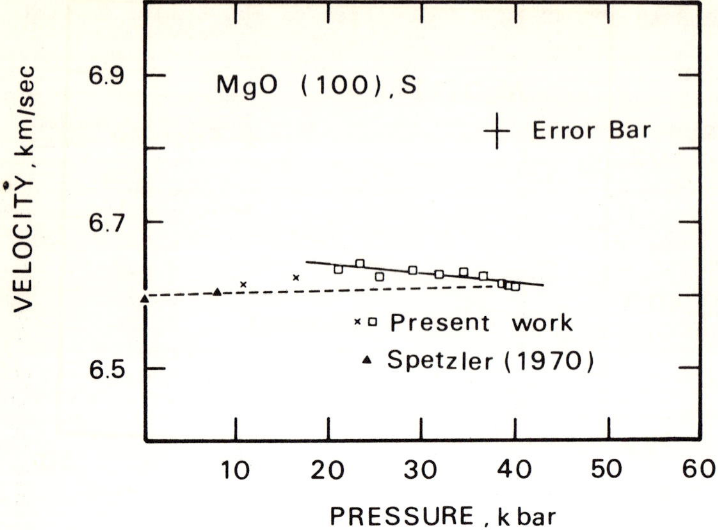

Fig. 8. Shear wave velocities of MgO along [100] direction as a function of pressure. Two triangles and the dotted line show Spetzler's [1970] data up to 8 kbar and their extrapolation. Error bar of the present measurement is also shown. The crosses are the data that were not used in the least squares calculation.

B. Polycrystalline Enstatite, $MgSiO_3$

A powder sample of $MgSiO_3$ was synthesized by intimately mixing stoichiometric amounts of MgO and SiO_2 and heating by a carbon arc in air. The resultant powder, proto-enstatite, was then hot-pressed and transformed to ortho-enstatite at $P = 40$ kbar and $T = 1000°C$ for 1 hour in the sliding-type, cubic-anvil, high-pressure apparatus. The crystallographic structure of the sample recovered at room condition was confirmed by X-ray diffraction technique.

The compressional and shear wave velocities in the recovered sample are shown in Figures 10 and 11, respectively, and tabulated in Table 3.

Compared with *Mizutani's* [1971] low pressure data, the present data seem to give intrinsic velocity values that are free from an effect of cracks or pores in the sample. The least-squares fit of the data between $P = 20 - 40$ kbar are

$$V_p(\text{km/sec}) = 8.237 + 0.0221\ P\ (\text{kbar}) \tag{5}$$

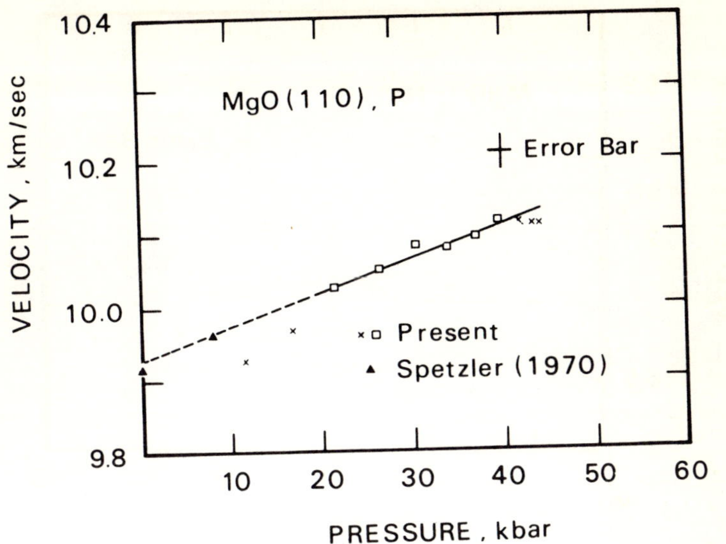

Fig. 9. Compressional wave velocities of MgO along [110] direction as a function of pressure. Two triangles show Spetzler's [1970] data up to 8 kbar. The crosses are the data that were not used in the least-squares calculation.

$$V_s\text{(km/sec)} = 4.943 + 0.0046\ P\ \text{(kbar)} \tag{6}$$

Using these pressure derivatives of compressional and shear wave velocities, the pressure derivative of the adiabatic bulk modulus and rigidity were calculated as

$$\frac{\partial K_S}{\partial P} = 10.7 \qquad \frac{\partial \mu}{\partial P} = 2.1 \tag{7}$$

The above values of $\partial V_p/\partial P$, $\partial V_S/\partial P$, $\partial K_S/\partial P$, and $\partial \mu/\partial P$ are in good agreement with those of VRH averages of the single crystalline-bronzite $(Mg_{0.8}Fe_{0.2})SiO_3$ data by *Frisillo and Barsch* [1972] which are

$$\frac{\partial V_p}{\partial P} = 0.02057\ \text{(km/sec/kbar)} \tag{8a}$$

$$\frac{\partial V_s}{\partial P} = 0.00516\ \text{(km/sec/kbar)} \tag{8b}$$

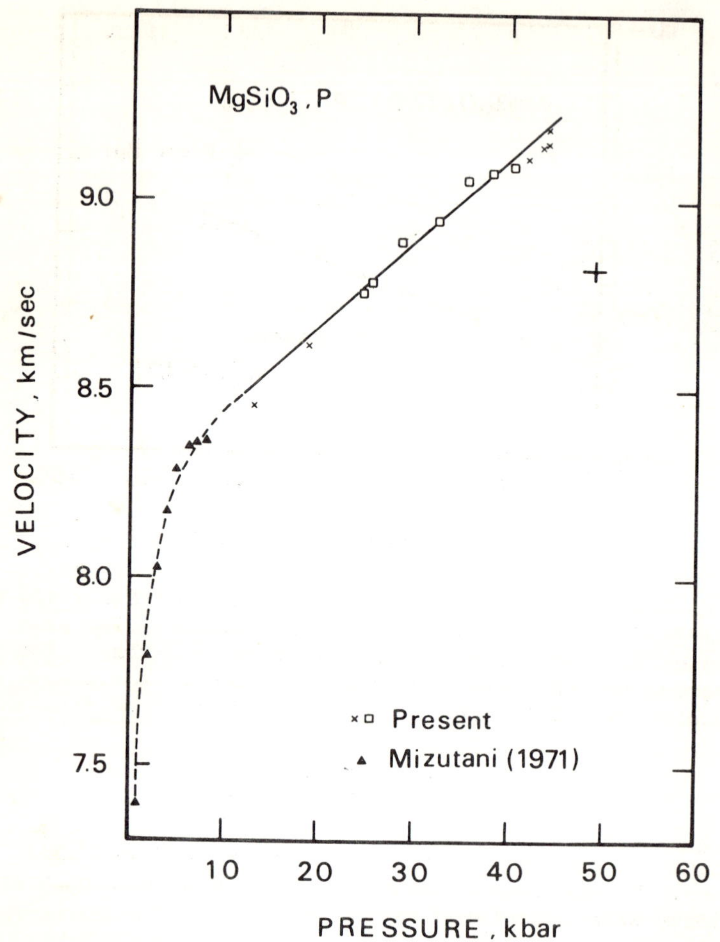

Fig. 10. Compressional wave velocities of enstatite. Triangles are low-pressure data by Mizutani [1971]. The crosses are the data that were not used in the least-squares calculation.

$$\frac{\partial K_S}{\partial P} = 9.59 \qquad \frac{\partial \mu}{\partial P} = 2.38 \tag{8c}$$

The present compressional wave pressure derivative of $MgSiO_3$, however, is significantly larger than that for bronzitite and pyroxenites reported by *Christensen* [1974] and *Wang* [1974]: their values are $\partial V_p/\partial P = 0.013 \sim 0.014$ (km/sec/kbar). There is at

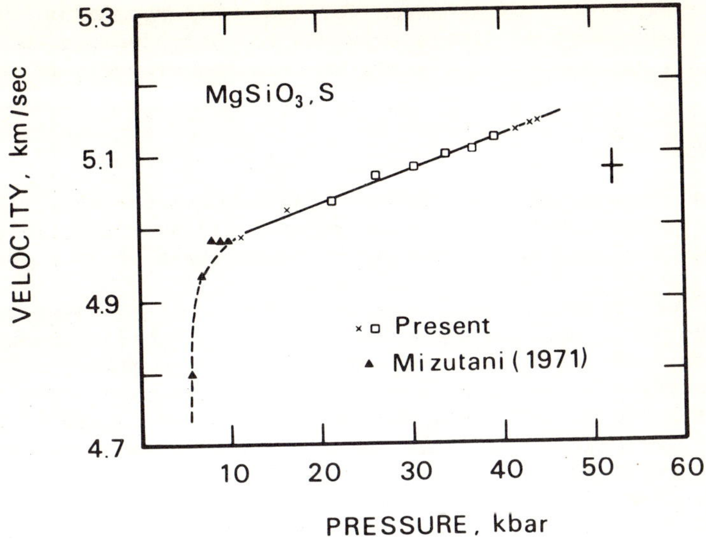

Fig. 11. Shear wave velocities of enstatite. Solid triangles are low-pressure data by Mizutani [1971].

TABLE 3. Compressional and Shear Wave Velocities of Polycrystalline Enstatite ($MgSiO_3$)

P-wave		S-wave	
Pressure[a]	V_p[b]	Pressure[a]	V_s[b]
13.0	8.465[c]	11.5	4.988[c]
19.0	8.613[c]	16.5	5.024[c]
24.5	8.762	21.5	5.035
25.5	8.785	26.5	5.072
28.5	8.894	30.5	5.085
32.5	8.950	34.0	5.099
35.5	9.063	37.0	5.108
38.0	9.078	39.5	5.124
40.5	9.099	41.5	5.134[c]
42.0	9.116[c]	43.0	5.138[c]
43.5	9.140[c]	44.0	5.143[c]
44.0	9.152[c]		
44.0	9.194[c]		

a. Pressure in kbar

b. Velocity in km/sec.

c. Not used in the calculation of pressure derivatives.

present no clear-cut explanation for these discrepancies. More precise measurement on single crystals and polycrystals of (Mg,Fe)SiO_3 are needed to clarify the elastic property of pyroxenes.

C. Dunite

As an example of simultaneous high-temperature and high-pressure measurements, shear wave velocity of a Twin Sisters dunite was measured as a function of temperature at 48.5 kbar pressure.

The Twin Sisters dunite sample is fine grained (average grain size ∿ 1.0 mm) and seems very isotropic [*Christensen*, 1974].

First, pressure was increased at room temperature, and shear wave velocity was measured as a function of pressure. The results for the Twin Sisters dunite are shown in Figure 12 and Table 4. The present value $\partial V_S/\partial P = 4.6 \times 10^{-3}$ km/sec/kbar compares well with that of a single-crystal olivine $\partial V_S/\partial P = 3.60 \times 10^{-3}$ km/sec/kbar [*Kumazawa and Anderson*, 1969].

Then temperature of the sample was increased up to 600°C keeping the pressure (load) constant. The shear wave velocity was measured as a function of temperature (Fig. 13 and Table 5). If we fit the data by a linear regression, the temperature derivative of the shear wave velocity of Twin Sisters dunite is

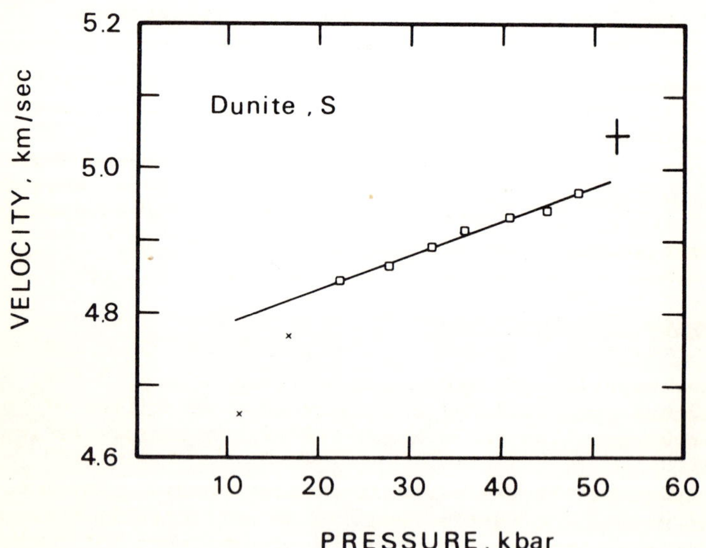

Fig. 12. Shear wave velocities of Twin Sisters dunite as a function of pressure at room temperature. The crosses are data that were not used in the least squares calculation.

TABLE 4. Shear Wave Velocities of Twin Sisters Dunite as a Function of Pressure

Pressure, kbar	Velocity, km/sec
11.5	4.659[a]
17.0	4.778[a]
22.5	4.845
28.0	4.866
32.5	4.891
36.5	4.915
41.0	4.933
45.0	4.939
48.5	4.967

a. Not used in the calculation of pressure derivatives. Data obtained at pressures above 40 kbar were used in a least-squares calculation because the pressure determination was more reliable than that with the sample assembly for the room-temperature measurement.

$$\left(\frac{\partial V_s}{\partial T}\right)_P = 48.5 \text{ kbar} = -\ 2.7 \times 10^{-4} \text{ km/sec/deg} \qquad (9)$$

which is somewhat smaller than that of the single crystal at zero pressure $(\partial V_S/\partial T)_P = 0 = -\ 3.40 \times 10^{-4}$ km/sec/deg by *Kumazawa and Anderson* [1969].

Comparison of the present data with *Kumazawa and Anderson's* [1969] data may indicate that $\partial^2 V_S/\partial T \partial P$ is negative, however, we feel that quantitative determination of $\partial^2 V_S/\partial T \partial P$ is premature at the present stage of the accuracy in our technique.

IV. SUMMARY AND CONCLUSIONS

A new experimental technique has been developed that enables us to measure compressional and shear wave velocities of solids at high pressures (up to 50 kbar) and high temperatures (up to 600°C).

In spite of the use of solid pressure media, pressure distribution around the sample seems to be sufficiently hydrostatic, especially since the transducers and samples were not broken and plastic deformation of the sample was negligible. Results on the single-crystal MgO show that the present technique is precise enough to determine the pressure derivative of elastic wave velocities of solids.

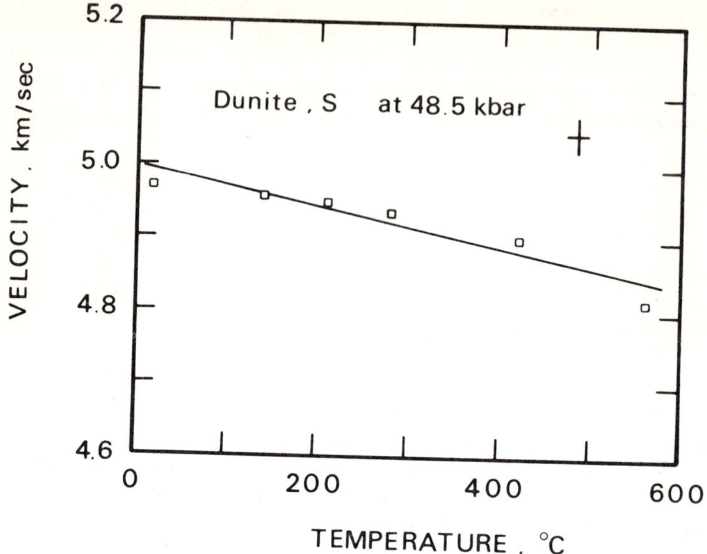

Fig. 13. Shear wave velocities of Twin Sisters dunite at 48.5 kbar as a function of temperature.

TABLE 5. Shear Wave Velocities of Twin Sisters Dunite at 48.5 kbar as a Function of Temperature

Temperature, °C	Velocity, km/sec
18	4.967
140	4.956
210	4.948
290	4.935
420	4.903
560	4.815

Measurement at high temperatures and high pressures should enable us to determine the pressure and temperature derivatives as well as the mixed derivatives of elastic wave velocities. Measurements on high-pressure minerals under their equilibrium conditions will also be possible. We think it would not be very difficult to extend the pressure and temperature range of these measurements up to 100 kbar and 1000°C with the present technique.

Acknowledgments. The authors are pleased to acknowledge the continuing guidance and technical assistance of M. Wakatsuki and T. Aoki of Toshiba Research and Development Center. Without their help, the present study would not have been possible. We express appreciation to R.C. Liebermann for reviewing the manuscript.

REFERENCES

Christensen, N. I., Compressional wave velocities in possible mantle rocks to pressures of 30 kilobars, *J. Geophys. Res.*, *79*, 407-412, 1974.

Cook, R. K., Variation of elastic constants and static strains with hydrostatic pressure: a method for calculation from ultrasonic measurements, *J. Acoust. Soc. Amer.*, *29*, 445-449, 1957.

Frisillo, A. L., and G. R. Barsch, Measurement of single-crystal elastic constants of bronzite as a function of pressure and temperature, *J. Geophys. Res.*, *77*, 6360-6384, 1972.

Fritz, I. J., Pressure and temperature dependences of the elastic properties of rutile (TiO_2), *J. Phys. Chem. Solids*, *35*, 817-826, 1974.

Fujisawa, H., and H. Kinoshita, Measurement of ultrasonic wave velocity of small mineral sample in solid-media high pressure apparatus, *Proc. 4th Internat. Conf. High Pressure*, 378-379, Kyoto, 1974.

Hall, T., Fixed points near room temperature, in *Accurate Characterization of the High Pressure Environment*, edited by E. C. Lloyd, Nat. Bureau of Standards, Spec. Publ. No. 326, 313-314, 1971.

Heydemann, P. L. M., and J. C. Houck, Ultrasonic and dilatometric measurements at very high pressures, in *Accurate Characterization of the High-Pressure Environment*, edited by E. C. Lloyd, Nat. Bureau of Standards, Spec. Publ. No. 326, 11-23, 1971.

Houck, J. C., and U. O. Hutton, Correlation of factors influencing the pressure generated in multi-anvil devices, in *High Pressure Measurement*, edited by A. A. Giardini and E. C. Lloyd, 221-245, Butterworth, Washington, D.C., 1963.

Ichinose, K., M. Wakatsuki, and T. Aoki, A new sliding type cubic anvil high pressure apparatus (in Japanese), *Pressure Engineering, J. Japan High Pressure Inst.*, *13*, 244-253, 1975.

Kumazawa, M., and O. L. Anderson, Elastic moduli, pressure derivatives, and temperature derivatives of single-crystal olivine and single-crystal forsterite, *J. Geophys. Res.*, *74*, 5961-5972, 1969.

Liebermann, R. C., Compressional velocities of polycrystalline olivine, spinel and rutile minerals, *Earth Planet. Sci. Lett.*, *17*, 263-268, 1972.

Lloyd, E. C., U. O. Hutton, and D. P. Johnson, Compact multi-anvil wedge-type high pressure apparatus, *J. Res. Nat. Bureau of Standards, 63C,* No. 1, 59-64, 1959.

Matsushima, S., K. Suito, and S. Kondo, Variation of the elastic properties accompanying the calcite I-II transition at high pressure and temperature, *Proc. 4th Internat. Conf. High Pressure,* 383-388, Kyoto, 1974.

Mizutani, H., Elasticity of mantle minerals and the constitution of the upper mantle, Ph.D. Thesis, Univ. of Tokyo, 1971.

Mizutani, H., Y. Hamano, and S. Akimoto, Elastic-wave velocities of polycrystalline stishovite, *J. Geophys. Res., 77,* 3744-3749, 1972.

Mizutani, H., H. Ito, and S. Akimoto, Measurement of ultrasonic-wave velocity of solids under very high pressure, *Proc. 4th Internat. Conf. High Pressure,* 380-382, Kyoto, 1974.

Montgomery, P. W., C. Montgomery, D. A. Wald, and J. L. S. Bellin, Sound velocity measurements at high pressures, *Rev. Sci. Instr., 38,* 1073-1076, 1967.

Peselnick, L., and R. M. Stewart, A sample assembly for velocity measurements of rocks at elevated temperatures and pressures, *J. Geophys. Res., 80,* 3765-3768, 1975.

Spetzler, H., Equation of state of polycrystalline and single-crystal MgO to 8 kilobars and 800°K, *J. Geophys. Res., 75,* 2073-2087, 1970.

Spetzler, H. A., and D. L. Anderson, Discrepancies in elastic constant data for MgO polycrystals and single crystals, *J. Amer. Ceram. Soc., 54,* 520-525, 1971.

Wang C., Pressure coefficient of compressional wave velocity for a bronzitite, *J. Geophys. Res., 79,* 771-772, 1974.

Warner, A. W., M. Onoe, and G. A. Coquin, Determination of elastic and piezoelectric constants for crystals in class (3m), *J. Acoust. Soc. Amer., 42,* 1223-1231, 1967.

GRÜNEISEN PARAMETER OF NaCl AT HIGH COMPRESSIONS

R. BOEHLER, I. C. GETTING, G. C. KENNEDY
Institute of Geophysics and Planetary Physics
University of California
Los Angeles, CA 90024

Abstract

A new method of determining the pressure dependence of the Grüneisen parameter is described. The measurements were carried out on NaCl to 45 kbar at room temperature using an end-loaded piston-cylinder apparatus. Both solid and fluid cell arrangements were used. Changes of sample temperature associated with small adiabatic pressure changes were measured and the Grüneisen parameter could be calculated from the thermodynamic relationship $\gamma = K_S/T\ (\partial T/\partial P)_S$ where K_S is the adiabatic bulk modulus. Our results are in excellent agreement with those reported by *Spetzler et al.* [1972] who calculated the pressure dependence of γ from ultrasonics up to 10 kbar and in excellent agreement with those reported by *Hardy et al.* [1975] who carried out a lattice-dynamical calculation.

STRAIN FIELDS ASSOCIATED WITH FRACTURE UNDER HIGH PRESSURE, VIEWED WITH HOLOGRAPHIC INTERFEROMETRY

H. SPETZLER, N. SOGA,[1] H. MIZUTANI,[2] and R. J. MARTIN III
Cooperative Institute for Research in Environmental Sciences
University of Colorado/NOAA
Boulder, Colorado 80309

Abstract

Holographic interferometry is used to observe the strain field and the development of an intensive deformation zone in Westerly Granite specimens under loading. Copper jacketed samples were subjected to differential stress under confining pressure up to 0.8 kbar. An elliptical bulging of the sample was observed to form at the onset of dilatancy. The ultimate failure plane and the major axis of the intense deformation zone are coincident. A bulge outlining the incipient failure plane becomes apparent below 90% of ultimate strength of the rock. In the post-failure regime, strain concentrations prior to frictional sliding of the sample are largely confined to this zone. The anelastic bulge in the intense deformation zone amounts to about twice that outside the zone at the ultimate failure of the sample.

I. INTRODUCTION

The physical and chemical mechanisms involved in deformation and failure of geologic materials are exceedingly complex. In order to predict failure, it is necessary to study several physical properties, such as changes in acoustic velocities [*Matsushima*, 1960], in magnetic properties [Martin and Wyss, 1976], and in electrical properties [*Brace et al.*, 1965; *Brace and Orange*, 1968], as well as the influence of chemically related properties [*Martin*, 1972; *Martin and Durham*, 1975]. In addition to the applied stress, the volumetric strain is perhaps that parameter which should be determined, along with any of the above, in a particular failure experiment.

---

[1]On leave from Department of Industrial Chemistry, Kyoto University, Kyoto, Japan.
[2]On leave from Geophysical Institute, University of Tokyo, Tokyo, Japan.

Brace et al. [1962] studied the volumetric strain of brittle rocks during fracture tests. They observed that above approximately half to two-thirds of the compressive strength the sample began to dilate and attributed the dilatancy to the opening of cracks parallel to the maximum compressive stress. Studies of crack distribution in sections of deformed samples showed that the cracks were not uniformly distributed but tended to cluster in highly crushed zones as the fracture stress was approached [*Wawersik and Brace*, 1971]. While these intensely deformed zones were not able to be delineated with strain gauges, they should be readily observable with holographic interferometry.

In geophysics we are usually concerned with anisotropic, inhomogeneous aggregates of crystals, i.e., rocks. To study the deformation of rocks and be able to apply the results to physical phenomena occurring within the earth, it is necessary to perform the experiments under confining pressure. We have adapted optical holographic interferometry to measure strain under confining pressure.

II. EXPERIMENTAL METHODS

Our particular approach to optical holographic interferometry enables quantitative interpretation of the holographic interferograms. A detailed description has been presented elsewhere and will not be repeated here [*Spetzler et al.*, 1974]. A hologram of an object is produced when laser light, scattered from that object, is photographically recorded together with a coherent reference beam from the same laser. When such an optically recorded interference pattern, or hologram, is reilluminated with the reference beam, a three-dimensional image of the original object is produced. This image is optically indistinguishable from the original object. If, in producing a hologram, the photographic plate is exposed twice and the object is strained between the exposures, both the strained and unstrained images are recorded on the plate. When the two images are viewed, interference fringes are produced according to path length differences between the two images and the observer. Using a parallel light beam to illuminate the object and, with the help of a beam splitter which makes the illuminating and viewing directions coincident, produces doubly exposed holograms whose interference fringes are easily interpreted. An interference pattern thus produced can be interpreted like a topographic map of the surface deformation of the object between the two exposures. The contour interval corresponds to one-half the wavelength of the illuminating laser light, where the wavelength must be taken in the medium where the change in optical path length occurs.

We have used the holographic technique as an aid in designing a large optical window for a pressure vessel [*Soga et al.*, 1976]. Figure 1 shows a schematic of our test arrangement. The

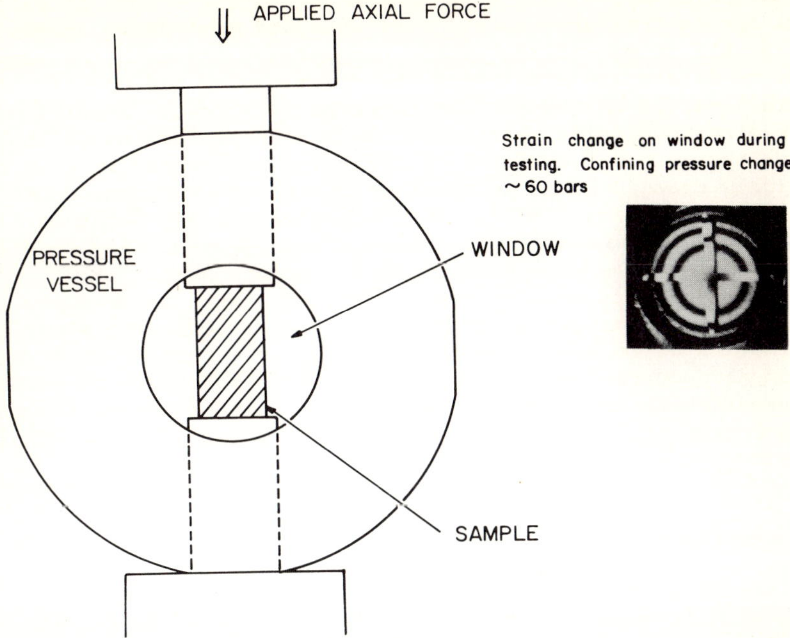

Fig. 1. Experimental arrangement showing the sample in the pressure vessel and the rams of the 100-ton hydraulic press which was used to apply the axial force. The insert is a photograph of a doubly exposed hologram of the window which provides for optical viewing of the sample during its deformation.

insert is a photograph of a doubly exposed hologram taken of the window as the confining pressure was changed by 60 bar. The conical window was painted white on its larger flat area (the high-pressure side) and had a white cross painted on the front surface (the smaller, flat surface). The three circular fringes show the strain on the high-pressure side, and the fringes on the cross show the front surface strain. Without a change in confining pressure between exposures, the optical path length is not changed by the window, and therefore no fringes appear.

In the present set of experiments, we examined the surface deformation of samples of Westerly Granite. The samples had a square cross-section and were cut and ground to 1.75 cm x 1.75 cm x 5.00 cm. Thin copper foil (0.01 cm thick) was used to jacket the samples in order to prevent the pressure fluid from penetrating the pores. Two strain gauges were mounted perpendicular to the compression axis as shown in Figure 2. The samples were

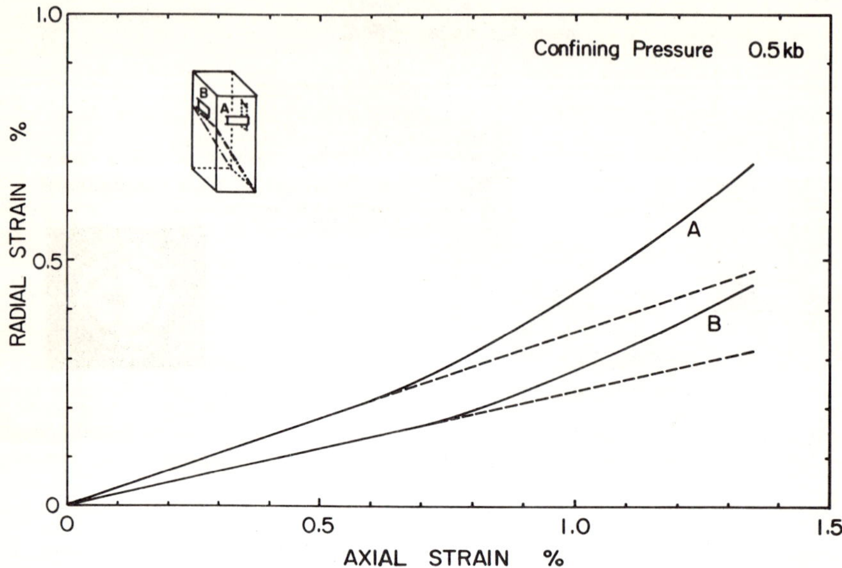

Fig. 2. Radial strain versus axial strain for the 0.5 kbar confining pressure experiment. The dashed lines are the extrapolations of the linear part of the curves giving the elastic part of the strain. The labels A and B refer to the strain perpendicular and parallel, respectively, to the ultimate failure plane.

placed within the pressure vessel, and an axial force was applied. The orientation of the samples is shown in Figure 3. In the experiment where the confining pressure was 0.8 kbar, two sample faces were visible through the window. In the 0.5 kbar experiment, the viewing direction was perpendicular to a sample face that ultimately contained the fault trace. Hydrostatic pressure was applied with Dow Corning #200 silicone oil. The axial force and strain were measured outside the pressure vessel with a force gauge and a linear variable differential transformer (LVDT), respectively.

During the course of an experiment, the hydrostatic pressure was held constant. The axial force, generated with a hydraulic system, was increased smoothly at a strain rate of $\sim 5 \times 10^{-7}$ sec^{-1} until failure of the specimen occurred. Figure 2 shows the strains normal to the loading axis recorded with the strain gauges on the prism faces labeled A and B, which are plotted as a function of axial strain. Initially, the strains change linearly with axial strain; that is, the sample behaves elastically. Above approximately 0.6 percent axial strain, the sample goes dilatant, and the radial strains become nonlinear with respect to the axial strain. The strain gauge attached on the

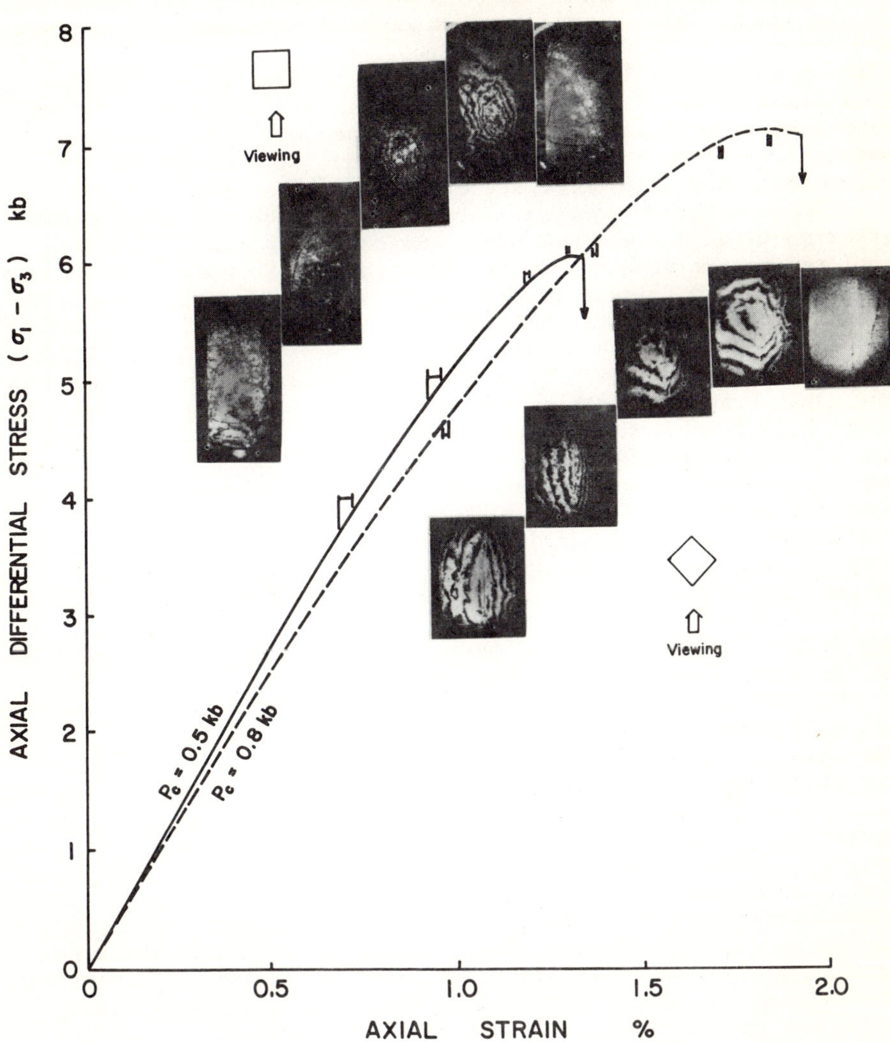

Fig. 3. Axial differential stress versus axial strain is plotted for the two experiments described in the text. Representative photographs of doubly exposed holograms are included. The upper set of photographs goes with the 0.5 kbar confining pressure experiment. The axial strain intervals between the two exposures of each hologram are also marked. One contour interval on a hologram corresponds to one-half of the wavelength ($\lambda/2$) of the illuminating laser (He-Ne, λ = 632.8 nm in vacuum) light.

sample face A, which is perpendicular to the ultimate failure plane (see Fig. 2), shows much larger nonelastic strain than that on the sample face B. This indicates that dilatancy occurs very anisotropically, as recently pointed out by *Hadley* [1975]. It is also at about that level of strain that holography illustrates the initiation of an elliptical bulge, indicating an intense deformation zone. The bulge develops and becomes more pronounced as the sample approaches failure. The major axis of this elliptical bulge coincides with the trace of the ultimate failure plane.

Representative holograms are shown in Figure 3 for the two experiments at 0.5 and 0.8 kbar. Since the specimen faces are inclined to the viewing direction in the 0.8-kbar experiment, the bulge is more difficult to visualize, while the radial strain is readily apparent. Note that the strain on the right face, which is perpendicular to the fracture plane, shows the greater strain, which is consistent with the strain gauge data shown in Figure 2. The fault trace is visible (on the far right picture) on the right-hand face from top right to bottom left. In the 0.5-kbar experiment, we oriented the specimen so that the viewing direction was perpendicular to the specimen face that would ultimately contain the fault trace. With this orientation, a fringe pattern visible on a doubly exposed hologram results only from the strain perpendicular to the viewing direction that has accumulated between the two exposures. The fault trace of the fractured sample is again visible on the far right picture. It goes from top left to bottom right. Figure 4 shows the strain accumulation on the bulge shortly before failure.

Figure 5 shows the deformation of the sample perpendicular to the surface within which the ultimate fault trace would lie. The data presented in this figure are a composite of that shown for strain gauge B in Figure 2 and the holograms of the 0.5-kbar experiment for which representative samples are shown again in Figure 6. The datum on the left vertical axis is the direct output of strain gauge B in Figure 2, while the solid line shows a cut through the bulge as calculated from the interference patterns produced by holography. The elastic strain as extrapolated in Figure 2 is also indicated. The shaded area above the elastic strain represents the dilatant contribution of strain. Note that the dilatant strain at the maximum of the bulge is more than twice the dilatant part furthest from the bulge.

III. SUMMARY

An experimental technique has been developed whereby the strain field of a sample being deformed under differential stress can be observed by holography under confining pressure. This technique is advantageous because it produces topographic strain maps over a surface rather than averaged values of strain as they are obtained with conventional techniques, such as strain gauges.

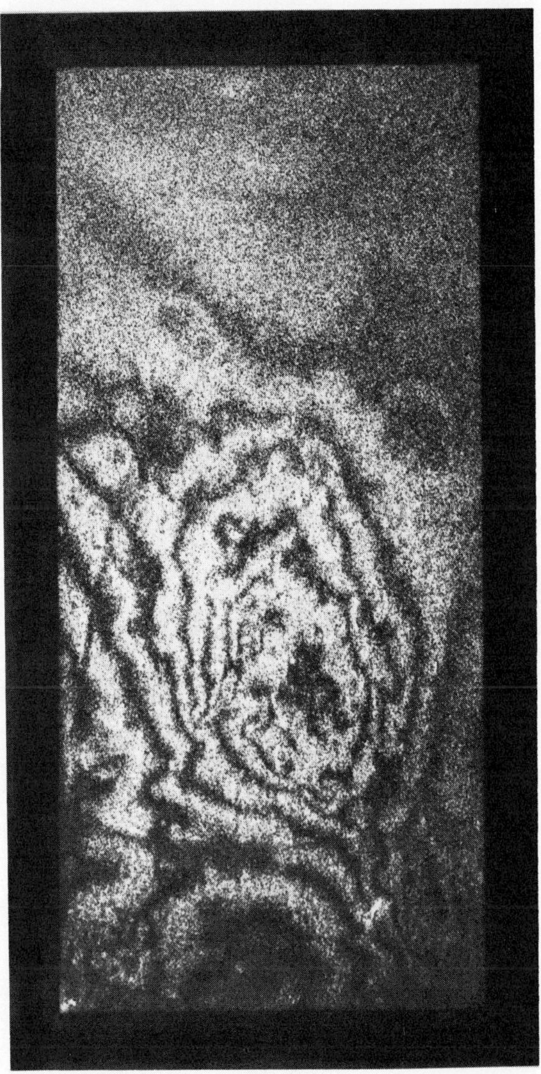

Fig. 4. See figure caption for Figure 3. The contours outlining the intense deformation zone are easily visible. The hologram was made shortly before failure, when the axial strain had reached 96% of its value at failure.

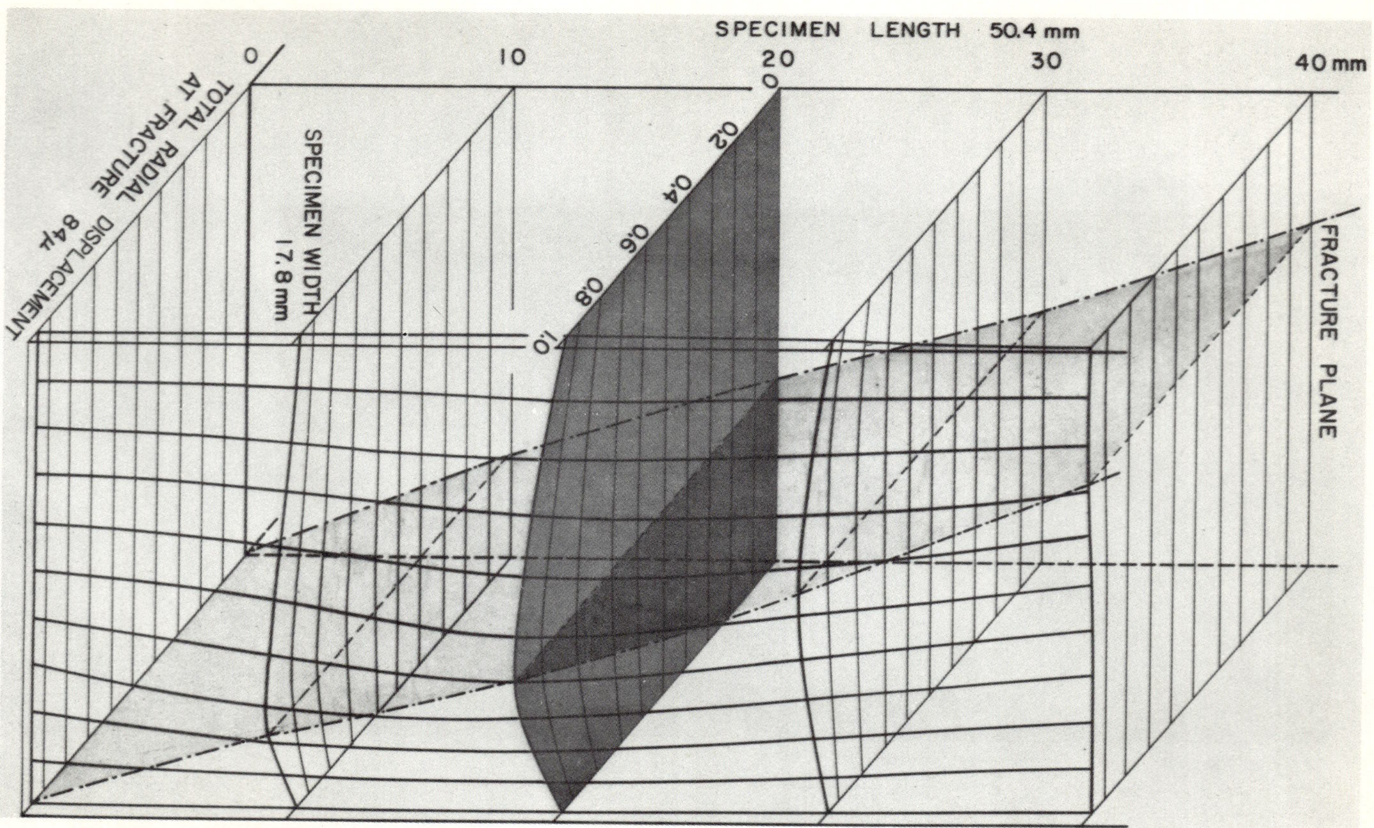

Fig. 5. A composite of the information contained in Figures 2 and 3. The progressive development of the dilatant bulge in relation to the ultimate failure plane is shown. Confining pressure = 0.5 kbar; axial differential stress at fracture = 5.7 kbar.

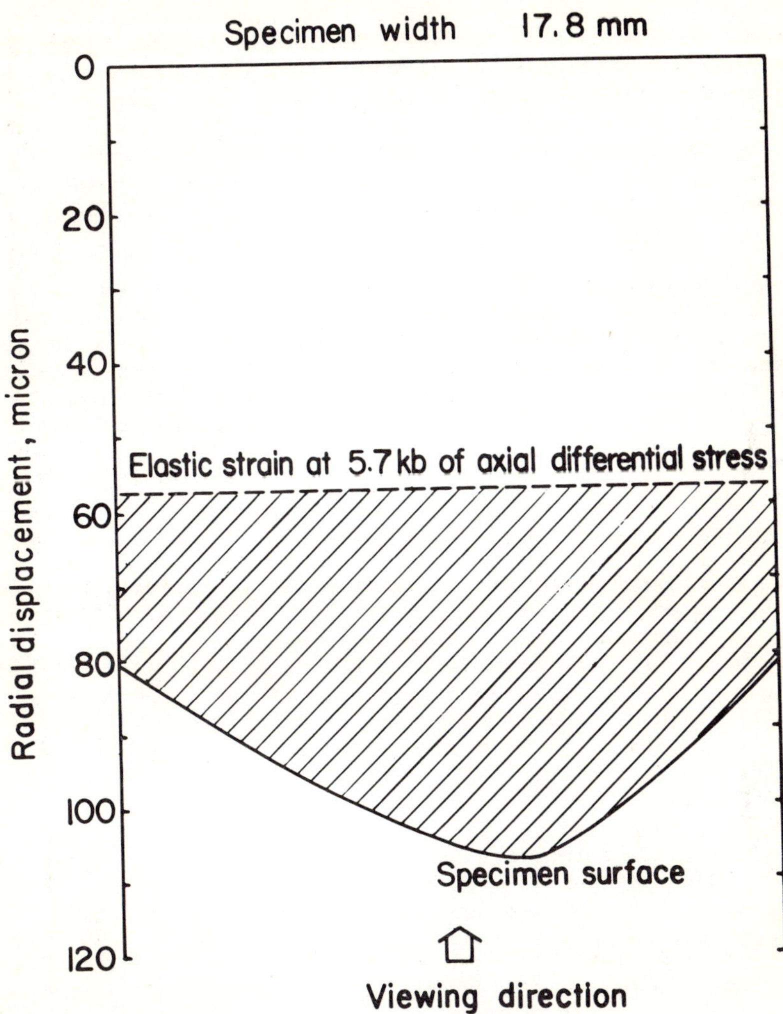

Fig. 6. The shaded horizontal plane in Figure 5 is shown in more detail. The elastic portion of the radial strain is approximately 57 μm. The dilatant portion of the strain is shaded. It varies across the sample by more than a factor of two.

We have applied this method to measure strain fields on samples of Westerly Granite as they are brought to failure under differential stress and confining pressure. At approximately 60% of the ultimate axial strain at failure, the sample becomes dilatant. An intense deformation zone forms at the center of the specimen. The dilatant strain within the zone is two times larger than the strain outside the zone. Therefore, when other physical properties (velocity, resistivity, etc.) are to be determined, the location of their measurement path in relation to the intense deformation zone becomes very important.

Acknowledgments. This work was in part supported by U.S. Geological Survey Contract #14-08-0001-15256 and National Aeronautics and Space Administration Grant #NAG 9036, for which the authors are very grateful. N. Soga is a Visiting Fellow at the Cooperative Institute for Research in Environmental Sciences, supported by a contract with Environmental Research Laboratories of NOAA.

REFERENCES

Brace, W. F., and A. S. Orange, Electrical resistivity changes in saturated rocks during fracture and frictional sliding, *J. Geophys. Res., 73,* 1433-1445, 1968.

Brace, W. F., A. S. Orange, and T. R. Madden, The effect of pressure on the electrical resistivity of water-saturated crystalline rocks, *J. Geophys. Res., 70,* 5669-5678, 1965.

Brace, W. F., B. W. Paulding, Jr., and C. H. Scholz, Dilatancy in the fracture of cyrstalline rocks, *J. Geophys. Res., 71*(16), 3939, 1966.

Hadley, K., Comparison of calculated and observed crack densities and seismic velocities in Westerly Granite, *J. Geophys. Res., 81,* 3484-3494, 1976.

Martin R. J., III, Time-dependent crack growth in quartz and its application to the creep of rocks, *J. Geophys. Res., 77,* 1406-1419, 1972.

Martin, R. J., III, and W. B. Durham, Mechanisms of crack growth in quartz, *J. Geophys. Res., 80,* 4837-4844, 1975.

Martin, R. J., III, and M. Wyss, Magnetism of rocks as a function of volumetric strain (abstract), *Trans. Amer. Geophys. Union, 57*(4), 331, 1976.

Matsushima, S., Variation of the elastic wave velocities of rocks in the process of deformation and fracture under high pressure, *Bull. 32,* 7 pp., Disaster Prevention Research Institute, Kyoto University, Japan, March, 1960.

Soga, N., D. Holcomb, and H. Spetzler, Determination of strains in optical windows of a high pressure chamber by holographic interferometry and finite element analysis, *Rev. Sci. Instrum.*, *47*, 46-49, 1976.

Spetzler, H., C. H. Scholz, and C.-P. J. Lu, Strain and creep measurements on rocks by holographic interferometry, *Pure Appl. Geophys.*, *112*, 571-581, 1974.

Wawersik, W. R., and W. F. Brace, Post failure of a granite and diabase, *Tectonophysics*, *3*, 61-85, 1971.

SUBJECT INDEX

E

F

G

H

I

J

K

L

M

N

S

T

U

V

W

X

Z

A
B 7
C 8
D 9
E 0
F 1
G 2
H 3
I 4
J 5